Sorghum: Technology and Production

Sorghum: Technology and Production

Editor:

Valeria Severino, Ph.D.

Delve Publishing

www.delvepublishing.com

Sorghum: Technology and Production

Editor: Valeria Severino, Ph.D.

708 3rd Avenue, 6th Floor
New York, NY 10017
United States of America
www.delvepublishing.com

ISBN: 978-1-68095-777-8

Library of Congress Control Number: 2016951533

About the Editor

Valeria Severino, Ph.D.

Valeria Severino obtained her PhD in Biological Processes and Biomolecules from the Second University of Naples in 2009. Her scientific interests concern the structural and functional characterization of proteins and peptides by using biochemical approaches. Fields of expertise comprises food chemistry, cellular biology, natural products and related applications. She published more than 40 papers on international peer-reviewed journals.

Contents

List of Contributors

Nadia Shakoor
Chromatin Inc, Chicago, Illinois, USA
Department of Biological Sciences, The University of South Carolina, 715 Sumter Street, Columbia, SC 29208, USA

Ramesh Nair
Chromatin Inc, Chicago, Illinois, USA

Oswald Crasta
Chromatin Inc, Chicago, Illinois, USA
Dow AgroSciences, Indianapolis, Indiana, USA

Morris
Department of Biological Sciences, The University of South Carolina, 715 Sumter Street, Columbia, SC 29208, USA

Alex Feltus
Department of Genetics and Biochemistry, Clemson University, Clemson, South Carolina, USA

Stephen Kresovich
Department of Biological Sciences, The University of South Carolina, 715 Sumter Street, Columbia, SC 29208, USA
Department of Genetics and Biochemistry, Clemson University, Clemson, South Carolina, USA

Jieqin Li
College of Agriculture, Anhui Science and Technology University, Fengyang, China

Feifei Fan
College of Agriculture, Anhui Science and Technology University, Fengyang, China

Lihua Wang
College of Agriculture, Anhui Science and Technology University, Fengyang, China

Qiuwen Zhan
College of Agriculture, Anhui Science and Technology University, Fengyang, China

Peijin Wu
College of Agriculture, Anhui Science and Technology University, Fengyang, China

Junli Du
College of Agriculture, Anhui Science and Technology University, Fengyang, China

Xiaocui Yang
College of Agriculture, Anhui Science and Technology University, Fengyang, China

Yanlong Liu
College of Agriculture, Anhui Science and Technology University, Fengyang, China

Alessandra Fracasso
Department of Sustainable Crop Production, Catholic University of the Sacred Heart, Via Emilia Parmense, 84, 29122 Piacenza, Italy

Luisa M. Trindade
Wageningen UR Plant Breeding, Wageningen University and Research Centre, 6708 PB Wageningen, The Netherlands

Stefano Amaducci
Department of Sustainable Crop Production, Catholic University of the Sacred Heart, Via Emilia Parmense, 84, 29122 Piacenza, Italy

Riyazaddin Mohammed
International Crops Research Institute for the Semi-Arid Tropics (ICRISAT), Patancheru, India
Department of Genetics, Osmania University, Hyderabad, India

Ashok Kumar Are
International Crops Research Institute for the Semi-Arid Tropics (ICRISAT), Patancheru, India

Rajendra Sudhakar Munghate
International Crops Research Institute for the Semi-Arid Tropics (ICRISAT), Patancheru, India

Ramaiah Bhavanasi
International Crops Research Institute for the Semi-Arid Tropics (ICRISAT), Patancheru, India

Kavi Kishor B. Polavarapu
Department of Genetics, Osmania University, Hyderabad, India

Hari Chand Sharma
International Crops Research Institute for the Semi-Arid Tropics (ICRISAT), Patancheru, India

Palakolanu Sudhakar Reddy
International Crops Research Institute for the Semi-Arid Tropics, Patancheru, India

Dumbala Srinivas Reddy
International Crops Research Institute for the Semi-Arid Tropics, Patancheru, India

Kaliamoorthy Sivasakthi
International Crops Research Institute for the Semi-Arid Tropics, Patancheru, India

Pooja Bhatnagar-Mathur
International Crops Research Institute for the Semi-Arid Tropics, Patancheru, India

Vincent Vadez
International Crops Research Institute for the Semi-Arid Tropics, Patancheru, India

Kiran K. Sharma
International Crops Research Institute for the Semi-Arid Tropics, Patancheru, India

Jieqin Li
College of Agriculture, Anhui Science and Technology University, Fengyang, China

Lihua Wang,
College of Agriculture, Anhui Science and Technology University, Fengyang, China

Qiuwen Zhan,
College of Agriculture, Anhui Science and Technology University, Fengyang, China

Yanlong Liu
College of Agriculture, Anhui Science and Technology University, Fengyang, China

Xiaocui Yang
College of Agriculture, Anhui Science and Technology University, Fengyang, China

Isabel Ruiz-Ballesta
Department of Plant Biology, Faculty of Biology, University of Sevilla, Avda Reina Mercedes No. 6, 41012 Sevilla, Spain

Guillermo Baena
Department of Plant Biology, Faculty of Biology, University of Sevilla, Avda Reina Mercedes No. 6, 41012 Sevilla, Spain

Jacinto Gandullo

Department of Plant Biology, Faculty of Biology, University of Sevilla, Avda Reina Mercedes No. 6, 41012 Sevilla, Spain

Liqun Wang
Shanghai Center for Plant Stress Biology, Chinese Academy of Sciences, 3888 Chenhua Road, Shanghai 201602, China
State Key Laboratory of Crop Genetics and Germplasm Enhancement, Soybean Research Institute, Nanjing Agricultural University, Nanjing, Jiangsu 210095, China

Yi-Min She
Shanghai Center for Plant Stress Biology, Chinese Academy of Sciences, 3888 Chenhua Road, Shanghai 201602, China

William Charles Plaxton
Department of Biology, Queen's University, Kingston, Ontario K7L 3N6, Canada

Cristina Echevarría
Department of Plant Biology, Faculty of Biology, University of Sevilla, Avda Reina Mercedes No. 6, 41012 Sevilla, Spain

Chukwuma C. Ogbaga
The University of Manchester, Faculty of Life Sciences, Oxford Road, Manchester, United Kingdom

Piotr Stepien
The University of Manchester, Faculty of Life Sciences, Oxford Road, Manchester, United Kingdom
Wroclaw University of Environmental and Life Sciences, Department of Plant Nutrition, ul. Grunwaldzka, Wroclaw, Poland

Beth C. Dyson
The University of Manchester, Faculty of Life Sciences, Oxford Road, Manchester, United Kingdom
Manchester Institute of Biotechnology, School of Chemistry, University of Manchester, United Kingdom

Nicholas J. W. Rattray
Manchester Institute of Biotechnology, School of Chemistry, University of Manchester, United Kingdom

David I. Ellis
Manchester Institute of Biotechnology, School of Chemistry, University of Manchester, United Kingdom

Royston Goodacre
Manchester Institute of Biotechnology, School of Chemistry, University of Manchester, United Kingdom

Giles N. Johnson
The University of Manchester, Faculty of Life Sciences, Oxford Road, Manchester, United Kingdom

Gangcheng Wu
Food Science and Technology Program, School of Public Health, Curtin University, Perth, WA, 6845, Australia

Stuart K. Johnson
Food Science and Technology Program, School of Public Health, Curtin University, Perth, WA, 6845, Australia

Janet F. Bornman
International Institute of Agri-Food Security (IIAFS), Curtin University, Perth, WA, 6845, Australia

Sarita J. Bennett
Department of Environment and Agriculture, School of Science, Curtin University, Perth, WA, 6845, Australia

Vijaya Singh
Queensland Alliance for Agriculture and Food Innovation, The University of Queensland, Brisbane, Qld, 4072, Australia

Azra Simic
National Measurement Institute, Perth, WA, 6151, Australia

Zhongxiang Fang
Food Science and Technology Program, School of Public Health, Curtin University, Perth, WA, 6845, Australia
Faculty of Veterinary and Agricultural Sciences, The University of Melbourne, Parkville, Victoria, 3010, Australia

Riyazaddin Mohammed
International Crops Research Institute for the Semi-Arid Tropics, Patancheru, India
Department of Genetics, Osmania University, Hyderabad, India

Ashok K. Are
International Crops Research Institute for the Semi-Arid Tropics, Patancheru, India

Ramaiah Bhavanasi
International Crops Research Institute for the Semi-Arid Tropics, Patancheru, India

Rajendra S. Munghate
International Crops Research Institute for the Semi-Arid Tropics, Patancheru, India

Polavarapu B. Kavi Kishor
Department of Genetics, Osmania University, Hyderabad, India

Hari C. Sharma
International Crops Research Institute for the Semi-Arid Tropics, Patancheru, India

Swapan Kumar Roy
Department of Crop Science, Chungbuk National University, Cheong-ju, Korea,

Seong-Woo Cho
Division of Rice Research, National Institute of Crop Science, Rural Development Administration, Suwon, Korea,

Soo Jeong Kwon
Department of Crop Science, Chungbuk National University, Cheong-ju, Korea,

Abu Hena Mostafa Kamal
Department of Chemistry and Biochemistry, University of Texas at Arlington, Arlington, Texas, United States of America,

Sang-Woo Kim
Department of Crop Science, Chungbuk National University, Cheong-ju, Korea,

Myeong-Won Oh
National Agrobiodiversity Center, National Academy of Agricultural Science, Rural Development Administration, Jeonju, Korea,

Moon-Soon Lee
Department of Industrial Plant Science & Technology, Chungbuk National University, Cheong-ju, Korea,

Keun-Yook Chung
Department of Environmental & Biological Chemistry, Chungbuk National University, Cheong-ju, Republic of Korea,

Zhanguo Xin
Plant Stress and Germplasm Development Unit, USDA-ARS, 3810 4th Street, Lubbock, TX, United States of America

Sun-Hee Woo
Department of Crop Science, Chungbuk National University, Cheong-ju, Korea,

DaoqianChen
State Key Laboratory of Soil Erosion and Dryland Farming on the Loess Plateau, Institute of Soil and Water Conservation, Northwest A&F University, Yangling, Shaanxi, 712100, China
College of Life Sciences, Northwest A&F University, Yangling, Shaanxi, 712100, China

BeibeiCao
College of Natural Resources and Environment, Northwest A&F University, Yangling, Shaanxi, 712100, China

ShiwenWang
State Key Laboratory of Soil Erosion and Dryland Farming on the Loess Plateau, Institute of Soil and Water Conservation, Northwest A&F University, Yangling, Shaanxi, 712100, China

College of Natural Resources and Environment, Northwest A&F University, Yangling, Shaanxi, 712100, China

Peng Liu
Department of Plant Protection, Shandong Agricultural University, Tai'an, Shandong, 271018, China

Xiping Deng
State Key Laboratory of Soil Erosion and Dryland Farming on the Loess Plateau, Institute of Soil and Water Conservation, Northwest A&F University, Yangling, Shaanxi, 712100, China
College of Life Sciences, Northwest A&F University, Yangling, Shaanxi, 712100, China

LinaYin
State Key Laboratory of Soil Erosion and Dryland Farming on the Loess Plateau, Institute of Soil and Water Conservation, Northwest A&F University, Yangling, Shaanxi, 712100, China

SuiqiZhang
State Key Laboratory of Soil Erosion and Dryland Farming on the Loess Plateau, Institute of Soil and Water Conservation, Northwest A&F University, Yangling, Shaanxi, 712100, China

GangchengWu
Food Science and Technology Program, School of Public Health, Curtin University, Perth, WA 6845, Australia

Stuart K. Johnson
Food Science and Technology Program, School of Public Health, Curtin University, Perth, WA 6845, Australia

Janet F. Bornman
International Institute of Agri-Food Security (IIAFS), Curtin University, Perth, WA 6845, Australia

Sarita J. Bennett
Department of Environment and Agriculture, School of Science, Curtin University, Perth, WA 6845, Australia

MichaelW. Clarke
Centre for Microscopy, Characterisation and Analysis, University of Western Australia, Perth, WA, 6009, Australia

Vijaya Singh
Queensland Alliance for Agriculture and Food Innovation, The University of Queensland, Brisbane, Qld 4072, Australia

Zhongxiang Fang
Food Science and Technology Program, School of Public Health, Curtin University, Perth, WA 6845, Australia
Faculty of Veterinary and Agricultural Sciences, The University of Melbourne, Parkville, Victoria 3010, Australia.

Geraldo Carvalho Jr
Embrapa Maize and Sorghum, Sete Lagoas, Minas Gerais, 35701-970 Brazil

Robert Eugene Schaffert
Embrapa Maize and Sorghum, Sete Lagoas, Minas Gerais, 35701-970 Brazil

Marcos Malosetti
Biometris, Wageningen University and Research Center, Wageningen, 6700AC The Netherlands

Joao Herbert Moreira Viana
Embrapa Maize and Sorghum, Sete Lagoas, Minas Gerais, 35701-970 Brazil

Cicero Bezerra Menezes
Embrapa Maize and Sorghum, Sete Lagoas, Minas Gerais, 35701-970 Brazil

Lidianne Assis Silva
Embrapa Maize and Sorghum, Sete Lagoas, Minas Gerais, 35701-970 Brazil
Department of Plant Breeding and Genetics, Federal University of Acre, Cruzeiro do Sul, AC, 69915-900 Brazil

Claudia Teixeira Guimaraes
Embrapa Maize and Sorghum, Sete Lagoas, Minas Gerais, 35701-970 Brazil

Antonio Marcos Coelho
Embrapa Maize and Sorghum, Sete Lagoas, Minas Gerais, 35701-970 Brazil

Leon V. Kochian
Robert W. Holley Center of Agriculture and Health, USDA-ARS, Cornell University, Ithaca, New York 14853

Fred A. van Eeuwijk
Biometris, Wageningen University and Research Center, Wageningen, 6700AC The Netherlands

Jurandir Vieira Magalhaes
Embrapa Maize and Sorghum, Sete Lagoas, Minas Gerais, 35701-970 Brazil

Ignacio A. Ciampitti
Department of Agronomy, Kansas State University, Manhattan, KS, USA

P. V. Vara Prasad
Department of Agronomy, Kansas State University, Manhattan, KS, USA

Caitlin S. Byrt
Australian Research Council Centre of Excellence in Plant Cell Walls, School of Agriculture, Food and Wine, University of Adelaide, Urrbrae, South Australia, Australia

Natalie S. Betts
Australian Research Council Centre of Excellence in Plant Cell Walls, School of Agriculture, Food and Wine, University of Adelaide, Urrbrae, South Australia, Australia

Hwei-Ting Tan
Australian Research Council Centre of Excellence in Plant Cell Walls, School of Agriculture, Food and Wine, University of Adelaide, Urrbrae, South Australia, Australia

Wai Li Lim
Australian Research Council Centre of Excellence in Plant Cell Walls, School of Agriculture, Food and Wine, University of Adelaide, Urrbrae, South Australia, Australia

Riksfardini A. Ermawar
Australian Research Council Centre of Excellence in Plant Cell Walls, School of Agriculture, Food and Wine, University of Adelaide, Urrbrae, South Australia, Australia

Hai Yen Nguyen
Australian Research Council Centre of Excellence in Plant Cell Walls, School of Agriculture, Food and Wine, University of Adelaide, Urrbrae, South Australia, Australia

Neil J. Shirley
Australian Research Council Centre of Excellence in Plant Cell Walls, School of Agriculture, Food and Wine, University of Adelaide, Urrbrae, South Australia, Australia

Jelle Lahnstein
Australian Research Council Centre of Excellence in Plant Cell Walls, School of Agriculture, Food and Wine, University of Adelaide, Urrbrae, South Australia, Australia

Kendall Corbin
Australian Research Council Centre of Excellence in Plant Cell Walls, School of Agriculture, Food and Wine, University of Adelaide, Urrbrae, South Australia, Australia

Geoffrey B. Fincher
Australian Research Council Centre of Excellence in Plant Cell Walls, School of Agriculture, Food and Wine, University of Adelaide, Urrbrae, South Australia, Australia

Vic Knauf
Arcadia Biosciences, Davis, CA, United States of America

Rachel A. Burton

Australian Research Council Centre of Excellence in Plant Cell Walls, School of Agriculture, Food and Wine, University of Adelaide, Urrbrae, South Australia, Australia

Hiroshi Teramura
Rokkodaicho, Nada-ku, Hyogo, Kobe 657-8501, Japan

Kengo Sasaki
Organization of Advanced Science and Technology, Kobe University, 1-1 Rokkodaicho, Nada-ku, Hyogo, Kobe 657-8501, Japan

Tomoko Oshima
Rokkodaicho, Nada-ku, Hyogo, Kobe 657-8501, Japan

Fumio Matsuda
Department of Bioinformatic Engineering, Graduate School of Information Science and Technology, Osaka University, 1-5 Yamadaoka, Osaka, Suita 565-0871, Japan
RIKEN Center for Sustainable Resource Science, 1-7-22 Suehiro-cho, Tsurumi-ku, Kanagawa, Yokohama 230-0045, Japan

Mami Okamoto
RIKEN Center for Sustainable Resource Science, 1-7-22 Suehiro-cho, Tsurumi-ku, Kanagawa, Yokohama 230-0045, Japan

Tomokazu Shirai
RIKEN Center for Sustainable Resource Science, 1-7-22 Suehiro-cho, Tsurumi-ku, Kanagawa, Yokohama 230-0045, Japan

Hideo Kawaguchi
Rokkodaicho, Nada-ku, Hyogo, Kobe 657-8501, Japan

Chiaki Ogino
Rokkodaicho, Nada-ku, Hyogo, Kobe 657-8501, Japan

Ko Hirano
Bioscience and Biotechnology Center, Nagoya University, 1 Furo-cho, Chikusa-ku, Nagoya 464-8601, Japan

Takashi Sazuka
Bioscience and Biotechnology Center, Nagoya University, 1 Furo-cho, Chikusa-ku, Nagoya 464-8601, Japan

Hidemi Kitano
Bioscience and Biotechnology Center, Nagoya University, 1 Furo-cho, Chikusa-ku, Nagoya 464-8601, Japan

Jun Kikuchi
RIKEN Center for Sustainable Resource Science, 1-7-22 Suehiro-cho, Tsurumi-ku, Kanagawa, Yokohama 230045, Japan
Graduate School of Medical Life Science, Yokohama City University, 1-7-29 Suehirocho, Tsurumi-ku, Yokohama 230-0045, Japan
Graduate School of Bioagricultural Sciences and School of Agricultural Sciences, Nagoya University, 1 Furo-cho, Chikusa-ku, Nagoya 464-8601, Japan

Akihiko Kondo
Rokkodaicho, Nada-ku, Hyogo, Kobe 657-8501, Japan
RIKEN Center for Sustainable Resource Science, 1-7-22 Suehiro-cho, Tsurumi-ku, Kanagawa, Yokohama 230-0045, Japan

Sunan Nuanpeng
Graduate School, Khon Kaen University, Khon Kaen 40002, Thailand
Department of Biotechnology, Faculty of Technology, Khon Kaen University, Khon Kaen 40002, Thailand

Sudarat Thanonkeo
Walai Rukhavej Botanical Research Institute, Mahasarakham University, Maha Sarakham 44150, Thailand

Mamoru Yamada
Department of Biological Chemistry, Faculty of Agriculture, Yamaguchi University, Yamaguchi 753-8515, Japan

Pornthap Thanonkeo
Department of Biotechnology, Faculty of Technology, Khon Kaen University, Khon Kaen 40002, Thailand

Fermantation Research Center for Value Added Agricultural Products, Khon Kaen University, Khon Kaen 40002, Thailand

Girma Getachew
Department of Plant Sciences, University of California, Davis, CA, USA

Daniel H. Putnam
Department of Plant Sciences, University of California, Davis, CA, USA

Christopher M. De Ben
Department of Plant Sciences, University of California, Davis, CA, USA

Edward J. De Peters
Department of Animal Science, University of California, Davis, CA, USA

Preface

Sorghum is a genus belonging to the subfamily Panicoideae and the tribe of Andropogoneae. It comprises many species and sub-species, presenting several kind of sorghum, including grain sorghums, also known as Sorghum bicolor, grass sorghums, sweet sorghums and others. With wheat, corn, rice, and barley, sorghum is one of the world's most important crops, according to the U.S. Grains Council. The plants of sorghum are cultivated worldwide in warm climates, especially in Africa, Australia, Asia and Mesoamerica, being water availability the main limit for production. Sorghum varieties are studied and cultivated for several purposes, including human consumption and to brew beer, for feeding livestock, and recently for biofuel production. Sorghum breeding is based on the use of either inbred or hybrid cultivars and the study of genotype of sorghum varieties is a very hot topic, in order to improve germplasms. The first section of this book is focused exactly on recent development in sorghum genotyping and genetic studies. Analyses of genes involved in resistance to drought, biotic and abiotic constrains are proposed, along with biochemical characterizations. In the second section of this book the role of some important factors (e.g. silicon, ions, growth temperature,and soil) on sorghum production are investigated. Recent innovative proposal to use sorghum for biofuel production and as an alternative to corn forage are finally presented. The demand for dryland crops in sub-Saharan Africa and South Asia, such as sorghum and millets, is growing as a number of major global issues continue to impact the world's food security. Trends include: extreme climate variabilities and increasing drought; the rise in global demand for livestock feed; the growing use of smart foods with high levels of iron, fiber and calcium for weaning children, nursing mothers, and the gluten intolerant; and an increasing global population requiring more food.

Editor
Valeria Severino, Ph.D.

INTRODUCTION

Sorghum is a genus of grass with many species and subspecies, and there are various types, including grain sorghum, grass sorghum (for pasture and hay), sweet sorghum (for syrup), and broomcorn. Sorghum is part of the subfamily Panicoideae and part of the tribe of grasses that include sugar cane and big bluestem. One of the most important species of this family is sorghum bicolor (sorghum vulgare) which is used for food and grain (including syrup or molasses), fodder, alcohol, ethanol, and many other uses as well.

The color of sorghum kernels vary from white and gold to red, purple and brown; white, red, and brown being the most common. Because of its hardy nature, sorghum is actually one of the cheapest grains to grow with a very low cost per acre. For many growers of sorghum, they simply plant the crop, then harvest it in the fall with very little tending or maintenance.

Sorghum is one of the main coarse cereal crops of India. India is second largest producer of sorghum in the world. This crop is ideally suited for semi-arid agro climatic regions of the country and, it gives reasonably good yield with minimal requirement of irrigation and fertilizers as compare to other cereals such as Wheat & rice. Mostly cultivated in the semi-arid regions for fodder to feed the large cattle population of the country as well as to meet the demand of industrial applications.

Sorghum is mainly used in the form of flour or paste processed into two main dishes."Tuwo" a thick porridge and OGI or KAMU a thin diet or porridge. Other dishes that are sometimes made from sorghum include a number of deep fried snacks, steamed dumplings and other boiled or roasted snack foods. The production constraint of sorghum include farmers at the margin of subsistence level do not invest much in fertilizer and improve varieties, rising labor cost, consumer change in food preference due to income increase, birds attacks and parasitic weed such as striga.

Grain moulds causes significant losses in both grain yield and quality, particularly in areas where improved cultivars have been adopted. Other important diseases include anthracnose, ergot and leaf blight. Insect pests constrain production in many areas. Stem borers, head bugs, midge and shoot fly are among the most important.

Another major problem is that variable rainfall leads to large fluctuation in production. Price fall abruptly in good years, leaving traders reluctant to enter the market, especially since stockholding infrastructure is usually inadequate. This increases the price risk that sorghum producer's face, and their unwillingness to invest in commercial sorghum production.

Production Inputs

Production inputs include seed, fertilizer, herbicides, insecticides and other chemicals used in sorghum production.

Seed

Seed is an important input in sorghum production. In order to have good germination, farmers should carefully select seeds that are not damage.

Fertilizer

Fertilizer can be divided into organic and inorganic which are both important for sorghum production. A detail of rates and application methods is presented under cultural practices.

Herbicides

Weed can be controlled in cropland through cultural, mechanical and chemical means. Wise use of these individual methods or a combination of them can manage weeds effectively without causing economic loss or harming the environment. Deciding which practice to use will depend largely on the type of weed(s) to be control and the infestation level. Also, the crop being planted will play a major role in determining the timeliness of mechanical measures.

Processing Inputs

The processing inputs are divided into traditional and modern. The traditional ones include machete, knives, ropes, empty sacks/ tarpaulin, mortar and pestle, beating stick, winnower/calabashes/bowls and grinding stones while the modern ones are different type of threshers, dryers, dehuller, milling machines, etc.

Production

Cultural Practices

Sorghum is a cereal crop that can grow in a wide variety of soils ranging from heavy clay in the Southern Guinea savanna to sandy loam in the Sudan /Sahel savanna ecologies. It does best in soils with high moisture retention capacity, well drained and fertile clay loam in nature. It is fairly tolerant to alkalinity and salinity. It can tolerate some drought situations that is why It is described as a hardy crop.

Land clearing

Land clearing is usually done just before the on-set of the rains. Crop residue of the previous crop grown on the area including shrubs are cut and burnt, while bushy trees are pruned to reduce their shading effect on the crop.

Land Preparation

Sorghum is most important in the northern Guinea and the Sudan savanna ecologies where the soils are light and sandy; this is also similar in the Sahel savanna where minimum land preparation through tiling is required. The seeds are planted in old furrows by farmers in such areas. In heavier soils such as in the southern Guinea savanna ecology proper land preparation is necessary. Where a tractor is available the land should be ploughed and left for two weeks. It is then harrowed and ridged 75 to 90cm apart.

With the ox-drawn mould board, bulls are trained to break or split old ridges to form new ones in the old furrows. After the land preparation the seed bed should be weed-free and the soil well pulverized.

Choice of varieties: There are three main ecologies with seasonal variations, these include the Southern Guinea Savanna (SGS) with a long growing period of about seven to eight (7-8)months of rainfall, the varieties to grow here should be long duration or late maturing varieties. In the Northern Guinea Savanna (NGS) the growing season is shorter than in the SGS but longer than in Sudan/Sahel savanna. Varieties of sorghum to be grown here are medium maturing. The rainfall here last between six and seven months (6-7 months). In the Sudan/Sahel savanna, the growing period is short with the rainfall lasting between four and five months (4-5 months) Varieties of sorghum grown here are short duration or early maturing in order to escape the drought that is frequently experienced at the end of the growing season.

Seed rate

To plant a sole crop of sorghum in one hectare of land, you need 10-15kg of seeds. Seeds are planted 3-5 seeds/hole and at 25-30cm between stands for good germination to occur.

For intercropping

Sorghum: 8kg of seed per hectare. Planted 3-5 seed/hole and thin to 2-3 plant per stand after 10 days of emergence.

Seed dressing

Most soils are contaminated with soil borne diseases like bacteria and fungi. In some cases insect pests and soil pests such as root-knot nematode may be found. Before planting your seed, dress all seeds with seed dressing chemicals. Use one sachet of Aprons star or Apron plus 50 DS to dress 1.0kg (1 mudu) of seeds.

Planting

Sorghum seed are plant on a well prepared seed bed as soon as the rains is established. To enhance good germination plant after a good rain. Poor seed germination, will eventually affect the yield.

Thinning

Thinning is done 2-3 weeks after planting. The seedlings are thinned to two plants per stand, where gaps exist you need to transplant when the soil is wet and preferably in the evening. The transplants should carry as much root as possible and the foliage should be slightly pruned to reduce vapor-transpiration and shock. The transplants are planted upright. In both the planting and the transplants the soil around crop should be firmed.

Fertilizer application

Nigerian soils are deficient in the major nutrient elements-nitrogen, phosphorus and potassium and some essential micro-nutrients.

Rate

For good sorghum growth and yield, 64kg Nha^{-1} 30 kg P_20_5 ha^{-1} and 30kg K_2Oha^{-1} are required. Phosphorus and potassium are applied during land preparations and nitrogen is applied in split application. Half the nitrogen rate is either applied at planting or three weeks after sowing and the balance banded or spot application at about 25cm from the plant at six weeks after so wing. Fertilizer rates can be satisfied using the following:

- 4 bags of 15:15:15 NPK compound fertilizer during land preparation or at planting or at 3 weeks after sowing. This should be followed by 1.5 bags Urea 6 weeks after sowing.
- 1.5 bags of urea plus 3.3 bags of SSP (18% P205) at planting followed by 1.3 bags of urea 6 weeks after sowing.

Where the land is prepared with tractor the first dose of the fertilizer or combination of fertilizers should be applied after harrowing and before ridging. It is more economical to spot apply the first dose of fertilizer about 3 weeks after sowing. This increases the crop vigor and controls weeds. There is need for molding up the ridges after the application of the second dose of nitrogen, this helps in covering the fertilizer up and weed control.

Weed Control

Weed infestation can reduce crop yields by as much as 30-50% if not controlled, for instance Striga, a parasitic weed, had been found to cause total crop failure where susceptible sorghum varieties have been planted in highly infested fields.

To control the non-parasitic weeds, two weedings at 3 weeks and 6 weeks after sowing and ridge moulding (8 week after sowing) are required. Where parasitic weeds are a problem, apart from choosing a striga tolerant or resistant sorghum variety, all emerged striga plants should be hand pulled and burnt. Increasing the nitrogen fertilizer rate from 64 kg N to 90kg N ha^{-1} will help the crop growth vigor to withstand the attack by striga. Recommended herbicides for weed control include Atrazine, premextragold pendimenthal in each at 4L/ha and also apply pre-emergence (0-1 DAS).

Harvesting

Time of harvest varies with the duration of the crop variety. Early maturing varieties mature early and are ready for harvest in Oct/Nov. Sorghum that is mature and ready for harvest will be well filled and the grain will be hard. The leaves will turn yellow to brown with the lower leaves drying up completely. As soon as the crop is mature, harvest should commence immediately as some of the varieties tend to shatter and you will loose much seed (yield).For tall and Medium-dwarf varieties, the plant is cut down before the panicle(s) (head) is cutoff. For dwarf varieties the panicles are cut off directly from the plant and are packed in a basket or bag. In both cases the heads are either allowed to further dry in the field or taken home straight.

Post-Harvest Processing and Storage

Where the sorghum is to be stored in heads, the heads should be well dried for good keeping. If the grains are to be stored, the heads are dried thoroughly before they are threshed. The panicles are pounded in a mortar or beaten with sticks on a smooth floor preferably cement floor to avoid sand particles.

Whether the heads or grains are to be stored, the store should be cleaned fumigated with phostoxin or actellic dust (11 mL/L of water and apply 1 L of diluted mixture for every 20m2 of surface area) so that it will be insect-free. If the crop is to be stored in bags, the bags should be air tight and fumigated with phostoxin at the rate of two tablets per bags of 100 kg seed.

Crop Protection

Over 32 diseases caused by fungi, bacteria, viruses and nematodes have been reported. A few of these diseases are currently of major importance, some have potential as emerging, while many are limited in occurrence. A large number of insect species have been reported on sorghum in Nigeria. However, only a few are of economic importance.

Uses of Sorghum

Sorghum actually has several uses and is one of the most versatile of crops. Since there are varieties of sorghum that are heat resistant and tolerant of drought conditions, it has become a staple crop for those living in very harsh conditions.

Food: In most areas where sorghum is grown, it is an important food crop. This is especially true in Central America, Africa and South Asia where it is used as a cereal grain. This is a highly versatile crop being included in several different types of foods.

Forage or Fodder: Often used as a fodder plant, sorghum is either harvested or used as part of the pasture for silage. Since the plant can grow under many different conditions, it is often left to grow wild in pastures where farm animals can feast on the crop.

Alcoholic Beverages: As with other cereal grain crops, sorghum is also used in conjunction with the creation of alcoholic beverages. Sorghum beer, wine, and spirits are produced all the world over, with a bulk of the US grain going to mainland China to supply the large Asian spirit and wine production.

Ethanol: Because sorghum is a stable, plentiful crop, it is one that is often used in creating biofuels such as ethanol. Because the crop grows fairly quickly and is cheap, sorghum makes for an effective ethanol product, although there are issues with transportation. However, given that it is cheaper to plant, grow and cultivate an acre of sorghum than it is to merely buy the seeds for corn, sorghum is being used more and more as a biofuel.

There is little doubt about the versatility of this crop, being one of the heartiest of the cereal grains. Sorghum will continue to be used in a variety of means to meet the needs of people around the world.

Growth Requirements and Adaptation

The optimum growth requirements of sorghum plants, in order to exploit its inherit yield potential, are a deep well-drained fertile soil, a medium to good and fairly stable rainfall pattern during the growing season, temperate to warm weather (20 – 30 °C) and a frost-free period of approximately 120 to 140 days.

Soil requirements

Sorghum is mainly grown on low potential, shallow soils with high clay content, which usually are not suitable for the production of maize. Sorghum usually grows poorly on sandy soils, except where a heavy textured subsoil is present. Sorghum is more tolerant of alkaline salts than other grain crops and can therefore be successfully cultivated on soils with a pH (KCl) between 5,5 and 8,5. Sorghum can better tolerate short periods of waterlogging compared to maize. Soils with a clay percentage of between 10 and 30 % are optimal for sorghum production.

Climatic requirements

The climatic requirements for the production of sorghum are divided into temperature, day length and water needs.

Temperature: Sorghum is a warm-weather crop, which requires high temperatures for good germination and growth. The minimum temperature for germination varies from 7 to 10 °C. At a temperature of 15 °C, 80 % of seed germinate within 10 to 12 days. The best time to plant is when there is sufficient water in the soil and the soil temperature is 15 °C or higher at a depth of 10 cm. Temperature plays an important role in growth and development after germination. A temperature of 27 to 30 °C is required for optimum growth and development. The temperature can, however, be as low as 21 °C, without a dramatic effect on growth and yield.

Exceptionally high temperatures cause a decrease in yield. Flower initiation and the development of flower primordia are delayed with increased day and night temperatures. Plants with four to six mature leaves that are exposed to a cold treatment (temperatures less than 18°C) will form lateral shoots. However, in plants in or beyond the eight-leaf stage, apical dominance will prevent the formation of lateral shoots. Temperatures below freezing are detrimental to sorghum and may kill the plant. At an age of 1 to 3 weeks, plants may recover if exposed to a temperature of 5°C below freezing point, but at 7 °C below freezing, plants are killed. Plants older than 3 weeks are less tolerant to low temperatures and may die off at 0°C.

Day length: Sorghum is a short-day plant, which means that the plant requires short days (long nights) before proceeding to the reproductive stage. The optimum photoperiod, which will induce flower formation, is between 10 and 11 hours. Photoperiods longer than 11 to 12 hours stimulate vegetative growth. The tropical varieties are usually more sensitive to photoperiod than the quick, short-season varieties. Sorghum plants are most sensitive to photoperiod during flower initiation.

Water requirements: Sorghum is produced in South Africa on a wide range of soils, and under fluctuating rainfall conditions of approximately 400 mm in the drier western parts to about 800 mm in the wetter eastern parts.

Drought Tolerance

Sorghum is able to tolerate drought better than most other grain crops. This can be attributed to:

- An exceptionally well-developed and finely branched root system, which is very efficient in the absorption of water.
- It has a small leaf area per plant, which limits transpiration.
- The leaves fold up more efficiently during warm, dry conditions than that of maize.
- It has an effective transpiration ratio of 1:310, as the plant uses only 310 parts of water to produce one part of dry matter,

compared to a ratio of 1:400 for maize.

- The epidermis of the leaf is corky and covered with a waxy layer, which protects the plant form desiccation.
- The stomata close rapidly to limit water loss. During dry periods, sorghum has the ability to remain in a virtually dormant stage and resume growth as soon as conditions become favorable. Even though the main stem can die, side shoots can develop and form seed when the water supply improves.

CHAPTER 1

A SORGHUM BICOLOR EXPRESSION ATLAS REVEALS DYNAMIC GENOTYPE-SPECIFIC EXPRESSION PROFILES FOR VEGETATIVE TISSUES OF GRAIN, SWEET AND BIOENERGY SORGHUMS

Nadia Shakoor, Ramesh Nair, Oswald Crasta, Geoffrey Morris, Alex Feltus and Stephen Kresovich

Keywords: *Sorghum bicolor,* Gene atlas, Transcriptome, Gene expression, Functional genomics, Microarray

BACKGROUND

Sorghum [*Sorghum bicolor* (L.) Moench] is a staple cereal crop for millions of people in the marginal, semi-arid environments of Africa and South Asia. Its unique and advanced ability to grow in regions of low and variable rainfall highlight its potential to impact agricultural productivity in widespread water-limited environments [1,2]. Originating and evolving across the diverse environmental landscape of Africa, morphological and physiological adaptation strategies has advanced sorghum as a naturally heat and drought-tolerant warm season C_4 grass that is more efficient at utilizing water, nitrogen and energy resources with respect to other major crops, including maize and wheat [1, 3, 4]. Occupying seven million hectares of farmland, the United States is currently the world's top sorghum producer (8.8 million annual metric tons), followed by India (7.0), Mexico (6.9), and Nigeria (4.8) (http://cgiar.org/sorghum). Cultivated in diverse climates and environmental conditions, the challenges of increasing performance and yield on marginal lands and cooler climates remains at the forefront of sorghum improvement efforts worldwide [5, 6].

Sorghum is globally established as an important source of food, feed, sugar and fiber, and recent interest in bioenergy feedstocks also spotlights sorghum as an attractive alternative for sustainable biofuel production [4]. Framed upon the 2009 sorghum reference genome [7], translational genomic resources have been developed that directly impact research in other closely related C_4 feedstock grasses, including switchgrass and *Miscanthus*[8, 9]. Comprehensive understanding of the genetic and molecular mechanisms that regulate metabolite biosynthesis, transport and storage in these species is essential for the efficient development of biofuel feedstocks.

Global transcriptome profiling further provides a means to access gene networks for the discovery of functional connections between genes, mRNAs and their regulatory proteins, and complex traits expressed through coordinated and dynamic gene networks across different tissues and developmental stages [10]. Over the last decade, microarray-based expression profiling has provided a reliable high-throughput platform for genome-wide analysis of gene expression in many organisms. Microarrays offer substantial advantages for functional genomics, as they are increasingly cost-effective, provide a comparable accuracy of expression profiling to RNA-sequencing, and have been shown to provide comprehensive expression data (up to 90% of the transcriptome) in a given tissue [11]. Well-established microarray data analysis tools are also available for querying, visualizing and analyzing the genomes and predicted genes [12, 13], as well as for analyzing the transcriptome profiling data and integrating with other public datasets [14, 15, 16, 17].

To provide insight into the sorghum transcriptome, we generated a record of gene expression in a set of seven tissues

Citation: Nadia Shakoor, Ramesh Nair, Oswald Crasta, Geoffrey Morris, Alex Feltus and Stephen Kresovich, A Sorghum bicolor expression atlas reveals dynamic genotype-specific expression profiles for vegetative tissues of grain, sweet and bioenergy sorghums, DOI: 10.1186/1471-2229-14-35.

and six diverse sorghum genotypes. The choice of samples reflects our aim to develop and enrich the current sorghum transcriptome literature. Previous studies have predominantly focused on reproductive tissues, and the majority of these reports do not represent the complete sorghum transcriptome. Several of these studies have also been limited to the reference genome (BTx623) or Keller, a recently resequenced sweet sorghum variety [18, 19, 20, 21, 22].

Comparable whole plant transcriptome maps are available for a number of other model species, including*Arabidopsis thaliana*[23], maize (*Zea mays)*[24], barley (*Hordeum vulgare*) [25], rice (*Oryza sativa*) [26, 27], and soybean (*Glycine max*) [28]. These recent transcriptome surveys were constructed with only one genotype or line/accession for their respective species of interest, whereas the present study aims to highlight the practical importance of examining expression profiles across diverse tissue types, developmental stages, as well as genotypes in order to accurately target genes and metabolic pathways for the efficient development of improved feedstocks.

Fundamental understanding of sorghum genomics is necessary for improving sorghum for agronomic and compositional traits. Specifically, genotypes with high biomass and increased levels of fermentable stem sugars are ideal for developing feedstocks for the biofuel industry. We developed this genomic resource, the whole-transcriptome array as well as the vegetative transcriptome in diverse genotypes and tissues, in order to facilitate the characterization of molecular networks and regulatory mechanisms governing important metabolic pathways including, but not limited to, cell wall biosynthesis for lignocellulosic biomass as well as synthesis, translocation, and storage of fermentable photosynthates for energy content. The relevance of our dataset is demonstrated by genotype and tissue-specific expression of the phenylpropanoid and lignin biosynthetic pathway genes.

Intended as readily available public resource for functional gene characterization, the transcriptome data presented here is available through NCBI's Gene Expression Omnibus (GEO) under accession number GSE49879, and the Sorghum Genome Array is available through Affymetrix (http://affymetrix.com).

RESULTS AND DISCUSSION

Generation and quality assessment of data

A whole-transcriptome exon array for *Sorghum bicolor* was custom-designed by Chromatin, Inc. (http://chromatininc.com) and Affymetrix: *Sorgh-WTa520972F*. This genechip contains 1,026,373 probes covering 149,182 exons (27,577 genes) across the *Sorghum bicolor* genome (10 chromosomes), chloroplast and mitochondria. The sequences used to construct the probesets included all identified *Sorghum bicolor*exons from the Sbi1 assembly (http://www.phytozome.net). Multiple probes were chosen for each exon, with a minimum of one probe per exon and 25 probes per gene. In addition to standard Affymetrix controls, positive controls in the microarray design included probes for constitutively expressed *Sorghum bicolor* genes (actin, ubiquitin and eIF4a1). Probes for intronic regions of actin and ubiquitin were also utilized to determine background expression levels. To study the sorghum transcriptome and build a gene expression atlas, we collected 78 diverse samples from various developmental stages and tissue types. In order to broadly capture sorghum genetic diversity, we included genotypes representing three major ideotypes, including grain, sweet, and bioenergy sorghums. Our study includes R159, an elite grain sorghum characterized by the valuable agronomic traits of uniform growth and disease resistance [29]. Grain sorghum is cultivated primarily for its high starch content, applications in human/animal health and nutrition, and as biofuel feedstock for ethanol production [5]. We also included two sweet sorghums, Fremont and Atlas, that produce increased biomass and accumulate high levels of fermentable carbohydrates in the stem. Additionally, Fremont is drought resistant and flowers early, while Atlas is less susceptible to lodging (due to a stiff stalk phenotype) and flowers later [30]. We also selected three bioenergy or high biomass lines, PI455230, PI152611, and AR2400 that produce increased levels of cellulosic material and are photoperiod sensitive, which allows the plant to produce higher amounts of vegetative matter under long day conditions. PI152611 is specifically a forage line, a fast-growing, highly digestible grass utilized for livestock feed [5, 29].

The primary goal of this study was to obtain relevant and applicable data for the research community developing sorghum as a global feedstock; this research interest guided our sample selection towards vegetative tissues, with a strong bias for stem tissues. A comprehensive trancriptomic profile of sorghum inflorescence and leaf data was recently made available to the community [19]. We compared the leaf RNA sequencing dataset with the present leaf dataset to demonstrate and confirm that our microarray analysis approach towards transcriptome profiling was appropriate. The Spearman correlation of the transcriptome across technologies is 0.61, which is consistent with several studies comparing RNA-seq and microarray methods for genome-wide transcriptome profiling [31, 32, 33]. The present comparison corroborates these studies and demonstrates that the microarray platform for expression profiling correlates well with current sequencing methods. With a common goal of crop improvement, complementary datasets such as these generate a core of information that can be explored for the functional characterization of genes and genetic pathways.

We assessed data quality for hybridization by comparing normalized signals of all probe sets between biological replicates using Pearson's correlation analysis. The biological replicates were highly correlated, with an average Pearson's correlation coefficient of 0.99. The highly reproducible results of the replicate data further validate the quality of the microarray platform and present dataset. Previous studies have consistently established strong correlations between qRT-PCR data and microarray data processed using robust multi-array analysis (RMA) [34, 35]. However, we also tested a small subset of these genes via qRT-PCR to validate the array-generated expression data and expression patterns across multiple tissue types.

To further assure data quality, we also examined the general expression patterns of well-characterized genes that have been highlighted for tissue-specific expression in previous studies. In microarray experiments with RNA isolated from shoot tips, we observed high expression levels for homologs of SPATULA, a shoot tip transcription factor that is strongly expressed in shoot tips and young leaf primordia [36]. Similarly, the sorghum homolog for TIP2-3, a root-specific aquaporin gene [37], was also expressed at higher levels in our study using root-isolated RNA.

Global gene expression patterns

We detected the expression of 19,354 genes in at least one of the 78 samples, representing 70.2% of all genes on the array (27,577 genes). The number of expressed transcripts detected in the various tissues ranged between 10,850 and 11,587 (representing 56 to 60% of all expressed genes on the array). Expressed genes were determined following established methods [24], and with a conservative and arbitrary expression threshold cutoff of 320 (five times the mean normalized signal from intronic gene probes used as controls), we found that 15.4% of genes on the array were detected in all tissues (4256/27,577). Gene ontology (GO) annotation analysis of these constitutively expressed genes reveals that most are involved in basic biological processes including development, protein synthesis/modification, and signal transduction. Similar to published work in maize, expression of constitutive genes varied among the samples, with the coefficient of variation (CV) ranging from 5% to 129%. With a CV of 10.4%, we identified a ubiquitin-conjugating enzyme, Sb09g023560, as one of the most stably expressed genes. This class of genes was also identified in the maize atlas as the most stably expressed among variable tissues [24].

A diverse range of plant tissues was sampled in this study; however, 29.8% of the probesets were not detected above our designated expression threshold level. Several plausible explanations can account for this incomplete expression coverage, including gene expression from specific tissues and/or developmental stages not included in this study, false positive gene models, and levels of expression below detection threshold limits. Further, the arrays were developed utilizing the BTx623 reference sequence and do not capture polymorphisms, copy number variation and presence-absence variation across all the sampled genotypes.

Transcriptome-based classification of sorghum tissues

A Pearson's distance correlation matrix was constructed to compare and evaluate the transcriptome data from each sample (Figure 1). This data shows strong correlations among and within the individual tissue types. The associated dendrogram reveals clustering according to tissue type as well as genotype, highlighting the significance of genotype-specific expression in this study (Figure 2). Utilizing GO categories, functional analysis of the identified gene sets revealed enrichment of known tissue-specific biological processes. For example, the leaf and shoot-associated gene sets were enriched for photosynthetic genes relative to the roots, as expected. We found that components of protein synthesis were overexpressed in the seedling roots and shoots, whereas genes involved in metabolism were over-represented in the shoot tip and stem tissues (Figure 3). These data identify core sets of genes associated with various biological processes and are clear targets for future study aimed to definitively characterize their functions in specific tissues.

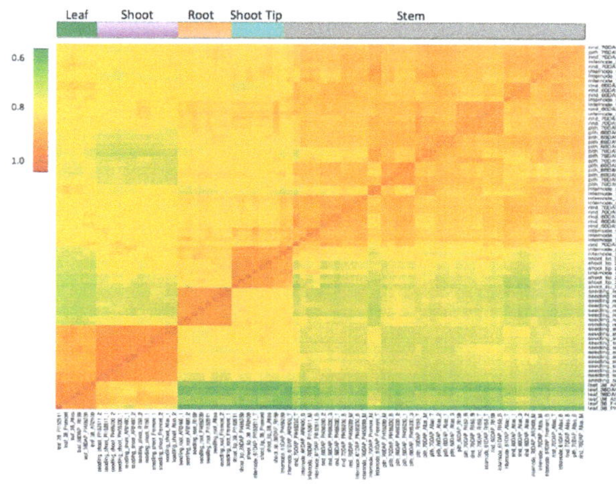

Figure 1: Pearson's correlation matrix of the whole dataset. Pair-wise Pearson correlation coefficients were calculated from the gene expression values of the whole transcriptome (27,577 genes) in all 78 samples. The hierarchical clusters were obtained based on Euclidian distance and are indicated by the color bar on the top side of the figure. The color scale indicates the degree of correlation.

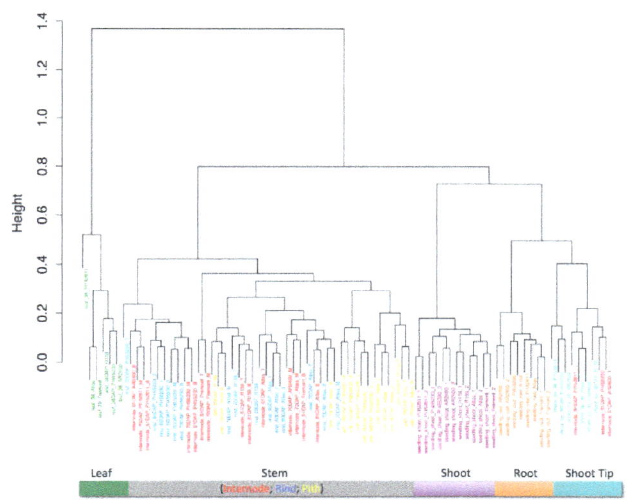

Figure 2: Cluster dendrogram of the whole dataset (78 samples). The hierarchical clusters of organs were grouped based on Euclidian distance. The 5 clusters are indicated by the color bar on the bottom side of the figure.

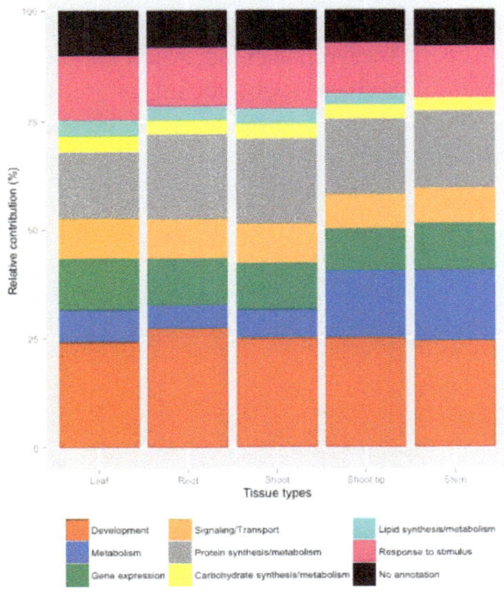

Figure 3: Functional category distribution of tissue-specific transcripts. Expression levels of select Gene Ontology categories across tissue types. The Sbi1.4 version of the sorghum annotation allowed for the identification of ~85% of expressed genes across all tissue types. The transcripts were manually verified and grouped into 7 functional categories based on Plant GO slim classifications.

Differential transcriptomes of developmentally distinct vegetative tissues were also apparent from the principal component analysis (PCA) (Figure 4). The PCA reveals clustering of functionally related tissue types, and the first two principal components (PC) of this analysis explain 68% of the variance among samples (PC1=48%, PC2=20%). Apical meristematic zones of the roots and shoot tips clustered together and weakly clustered with leaves, shoots and stem tissues. The large group of stem tissues (46 samples) including internode, pith, and rind strongly clustered together and weakly with the remaining tissues. These results are consistent with previous studies in maize and *P. halli* crop models, that show core similarities among stem-associated tissues and subsequent divergence of root and leaf samples [24, 38].

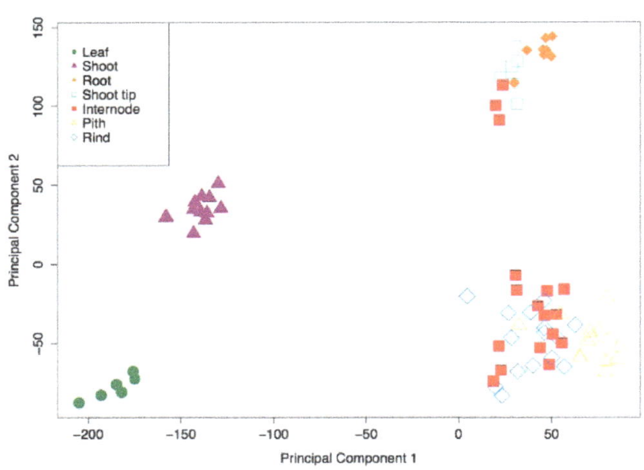

Figure 4: Classes sharing similar expression patterns. Principal component analysis was applied to 78 tissue samples, based on expression of 29,065 probe sets (27,577 genes, 654 controls and 834 small RNA probe sets). Each symbol represents a single sample. Tissue types are indicated by color and shape of symbol.

Interestingly, three out of 46 stem samples clustered near the group of meristematic tissues (roots and shoot tips). All three of these ‹outlier' samples were collected at the top internode, 61 days after planting (DAP) in three of the six sampled genotypes (PI455230, PI152611, and AR2400). At 70DAP, the stem samples from same genotypes clustered with the other stem samples. These lines are characterized as high biomass genotypes, whereas the remaining three genotypes can be characterized as either grain or sweet lines (R159, Atlas, and Fremont). The PCA indicates that at 61DAP, the patterns of gene expression in the stem of the high biomass lines are more related to meristematic regions, or regions of active growth. While it is possible that these three stem samples were collected too close to the meristematic shoot tip region, further study may indicate that the differential transcriptome in the stems of these lines capture a transition zone of gene expression in which sorghum commits to post-reproductive pathways of sugar production and grain fill versus continued biomass production. This result further demonstrates the importance of examining genotype, tissue type, as well as temporal expression patterns when targeting transcriptional programs of interest.

Tissue and genotype-specific patterns of gene expression

To identify tissue-specific genes, we created genotype-specific datasets for PI152611, Fremont, and AR2400, each representing one of three major classes of sorghum: forage, sweet, and high biomass types respectively. Excluding replicate tissues from the same major organ, we identified genes exclusively expressed in the leaf, shoot, root, shoot tip and stem (Figure 5). The leaf and meristematic shoot tips expressed the greatest number of tissue-specific transcripts across all three genotypes, whereas the seedling shoots expressed the fewest number of tissue-specific genes. Of particular interest in this dataset is the extent of variation observed across genotypes. For example, in stems, over 800 stem-specific genes are identified in representative examples of sweet and high biomass sorghum. Over 500 stem-specific genes are detected in forage sorghum; however, only 103 stem-specific genes are common among all three sorghum types. This lack of shared tissue-specific genes across genotypes is observed in all major tissue types. We also carried out this analysis for the small RNAs included on the array. Similar to gene expression, we observed both tissue and genotype-specific expression of the small RNAs. For purposes of functional crop improvement, these results highlight the significance of intra-species variation in sorghum and the importance of selecting the appropriate genotype for targeted changes to gene expression via transgenic and breeding approaches.

To illustrate the expression dynamics among tissues, we also calculated the relative gene expression levels (Z-scores) of each of the major tissues (Figure 6). Consistent with previous studies, tissues with a relatively higher number of tissue-specific genes (e.g. leaf, root, shoot tip, pith) had a wide distribution of genes deviating from their mean expression. Stem-associated tissues had similar expression profiles and gene expression was closer to the overall average across tissue types [24, 38].

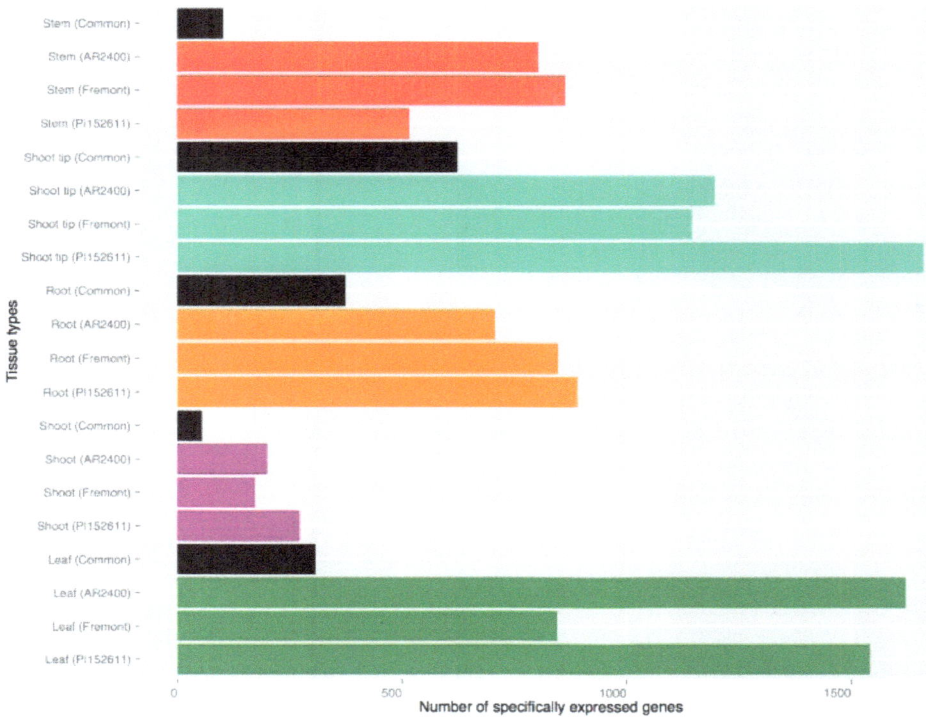

Figure 5: Number of tissue-specific genes in across sorghum ideotypes. AR2400: biomass sorghum; Fremont: sweet sorghum; PI152611: forage sorghum; Common: number of genes in common among all three ideotypes.

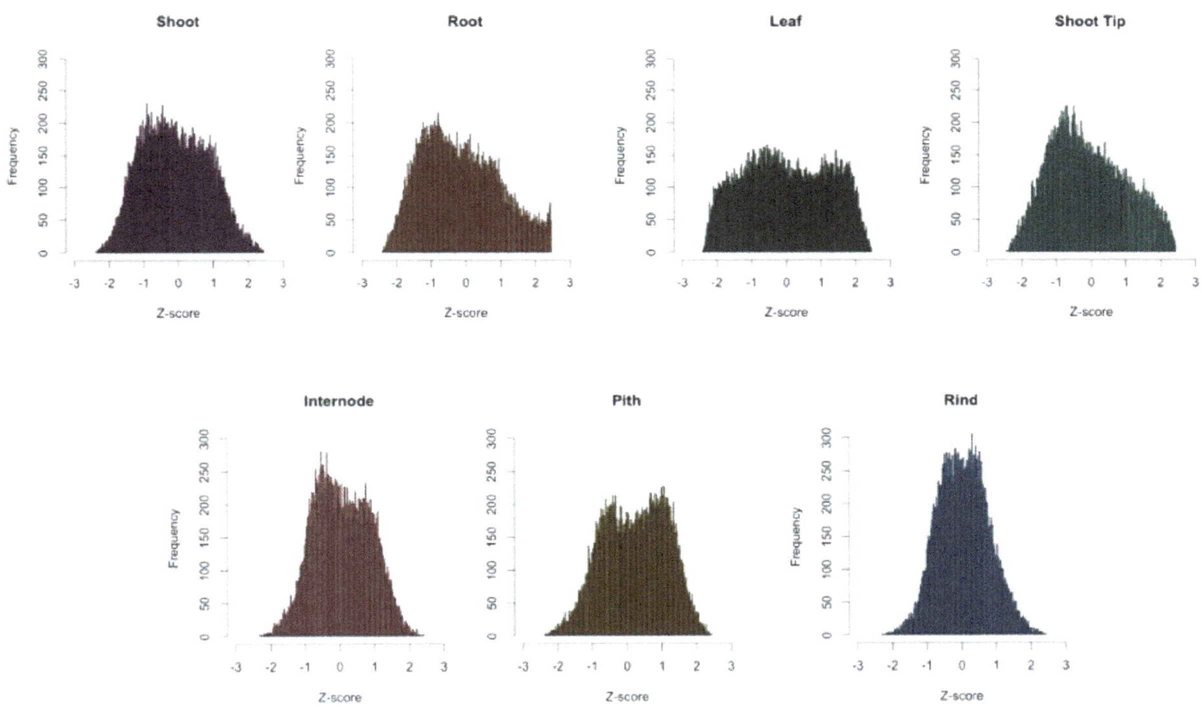

Figure 6: Distribution of global gene expression across sorghum tissue types. Histograms of relative expression levels (measured by Z-scores) in each tissue type. For each of these tissues, Z scores were calculated as follows: $Z=(X-X_{mean})/SD$, where X is the average expression of a given gene in a tissue, and X_{mean} and SD are the mean expression and standard deviation respectively of that gene across all the selected tissues.

We next attempted to determine whether functional gene classes were over-represented in specific genotypes. GO analysis did not reveal statistical differences in the enrichment of GO slim terms using agriGO (Fisher's exact test and the Yekutieli (false-discovery rate under dependency) multi-test adjustment method) [39]. However, this can partially be attributed to the incomplete annotation of the sorghum genome, as well as stage and tissue-specific expression not captured in our sample collection. To identify genotype-specific expression patterns, we examined the expression of several known sugar metabolizing enzymes and sucrose transporters in sorghum with the hypothesis that differential expression of these genes would be observable across genotypes. Differential expression between sweet and grain sorghum has recently been shown [21, 40], and our results further validate this observation, with the majority of sugar-related genes showing differential expression among tissues and genotypes. For example, sweet and high biomass varieties showed consistently higher expression of SPS2 and SPS5, sugar phosphate enzymes thought to play significant roles in sucrose biosynthesis, compared to grain varieties (Figure 7). A comprehensive gene list and more detailed expression analysis of sugar related genes across genotypes may provide insight into the mechanisms governing trade-offs in sorghum grain yield and stem sugar content.

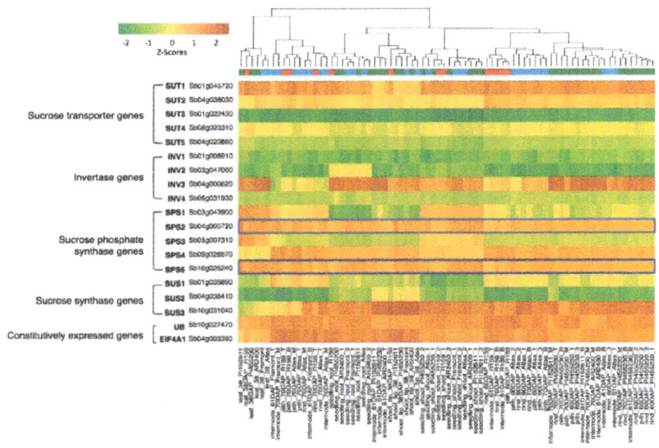

Figure 7: Hierarchical clustering of samples based on expression of sucrose metabolizing enzymes and sucrose transporter genes. Color bar key: Blue: sweet sorghum; red: grain sorghum; green: high biomass sorghum. Outlined in blue, the expression of sucrose phosphate synthase genes, SPS2 and SPS5, is consistently lower in grain types sweet and high biomass lines. Sugar metabolism gene list is appropriated from current literature [40].

We further analyzed tissue-specific transcripts to identify shared and specifically expressed genes in multiple tissues (Figures 8 and 9). To avoid variation in gene expression due to genotypic differences, we chose samples from the genotype Atlas for this analysis. We identified 587, 489, and 698 genes that are specifically expressed in leaf, stem and root and 232 and 688 unique genes that are expressed in shoot and shoot tips, respectively (Figure 8). We also identified 960 genes that are specifically expressed in stem rind (predominantly lignified sclerenchymatous cells) as compared to 928 genes that are specifically expressed in stem pith (predominantly non-lignified parenchymatous cells; Figure 9). This dataset provides a unique opportunity to discover target sets of genes in core sorghum varieties that may be useful for modulating gene expression in a tissue-dependent manner. For example, these rind and pith-specific genes can be studied as potential candidate genes for biomass content and targets for compositional modification of biofuel feedstocks. Further, identification of promoter elements and corresponding DNA-binding regulatory proteins that regulate tissue-specific expression of genes could be identified from these data. As a direct application of this study, we are currently analyzing the promoter regions of candidate genes that are differentially expressed in the rind versus pith region of stem tissues.

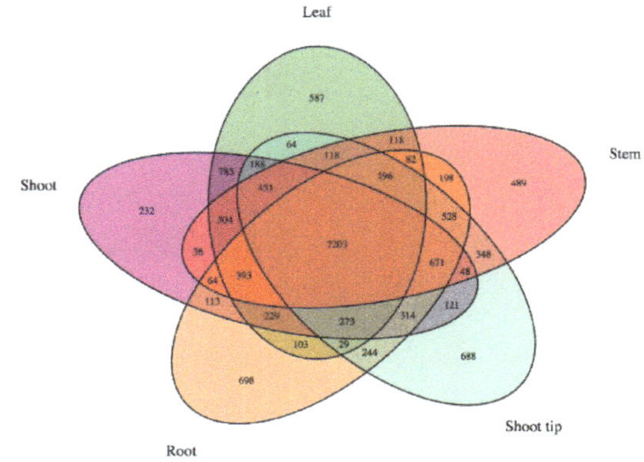

Figure 8: Number of shared and specific expression profiles of genes expressed in multiple tissue types (Atlas genotype).

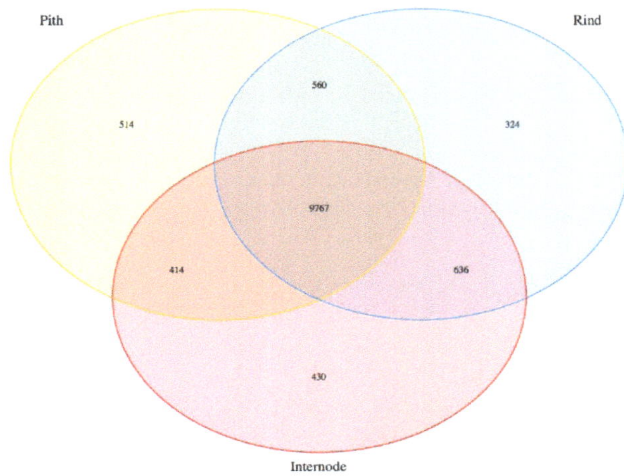

Figure 9: Number of shared and specific expression profiles of genes expressed in multiple stem tissues (Atlas genotype).

Tissue-specific expression of genes involved in the phenylpropanoid-monolignol pathway

To exemplify the functional utility of this data, we highlighted the expression data of 10 key enzymes associated with the phenylpropanoid-monolignol biosynthesis pathway. Currently, one of the primary strategies for bioenergy feedstock improvement is through lignin modification. Alterations in lignin content and composition aim to improve the digestibility of forage and saccharification efficiency of lignocellulosic biofuels [41, 42]. Thus, modifying the expression of genes in the lignin biosynthesis pathway is an attractive approach to achieving this goal.

Annotated in several databases, the majority of known and putative genes and homologs were analyzed for: phenylalanine ammonia-lyase (PAL, 9 sequences), coumaroyl shikimate 3'-hydroxylase (C3'H, 1), ferulate 5-hydroxylase (F5H, 3), cinnamate 4-hydroxylase (C4H, 3), 4-coumarate:CoA ligase (4CL, 5), cinnamoyl CoA reductases (CCR, 3), hydroxycinnamoyl CoA:shikimate hydroxycinnamoyl transferase (HCT, 1), caffeoyl-CoA 3-O-methyltransferase (CCoAMOT, 6), caffeic acid 3-O-methyltransferase (COMT,

1), and cinnamyl alcohol dehydrogenase (CAD, 1). Similar to previous studies in maize and switchgrass, the highest expression of these genes was found in the roots and stems [8, 43]. Further, hierarchical clustering reveals that the expression of lignin biosynthesis genes varies with developmental stage, as well tissue type and genotype (Figure 10). Distinct expression signatures of gene homologs as well as clustering of above-ground vegetative tissues according to developmental stage has precedence in maize and, in general, most of the lignin genes showed organ-specific expression patterns consistent with studies in related species [24, 38].

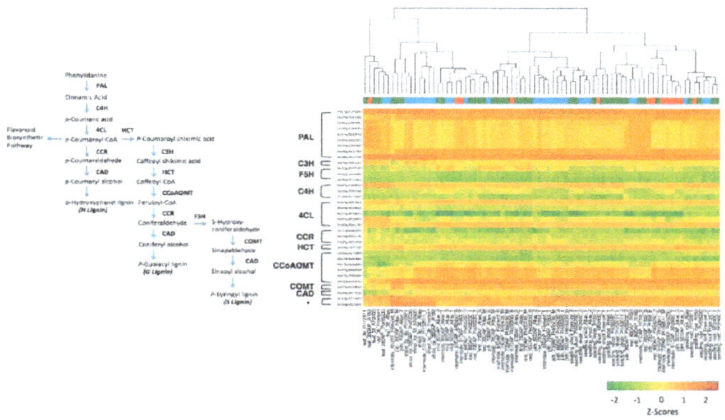

Figure 10: Hierarchical clustering of tissues based on expression of phenylpropanoid-monolignol biosynthesis pathway genes. *Constitutively expressed genes: Ubiquitin: Sb10g027470; EIF4A1: Sb04g003390. Color bar key: Blue: sweet sorghum; red: grain sorghum; green: high biomass sorghum. The color scale indicates the relative gene expression (Z-scores). Red, yellow, and green represent high, medium, and low levels of gene expression, respectively. The phenylpropanoid-monolignol pathway and enzyme nomenclature is appropriated from current literature [44].

CONCLUSIONS

Comprehensive transcriptome profiling provides a global over of gene networks and allows for the discovery of functional connections between genes, mRNAs and their regulatory proteins. In the present study, we constructed a gene expression atlas covering an array of tissues, developmental stages and genotypes using the first commercially available sorghum microarray (Sorgh-WTa520972F). We observed tissue and genotype-specific expression patterns of relevant metabolic pathways that highlight the significance of intra-species variation in sorghum.

Developed as a new resource for crop breeding and genomic discovery, Sorgh-WTa520972F is produced by Affymetrix and is available to the public research community. We are currently utilizing this microarray to identify differential gene expression related to key metabolic processes (*e.g.,* starch/lignin biosynthesis) for the identification of regulatory regions. Additional avenues for future study with this array are wide-ranging and can include gene expression profiling during abiotic/biotic stress, plant infection and disease establishment to investigate genetic mechanisms and applications to plant breeding and crop improvement. Detailed expression analysis of small RNAs included in the array design may also reveal key insights in diverse biological processes, including RNA-guided gene regulation. Sorgh-WTa520972F can also be utilized in quantitative trait locus (QTL) mapping and validation methods (*e.g.,* identify differentially expressed genes from 'tolerant' versus 'sensitive' varieties). Minimal costs associated with microarray analysis allow for the generation of high-throughput expression profiles or combinations of profiles of elite breeding lines for accelerated crop-breeding efforts. Applications of this resource can target numerous agronomic traits in sorghum as well as provide insight in closely related grasses (*e.g.,* sugarcane, switchgrass, *Miscanthus x giganteus*) for improved feedstock development.

METHODS

Tissue collection

To study the sorghum transcriptome and build the present gene expression atlas, we collected 78 samples from various developmental stages and tissue types. Six diverse sorghum genotypes were grown in Chromatin's greenhouse and field sites (Champaign, IL). These six genotypes were chosen to represent ideotypes of sorghum cultivation, including sweet, grain and high biomass sorghum varieties. Greenhouse grown seedling shoot and root samples were collected at 10DAP, which is roughly five days after plant emergence. Whole leaf and meristematic shoot tip samples were collected at 38DAP. This time-point captures the active growth phase of vegetative structures, including leaves, shoots and tillers. The stem tissue samples were collected at two time points: 61 and 70DAP. At 61DAP, the stem is fully formed in both flowering and non-flowering types. In flowering types, the head is also fully formed, and the period between 61 and 70DAP is a stage of active metabolism, capturing the transition between flowering (61DAP) and active grain filling (70DAP) [45]. The stem tissue was further dissected into the pith and the rind. As a bioenergy crop, the majority of fermentable sugar available in sorghum is present in the pith. The majority of lignin, however, is found in the rind [46]. Two tissue types (shoot and root) were represented by two biological replicates.

Microarray design

A whole-transcriptome exon array for *Sorghum bicolor*: *Sorgh-WTa520972F* was designed and utilized for the present expression study. The array contains 1,026,373 probes covering 149,182 exons (27,577 genes) across the *Sorghum bicolor* nuclear, chloroplast and mitochondrial genome. The sequences used to construct the probesets included all identified *Sorghum bicolor* exons from the Sbi1 assembly and Sbi1.4 annotation (http://phytozome.net). We also added sequences for putative non-coding RNAs in *Sorghum bicolor* that may play a role in gene regulation (*e.g.,* rRNAs, tRNAs, snoRNAs and microRNAs). Confirmed functional small RNAs in closely related species (maize, sugarcane) were also included in our array design (http://bioinformatics.cau.edu.cn/PMRD, http://www.ncrna.org/frnadb).

RNA Isolation and hybridization

Total RNA from all tissue types was extracted using a NucleoSpin RNA Plant Kit (Maxherey-Nagel, Germany). RNA integrity, as indicated by the detection of discrete ribosomal subunits, was verified electrophoretically. The RNA quality and quantity was further validated with a NanoDrop spectrophotometer (NanoDrop Technologies, Wilmington, DE). Prior to hybridization, the total RNA profile was also analyzed with Agilent 2100 Bioanalyzer (Agilent technologies, Waldbronn, Germany). Synthesis of cDNA, probe labeling and hybridization was performed by Precision Biomarker (Precision Biomarker Resources, Inc. Evanston, Illinois)

Data extraction and evaluation of gene expression

Background correction and normalization were performed using a robust multi-chip average (RMA) algorithm in the Bioconductor *Affy* package [13]. Present calls for expressed genes were determined following established methods [24]. In brief, an expressed gene was identified by a RMA-normalized linear expression of >/= 320 in at least one of the 78 samples. The expression cut-off was five times the mean RMA-normalized signal from 576 negative-control oligos selected from the intronic regions of known constitutive genes (*e.g.,*actin, ubiquitin, and eIF4a1). A mean signal intensity of 64 was determined for the negative control oligos analyzed across all 78 slides. Constitutively expressed genes were identified by a RMA-normalized linear expression value of >/= 320 in all 78 samples.

Principal component analysis, hierarchical clustering and z-scores

To study the biological relatedness and identify expression trends among the samples, we utilized the cmdscale function and then plotted using R. We used RMA-normalized $\log_2$ normalized expression values in the PCA analysis. Hierarchical clustering was performed using RMA-normalized $\log_2$ normalized expression values and clustered using Pearson's correlation analysis. The Z scores were calculated as follows: $Z=(X-X_{mean})/SD$, where X is the average expression of a given gene in a tissue, and X_{mean} and SD are the mean expression and standard deviation respectively of that gene across all the selected tissues.

GO Slim enrichment analysis

We evaluated enrichment of GO slim terms of biological process category (http://geneontology.org/GO.slims) in agriGO (http://bioinfo.cau.edu.cn/agriGO/) by Fisher's exact test (p-value ≤0.05) and the Yekutieli (false-discovery rate under dependency) multi-test adjustment method [39].

qRT-PCR

The relative mRNA expression was measured using Peltier Thermal Cycler PTC-200 PCR machine (MJ Research, Waltham, MA, USA) and the SuperScript III Platinum SYBR Green One-Step qRT- PCR kit (Invitrogen, Carlsbad, CA). Three independent reverse transcription reactions were performed for each RNA sample, and qRT-PCR was carried out under the following conditions: 100 nanograms of each RNA sample was reverse transcribed at 60°C for 3 minutes, and reverse transcription was followed by initial activation at 95°C for 5 minutes, and 40 amplification cycles at 95°C for 15 s and 50°C for 30s. Results were analysed using MJ Opticon Monitor 3.1.32 software, and relative expression of mRNA was calculated by the comparative Ct method ($2^{-[\Delta][\Delta]Ct}$) [47]. Gene expression values across tissue types were normalized to ubiquitin expression.

Availability of supporting data

The transcriptome dataset supporting the results of this article is available through NCBI's Gene Expression Omnibus (GEO) under accession number GSE49879, and the Sorghum Genome Array is available through Affymetrix (http://affymetrix.com).

DECLARATIONS

Acknowledgments

We would like to thank Otto Folkerts and Jennifer Mach for critically reing and editing this manuscript. We would also like to thank Shu-wen Huang for her contribution to data procurement. For their contributions to sample collection and RNA preparation, we also thank Zhanmin Fan, Piero Barone, Soumya Bhattacharya, Michael Root, David Sobczak, Kaushal Patel, Duyen Tran, and Shivani Verma. This project is partially funded through ARPA-E Cooperative Agreement DE-AR0470-1541.

Authors' Contributions

NS wrote the manuscript, contributed for microarray design, carried out RNA isolations, validation assays, data analysis and interpretation of the results. RN and OC contributed to experimental and microarray design and participated in tissue sampling and phenotyping. GM and AF participated in data analysis. SK participated in analysis and interpretation of the results. All authors read, revised and approved the final manuscript.

REFERENCES

1. Rosenow D, Quisenberry J, Wendt C, Clark L: Drought tolerant sorghum and cotton germplasm. Agric Water Manage. 1983, 7 (1): 207-222.
2. Mann J, Kimber C, Miller F: The origin and early cultivation of sorghums in Africa. Texas Agric Exp Station. 1983, 1: 128074-Available electronically from http://hdl.handle.net/1969.1/1280
3. de Vries SC, van de Ven GW, van Ittersum MK, Giller KE: Resource use efficiency and environmental performance of nine major biofuel crops, processed by first-generation conversion techniques. Biomass and Bioenergy. 2010, 34 (5): 588-601. 10.1016/j.biombioe.2010.01.001.
4. Kimber CT, Dahlberg JA, Kresovich S: The gene pool of Sorghum bicolorand its improvement. In Genomics of the Saccharinae. New York: Springer;2013:23–41.
5. Shoemaker C, Bransby DI: The role of sorghum as a bioenergy feedstock.In Sustainable alternative fuel feedstock opportunities, challenges androadmaps for six US regions. Ankeny, IA: Soil and Water ConservationSociety; 2010:149–159.
6. Burow G, Burke JJ, Xin Z, Franks CD: Genetic dissection of early-season cold tolerance in sorghum (Sorghum bicolor (L) Moench). Mol Breeding. 2011, 28 (3): 391-402. 10.1007/s11032-010-9491-4.

7. Paterson AH, Bowers JE, Bruggmann R, Dubchak I, Grimwood J, Gundlach H, Haberer G, Hellsten U, Mitros T, Poliakov A: The Sorghum bicolor genome and the diversification of grasses. Nature. 2009, 457 (7229): 551-556. 10.1038/nature07723.

8. Zhang J-Y, Lee Y-C, Torres-Jerez I, Wang M, Yin Y, Chou W-C, He J, Shen H, Srivastava AC, Pennacchio C: Development of an integrated transcript sequence database and a gene expression atlas for gene discovery and analysis in switchgrass (Panicum virgatum L). Plant J. 2013, 74 (1): 160-173. 10.1111/tpj.12104.

9. Swaminathan K, Alabady MS, Varala K, De Paoli E, Ho I, Rokhsar DS, Arumuganathan AK, Ming R, Green PJ, Meyers BC: Genomic and small RNA sequencing of Miscanthus x giganteus shows the utility of sorghum as a reference genome sequence for Andropogoneae grasses. Genome Biol. 2010, 11 (2): R12-10.1186/gb-2010-11-2-r12.

10. Komili S, Silver PA: Coupling and coordination in gene expression processes: a systems biology . Nat Rev Genet. 2008, 9 (1): 38-48. 10.1038/nrg2223.

11. Mane S, Evans C, Cooper K, Crasta O, Folkerts O, Hutchison S, Harkins T, Thierry-Mieg D, Thierry-Mieg J, Jensen R: Transcriptome sequencing of the Microarray Quality Control (MAQC) RNA reference samples using next generation sequencing. BMC Genomics. 2009, 10 (1): 264-10.1186/1471-2164-10-264.

12. Goodstein DM, Shu S, Howson R, Neupane R, Hayes RD, Fazo J, Mitros T, Dirks W, Hellsten U, Putnam N: Phytozome: a comparative platform for green plant genomics. Nucleic Acids Res. 2012, 40 (D1): D1178-D1186. 10.1093/nar/gkr944.

13. Gentleman RC, Carey VJ, Bates DM, Bolstad B, Dettling M, Dudoit S, Ellis B, Gautier L, Ge Y, Gentry J: Bioconductor: open software development for computational biology and bioinformatics. Genome Biol. 2004, 5 (10): R80-10.1186/gb-2004-5-10-r80.

14. Barrett T, Troup DB, Wilhite SE, Ledoux P, Rudnev D, Evangelista C, Kim IF, Soboleva A, Tomashevsky M, Edgar R: NCBI GEO: mining tens of millions of expression profiles–database and tools update. Nucleic Acids Res. 2007, 35 (suppl 1): D760-D765.

15. Zimmermann P, Laule O, Schmitz J, Hruz T, Bleuler S, Gruissem W: Genevestigator transcriptome meta-analysis and biomarker search using rice and barley gene expression databases. Mol Plant. 2008, 1 (5): 851-857. 10.1093/mp/ssn048.

16. Joung J-G, Corbett AM, Fellman SM, Tieman DM, Klee HJ, Giovannoni JJ, Fei Z: Plant MetGenMAP: an integrative analysis system for plant systems biology. Plant Physiol. 2009, 151 (4): 1758-1768. 10.1104/pp.109.145169.

17. Katari MS, Nowicki SD, Aceituno FF, Nero D, Kelfer J, Thompson LP, Cabello JM, Davidson RS, Goldberg AP, Shasha DE: VirtualPlant: a software platform to support systems biology research. Plant Physiol. 2010, 152 (2): 500-515. 10.1104/pp.109.147025.

18. Buchanan CD, Lim S, Salzman RA, Kagiampakis I, Morishige DT, Weers BD, Klein RR, Pratt LH, Cordonnier-Pratt M-M, Klein PE: Sorghum bicolor's transcriptome response to dehydration, high salinity and ABA. Plant Mol Biol. 2005, 58 (5): 699-720. 10.1007/s11103-005-7876-2.

19. Davidson RM, Gowda M, Moghe G, Lin H, Vaillancourt B, Shiu SÄ, Jiang N, Robin Buell C: Comparative transcriptomics of three Poaceae species reveals patterns of gene expression evolution. Plant J. 2012, 71 (3): 492-502.

20. Salzman RA, Brady JA, Finlayson SA, Buchanan CD, Summer EJ, Sun F, Klein PE, Klein RR, Pratt LH, Cordonnier-Pratt M-M: Transcriptional profiling of sorghum induced by methyl jasmonate, salicylic acid, and aminocyclopropane carboxylic acid reveals cooperative regulation and novel gene responses. Plant Physiol. 2005, 138 (1): 352-368. 10.1104/pp.104.058206.

21. Jiang S-Y, Ma Z, Vanitha J, Ramachandran S: Genetic variation and expression diversity between grain and sweet sorghum lines. BMC Genomics. 2013, 14 (1): 18-10.1186/1471-2164-14-18.

22. Dugas DV, Monaco MK, Olson A, Klein RR, Kumari S, Ware D, Klein PE: Functional annotation of the transcriptome of Sorghum bicolor in response to osmotic stress and abscisic acid. BMC Genomics. 2011, 12 (1): 514-10.1186/1471-2164-12-514.

23. Schmid M, Davison TS, Henz SR, Pape UJ, Demar M, Vingron M, Schölkopf B, Weigel D, Lohmann JU: A gene expression map of Arabidopsis thaliana development. Nat Genet. 2005, 37 (5): 501-506. 10.1038/ng1543.

24. Sekhon RS, Lin H, Childs KL, Hansey CN, Buell CR, de Leon N, Kaeppler SM: Genome-wide atlas of transcription during maize development. Plant J. 2011, 66 (4): 553-563. 10.1111/j.1365-313X.2011.04527.x.

25. Druka A, Muehlbauer G, Druka I, Caldo R, Baumann U, Rostoks N, Schreiber A, Wise R, Close T, Kleinhofs A: An atlas of gene expression from seed to seed through barley development. Funct Integr Genomics. 2006, 6 (3): 202-211. 10.1007/s10142-006-0025-4.

26. Jiao Y, Tausta SL, Gandotra N, Sun N, Liu T, Clay NK, Ceserani T, Chen M, Ma L, Holford M: A transcriptome atlas of rice cell types uncovers cellular, functional and developmental hierarchies. Nat Genet. 2009, 41 (2): 258-263. 10.1038/ng.282.

27. Wang L, Xie W, Chen Y, Tang W, Yang J, Ye R, Liu L, Lin Y, Xu C, Xiao J: A dynamic gene expression atlas covering the entire life cycle of rice. Plant J. 2010, 61 (5): 752-766. 10.1111/j.1365-313X.2009.04100.x.

28. Libault M, Farmer A, Joshi T, Takahashi K, Langley RJ, Franklin LD, He J, Xu D, May G, Stacey G: An integrated transcriptome atlas of the crop model Glycine max, and its use in comparative analyses in plants. Plant J. 2010, 63 (1): 86-99.

29. USDA, ARS, National Genetic Resources Program: Germplasm ResourcesInformation Network - (GRIN). Beltsville, Maryland: National GermplasmResources Laboratory; 2014.

30. Martin JH, Stephens JC: The culture and use of sorghums for forage.Washington D.C: US Department of Agriculture; 1955.

31. Agarwal A, Koppstein D, Rozowsky J, Sboner A, Habegger L, Hillier LW, Sasidharan R, Reinke V, Waterston RH, Gerstein M: Comparison and calibration of transcriptome data from RNA-Seq and tiling arrays. BMC Genomics. 2010, 11 (1): 383-10.1186/1471-2164-11-383.

32. Kogenaru S, Yan Q, Guo Y, Wang N: RNA-seq and microarray complement each other in transcriptome profiling. BMC Genomics. 2012, 13 (1): 629-10.1186/1471-2164-13-629.

33. Marioni JC, Mason CE, Mane SM, Stephens M, Gilad Y: RNA-seq: an assessment of technical reproducibility and comparison with gene expression arrays. Genome Res. 2008, 18 (9): 1509-1517. 10.1101/gr.079558.108.

34. Dallas PB, Gottardo NG, Firth MJ, Beesley AH, Hoffmann K, Terry PA, Freitas JR, Boag JM, Cummings AJ, Kees UR: Gene expression levels assessed by oligonucleotide microarray analysis and quantitative real-time RT-PCR - how well do they correlate?. BMC Genomics. 2005, 6 (1): 59-10.1186/1471-2164-6-59.

35. Morey JS, Ryan JC, Van Dolah FM: Microarray validation: factors influencing correlation between oligonucleotide microarrays and real-time PCR. Biol Proced Online. 2006, 8 (1): 175-193. 10.1251/bpo126.

36. Ichihashi Y, Horiguchi G, Gleissberg S, Tsukaya H: The bHLH transcription factor SPATULA controls final leaf size in Arabidopsis thaliana. Plant Cell Physiol. 2010, 51 (2): 252-261. 10.1093/pcp/pcp184.

37. Lopez F, Bousser A, Sissoéff I, Hoarau J, Mahé A: Characterization in maize of ZmTIP2-3, a root-specific tonoplast intrinsic protein exhibiting aquaporin activity. J Exp Bot. 2004, 55 (396): 539-541. 10.1093/jxb/052.

38. Meyer E, Logan TL, Juenger TE: Transcriptome analysis and gene expression atlas for Panicum hallii var. filipes, a diploid model for biofuel research. Plant J. 2012, 70 (5): 879-890. 10.1111/j.1365-313X.2012.04938.x.

39. Du Z, Zhou X, Ling Y, Zhang Z, Su Z: agriGO: a GO analysis toolkit for the agricultural community. Nucleic Acids Res. 2010, 38 (suppl 2): W64-W70.

40. Qazi HA, Paranjpe S, Bhargava S: Stem sugar accumulation in sweet sorghum - activity and expression of sucrose metabolizing enzymes and sucrose transporters. J Plant Physiol. 2012, 169 (6): 605-613. 10.1016/j.jplph.2012.01.005.

41. Zhao Q, Tobimatsu Y, Zhou R, Pattathil S, Gallego-Giraldo L, Fu C, Jackson LA, Hahn MG, Kim H, Chen F: Loss of function of cinnamyl alcohol dehydrogenase 1 leads to unconventional lignin and a temperature-sensitive growth defect in Medicago truncatula. Proc Natl Acad Sci USA. 2013, 110 (33): 13660-13665. 10.1073/pnas.1312234110.

42. Saballos A, Sattler SE, Sanchez E, Foster TP, Xin Z, Kang C, Pedersen JF, Vermerris W: Brown midrib2 (Bmr2) encodes the major 4 coumarate: coenzyme A ligase involved in lignin biosynthesis in sorghum (Sorghum bicolor (L.) Moench). Plant J. 2012, 70 (5): 818-830. 10.1111/j.1365-313X.2012.04933.x.

43. Jouanin L, Lapierre C: Lignins: biosynthesis. Biodegradation andBioengineering: Elsevier Science; 2012.

44. Sattler SE, Funnell-Harris DL: Modifying lignin to improve bioenergy feedstocks: strengthening the barrier against pathogens?. Front Plant Sci. 2013, 4: 70.

45. Vanderlip R, Reeves H: Growth stages of sorghum [Sorghum bicolor (L.) Moench]. Agronomy J. 1972, 64 (1): 13-16. 10.2134/agronj1972.00021962006400010005x.

46. Whitfield MB, Chinn MS, Veal MW: Processing of materials derived from sweet sorghum for biobased products. Ind Crops Products. 2012, 37 (1): 362-375. 10.1016/j.indcrop.2011.12.011.

47. Livak KJ, Schmittgen TD: Analysis of relative gene expression data using real-time quantitative PCR and the 2(-Delta Delta C(T)) Method. Methods. 2001, 25 (4): 402-408. 10.1006/meth.2001.1262.

CHAPTER 2

CLONING AND EXPRESSION ANALYSIS OF CINNAMOYL-COA REDUCTASE (CCR) GENES IN SORGHUM

Jieqin Li, Feifei Fan, Lihua Wang, Qiuwen Zhan, Peijin Wu, Junli Du, Xiaocui Yang and Yanlong Liu

Keywords: Sorghum, Gene cloning, qRT-PCR, CCR, Phylogenetic analysis

INTRODUCTION

Lignin is a complex aromatic polymer present mainly in the secondary cell walls of vascular plants. It plays an important role in specialized conducting and supporting tissues of plants, facilitating water transport, providing mechanical strength, and defending against biotic and abiotic stress (Bhuiyan et al., 2009; Vermerris, Sherman & McIntyre, 2010; Jin et al., 2014). The lignin biosynthetic pathway has attracted researchers' attention because lignin is a limiting factor in a number of agro-industrial processes, such as chemical pulping, forage digestion, and the conversion of lignocellulosic plant biomass to bioethanol (Poovaiah et al., 2014; Tang et al., 2014).

Lignin is derived from the polymerisation of monomeric subunits known as monolignols. The three main monolignols in grasses are *p*-coumaryl, coniferyl and sinapyl alcohols, which give rise to *p*-hydroxyphenyl (H), guaiacyl (G) and syringyl (S) residues, respectively (Boerjan, Ralph & Baucher, 2003; Vermerris, Sherman & McIntyre, 2010). The biosynthesis of monolignols requires the action of a series of enzymes.

Abbreviations: *PAL*, phenylalaine ammonia-lyase; *C4H*, cinnamate 4-hydroxylase; *4CL*, 4-coumarate: CoA ligase; *HCT*, *p*-hydroxycinnamoyl-CoA: quinate shikimate *p*-hydroxycinnamoyltransferase; *C3H*, *p*-coumarate 3-hydroxylase; *CCoAOMT*, caffeoyl-CoA *O*-methyltransferase; *CCR*, cinnamoyl-CoA reductase; *CAD*, cinnamyl alcohol dehydrogenase.

Citation: Li J, Fan F, Wang L, Zhan Q, Wu P, Du J, Yang X, Liu Y. (2016) Cloning and expression analysis of cinnamoyl-CoA reductase (CCR) genes in sorghum. PeerJ4:e2005 https://doi.org/10.7717/peerj.2005.

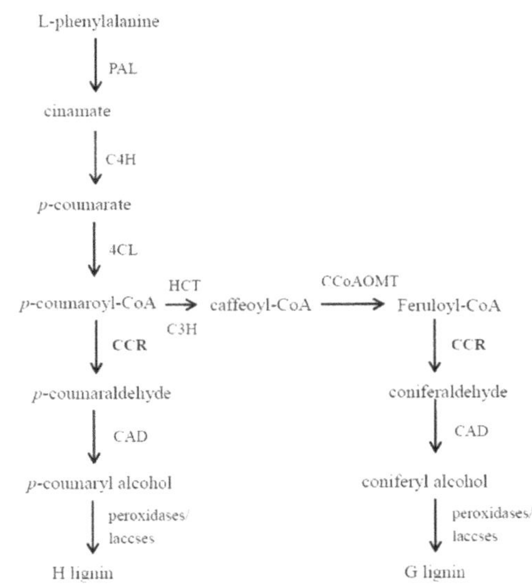

Figure 1: The outline of the lignin biosynthetic pathway.

Cinnamoyl-CoA reductase (CCR) is the first enzyme in the monolignol-specific branch of the lignin biosynthetic pathway, where it converts feruloyl-CoA to coniferaldehyde (Fig. 1) (Leple et al., 2007). Genes encoding CCR proteins have been studied in many species including Arabidopsis (Lauvergeat et al., 2001; Xue et al., 2015), rice (Kawasaki et al., 2006), tobacco (Chabannes et al., 2001), soybean (Luderitz & Grisebach, 1981), poplar tree (Leple et al., 2007), maize (Pichon et al., 1998; Tamasloukht et al., 2011), dallisgrass (Giordano et al., 2014) and wheat (Ma & Tian, 2005). The down regulation of CCR in transgenic Arabidopsis and tobacco leads to significant reduction of lignin content. The CCR gene family is very diverse in plants (Barakat et al., 2011). Multiple homologs of CCR genes involved in different functions can be present in a same plant. For instance, *AtCCR1* is involved in developmental lignification, while *AtCCR2* is for stress and elicitor response (Lauvergeat et al., 2001). It has been demonstrated that different CCR genes play different roles in plant's development. Although many studies have been done on CCR genes in different plants, to date, little work on CCR genes in sorghum has been done.

In this study, three CCR genes were cloned from the cDNA in sorghum seedlings by homology cloning. The structures and functions of their deduced proteins were analyzed by a bioinformatics method. To understand the expression of the three CCR genes in different tissues and treatments, expression analyses were performed. The findings provide a basis for further revealing the roles of the three CCR genes in sorghum.

MATERIALS AND METHODS

Plant materials and gene cloning

Sorghum Tx623B was used for gene cloning and expression analysis. Seeds of Tx623B were pretreated with 75% alcohol for 2 min and then washed three times with distilled water. The sterilized seeds were planted in pots. After germination, the seedlings were grown at 28 °C under 12 h light/12 h dark for two weeks, taken and ground in liquid N_2 for RNA extraction. Total RNA of the seedling samples was extracted using an RNA Prep Pure Plant kit (Tiangen Co., Beijing, China) and was reverse transcribed using a SuperScript II kit (TaKaRa Biomedical, Shiga, Japan).

Table 1: The primers used in the research

Primer name	Forward primer	Reverse primer	Usage
GSbCCR1	CATCACTCGACCGCACATAC	CAGCCAGCGAACAAACACTA	Gene cloning
GSbC-CR2-1	TCAGACTAATAACCCGCCTAG	GCAGTATCAGCGTTGGAAA	
GSbC-CR2-2	TCTTTCCGCTTCCACCGAT	GACTGCCAAAATTTAATAAC-CAA	
SSbCCR1	CGGACTAGTATGACCGTCGTC-GAC	TCCCCCGGGCGCACGGATG-GCGAT	Subcellular localiza-tion
SSbC-CR2-1	CGGACTAGTATGCCAACAG-CAGAG	CGCGGATCCTGATTTGTG-GAGTTG	
SSbC-CR2-2	CGGACTAGTATGGCCGTCGTC-GTG	TCCCCCGGGAAGTTTTGAAAT-CAA	
YSbCCR1	ATGCTGCTCGAGAAGGGATA-CAC	GTTCTTCGGGTCATCTGGGTTC	qRT-PCR
YSbC-CR2-1	ACCCACAAAGTGCAAGGAC-GAC	CCGCTGGTTCGTGAACTTG-TATCC	
YSbC-CR2-2	GGAGTACCCTATTCCGA-CAAGGTG	GCACTGGCGTGAACTTGATTCC	

Cloning primers for the three CCR genes (i.e., *Sb07g021680* named*SbCCR1*, *Sb02g014910* named *SbCCR2-1*, *Sb04g005510* named *SbCCR2-2*) were designed according to the CCR gene sequences in the National Center for Biotechnology Information (NCBI) database. The amplification primers are listed in Table 1. Polymerase chain reaction (PCR) was performed using KOD FX polymerase (Toyobo, Osaka, Japan). The amplified fragments were separated on 1% agarose gels and purified by DNA Gel Extraction Kit (AxyPrep DNA Gel Extraction kit; Axygen, USA). Then, the purified fragments were linked to pMD18-T vector (Takara, China) and transformed into *E. coli* DH5 . The recombinant plasmids were verified by PCR; the positive clones were sent to Genscript Company (Nanjing, China) for sequencing.

Drought and pest defense experiments

For the drought defense experiment, Tx623B seeds were sterilized as described above and germinated on wet paper. After germination, the seedlings were grown at 28 °C under 12 h light/12 h dark for 2 weeks. Then half of the seedlings were treated with 20% polyethylene glycol (PEG) 6000 as drought treatment and sampled at 24 h and 48 h. The other half of the seedlings were untreated with PEG 6000 but sampled at the same times as control. For the pest defense experiment, Tx623B seeds were sterilized, germinated and grown as in the drought defense experiment. Then, for half of the seedlings, 4 pest-attacked sorghum leaves with 2–4 sorghum aphids [*Melanaphis sacchari* (Zehntner)] on each were put close to each seedling so that the aphids would migrate to the seedlings; subsequently, the seedlings were placed in a cage. After the sorghum seedlings were attacked by the aphids, their pest-attacked leaves were sampled at 24 h and 48 h for RNA extraction. The other half of the seedlings was used as control and leaves were sampled at the same times for RNA extraction.

Sequence analysis and phylogenetic analysis

Nucleotide sequences were translated into protein sequences which were then aligned using BioEdit. Phylogenetic analyses were performed using the neighbor joining (NJ) method with Mega 5.1 software. Branch support was assessed with 1,000 bootstrap replicates.

Subcellular localization

The primer pairs for the three CCR genes were used to amplify the cDNA fragments encoding the full-length CCR proteins.

The PCR fragments for each CCR gene were inserted into the vector 1305GFP at the N-terminus of the green fluorescent protein (GFP) under the control of cauliflower mosaic virus 35S promoter. An *Agrobacterium tumefaciens* strain carrying the*35S::CCR1-GFP*, *35S::CCR2-1-GFP*, *35S::CCR2-2-GFP* or *35S::GFP*plasmid was infiltrated into *Nicotiana benthaminana* leaves and analyzed with confocal microscopy 48 h after agroinfiltration as described previously (Goodin et al., 2002). Fluorescence of GFP was observed with a Leica LSM710 confocal laser scanning microscope.

Quantitative real-time PCR

Quantitative real-time PCR (qRT-PCR) was performed using a SYBR Green supermix (Bio-Rad, USA) on an ABI prism 7900 real-time PCR system. The primers for the three CCR genes are listed in Table 1. The sorghum *eIF4a1*gene (*Sb04g003390*) was used as the endogenous control in the experiment. All reactions were run in three replicates. The $2^{-\Delta\Delta CT}$ method was used to analyze relative changes in gene expression (Livak & Schmittgen, 2001).

Absolute qRT-PCR

The three genes' coding fragments were aligned with pMD18-T vector. Fresh 10-fold serial dilutions were created from the three constructs. The concentration of the plasmid preps was measured using a Biodrop Touch spectrophotometer. The corresponding gene copy number in plasmid was calculated with the following equation:

$$\text{gene copy number} = \frac{6.02\times10^{23}(\text{copy/mol})\times\text{DNA amount (g)}}{\text{DNA length (bp)}\times660\,(\frac{\text{g}}{\text{mol}}/\text{bp})}$$

where DNA length stands for the combined length of plasmid and insert, in base pairs, and DNA amount equals the corresponding plasmid concentration times the volume (Gotia et al., 2016). The linear regression equation and R^2 were obtained by comparing the log values of the DNA concentration and the Ct values. Then, gene copy number was calculated with the corresponding linear regression equation.

RESULTS

Cloning and structure analysis of *SbCCR1, SbCCR2-1* and *SbCCR2-2*

Based on data obtained from the sorghum genome sequence and research results (Giordano et al., 2014), 3 CCR genes were identified by PCR and sequencing. Sequence annotation and protein chemical characterization of the three CCR genes are shown in Table 2. The sequencing results showed that the cDNA sequences of the three CCR genes were completely consistent with those in NCBI database. Molecular analysis of the deduced proteins showed that SbCCR1, SbCCR2-1 and SbCCR2-2 contained 374, 346 and 343 amino acids, respectively. Multiple alignments showed that the highest similarity between SbCCR1 and ZmCCR1 was 89.3% (Fig. 2). The similarity between SbCCR2-1 and SbCCR2-2 (59.8%) was higher than that between SbCCR1 and SbCCR2-1 (58.3%) and that between SbCCR1 and SbCCR2-2 (49.3%).

Table 2: Molecular characteristics of CCR genes in *Sorghum bicolor*

Name	Gene number	CDS length	Peptide residue	Theoretical Mw (kDa)	Theoretical PI
SbCCR1	Sb07g021680	1,549	374	40.24	5.47
SbCCR2-1	Sb02g014910	1,268	346	38.21	6.50
SbCCR2-2	Sb04g005510	1,378	343	37.95	8.79

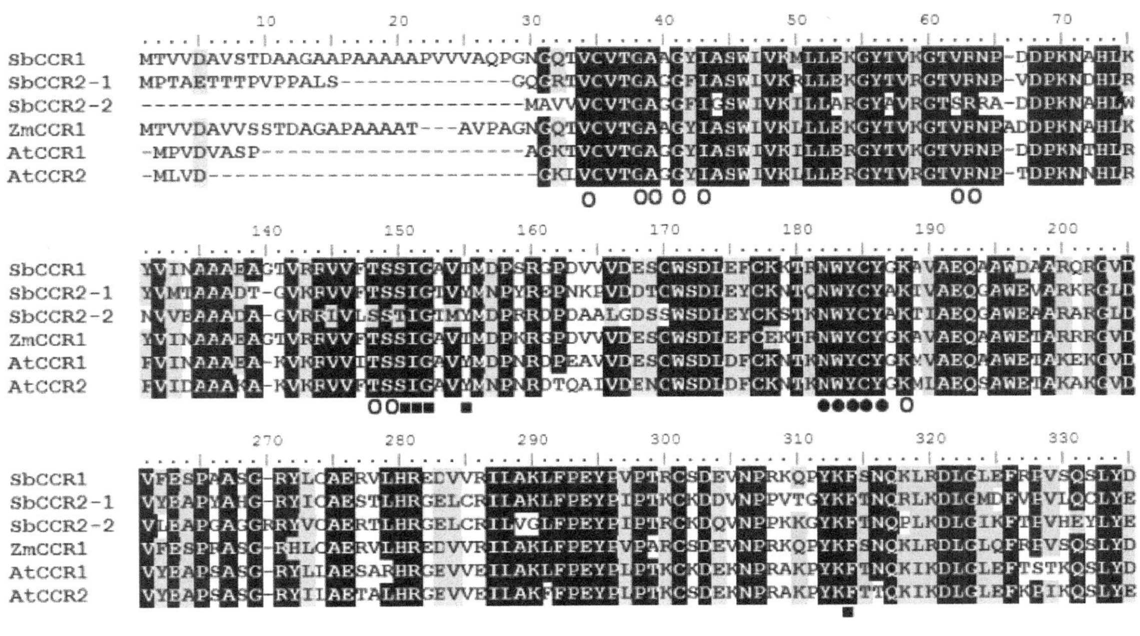

Figure 2: Multiple alignment of the protein sequences of sorghum CCR proteins and other plant CCR proteins.

The alignment includes sequences from *Arabidopsis* (*AtCCR1* and *AtCCR2*) and maize (*ZmCCR1*). The amino acids believed to be part of the active sites are shown below the alignment with the following codes: ●, catalytic activity domain; ○, NADP binding domain; ■, substrate binding domain.

Commonly, a CCR protein contains an NADP binding domain and a catalytic activity domain (NWYCY) (Pan et al., 2014). The multiple alignments showed that SbCCR1, SbCCR2-1 and SbCCR2-2 contained an NADP binding domain, an NWYCY domain and a substrate domain (Fig. 2). These results demonstrated that the three CCR genes contained the functional domains of CCR, suggesting that the three CCR proteins should be of CCR activities.

Phylogenetic analysis of the CCR proteins

Phylogenetic analysis of the CCR proteins encoded in sorghum and other plant genomes showed that there were two major clades (Fig. 3). One clade was mainly comprised of CCR1 proteins; the other was mainly comprised of CCR2 proteins. The two clades were supported with moderate bootstrap values. SbCCR1 was included in the first clade; while SbCCR2-1 and SbCCR2-2 were included in the second clade. The clade containing mainly CCR1 proteins was comprised of two major sub-clades: one consisted of dicotyledons (e.g., *Glycine max* and *Arabidopsis thaliana*), the other monocotyledons (e.g., *Sorghum bicolor*, *Hordeum vulgare*, *Lolium perenne*,*Zea mays* and *Oryza sativa*) and both were supported with high bootstrap values.

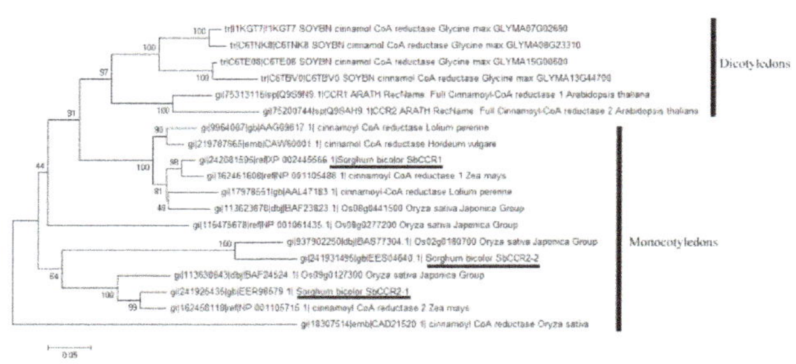

Figure 3: Phylogenetic analyses of sorghum CCR proteins and other plant CCR homologues.

Subcellular localization of CCR proteins

CCR genes perform their functions in cytoplasm. Therefore, the three CCR proteins were further analyzed with the protein subcellular localization prediction program WoLF PSORT (http://wolfpsort.org/). All of these proteins were predicted to be in cytoplasm. To verify the predicted results, a transient expression assay for the three CCR proteins was performed in *N. benthamiana* leaves. GFP alone was expressed as control, SbCCR1, SbCCR2-1 and SbCCR2-2 in full-length were fused to the N-terminus of GFP. The results showed that free GFP dispersed throughout the cytoplasm in the *N. benthamiana* epidermal cells; and the green fluorescent signal of GFP was not co-located with the auto-fluorescence of chlorophylls in chloroplasts. The localization patterns of SbCCR1-GFP, SbCCR2-1-GFP and SbCCR2-2-GFP were similar to that of free GFP (Fig. 4), indicating that the three CCR proteins were targeted to the cytoplasm.

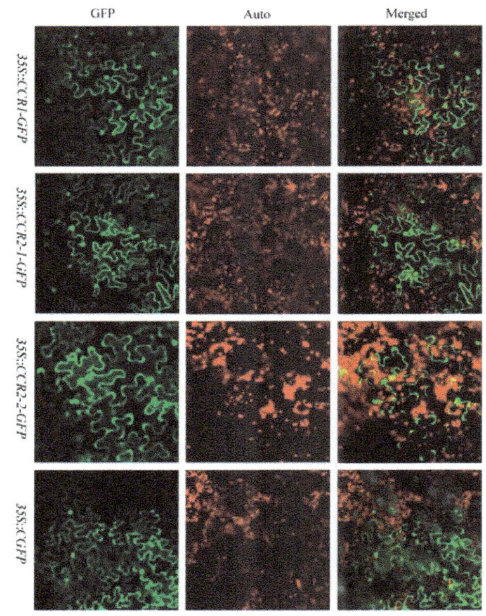

Figure 4: The subcellular localization of sorghum CCR genes in *N. benthamiana*leaves.

Transient expression of SbCCR1–GFP, SbCCR2-1-GFP and SbCCR2-2-GFP fusion proteins and GFP in *N. benthamiana* leaves; GFP, fluorescence of SbCCR1-GFP, SbCCR2-1-GFP, SbCCR2-2-GFP and GFP; Auto, Chl auto-fluorescence; Merged, merged images of GFP and Auto ones in bright.

Expression analysis of CCR genes in drought and pest treatments

To understand the functions of the three CCR genes in sorghum, expression analysis was performed in drought treatment (Fig. 5). The results showed that the expression levels of *SbCCR*1, *SbCCR2-1* and *SbCCR2-2* were significantly higher in the drought treatment than in CK at 24 h, indicating that all of the three CCR genes were induced by drought. But at 48 h, there were no differences in the expression levels of *SbCCR1* and *SbCCR2-2*between CK and the drought treatment; the expression level of *SbCCR2-1*was higher in the drought treatment than in CK. The results suggested that all of these genes were involved in the process of drought defense. *SbCCR1*and *SbCCR2-2* mainly responded in the first phase of drought defense, while the *SbCCR2-1* gene responded during the whole drought defense period.

To identify the functions of the three CCR genes in pest defense, expression analysis was performed. The expression levels of the *SbCCR1* and*SbCCR2-1* genes were not different between CK and the pest treatment at 24 h or 48 h (Fig. 6). But the expression levels of the *SbCCR2-2* gene at 24 h and 48 h in the pest treatment were significantly higher than in CK, indicating that *SbCCR2-2* was involved in pest defense.

Spatial expression patterns of CCR genes

The linear regression equations and R^2 values of the three CCR genes were obtained by comparing the log values of DNA concentration and the Ct values. The slopes of the standard quantification curves for *SbCCR1, SbCCR2-1* and *SbCCR2-2* were −3.25, −3.56 and −3.39, respectively. TheR^2 values of the three equations were 0.99. The gene copy numbers of*SbCCR1*, *SbCCR2-1* and *SbCCR2-2* in various plant organs, including leaf, spikelet, stem and root, were quantified using their corresponding linear regression equations. The results demonstrated that the three CCR genes were expressed in allthe tested organs (Fig. 7). The expression patterns of the three CCR genes were consistent with the deposition patterns of lignin in these organs. The highest expression levels of the three CCR genes were found in stem. The expression levels of *SbCCR1* and *SbCCR2-2* in stem and root were obviously higher than in spikelet and leaf, but those of*SbCCR2*-1 in stem and spikelet were higher than in leaf and root. The gene copy number of *SbCCR1* was significantly higher than those of *SbCCR2-1*and *SbCCR2-2* in root, leaf, stem and spikelet, especially in stem. This indicated that the expression level of *SbCCR1* was obviously higher than those of *SbCCR2-1* and *SbCCR2-2* in the tested sorghum tissues.

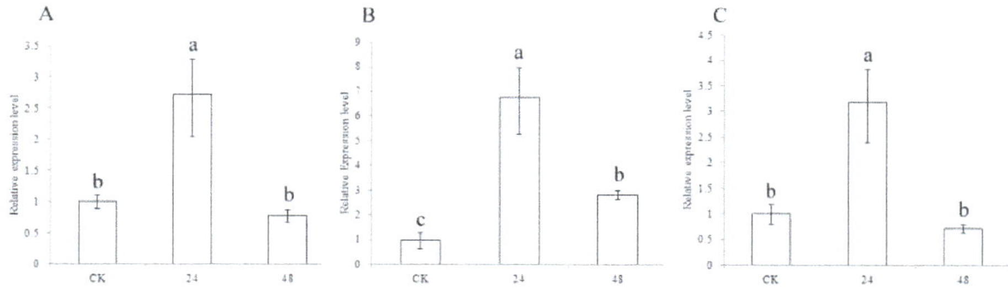

Figure 5: Expression of sorghum CCR genes in CK and the drought treatment at different times.

(A), *SbCCR1*; (B), *SbCCR2-1*; (C), *SbCCR2-2*; variance analysis was used for statistical test; different lowercase letters represent significant difference at 0.05 probability level.

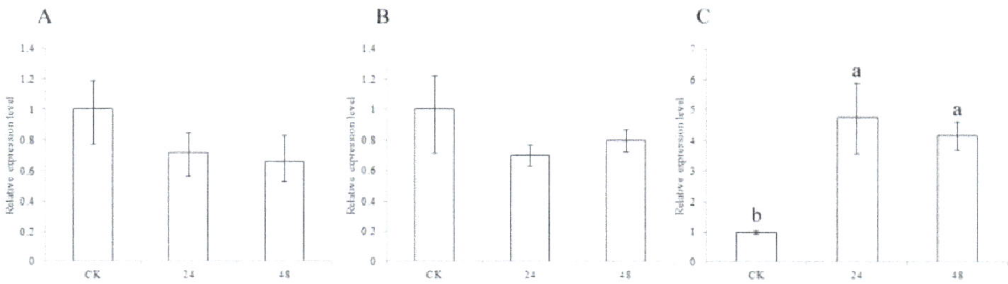

Figure 6: Expression of sorghum CCR genes in CK and the pest treatment at different times.

Pest used was sorghum aphids (*Melanaphis sacchari* (Zehntner)); A, *SbCCR1*; B,*SbCCR2-1*; C, *SbCCR2-2*; variance analysis was used for statistical test; different lowercase letters represent significant difference at 0.05 probability level.

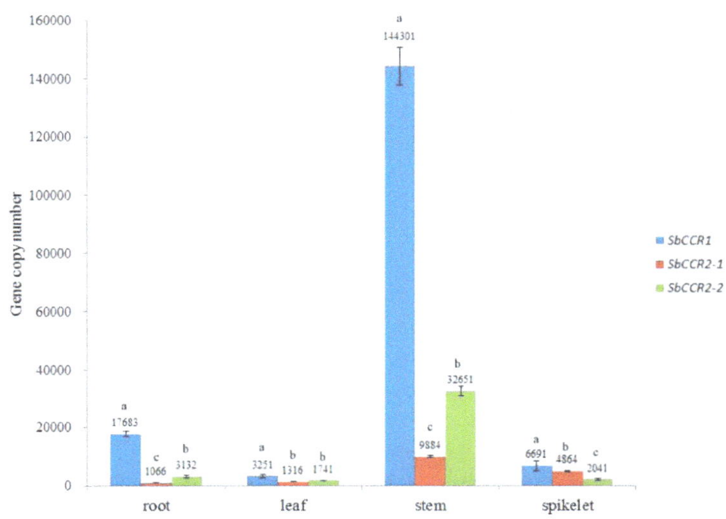

Figure 7: The copy numbers of CCR genes in different sorghum tissues.

Variance analysis was used for statistical test; different lowercase letters represent significant difference at 0.05 probability level.

DISCUSSION

The biosynthesis of lignin begins with the common phenylpropanoid pathway starting with the deamination of phenylalanine and leading to cinnamoyl-CoA esters which are then channeled into the lignin branch pathway via CCR (Pichon et al., 1998). CCR genes were first cloned and characterized in*Eucalyptus*, providing insight into the regulation of the lignin biosynthetic pathway in developmental and defense processes (Lacombe et al., 1997). In this research, the molecular and expression characteristics of three CCR genes in sorghum were reported. The results showed that the three CCR genes exhibited different biochemical properties and differential expression patterns in response to drought and pest treatments.

In higher plants, there are a number of CCR and CCR-like genes (Barakat et al., 2011). But only one to two CCR genes (bona fide CCR genes) are involved in lignin biosynthesis during plant development. For example,*AtCCR1* is the only bona fide CCR gene in *Arabidopsis* (Mir et al., 2008). In this study, the *SbCCR1* gene showed high sequence similarity with the*ZmCCR1* gene which is the bona fide CCR gene in maize (Tamasloukht et al., 2011). Phylogenetic analysis revealed that SbCCR1 was closer to other CCR1 proteins involved in lignin biosynthesis in plant developmental processes. The expression analysis also showed that the *SbCCR1* gene was mainly expressed in stem where lignin was deposited at high content. Additionally, the expression level of *SbCCR1* was significantly higher than those of *SbCCR2-1* and *SbCCR2-2* in various tissues. This indicated that the*SbCCR1* gene should be mainly involved in the biosynthesis of lignin. All these results indicated that *SbCCR1* should be a bona fide CCR gene in sorghum.

The CCR2 gene is mainly involved in defense-related processes in plant (Fan et al., 2006). In this research, phylogenetic analysis revealed that*SbCCR2-1* and *SbCCR2-2* were in a same clade, indicating that the two genes should belong to CCR2 gene. Spatial expression analysis showed that the three CCR genes were mainly expressed in tissues with lignin deposited, but expression analysis of the CCR genes in drought and pest treatments showed that *SbCCR2-1* was mainly involved in the process of drought defense, while *SbCCR2-2* was involved in pest defense. All these results suggested that *SbCCR2-1* and *SbCCR2-2* mainly played a role in defense-related processes in sorghum.

In plants, the CCR1 and CCR2 genes show different expression patterns in response to biotic and abiotic stresses (Fan et al., 2006; Tamasloukht et al., 2011). For instance, *AtCCR1* and *AtCCR2* show different expression patterns in response to pathogen infection. *AtCCR1* transcript level is not significantly changed, but AtCCR2 highly accumulates in infected leaves with*Xanthomonas campestris* pv. *campestris* (Lauvergeat et al., 2001). In this research, it was also observed that the transcript levels of *SbCCR1* and*SbCCR2-1* were not significantly affected by pest. Whereas, the transcript level of *SbCCR2-2* was strongly affected by pest attack. It was showed that*SbCCR2-1* and *SbCCR2-2* played different roles in response to pest attack, suggesting a role of *SbCCR2-2* in pest resistance.

CONCLUSIONS

In this research, three CCR genes were cloned and characterized in sorghum. Phylogenetic analysis revealed that the *SbCCR1* gene belongs to one clade, while *SbCCR2-1* and *SbCCR2-2* belong to another clade. Subcellular localization indicated that the three CCR genes were localized in the cytoplasm. qRT-PCR demonstrated that the three CCR genes were induced by drought. But the expression pattern of *SbCCR2-1* was different from those of *SbCCR1* and *SbCCR2-2*. The expressions of the *SbCCR1* and*SbCCR2-1* genes were not induced by pest attack, but that of *SbCCR2-2*was significantly induced by pest. It was suggested that *SbCCR2-2* was involved in the process of pest defense. Absolute qRT-PCR analysis showed that the three CCR genes have similar spatial expression patterns. The gene copy number of *SbCCR1* was significantly higher than those of *SbCCR2-1*and *SbCCR2-2* in the tested tissues, especially in stem. The results from this study provide important information for understanding the roles of the three CCR genes in sorghum.

Author Contributions

Jieqin Li conceived and designed the experiments, analyzed the data, contributed reagents/materials/analysis tools, wrote the paper.

Feifei Fan performed the experiments, contributed reagents/materials/analysis tools.

Lihua Wang and Junli Du performed the experiments.

Qiuwen Zhan conceived and designed the experiments, reviewed drafts of the paper.

Peijin Wu and Xiaocui Yang performed the experiments, prepared figures and/or tables.

Yanlong YL.L Liu analyzed the data.

Funding

This work was supported by grants from the young talent project of Anhui province (gxyqZD2016217), the key project of natural science research of Anhui provincial education department (KJ2016A177), the National Natural Science Foundation of China (31301383), the Key-construction Subject Plan of Anhui Province (No: WanJiaoMiKe[2014]28), University outstanding youth talent support program (No: WanJiaoMiRen[2015]211), and the youth talent support program of Anhui Science and Technology University (No:XiaoRenFa[2015]69). The funders had no role in study design, data collection and analysis, decision to publish, or preparation of the manuscript.

REFERENCES

1. Barakat A, Yassin NB, Park JS, Choi A, Herr J, Carlson JE. 2011. Comparative and phylogenomic analyses of cinnamoyl-CoA reductase and cinnamoyl-CoA-reductase-like gene family in land plants. *Plant Science* 181:249-257
2. Bhuiyan **NH,** Selvaraj **G, Wei Y, King J.** 2009. Role of lignification in plant defense.*Plant Signaling & Behavior* 4:158-159
3. Boerjan **W, Ralph J,** Baucher **M.** 2003. Lignin biosynthesis. *Annual Review of Plant Biology* 54:519-546
4. Chabannes M, Barakate A, Lapierre C, Marita JM, Ralph J, Pean M, Danoun S,Halpin C, Grima-Pettenati J, Boudet AM. 2001. Strong decrease in lignin content without significant alteration of plant development is induced by simultaneous down-regulation of cinnamoyl CoA

reductase (CCR) and cinnamyl alcohol dehydrogenase (CAD) in tobacco plants. *The Plant Journal* 28:257-270

5. Fan L, Linker R, Gepstein S, Tanimoto E, Yamamoto R, Neumann PM. 2006.Progressive inhibition by water deficit of cell wall extensibility and growth along the elongation zone of maize roots is related to increased lignin metabolism and progressive stelar accumulation of wall phenolics. *Plant Physiology* 140:603-612

6. Giordano A, Liu Z, Panter SN, Dimech AM, Shang Y, Wijesinghe H, Fulgueras K,Ran **Y,** Mouradov **A,** Rochfort **S**+2 more. 2014. Reduced lignin content and altered lignin composition in the warm season forage grass Paspalum dilatatum by down-regulation of a Cinnamoyl CoA reductase gene. *Transgenic Research* 23:503-517

7. Goodin MM, Dietzgen RG, Schichnes D, Ruzin S, Jackson AO. 2002. pGD vectors: versatile tools for the expression of green and red fluorescent protein fusions in agroinfiltrated plant leaves. *The Plant Journal* 31:375-383

8. Gotia HT, Munro JB, Knowles DP, Daubenberger CA, Bishop RP, Silva JC. 2016.Absolute quantification of the host-to-parasite DNA ratio in Theileria parva-infected lymphocyte cell lines. *PLoS ONE* 11:e150401

9. Jin Y, Zhang C, Liu W, Qi H, Chen H, Cao S. 2014. The cinnamyl alcohol dehydrogenase gene family in melon (Cucumis melo L.): bioinformatic analysis and expression patterns. *PLoS ONE* 9:e101730

10. Kawasaki T, Koita H, Nakatsubo T, Hasegawa K, Wakabayashi K, Takahashi H,Umemura K, Umezawa T, Shimamoto K. 2006. Cinnamoyl-CoA reductase, a key enzyme in lignin biosynthesis, is an effector of small GTPase Rac in defense signaling in rice. *Proceedings of the National Academy of Sciences of the United States of America* 103:230-235

11. Lacombe E, Hawkins S, Van Doorsselaere J, Piquemal J, Goffner D,Poeydomenge O, Boudet AM, Grima-Pettenati J. 1997. Cinnamoyl CoA reductase, the first committed enzyme of the lignin branch biosynthetic pathway: cloning, expression and phylogenetic relationships. *The Plant Journal* 11:429-441

12. Lauvergeat V, Lacomme C, Lacombe E, Lasserre E, Roby D, Grima-Pettenati J.2001. Two cinnamoyl-CoA reductase (CCR) genes from *Arabidopsis thaliana* are differentially expressed during development and in response to infection with pathogenic bacteria. *Phytochemistry* 57:1187-1195

13. Leple JC, Dauwe R, Morreel K, Storme V, Lapierre C, Pollet B, Naumann A, KangKY, Kim H, Ruel **K**+16 more. 2007. Downregulation of cinnamoyl-coenzyme A reductase in poplar: multiple-level phenotyping reveals effects on cell wall polymer metabolism and structure. *The Plant Cell* 19:3669-3691

14. Livak **KJ,** Schmittgen **TD.** 2001. Analysis of relative gene expression data using real-time quantitative PCR and the 2(-Delta Delta C(T)) Method. *Methods* 25:402-408

15. Luderitz **T,** Grisebach **H.** 1981. Enzymic synthesis of lignin precursors. Comparison of cinnamoyl-CoA reductase and cinnamyl alcohol:NADP+dehydrogenase from spruce (Picea abies L) and soybean (*Glycine max* L) *European Journal of Biochemistry*119:115-124

16. Ma QH, Tian **B.** 2005. Biochemical characterization of a cinnamoyl-CoA reductase from wheat. *Biological Chemistry* 386:553-560

17. Mir DM, Sierra JB, Ruel K, Pollet B, Do CT, Thevenin J, Buffard D, Jouanin L,Lapierre C. 2008. Redirection of the phenylpropanoid pathway to feruloyl malate in Arabidopsis mutants deficient for cinnamoyl-CoA reductase 1. *Planta* 227:943-956

18. Pan H, Zhou R, Louie GV, Muhlemann JK, Bomati EK, Bowman ME, Dudareva N,Dixon RA, Noel JP, Wang X. 2014. Structural studies of cinnamoyl-CoA reductase and cinnamyl-alcohol dehydrogenase, key enzymes of monolignol biosynthesis. *The Plant Cell* 26:3709-3727

19. Pichon M, Courbou I, Beckert M, Boudet AM, Grima-Pettenati J. 1998. Cloning and characterization of two maize cDNAs encoding cinnamoyl-CoA reductase (CCR) and differential expression of the corresponding genes. *Plant Molecular Biology*38:671-676

20. Poovaiah CR, Nageswara-Rao M, Soneji JR, Baxter HL, Stewart CJ. 2014. Altered lignin biosynthesis using biotechnology to improve lignocellulosic biofuel feedstocks.*Plant Biotechnology Journal* 12:1163-1173

21. Tamasloukht B, Wong QLM, Martinez Y, Tozo K, Barbier O, Jourda C, Jauneau A,Borderies **G,** Balzergue **S,** Renou **JP**+7 more. 2011. Characterization of a cinnamoyl-CoA reductase 1 (CCR1) mutant in maize: effects on lignification, fibre development, and global gene expression. *Journal of Experimental Botany* 62:3837-3848

22. Tang HM, Liu S, Hill-Skinner S, Wu W, Reed D, Yeh CT, Nettleton D, Schnable PS.2014. The maize brown midrib2 (bm2) gene encodes a methylenetetrahydrofolate reductase that contributes to lignin accumulation. *The Plant Journal* 77:380-392

23. Vermerris **W, Sherman DM, McIntyre LM.** 2010. Phenotypic plasticity in cell walls of maize brown midrib mutants is limited by lignin composition. *Journal of Experimental Botany* 61:2479-2490

24. Xue **J,** Luo **D,** Xu **D, Zeng M, Cui X, Li L, Huang H.** 2015. CCR1, an enzyme required for lignin biosynthesis in Arabidopsis, mediates cell proliferation exit for leaf development. *The Plant Journal* 83:375-387

CHAPTER 3

DROUGHT STRESS TOLERANCE STRATEGIES REVEALED BY RNA-SEQ IN TWO SORGHUM GENOTYPES WITH CONTRASTING WUE

Alessandra Fracasso, Luisa M. Trindade and Stefano Amaducci

Keywords: RNA-Seq, Drought stress, *Sorghum bicolor*, Water Use Efficiency, Drought tolerance

BACKGROUND

Drought is the most important abiotic stress in terms of limiting crop productivity worldwide. Water availability is, therefore, of primary importance for a non-limiting crop production in the current changing global climate scenario. The slogan "more crop per drop" [1] was the track for crop improvement in water limited environments aiming to address the growing demand for water, food and commodities (such as energy) of the growing world population [2]. Among the C4 cereals, Sorghum bicolor is the species most suited to environments that are prone to drought. Its tolerance to drought is a consequence of morphological and anatomical characteristics (thick leaf wax, deep root system) and physiological responses (osmotic adjustment, stay green, quiescence) [3]. The high genetic variability among sorghum genotypes and the relatively small size of its genome make this cereal a good model for the identification of drought related genomic regions and genes valuable to unravel the high complexity of drought tolerance related traits [4, 5]. Several sorghum linkage maps, including high density maps [6], have been built using different types of DNA markers [7, 8]. Different genomic regions related to drought tolerance at pre-flowering and postflowering stage were identified [9] but it was the availability of the sorghum genome sequence [4] that has enabled the monitoring of the genome-wide gene expression profile at a single time in response to several abiotic stresses through microarray or RNA-Seq analysis [3, 10–12]. These studies resulted in the identification of drought stress responsive genes and their regulatory elements.

Several transcriptomics studies were carried out on sorghum using RNA-Seq analysis to monitor gene expression in response to osmotic stress and abscisic acid [3], to provide a S. bicolor expression atlas on the dynamic genotype-specific expression profiles [13], or to identify genome-wide SNPs that can potentially enhance genetic analysis and the application of molecular markers in sorghum genomics and breeding [14]. In addition to physiologic or agronomic approaches, genomics offer new opportunities for dissecting quantitative traits into their single determinants (quantitative trait loci, QTLs) paving the way to marker-assisted selection (MAS) or direct gene editing via genetic engineering [15]. Drought stress elicits a wide range of responses in plants [16]. It increases oxidative damage in chloroplasts [17, 18], reduces photosynthesis [19–21], limits metabolic reactions [22], triggers sugar catabolism, in order to provide osmotically active compound and signal molecules [23–25], and modifies cellular lipid composition [26]. To cope with drought stress, plants have developed various strategies, such as generation of larger and deeper root systems [27], regulation of stomatal closure to reduce water loss [28], accumulation of compatible solutes and protective proteins [29], and an increase in the level of antioxidants [30]. Identification of drought resistant traits was frequently labelled as "complex" although we already know the results of all the modifications adopted by plants to cope with drought stress [31].

Citation: Alessandra Fracasso, Luisa M. Trindade and Stefano Amaducci, Drought stress tolerance strategies revealed by RNA-Seq in two sorghum genotypes with contrasting WUE, DOI 10.1186/s12870-016-0800-x.

In this study we have furthered extended the knowledge on the drought response of two sorghum genotypes through transcriptomic analysis [32]. A massive parallel sequencing of RNA (RNA-Seq) on the Illumina platform was used to provide a thorough scenario on the whole sorghum transcriptome in response to drought stress. Several categories of key genes involved in drought response have been identified.

RESULTS

Physiological responses to drought stress

Twenty sorghum plants (ten per each genotype) were subjected to severe drought stress by withholding water from 26 DAE (Days After Emergence) until 34 DAE when 0.2 FTSW (Fraction of Transpirable Soil Water) was reached in all the stressed plants (Fig. 1, solid line, white dots). Subsequently the stressed plants were kept at 0.2 FTSW by irrigating daily for nine days, while the control plants were kept at FTSW values higher than 0.6 for the entire duration of the experiment (Fig. 1, solid line, full dots). The daily transpired water (DTW) was under 400 gr for the stressed plant, while it was up to 1000 gr for the control plants (Fig. 1, dotted lines).

Leaf area, chlorophyll fluorescence parameters (maximum quantum yield, Fv/Fm, the photosystem II efficiency, ΦPSII, and non-photochemical quenching, qNP) and gas exchange measurements (photosynthetic rate, Pn, and transpiration E) were quantified for the entire duration of the experiment (data not shown). The decreased FTSW led to a reduction in RWC (Relative Water Content) values and these changes were greater in the sensitive genotype IS20351 than in the tolerant genotype IS22330 (Table 1). Drought stress also dramatically reduced chlorophyll fluorescence and photosynthetic rate. Under stress conditions the tolerant genotype IS22330 showed a significantly higher value of Fv/Fm than the sensitive genotype IS20351 (Table 1). The same trend was observed for ΦPSII: 0.36 and 0.28 for the tolerant and the sensitive genotype, respectively. In contrast, the qNP under drought stress was higher in the sensitive genotype IS20351 than in the tolerant genotype IS22330 (Table 1).

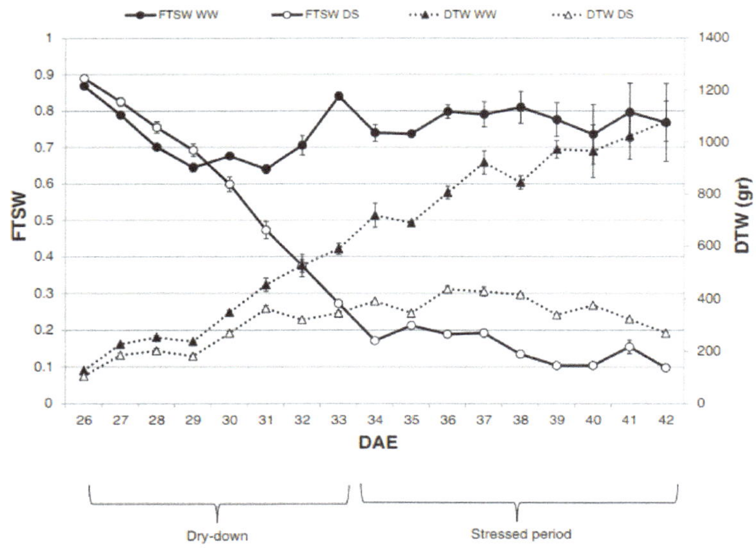

Figure 1: Trend of FTSW and daily transpired water during the dry-down experiment. On the left axis with circles symbols the trend of FTSW during the dry-down: with full circles the WW plants and with the empty circles the DS ones. On the right axis with triangles the daily transpired water: full triangles for the WW plants and empty triangles for the DS ones. DAE = days after emergence. Mean of 10 plants ± SE.

Drought stress affected Pn in both the genotypes differently; the sensitive genotype IS20351 had a greater reduction in Pn (36.5 %) while the tolerant genotype IS22330 showed a Pn reduction of 20.7 %. Transpiration (E) did not differ between the WW (Well-Watered) and DS (Drought-Stressed) plants of the tolerant genotype IS22330, while there was a statistically significant difference between the WW and DS plants of the sensitive genotype IS20351. The intrinsic water use efficiency (WUE_i) decreased linearly for the DS plants of both genotypes from the beginning of the experiment (26 DAE) until harvest (42 DAE), while the WW plants kept their WUE_i close to 6 μmol $mmol^{-1}$ (Fig. 2). WUE_i of DS plants of the tolerant genotype IS22330 was significantly higher than that of DS plants belonging to the sensitive genotype IS20351 during the stress period ($p < 0.05$) (Fig. 2). The agronomic water use efficiency (WUE_a), calculated at harvest, was higher for the tolerant genotype IS22330 (4.23 g/l) than for the sensitive genotype IS20351 (3.26 g/l), thereby confirming the trend highlighted by WUE_i.

Drought stress reveals different intergenic transcripts and novel splice sites

Transcription profiles of IS20351 and IS22330 under wellwatered (WW) and drought-stressed (DS) conditions were explored using the Illumina Genome Analyzer deep sequencing. Three biological replicates were analysed for each condition, resulting in twelve samples. In total, 0.56 billion clean reads, each 100 nucleotides long, were generated, with approximately 47 million clean reads from each sample. The reads mapping to the reference genome were categorised into two classes: uniquely mapped reads, that are reads that map to only one position in the reference genome, and multi-position match, that are reads mapping to more than one position in the reference genome (Table 2). The assembled transcripts were mapped on the genome: on average 72 % were known transcripts, 10 % were novel transcripts and 18 % were intergenic transcripts (Table 3). Drought stress induced alternative splicing events (ASE) in the two genotypes (Table 3): in the sensitive genotype IS20351 no difference in ASE were found, while in the tolerant genotype IS22330 the ASE were increased by 18 %.

Drought stress triggers differential expression of particular genes and GO classes

Each condition was represented by three biological replicates, resulting in eighteen pairwise comparisons between control and stressed plants of the two genotypes. The transcript abundance of each gene was calculated as reads per kilobase transcriptome per million mapped reads (RPKM) (Fig. 3a). This value was used to determine the differential expression analysis as Log_2 ratio between DS and WW plants per genotype and between the two genotypes under WW and DS conditions. Four comparisons were analysed in this study: i) the genotypes IS20351 and IS22330 under WW conditions (WW IS22-IS20 in yellow), ii) the genotypes IS20351 and IS22330 under DS conditions (DS IS22-IS20 in green), iii) the genotype IS20351 in response to DS conditions (IS20 DS-WW in blue), iv) the genotype IS22330 in response to DS conditions (IS22 DS-WW in red).

Table 1: Physiological responses of sorghum genotypes to drought stress

Genotype	Condition	FTSW	RWC	Chlorophyll fluorescence			Gas exchange			
				Fv/Fm	ΦPSII	qNP	Pn	E	WUE_i	WUE_a
		%	%				μmol m^{-2} s^{-1}	mmol m^{-2} s^{-1}	μmol mol^{-1}	g/l
IS20351	WW	0.70	92.7	0.803^a	0.50^a	0.18^a	31.2^a	4.58^b	6.38^a	3.31^b
	DS	0.2	78.4	0.779^c	0.28^c	0.15^b	19.8^c	5.56^a	3.56^c	3.26^b
IS22330	WW	0.80	92.9	0.804^a	0.52^a	0.11^b	30.4^a	4.51^b	6.74^a	3.74^{ab}
	DS	0.2	88.4	0.791^b	0.36^b	0.08^c	24.1^b	4.93^b	4.88^b	4.23^a
$LSD_{(0.05)}$			7.76	0.006	0.08	0.002	2.4	0.58	0.28	0.59
P		<0.0001	<0.01	<0.0001	<0.05	<0.0001	<0.0001	<0.001	<0.01	<0.05

Analysis of relative water content (RWC), chlorophyll fluorescence (Fv/Fm, FPSII and qNP), gas exchange (Photosynthetic rate, Pn, and Transpiration, E), intrinsic (WUEi) and agronomic WUE (WUEa) in sorghum plants in well-watered (WW) and drought stress (DS) conditions at vegetative stage of 9th leaf. Values followed by the same letter are not statistically significant at LSD test $p < 0.05$ performed on the interaction genotypex irrigation.

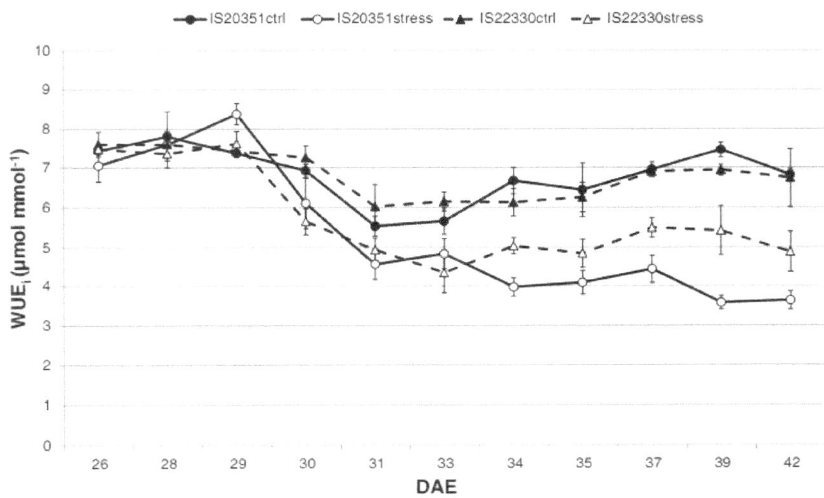

Figure 2: Trend of WUEi calculated during the dry down experiment. Circles represent the sensitive genotype IS20351 and triangles the tolerant IS22330. For both the genotypes the full symbols represents the WW plants whilst the empty symbols represent the DS ones. Mean of 10 plants ± SE.

After applying a stringent cut-off (see Methods section), the comparison of genotypes IS20351 and IS22330 under WW conditions identified 1643 differentially expressed genes (DEGs), and the comparison of genotypes IS20351 and IS22330 under DS conditions identified 1845 DEGs. 1599 genes were differentially expressed in IS20351 in response to drought stress, whilst only 636 were differentially expressed in IS22330 (Fig. 3b). Venn diagrams highlight the overlap of DEGs between each pairwise comparison (Fig. 3c).

Comparison between IS22330 and IS20351 under WW conditions (Fig. 3c in yellow) resulted in 1030 up-regulated genes and 613 down-regulated genes. Only 340 genes were uniquely up- and 160 genes down-regulated in IS22330 in these conditions. The singular enrichment analysis (SEA), carried out with AgriGO software (http://bioinfo.cau.edu.cn/agriGO/index.php) on the 340 up-regulated genes, highlighted 34 GO terms significantly enriched: "aromatic compound biosynthetic process" (GO:0019438), "secondary metabolic process" (GO:0019748), and "flavonoid biosynthetic process" (GO:0009812) in the cellular processes category; "glutathione transferase activity" (GO:0004364), "oxygen binding" (GO:0019825), "UDP-glucosyltransferase activity" (GO:0035251) in molecular functions category. "Apoptosis" (GO:0006915) and "oxidoreductase activity" (GO:0016491) were the most enriched GO terms in the biological processes and molecular function categories among the 160 uniquely downregulated genes expressed in WW conditions in IS22330.

Table 2: Number of reads sequenced and mapped with SOAPaligner/SOAP2

Genotype	Treatment	Total Reads	Total Unmapped Reads	Total Mapped Reads	Unique match	Multi-position match
IS20351	WW	47090292	11993194	35097098	32702251	2394847
	DS	46866452	11969692	34896760	32188700	2708060
IS22330	WW	47504944	11856843	35648101	32660622	2987479
	DS	46840269	11086330	35753938	32549143	3204795

The numbers of unique mapped reads plus the multi-position match equals the total number of mapped reads in well-watered (WW) and drought stress (DS) conditions

Table 3: Classification of transcript produced in sorghum leaves under well-watered (WW) and drought stress (DS) conditions

Genotype	Treatment	Total Mapped Reads	Match to known transcripts	Intergenic transcripts	Novel Transcripts	Alternative Splicing Events
IS20351	WW	75 %	73 %	18 %	9 %	24178
	DS	74 %	71 %	20 %	9 %	24367
IS22330	WW	75 %	72 %	18 %	10 %	20498
	DS	76 %	72 %	18 %	10 %	24304

Percentage of total mapped reads on the reference genome, percentage of match with known transcripts, with intergenic transcripts and novel transcripts identified, and alternative splicing events identified

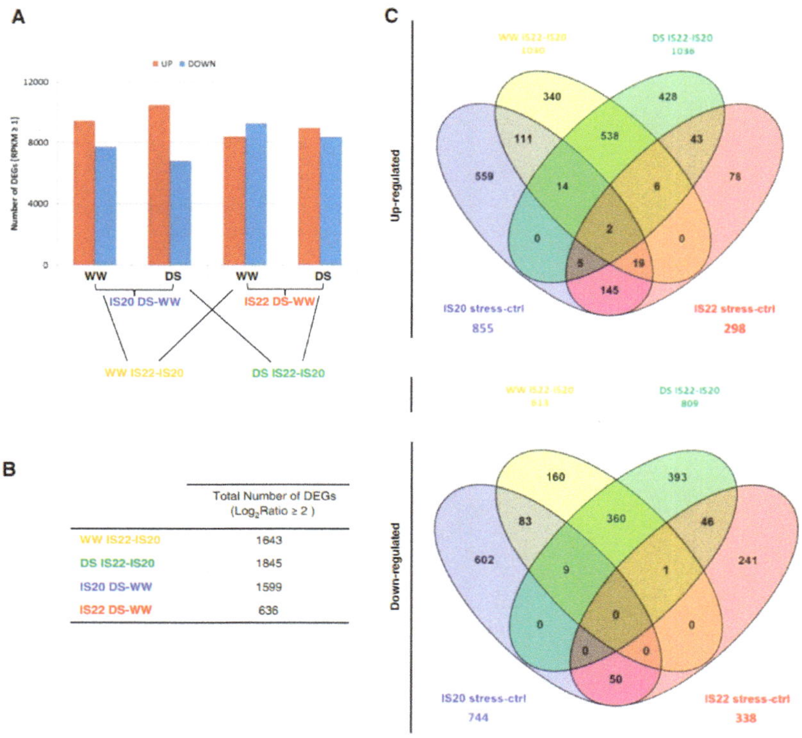

Figure 3: Comparison under study. a Number of DEGs (RPKM) in each pairwise comparison. Blue and red bar are up- an down-regulated genes respectively expressed in well-watered (WW) and drought stressed (DS) conditions in the genotypes IS20351 (IS20) and IS22330 (IS22). b Total number of DEGs that passed the cut-off of Log2 FC >2 in each comparison. In yellow the number of DEGs resulting from the comparison between IS20351 and IS22330 in wellwatered (WW) conditions, in green the number of DEGs resulting from the comparison between the two genotypes under drought stress (DS) conditions; in blue the numbers of DEGs in response to drought stress in IS20351 and in red the number of DEGs in response to drought stress in IS22330. c Venn diagram showing the numbers of up- and down- regulated genes resulted from the four comparison performed. The number of up- or down- regulated genes shared among the four comparison is represented by overlapping circles.

The comparison between the two genotypes under DS conditions resulted in 1036 up- and 809 down-regulated genes. Among these genes, only 428 and 393 were uniquely up- and down- regulated in the genotype IS22330 in comparison to IS20351. "Regulation of DNA replication" (GO: 0006275), "cell death" (GO:0008219), "regulation of cell growth by extracellular stimulus" (GO:0001560), "secondary metabolic processes" (GO:0019748) including "terpenoids biosynthetic process" (GO:0016114), "glutathione transferase activity" (GO:0004364) and "prereplicative complex" (GO:0005656) were the most enriched GO terms among the 75 identified after SEA of the 428 up-regulated genes. Among the 393 down-regulated genes 24 GO terms were significantly enriched: "lipid localization" (GO:0010876), "apoptosis" (GO:0006915), "flavonol biosynthetic process" (GO:0051555), "electron carrier activity" (GO:0009055) and "heme binding" (GO:0020037).

The main difference between the two genotypes was in the total number of genes differentially expressed in response to drought stress: 1599 for the sensitive IS20351 and 636 for the tolerant IS22330. The SEA analysis, performed on all the 1599 and 636 DEGs expressed in response to drought in the genotypes IS20351 and IS22330, showed 197 significantly enriched GO terms (p-value < 0.05) in the sensitive genotype IS20351 while 34 in the tolerant IS22330. Twenty GO terms were enriched in both the genotypes in response to drought stress and are represented in the heat map (Fig. 4). "Response to heat", "RNA modification", "cytosolic part" and "ribosomal subunit" GO terms were enriched with the same extent in both the genotypes. Different GO enrichment was recorded between IS203351 and IS22330 for "oxidation-reduction process", "response to abiotic stimulus", "oxidoreductase activity", "response to chemical stimulus", "small molecule metabolic process", "response to stress", "chloroplast", "single-organism metabolic process" and "cytoplasm component". All these GO terms were more enriched in IS20351 than in IS22330.

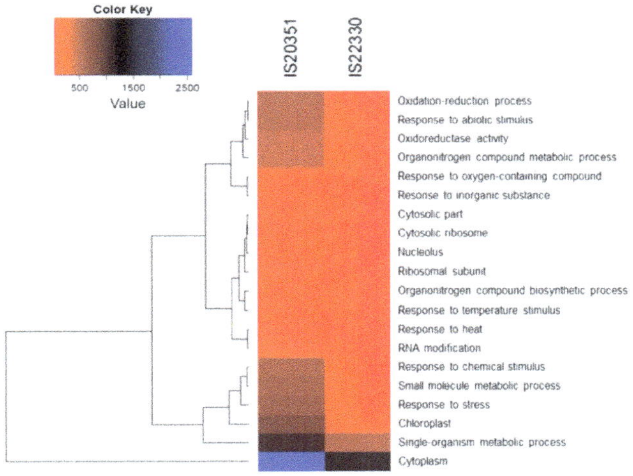

Figure 4: Heat map showing the 20 common GO terms enriched under drought stress in sorghum leaves of IS20351 and IS22330. The cluster frequency was used as a parameter for the parametric analysis of gene enrichment analysis. The figure was generated using R software, Limma package.

Between the two genotypes there were 145 common upregulated genes in response to drought stress and 50 common down-regulated genes (Fig. 3c). The SEA performed on these common DEGs highlighted 11 enriched GO terms belonging to biological processes: "response to abscisic acid stimulus" (GO:0009737), "response to water deprivation" (GO:0009414), "photosynthesis, light reaction" (GO:0019684) were the most enriched GO. The SEA analysis performed with AgriGO on the unique up-regulated genes of IS20351and IS22330 (respectively 559 and 78 genes) highlighted 74 enriched GO terms in IS20351 and and 6 enriched GO terms IS22330. The cross comparison of SEA (http://bioinfo.cau.edu.cn/agriGO/analysis.php?method=compare) highlighted 6 common GO terms. The SEA analysis performed on the unique down-regulated genes (602 and 241 for IS20351 and IS22330 respectively) highlighted 166 and 32 significantly enriched GO terms in IS20351 and IS22330 respectively; after the cross comparison of SEA only 6 resulted as being common to both genotypes.

Drought stress affects different pathways

The KEGG pathway analysis was performed to assign the related biological pathways in which DEGs were involved. One-hundred and seventy-one genes, uniquely expressed in response to drought stress in both the genotypes, were assigned to 112 different KEGG pathways belonging to 24 clades under five major KEGG categories including 'organismal system' (I), 'cellular process' (II), 'environmental information processing' (III), 'genetic information processing' (IV), and 'metabolism' (V) (Fig. 5). Gene-set enrichment analysis showed that translation, signal transduction and carbon metabolism were the top three up-regulated pathways represented by the genes uniquely expressed in response to drought stress; metabolism pathways (V) and signal transduction were, on the other hand, the most enriched down-regulated pathways (Fig. 5).

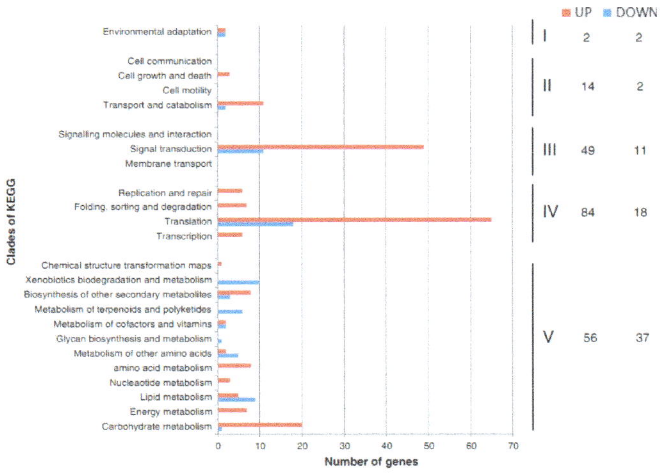

Figure 5: Number of up- and down-regulated genes in each clade of the KEGG pathway maps. The 171 unigenes were assigned 112 KEGG pathways within 24 clades under five major categories: "organismal systems" (I), "cellular processes" (II), "environmental information processing" (III), "genetic information processing" (IV), "metabolism" (V). Per each clades are shown the up- (in red) and the down- (in blue) regulated genes.

KEGG pathway analysis was also performed on the genes that were uniquely up- and down-regulated in response to drought stress in both genotypes (Fig. 6). Transcription factors, 'environmental information processing' pathways, and pathways related to 'cellular processes' and 'organismal system' remained unchanged among the uniquely upregulated genes (Fig. 6 in red). The most striking differences in the transcriptomic profiles of the two genotypes in response to drought were mainly in the 'metabolism' pathways (that were up-regulated by 36 % in IS20351 and 22 % in IS22330), in the 'genetic information processing' pathway (that was up-regulated to a greater extent in IS20351) and in the number of genes not assigned to pathways (Fig. 6 in red). Focusing on the up-regulated 'metabolism' pathways, the tolerant genotype IS22330 showed a two-fold (or greater) enrichment in the metabolism of other amino acids, the nucleotide metabolism, the glycan biosynthesis metabolism and the lipid metabolism compared to the sensitive genotypes IS20351 (Fig. 6 in red). Amino acid metabolism, carbohydrate metabolism and energy metabolism were more enriched in the sensitive genotype IS20351 than in the tolerant genotype IS22330 (Fig. 6 in red).

The 'metabolism' pathways of IS20351 and IS22330 were down-regulated to the same degree in response to drought stress (Fig. 6 in blue). 'Cellular processes' pathways represented 4 % of the down-regulated genes in IS20351 and 2 % in IS22330 (Fig. 6 in blue). 'Organismal system' pathways, 'genetic information processing' pathways and transcription factors were down-regulated to a greater extent in the tolerant genotype IS22330 (Fig. 6 in blue). Among the down-regulated 'metabolism' pathways, energy metabolism, nucleotide, cofactors and vitamins metabolism, glycan biosynthesis and metabolism, and carbohydrate metabolism pathways were down-regulated with a higher frequency in the sensitive genotype IS20351 than in the tolerant IS22330 (Fig. 6 in blue).

Drought stress response of sorghum transcriptome

The MapMan software (3.5.1R2) [33] was used to show a pathway overview of 1599 and 636 DEGs expressed in response to drought stress and it was selected for its capacity to show statistically significant drought mediated gene expression data for the sensitive genotype IS20351 (Fig. 7a) and the tolerant genotype IS22330 (Fig. 7b). Three main aspects were selected for a deeper evaluation of drought tolerant traits: the antioxidant and secondary metabolism pathways, light reaction and carbon fixation pathways, lipid and carbon metabolism.

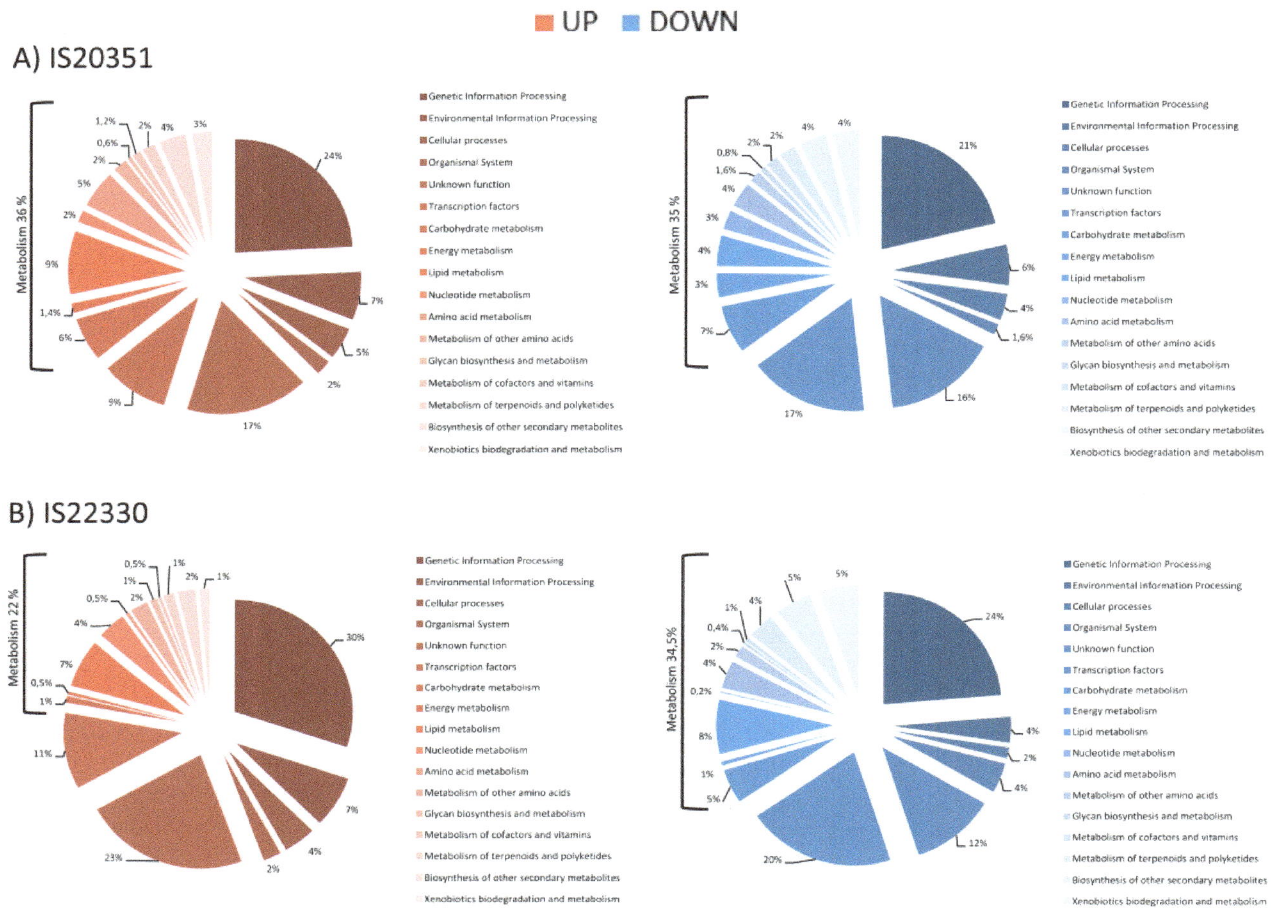

Figure 6: Distribution in KEGG pathways of the unique up- and down-regulated genes in response to drought for the genotype IS20351 and IS22330. Pie charts showing the percentage of genes up- (in red) and down- (in blue) regulated in response to drought stress for the genotypes IS20351 (a) and IS22330 (b).

Response of antioxidant and secondary metabolism related genes

DEGs related to antioxidant and secondary metabolism were analysed together because of the strong relationship between the capacity to scavenge ROS through antioxidant genes and metabolites derived from the secondary metabolism.

Seventeen DEGs were identified in the sensitive genotype IS20351 in response to drought: 5 were up-regulated and 12 down-regulated. In the tolerant genotype IS22330, in the same condition, only 4 DEGs were found and three of them were up-regulated. The sb09g025730.2 gene showed a peculiar behaviour; it was up-regulated in the tolerant genotype IS22330 and dramatically down-regulated in the sensitive IS20351. The sb06g001970.1 gene was up-regulated in the sensitive genotype IS20351 and remained unchanged in the tolerant IS22330. In contrast, the sb09g001690.1 gene was upregulated in the tolerant IS22330 and its expression remained unchanged in the sensitive IS20351.

Drought affected the secondary metabolism in both sorghum genotypes. Fifty DEGs were found in the sensitive genotype IS20351 and 27 in the tolerant IS22330. In the sensitive genotype IS20351, about the same number of genes were up- and down-regulated (25), whilst in the tolerant genotype IS22330 the down-regulated genes were more than the upregulated ones; 20 and 7, respectively. Among the down-regulated genes, the isoprenoids and phenylpropanoids metabolism was the most affected metabolism, with 20 genes in IS20351 and 10 in IS22330. The flavonoids pathway showed a peculiar behaviour being up-regulated by drought in the sensitive genotype IS20351 and down-regulated in the tolerant genotype IS22330. The changes in the secondary metabolism expression pattern, for example the change in the chlorophyll/carotenoids content, was reflected in the fluorescence parameters recorded.

Response of light reactions and carbon fixation pathways

The photosynthetic pathway was drastically affected by drought in the sensitive genotype IS20351, with 28 genes differentially expressed in response to drought: 19 belong to the light reaction pathway and 9 to the Calvin cycle.

Among the 19 DEGs belonging to the light reaction pathway, 15 genes were down-regulated in response to drought: 8 code for protein belonging to the light harvesting complex I or II (LHCI and LHCII), 6 code for protein related to photosystem I and II (PSI and PSII) and 1 codes for the gamma subunit of the ATP synthase. Two isoforms of PSII polypeptide subunits were strongly up-regulated together with the electron carrier ferrodoxin in the sensitive genotype IS20351 in response to drought. In the tolerant genotype IS22330 the light reaction pathway was also affected, but to a lower extent. Only three genes belonging to the light reaction pathway were up-regulated in response to drought: 2 implicated in PSII and one in photosynthetic electron transport, the ferrodoxin.

9 genes related to the carbon fixation pathway (Calvin cycle) were differentially expressed in the sensitive genotype IS20351: 6 were down-regulated by drought and 3 were up-regulated (Sb01g037510.1, Sb06g004280.1 and Sb05g027880.1). In the tolerant genotype IS22330 no genes were differentially expressed in response to drought.

Lipid and carbon metabolism in response to drought stress

In terms of DEGs the lipid metabolism was more greatly affected in the sensitive genotype IS20351. In this genotype fatty acid synthesis, elongation and lipid degradation via beta-oxidation cycle were all upregulated. Phospholipid and sphingolipid syntheses were down-regulated in response to drought. In the tolerant genotype IS22330 the steroids biosynthesis and phospholipase D were up-regulated.

Also the carbon metabolism was more greatly affected by drought in the sensitive genotype IS20351 than in the tolerant IS22330. In IS20351 drought highlighted 12 DEGs: 7 genes belonging to the degradation of starch and sucrose were up-regulated, and 5 genes were down-regulated. In the tolerant genotype IS22330 only 2 genes were down-regulated.

DISCUSSION

In plants exposure to drought triggers a wide range of responses, ranging from molecular expression, biochemical metabolism to ecosystem level, that involve lots of genes and pathways related to diverse mechanisms [16]. In this study we evaluated these mechanisms through RNA-Seq analysis of two sorghum genotypes subjected to the same extent of drought stress. The responses differed greatly between the sensitive IS20351 and the tolerant IS22330 genotypes in terms of the number of genes and pathways involved in drought stress response, but also in terms of the constitutive expression level of several pathways.

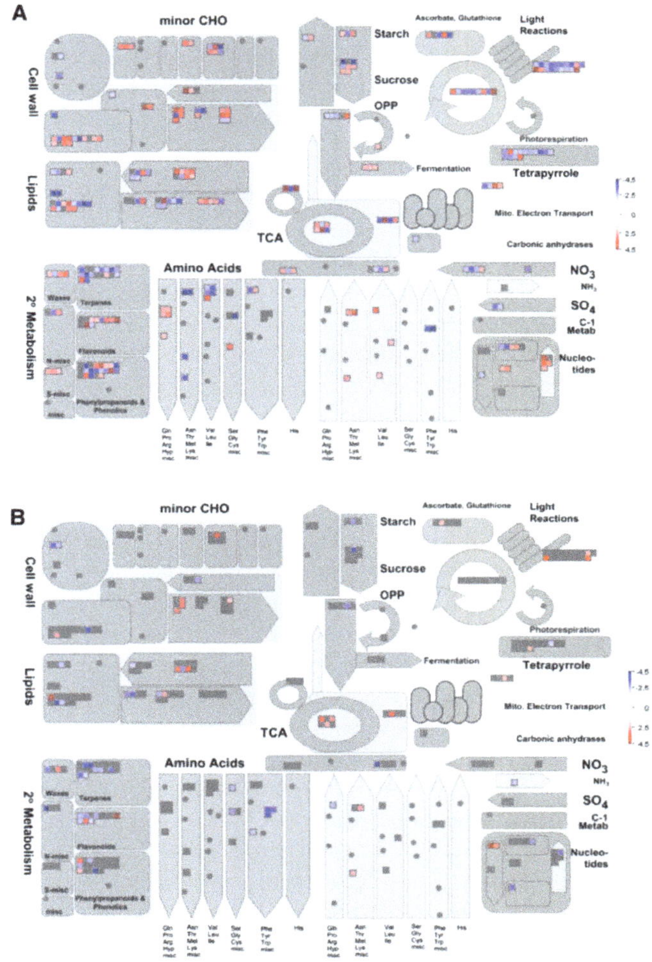

Figure 7: Distribution of up- (in red) and down- (in blue) regulated genes in metabolic pathways in response to drought stress for IS20351 and IS22330. Drought mediated expression changes in the metabolic pathways in leaves of IS20351 (a) and IS22330 (b). The figure was generated using MapMan and shows DEGs that passed the cut-off of Log_2 FC >2.

Constitutive drought tolerance trait

The trend of FTSW, together with the value of the daily transpiration rate, confirmed that the DS plants of both genotypes were subjected to the same environmental conditions and to the same extent of drought stress. In addition, transcriptomics analysis provided unequivocal evidence on RNA modifications triggered by drought stress. "Response to heat" (GO:0009408) and "RNA modification" (GO:0009451) GO terms were enriched to the same extent in both genotypes.

Although the drought stress level applied was equal (0.2 FTSW), the two genotypes responded differently; in IS20351 a significantly higher number of differentially expressed genes (DEGs) was observed than in the tolerant genotype IS22330, resulting in a greater enrichment of GO terms related to drought stress response in IS20351 than in IS22330. The up-regulation of genes under WW conditions of "secondary metabolic process" (GO:0019748), and related GO terms, in the genotype IS22330 confirm its intrinsic tolerance, previously only characterized from a physiological point of view [32]. In this genotype, the constitutive upper level of flavonoids and secondary metabolites led to increased drought tolerance traits according to Winkel-Shirley [34]. Furthermore the "glutathione transferase activity" (GO:000004364) was up-regulated in the tolerant genotype IS22330 confirming the role of the glutathione-S transferase family in improving environmental stress resistance in crops [35].

Drought tolerance strategies

Drought stress results in a massive production of reactive oxygen species (ROS) [17, 18] that cause oxidative stress. The sequence of events that occur in plant tissues in response to oxidative drought-induced stress was well described by Mano et al. [36]. The antioxidant enzymes constitute the "first line of defence" against ROS and oxidative stress generated by different abiotic and biotic injuries [37, 38]. The activity of these enzymes can be enhanced or repressed depending on the species, genotype, stress duration and severity [39–41]. In the "response to abiotic stimulus" (GO:0009628), "oxido-reductase activity" (GO:0016491) and "response to stress" (GO:0009628) gene ontology categories, genes were more greatly downregulated by drought in the sensitive genotype IS20351 than in the tolerant IS22330, enabling us to speculate that the tolerant IS22330 had a constitutively higher expression of antioxidant genes that is not affected by drought stress. Experimental evidence showed that the antioxidant enzyme activity might be depressed in excess-light conditions, especially when plants are faced with additional stresses such as drought or temperature [42]. To cope with the oxidative stress caused by drought, genes coding for secondary metabolites such as phenylpropanoids, phenolic compounds and flavonoids, are overexpressed [43]. Phenylpropanoids have the greatest potential to reduce ROS, the polyphenols act as antioxidants to protect plants against oxidative stress [44], flavonoids play different molecular functions, including stress protection in plants [34], and also flavanols were found to be oxidated in response to severe drought in tea plants, suggesting their involvement in plant protection [45]. All these compounds are widely synthetized in response to several abiotic stresses, including drought [46–50]. In wheat and willow leaves an increase in flavonoid and phenolic acids content was observed together with an induction of genes involved in the flavonoid biosynthetic pathway in response to various stresses, including drought [51, 52]. With our study, we confirm that under drought stress the up-regulation of these genes in the sensitive genotype IS20351 was higher than in the tolerant genotype IS22330, whilst a constitutively higher expression of these genes in the tolerant IS22330 under control conditions led to a lower synthesis of stress induced compounds. The accumulation of these compounds and the differential expression of the above mentioned genes remains genotype dependent [53].

Only in the last decade was it hypothesized that flavonoids might also play a role as antioxidant in response to severe excess of light complementing the role of antioxidant enzymes [54–57]. Agati et al. [42] found that flavonoid genes were up-regulated in response to drought in the sensitive genotype IS20351 whilst they were mostly downregulated in the tolerant IS22330. The biosynthesis of "antioxidant" flavonoids, in fact, increases more in stress sensitive species than in stress tolerant ones [42]. The reason for this lies in the fact that stress sensitive species display a less efficient "first line" of defence against ROS in conditions of stress and they are therefore exposed to a more severe oxidative stress [58, 59]. In any case, the relationship between antioxidant enzymes and flavonoids in response to abiotic and biotic stress it is not yet well clarified [42].

Drought stress induces a decrease in the chlorophyll content, a consequential change in the chlorophyll/carotenoid ratio [60] and an increase in the ratio of violaxanthin-cycle pigment. This results in a reduction of light absorption centres, an enhancement of nonphotochemical quenching in order to dissipate the excess of light, and a reduction in photosynthetic rate [19–21]. All these stress-induced physiological modifications (qNP and Pn) were observed to a greater extent in the sensitive genotype IS20351. The physiological response is supported by the observation that a high number of genes involved in the terpenoids and carotenoids biosynthesis were down-regulated in IS20351 and not in IS22330, in agreement with the decreased concentration of some carotenoids under severe drought stress [17, 38, 61].

The down-regulation of genes related to carotenoids and chlorophyll biosynthetic pathways leads to the down-regulation of light reaction and carbon fixation pathways, that in fact were dramatically affected by drought in the sensitive genotype IS20351. The decreased expression pattern mainly involved the light harvesting complex I and II and polypeptide subunits of the photosystems (I and II). In particular, the lightharvesting chlorophyll a/b-binding proteins (LHCBs) were extremely down-regulated in the sensitive genotype IS20351 according to several studies in which the down-regulation of LHCBs reduces plant tolerance [62–65]. The LHCBs, complexed with chlorophyll and xanthophylls, form the antenna complex [66] and play an important role in adaptation to environmental stress [63–65]. Their expression is regulated by multiple environmental factors including light [67], oxidative stress [68, 69] and abscisic acid (ABA) [70]. Also the genes involved in the "carbon fixation" were more greatly down-regulated in the sensitive genotype IS20351 rather than in the tolerant one. The upregulation of Sb03g040610.1 was the main exception in the expression pattern of this genotype; this gene codes for the electron carrier ferrodoxin. Comparing the Log2 values of this gene in the two genotypes, it appears that this gene was more up-regulated in the sensitive genotype than in the tolerant one (5.2 and 3.4 for IS20351 and IS22330, respectively). This result indicates that the tolerant genotype IS22330 could better cope with the excess of light during drought stress. This is further supported at a physiological level by the low qNP value recorded. Conversely, the sensitive genotype IS20351 over expressed this gene so that it can dispose the excess of electrons and consequently waste the excess of light in nonphotochemical reactions.

According to literature, under drought stress starch (inactive osmotically) content decreases, whilst content of soluble sugars (osmotically active) increases, assuring the maintenance of leaf water status and plant growth [23–25]. In the sensitive genotype IS20351, starch synthases were down-regulated and enzymes involved in the degradation of starch and sucrose up-regulated. According to Sturm and Tang [71] invertases play a role in several processes ranging from phloem loading to response to abiotic and biotic stresses [23, 72]. Exogenous ABA applied in soybean green beans

[73] and maize leaves exposed to drought [74] showed an increase in invertase activity. Gazarrani and McCourt [75] also highlighted that hexose-based signals originating from sucrose cleavage are implicated in the regulation of ABA biosynthetic genes. It is well known that sucrose plays a crucial role as a key molecule in energy transduction and as a regulator of cellular metabolism [76–78].

Furthermore, sucrose and other sugars are energy and carbon sources required for defence response and are necessary for plant survival under drought stress conditions [79]. Like hormones, sucrose can act as primary messenger controlling the expression of several genes involved in sugar metabolism.

Lipids are important membrane components and, under drought stress, significant modifications of the lipid membranes occur. For this reason our investigation also focused on this metabolic pathway. The fatty acid elongation is considered to be the rate-limiting step in cuticular wax biosynthesis [80, 81]. The accumulation of wax has a key role in limiting water losses from plants [82]. It is widely accepted that drought stress can increase the amount of wax in several species [83–87] and that this increase is associated with an improved drought tolerance [88]. According to our results, the sensitive genotype IS20351 up-regulated these genes in response to drought; on the contrary, the drought tolerant genotype IS22330 remained unchanged. The hypothesis is that the tolerant genotype IS22330 has a constitutively higher expression level of genes related to drought tolerance, such as genes involved in cuticular wax synthesis and fatty acid desaturation. This hypothesis is also confirmed by the observation that, according to Torres-Martin et al. [89], no changes in omega-3 desaturase expression were highlighted in response to drought in the tolerant genotype IS22330. On the contrary, the omega-3 desaturases were downregulated in the sensitive genotype IS20351 [89].

The first evidence of the involvement of sphingolipids in the signal-transduction pathways in plants, including in response to drought, was provided by Ng et al. [90]. Until that moment only the implication of sphingolipids in conferring stability to plant membranes, contributing to acclimation to drought stress had been hypothesized [91]. Spiegel and Milstien [92] afterwards explored the link between the sphingosine-1-phosphate and the drought hormone abscisic acid in the release of calcium from the vacuole. RNA-Seq results highlighted the ineffective response of the drought sensitive genotype IS20351 that down-regulated sphingolipids in response to drought, except for a ceramidase (sb03g028410.1). In cowpea leaves a massive breakdown of membrane lipids was observed in response to drought with a more severe degradation in the sensitive plants [93]. The main enzyme responsible for the drought-induced degradation of membrane phospholipids is phospholipase D (PLD) [94]. According to El Masouf et al. [95], the drought sensitive genotype IS20351 strongly up-regulated the PLD expression, whilst in the drought tolerant IS22330 the expression was only slightly up-regulated. Recently, PLD up-regulation was associated to drought and salt stress tolerance [96–99] and the product of its activity, the phosphatidic acid, is involved in ABA signalling in stomatal movement [100]. PLDa1, in particular, is the most predominant PLD in plants activated by ABA [101].

Some interesting genes provided insight into the drought tolerance of the genotypes analysed. The Sb06g014320 gene, encoding for a glycerophosphodiester phosphodiesterase, found to be up-regulated in response to drought in sorghum leaves [12], was strongly down-regulated in response to drought in the sensitive genotype IS20351. The Sb07g027910 gene, encoding for a monogalactosyldiacylglycerol (MGDG) synthase, found to map to a stay green QTL [102] and to be overexpressed in response to drought in sorghum leaves, was down regulated in the sensitive genotype IS20351. Since these genes are involved in drought tolerance related pathways, the first in choline biosynthesis and the second in phosphatidylinositol biosynthesis, a down regulation in response to drought is proof of sensitivity to drought stress for the sensitive genotype IS20351. A confirmation of the drought tolerance of IS22330 was the overexpression of genes related to the phosphatidylinositol biosynthesis, such as sb08g016610, sb08g022520 and sb05g026855.

CONCLUSION

RNA-Seq analysis, performed in this study, proved to be a good method to investigate complex traits in different genotypes. The sorghum transcriptome analysed in response to drought conditions revealed unequivocal traits of sensitivity and tolerance in the two sorghum genotypes studied.

The first evidence of sensitivity to drought of the genotype IS20351 was represented by the physiological measurements (gas exchange and chlorophyll fluorescence) that drought dramatically affected. This evidence was confirmed at a transcriptomic level by the higher number of differentially expressed genes (DEGs) observed in the sensitive genotype IS20351 and not in the tolerant genotype IS22330. The sensitivity to drought of IS20351 was further confirmed by the lower constitutive expression level of "secondary metabolic process" (GO:0019748) and "glutathione transferase activity" (GO:000004364) observed under well-watered conditions in IS20351 in comparison with the tolerant genotype IS22330. In addition, the enriched GO terms analysis highlighted the differences existing between the two genotypes in coping with drought stress and the strategies adopted. The sensitive genotype hydrolysed carbohydrates and sugars, while the tolerant IS22330 activated the synthesis of other amino acids (glycinbetaine, glutathione) to cope with drought stress. In conclusion, we can confirm that the sensitive genotype IS20351 perceived the drought stress imposed (0.2 FTSW) to a greater extent than the tolerant genotype IS22330, showing an overactive genetic response. IS22330, on the other hand, being generally less affected by drought in all the analysed pathways, could be used as a genetic donor to further improve the sorghum germoplasm with drought tolerance traits.

METHODS

Plant material, drought stress conditions and physiological measurements

Two sorghum genotypes of the durra race, IS20351 and IS22330, were cultivated in pots in July 2013 in a dry down experiment in open field condition in the experimental station of Università Cattolica del Sacro Cuore, Piacenza, Italy. The genotypes are part of germplasm collection of CIRAD and were provided by the CRB-T (Centre de Resources Biologiques Tropicales) CIRAD Montpellier. IS20351 and IS22330 were previously characterized in 2012 for their contrasting tolerance to drought [32]. According to Fracasso et al. [32], germination of seeds was carried out in Petri dishes at 25 °C and in dark conditions for 3 days. Five germinated seeds were planted in plastic pots (16 L capacity), filled with a base layer of sand to guarantee drainage and 8 kg of a soil mixture (24 % clay, 64 % silt, and 12 % sand), that had been previously sieved, dried and homogenized. At the 4th leaf stage, plants were thinned in order to have one healthy plant per pot.

The Fraction of Transpirable Soil Water (FTSW) was determined as the ratio of Available Soil Water Content (ASWC) divided by the Total Transpirable Soil Water (TTSW) as follows:

$$FTSW = \frac{ASWC}{TTSW} = \frac{SWC-WP}{FC-WP}$$

Where ASWC represent the Available Soil Water Content for the plant, derived from the actual soil water content calculated as difference between the Soil Water Content (SWC) and the soil water content at Wilting Point (WP), and TTSW as the difference between the soil water content at Field Capacity (FC) and the water content at WP. Both FC and WP were determined in a short previous experiment (data not shown).

Plants were grown under well-water conditions until they reached the 6th leaf stage. At this moment, all the plants were irrigated until FC, the soil surface was covered by a thin layer of perlite, and the top of the pot was covered with PVC bags. A little slit was made in the bottom of the plastic bag to allow the sorghum plant to grow through. The slit was sealed with adhesive packing tape to minimise water loss by evaporation. Following the protocol of the dry-down experiment [32] a decrease of pot weight between two consecutive weight determinations is only attributed to plant transpiration.

Forty plants were divided in two groups: the wellwatered (WW) and the drought stressed (DS) plants. Irrigation was withheld for half of them (the DS ones) till the FTSW value reached 0.2. This value was kept constant for nine days by re-integrating water losses of the DS plants day by day, while the WW plants were irrigated daily to maintain soil water content close to 0.7 FTSW. After nine days had passed, the plants were harvested in order to perform physiological and transcriptomic analysis.

Leaf area, chlorophyll fluorescence parameters (maximum quantum yield, Fv/Fm, photosystem II efficiency, ΦPSII, and non photochemical quenching, qNP), gas exchange measurements (photosynthetic rate, Pn, and transpiration, E) were measured for the entire duration of the experiment every two days. Pn and E data were used to calculate the intrinsic WUE (WUEi) as the ratio between photosynthetic activity and the transpiration rate (μmol mol^{-1}). At the destructive sampling date (on the 42nd day after emergence, DAE), leaves samples were collected in order to perform transcriptomic analysis. At the same moment, relative water content (RWC) and biomass production were determined in order to calculate the agronomic water use efficiency (WUEa) the ratio between dry biomass production (g) and the total transpired water (L) according to Mastrorilli et al. [104].

Measurements of chlorophyll fluorescence parameters Fv/Fm, qNP and ΦPSII were carried out with a portable chlorophyll fluorometer (Fluorescence Monitoring System, Hansatech instruments, Norfolk, England) on the youngest fully expanded leaf. The value of minimal fluorescence was determined through predawn measurements by applying weak modulated light (0.4 μmol m^{-2} s^{-1}) and maximal fluorescence (Fm) was induced by a short pulse (0.7 s) of saturating light (15300 μmol m^{-2} s^{-1}). The measurements were recorded between 12 and 2 pm for ΦPSII and before dawn for Fv/Fm.

Photosynthetic rate (Pn) was measured on the same leaf used for chlorophyll fluorescence measurements using a portable infrared gas analyser (CIRAS-2, PP System, Amesbury, USA): leaf surface area 4,5 cm^2 , saturated CO_2 concentration of 400 $\mu molmol^{-1}$, and PPFD 2000 μmol m^{-2} s^{-1} . Photosynthetic rate (Pn) was recorded between 12 and 2 pm.

Relative Water Content (RWC) was determined at the destructive sampling time according to the methodology described by Barr and Weatherley [103]. Twelve leaf disks of 20 mm of diameter were collected from each plant for the RWC determination. The disks were weighed, then soaked in distilled water for 24 h at 4 °C in the dark to determine the turgid weight. The dry weight was determined after drying the leaves for 72 h at 95 °C. The relative water content was then calculated using the following equation:

$$RWC = \frac{(FM-DM)}{(TM-DM)} * 100$$

where FW is the fresh weight, TW the turgid weight after the rehydration in distilled water and DW the dry weight after drying.

RNA extraction, cDNA library construction and sequencing

Three biological replicates were used for all RNA-Seq experiments from each genotypes and water treatment. The total RNA from the leaf meristem was extracted using Trizol reagent (Invitrogen, Carlsbad, CA) and purified using the RNeasy Plant Mini kit (Qiagen, Valencia, CA). On column DNase digestion was performed according to the manufacturer's protocol (Qiagen, Valencia, CA). RNA quality and integrity was verified using a 2100 Bioanalyzer RNA Nanochip (Agilent, Santa Clara, CA) and all three samples had RNA Integrity Number (RIN) value more than 8.5. The quantification of the total RNA was checked by a NanoDropND-1000 Spectrophotometer (Nano-Drop, Wilmington, DE) and agarose gel electrophoresis.

Illumina sequencing using the GAII platform was performed at Beijing Genomics Institute (BGI-Shenzhen, Shenzhen, China http://www.genomics.cn/en/index) according to the manufacturer's instructions (Illumina, San Diego, CA). Briefly, poly-A RNA was isolated from 20 μg of total RNA using Magnetic Oligo (dT) Beads (Illumina) and digested in short fragment. First and second strand synthesis were followed by end repair, and adenosines were added to the 3' ends. Adapters were ligated to the cDNA and fragments (200 ± 25 bp) were purified by agarose gel electrophoresis and amplified by PCR. Finally, after validating on an Agilent Technologies 2100 Bioanalyzer using the Agilent DNA 1000 chip kit, the cDNA library was sequenced on a PE flow cell using Illumina Genome Analyzer IIx, and the workflow was as follows: template hybridization, isothermal amplification, linearization, blocking, sequencing primer hybridization, and sequencing on the sequencer for Read 1.

Data processing and analysis

The RNA-seq reads generated by the Illumina Genome Analyzer were initially processed to remove the adapter sequences, reads in which unknown bases are more than 10 % and low-quality reads. After filtering, the remaining reads, so called "clean reads", were used for downstream bioinformatics analysis. In the pipeline, clean reads are aligned to the reference sequence (ftp://ftp.ensemblgenomes.org/pub/ plants/release-20/fasta/sorghum_bicolor/dna/) by using SOAPaligner/SOAP2. No more than 5 mismatches are allowed in the alignment. A quality control step was performed after that step and the distribution of reads on reference genes was analysed. Gene coverage was calculated as the percentage of a gene covered by reads. This value is equal to the ratio of the base number in a gene covered by unique mapping reads to the total base number of coding region in that gene. The expression level was, on the other hand, calculated using RPKM (Reads per Kilobase transcriptome per Million mapped reads) method [105], according to the following formula:

$$RPKM = \frac{10^6 C}{NL/10^3}$$

where C is the uniquely mapped counts determined from the high quality category, L is the cDNA length for the longest splice variant for a particular gene and N is the number of total mappable reads which was determined as the sum of the high quality reads and the highly repetitive reads. This method is able to eliminate the influence of different gene length and sequencing discrepancy on the calculation of gene expression. Log_2 transformations of this normalization were performed.

Screening, expression pattern, gene ontology analysis and pathway enrichment of DEGs

A strict algorithm was developed to identify differentially expressed genes between two samples and false positive and false negative errors are performed using Benjamini and Yekutieli [106] FDR method. We used FDR $\leq$0.001and the absolute value of Log_2Ratio $\geq$2 as the threshold to judge the significance of gene expression difference. Gene Ontology (GO) enrichment was based on AgriGO software [107] with hypergeometric statistical test and Hocberg (FDR).

Pathway enrichment analysis of DEGs was performed using the Kyoto Encyclopedia of Genes and Genome (KEGG, http://www.genome.jp/kegg/). This analysis allows to identify enriched metabolic pathways or signal transduction pathways in DEGs comparing with the whole genome background. A strict algorithm was used for the analysis:

$$P = 1 - \sum_{i=0}^{m-1} \frac{\binom{M}{i}\binom{N-M}{n-i}}{\binom{N}{m}}$$

Where N is the number of all genes with KEGG annotation; n is the number of DEGs in N, M is the number of all genes annotated to specific pathway. Pathways with Qvalue $\leq$0.05 are significantly enriched in DEGs.

Novel transcript prediction and alternative splicing analysis

The assembled transcripts were compared with the annotated genomic transcripts from the reference sequences in order to discover novel transcribed regions. Three requirements are needed: the transcript must be at least 200 bp away from annotated gene, the length of the transcript must be over 180 bp, the sequencing depth must be no less than 2. The Coding Potential Calculator (CPC: http://cpc.cbi.pku.edu.cn/) was used to assess the proteincoding potential. TopHat software [108] was used to detect alternative splicing events (ASE).

AUTHORS' CONTRIBUTIONS

AF designed the experiment, performed the physiological measurements and the RNA-Seq data meaning; LT and SA supervised this work and assisted with editing the manuscript. All authors read and approved the final manuscript.

ACKNOWLEDGEMENTS

We sincerely acknowledge two anonymous reviewers for their valuable comments and guidance. The authors kindly thank the CRB-T (Centre de Resources Biologiques Tropicales) CIRAD Montpellier that provided the plant material. Special thanks to Dr. Jamila Bernardi for constant support and discussion.

FUNDING

This work was supported by grants from European Project "SWEETFUEL-Sweet sorghum: an alternative energy crop", 7th Framework Programme of the European Union – grant agreement 227422, and from the project BIOSEA, funding by Italian Ministry of Agriculture.

REFERENCES

1. Zhang HP. Improving water productivity through deficit irrigation: examples from Syria, the North China Plain and Oregon, USA. In: Kijne J, Barker R, Molden D, editors. Water productivity in agriculture: limits and opportunities for improvement. Wallingford: CABI; 2003. p. 332.
2. Rooney WL, Blumenthal J, Bean B, Mullet J. Designing sorghum as a dedicated bioenergy feedstock. Biofuel, Bioprod Bioref. 2007;1:147–57.
3. Dugas DV, Monaco MK, Olsen A, Klein RR, Kumari S, Ware D, et al. Functional annotation of the transcriptome of Sorghum bicolor in response to osmotic stress and abscisic acid. BMC Genomics. 2011;12:514–19.
4. Paterson AH, Bowers JE, Bruggmann R, Dubchak I, Grimwood J, Gundlach H, et al. The Sorghum bicolor genome and the diversification of grasses. Nature. 2009;457(7229):551–56.
5. Sanchez AC, Subudhi PK, Rosenow DT, Nguyen HT. Mapping QTLs associated with drought resistance in sorghum (Sorghum bicolor L. Moench). Plant Mol Biol. 2002;48(5–6):713–26.
6. Ashraf M. Inducing drought tolerance in plants: recent advances. Biotechnol Adv. 2010;28(1):169–83.
7. Mace ES, Rami JF, Bouchet S, Klein PE, Klein RR,

Kilian A, et al. A consensus genetic map of sorghum that integrates multiple component maps and high-throughput Diversity Array Technology (DArT) markers. BMC Plant Biol. 2009;9:13–27.

8. Rami JF, Dufour P, Trouche G, Fliedel G, Mestres C, Davrieux F, et al. Quantitative trait loci for grain quality, productivity, morphological and agronomical traits in sorghum (Sorghum bicolor L. Moench). Theor Appl Genet. 1998;97(4):605–16.

9. Kebede H, Subudhi P, Rosenow D, Nguyen H. Quantitative trait loci influencing drought tolerance in grain sorghum (Sorghum bicolor L. Moench). Theor Appl Genet. 2001;103:266–76.

10. Buchanan CD, Lim S, Salzman RA, Kagiampakis I, Morishige DT, Weers BD, et al. Sorghum bicolor's transcriptome response to dehydration, high salinity and ABA. Plant Mol Biol. 2005;58(5):699–720.

11. Yazawa T, Kawahigashi H, Matsumoto T, Mizuno H. Simultaneous transcriptome analysis of Sorghum and Bipolaris sorghicola by using RNA-seq in combination with de novo transcriptome assembly. PloSone. 2013;8(4):e62460.

12. Pasini L, Bergonti M, Fracasso A, Marocco A, Amaducci S. Microarray analysis of differentially expressed mRNAs and miRNAs in young leaves of sorghum under dry-down conditions. J Plant Physiol. 2014;171(7):537–48.

13. Shakoor N, Nair R, Crasta O, Morris G, Feltus A, Kresovich S. A Sorghum bicolor expression atlas reveals dynamic genotype-specific expression profiles for vegetative tissues of grain, sweet and bioenergy sorghums. BMC Plant Biol. 2014;14:35–43.

14. Bekele WA, Wieckhorst S, Friedt W, Snowdon RJ. High-throughput genomics in sorghum: from whole-genome resequencing to a SNP screening array. Plant Biotech J. 2013;11(9):1112–25.

15. Tuberosa R, Salvi S. Genomics-based approaches to improve drought tolerance of crops. Trends Plant Sci. 2006;11(8):405–12.

16. Chaves MM, Maroco JP, Pereira JS. Understanding plant response to drought-from genes to the whole plant. Funct Plant Biol. 2003;30: 239–64.

17. Munne-Bosch S, Jubany-Mari T, Alegre L. Drought-induced senescence is characterized by a loss of antioxidant defences in chloroplasts. Plant Cell Environ. 2001;24:1319–27.

18. Jubany-Mari T, Munne-Bosch S, Alegre L. Redox regulation of water stress responses in field-grown plants. Role of hydrogen peroxide and ascorbate. Plant Physiol Biochem. 2010;48:351–8.

19. Kyparissis A, Petropoulou Y, Manetas Y. Summer survival of leaves in a softleaved shrub (Phlomis fruticosa L., Labiatae) under Mediterranean field conditions: avoidance of photoinhibitory damage through decreased chlorophyll contents. J Exp Bot. 2000;46:1825–31.

20. Kyparissis A, Drilias P, Manetas Y. Seasonal fluctuations in photoprotective (xanthophyll cycle) and photoselective (chlorophylls) capacity in eight Mediterranean plant species belonging to two different growth forms. Funct Plant Biol. 2000;27:265–72.

21. Havaux M, Tardy F. Loss of chlorophyll with limited reduction of photosynthesis as an adaptive response of Syrian barley landraces to highlight and heat stress. Aust J Plant Physiol. 1999;26:569–78.

22. Beck PA, Hutchison S, Gunter AS, Losi CT, Stewart BC, Capps KP, et al. Chemical composition and in situ dry matter and fibre disappearance of sorghum x Sudangrass hybrids. J Animal Sci. 2007;85:545–55.

23. Pelleschi S, Rocher JP, Prioul JL. Effect of water restriction on carbohydrate metabolism and photosynthesis in mature maize leaves. Plant Cell Environ. 1997;20:493–503.

24. Pinheiro C, Chaves MM, Ricardo CP. Alterations in carbon and nitrogen metabolism induced by water deficit in the stems and leaves of Lupinus albus L. J Exp Bot. 2001;52:1063–70.

25. Yang J, Zhang J, Wang Z, Zhu Q, Wang W. Remobilization of carbon reserves in response to water-deficit during grain filling of rice. Field Crops Res. 2001;71:47–55.

26. Toumi I, Gargouri M, Nouairi I, Moschou P, Salem-Fnayou A, Mliki A, et al. Water stress induced changes in the leaf lipid composition of four grapevine genotypes with different drought tolerance. Biol Plant. 2008;52:161–4.

27. Passioura J. Increasing crop productivity when water is scarce- from breeding to field management. In: 'New directions for a diverse planet'. Proceedings of 4th International Crop Sciences Congress, Brisbane, Australia. 2004

28. Cornic G. Drought stress inhibits photosynthesis by decreasing stomatal aperture – not by affecting ATP synthesis. Trends Plant Sci. 2000;5(5):187–88.

29. Chen TH, Murata N. Enhancement of tolerance of abiotic stress by metabolic engineering of betaines and other compatible solutes. Curr Opin Plant Biol. 2002;5:250–57.

30. Zhang J, Kirkham B. Antioxidant responses to drought in sunflower and sorghum seedlings. New Phytol. 1996;132:361–73.

31. Blum A. Drought resistance – is it really a complex trait? Func Plant Biol. 2011;38:753–57.

32. Fracasso A, Trindade L, Amaducci S. Drought tolerance strategies highlighted by two Sorghum bicolor races in a dry-down experiment. J Plant Phys. 2016;190:1–14.

33. Usadel B, Poree F, Nagel A, Lohse M, Czedik-Eysenberg A, Stitt M. A guide to using MapMan to visualize and compare Omics data in plants: a case study in the crop species, Maize. Plant Cell Env. 2009;32:1211–29.

34. Winkel-Shirley B. Biosynthesis of flavonoids and effects of stress. Curr Opin Plant Biol. 2002;5:218–23.

35. Ji W, Zhu L, Li Y, Yiang L, Zhao X, Cai H, et al. Overexpression of a glutathione S-transferase gene, GsGST, from wild soybean (Glycine soja) enhances drought and salt tolerance in transgenic tobacco. Biothecnol Lett. 2010;32:1173–79.

36. Mano J. Early events in environmental stresses in plants

induction mechanisms of oxidative stress. In: Inze D, Montago MV, editors. Oxidative stress in plants. New York: Taylor and Francis Publishers; 2002. p. 217–45.

37. Smirnoff N. Plant resistance to environmental stresses. Curr Opin Plant Biol. 1998;9:214–9.

38. Apel K, Hirt H. Reactive oxygen species: metabolism, oxidative stress and signal transduction. Annu Rev Plant Biol. 2004;55:373–99.

39. Schwanz P, Polle A. Differential stress responses of antioxidative systems to drought in pedunculate oak (Quercus robur) and maritime pine (Pinus pineaster) grown under high CO2 concentrations. J Exp Bot. 2001;52:133–43.

40. Ren J, Dai W, Xuan Z, Yao Y, Korpelainen H, Li C. The effect of drought and enhanced UV-B radiation on the growth and physiological traits of two contrasting poplar species. Forest Ecol Manag. 2007;239:112–9.

41. Liu C, Liu Y, Guo K, Fan D, Li G, Zheng Y, et al. Effect of drought on pigments, osmotic adjustment and antioxidant enzymes in six woody plant species in karst habitats of south western China. Environ Exp Bot. 2011;71:174–83.

42. Agati G, Azzarello E, Pollastri S, Tattini M. Flavonoids as antioxidants in plants: location and functional significance. Plant Sci. 2012;196:67–76.

43. Ramakrishna A, Ravishankar GA. Influence of abiotic stress signals on secondary metabolites in plants. Plant Signal Behavior. 2011;6(11):1720–31.

44. Grace S. Phenolics as antioxidants. In: Smirnoff N, editor. Antioxidants and reactive oxygen species in plants. Oxford: Blackwell Publishing Limited; 2005. p. 141–68.

45. Hernandez I, Alegre L, Munne-Bosch S. Enhanced oxidation of flavan-3-ols and proanthocyanidin accumulation in water-stressed tea plants. Phytochemistry. 2006;67:1120–6.

46. Hernández I, Alegrel L, Munné-Bosch S. Drought-induced changes in flavonoids and other low molecular weight antioxidants in Cistus clusii grown under Mediterranean field conditions. Tree Physiol. 2004;24:1303–11.

47. Tattini M, Galardi C, Pinelli P, Massai R, Remorini D, Agati G. Differential accumulation of flavonoids and hydroxycinnmates in leaves of Ligustrum vulgare under excess light and drought stress. New Phytol. 2004;163:547–61.

48. Castellarin SD, Pfeiffer A, Sivilotti P, Degan M, Peterlunger E, Di Gaspero G. Transcriptional regulation of anthocyanin biosynthesis in ripening fruits of grapevine under seasonal water deficit. Plant Cell Environ. 2007;30:1381–99.

49. Agati G, Tattini M. Multiple functional roles of flavonoids in photoprotection. New Phytol. 2010;186:786–93.

50. Pollastri S, Tattini M. Flavonols: old compounds for old roles. Ann Bot. 2011; 108:1225–33.

51. Ma D, Sun D, Wang C, Li Y, Guo T. Expression of flavonoid biosynthesis genes and accumulation of flavonoid in wheat leaves in response to drought stress. Plant Phys Biochem. 2014;80:60–6.

52. Larson RA. The antioxidants of higher plants. Phytochemistry. 1988;27:969–78.

53. Ithal N, Reddy AR. Rice flavonoid pathway genes, OsDfr and OsAns, are induced by hydration, high salt and ABA, and contain stress responsive promote elements that interact with the transcription activator, OsC1-MYB. Plant Sci. 2004;166:1505–13.

54. Hatier JHB, Gould KS. Foliar anthocyanins as modulators of stress signals. J Theor Biol. 2008;253:625–7.

55. Agati G, Stefano G, Biricolti S, Tattini M. Mesophyll distribution of 'antioxidant' flavonoids glycosides in Ligustrum vulgare leaves under contrasting sunlight irradiance. Ann Bot. 2009;104:853–61.

56. Agati G, Biricolti S, Guidi L, Ferrini F, Fini A, Tattini M. The biosynthesis of flavonoids is enhanced similarly by UV radiation and root zone salinity in L. vulgare leaves. J Plant Physiol. 2011;168:204–12.

57. Hernandez I, Alegre L, van Breusegem F, Munne-Bosch S. How relevant are flavonoids as antioxidants in plants? Trends Plant Sci. 2009;14:125–32.

58. Tattini M, Remorini D, Pinelli P, Agati G, Saracini E, Traversi ML, et al. Morphoanatomical and biochemical adjustments in response to root zone salinity stress and high solar radiation in two Mediterranean evergreen shrubs, Myrtus communis and Pistacia lentiscus. New Phytol. 2006;170:779–94.

59. Wolf L, Rizzini L, Stracke R, Ulm R, Rensing SA. The molecular and physiological responses of Physcomitrella patens to ultraviolet-B radiation. Plant Physiol. 2010;153:1123–34.

60. Anjum F, Yaseen M, Rasool E, Wahid A, Anjum S. Water stress in barley (Hordeum vulgare L.): on chemical composition and chlorophyll contents. Pak. J Agr Sci. 2003;40:45–9.

61. Krammer I, Beckett RP, Wornik S, Zorn M, Pfeifhofer HW. Revival of a resurrection plant correlates with its antioxidant status. Plant J. 2003;31:13–24.

62. Andersson J, Walters RG, Horton P, Jansson S. Antisense inhibition of the photosynthetic antenna proteins CP29 and CP26: implications for the mechanism of protective energy dissipation. Plant Cell. 2001;13: 1193–204.

63. Andersson J, Wentworth M, Walters RG, Howard CA, Ruban AV, Horton P, et al. Absence of the Lhcb1 and Lhcb2 proteins of the light-harvesting complex of the photosystem II: effects on photosynthesis, grana stacking and fitness. Plant J. 2003;35:350–61.

64. Ganeteg U, Kulheim C, Andersson J, Jansson S. Is each light-harvesting complex protein important for plant fitness? Plant Physiol. 2004;134:502–9.

65. Kovacs L, Damkjær J, Kereiche S, Ilioaia C, Ruban AV, Boekema EJ, et al. Lack of the lightharvesting complex CP24 affects the structure and function of the grana membranes of higher plant chloroplasts. Plant Cell. 2006;18:3106–20.

66. Jansson S. The light-harvesting chlorophyll a/ b-binding proteins. Biochim Biophys Acta. 1994;1184:1–19.

67. Humbeck K, Krupinska K. The abundance of minor

chlorophyll a/b-binding proteins CP29 and LHCI of barley (Hordeum vulgare L.) during leaf senescence is controlled by light. J Exp Bot. 2003;54:375–83.

68. Nott A, Jung HS, Koussevitzky S, Chory J. Plastid-tonucleus retrograde signaling. Ann Rev Plant Biol. 2006;57:739–59.

69. Staneloni RT, Rodriguez-Batiller MJ, Casal JJ. Abscisic acid, high-light, and oxidative stress down-regulate a photosynthetic gene via a promoter motif not involved in phytochrome-mediated transcriptional regulation. Mol Plant. 2008;1:75–83.

70. Weatherwax SC, Ong MS, Degenhardt J, Bray EA, Tobin EM. The interaction of light and abscisic acid in the regulation of plant gene expression. Plant Physiol. 1996;111:363–70.

71. Sturm A, Tang GQ. The sucrose-cleaving enzymes of plants are crucial for development, growth and carbon partitioning. Trends Plant Sci. 1999;4:401–7.

72. Zinselmier C, Westgate ME, Schussier JR, Jones R. Low water potential disrupts carbohydrate metabolism in Maize (Zea mays L.) ovaries. Plant Physiol. 1995;107:385–91.

73. Ackerson RC. Osmoregulation in Cotton in response to water stress. III. Effects of phosphorus fertility. Plant Physiol. 1985;77:309–12.

74. Trouverie J, Thévenot C, Rocher JP, Sotta B, Prioul JL. The role of abscisic acid in response of a specific vacuolar invertase to water stress in the adult maize leaf. J Exp Bot. 2003;54:2177–86.

75. Gazarrani S, McCourt P. Genetic interactions between ABA, ethylene and sugar signaling pathways. Curr Opin Plant Biol. 2001;4:387–91.

76. Smeekens S, Rook F. Sugar sensing and sugar-mediated signal transduction in plants. Plant Physiol. 1997;115:7–13.

77. Smeekens S. Sugar-induced signal transduction in plants. Ann Rev Plant Phys Plant Mol Biol. 2000;51:49–81.

78. Martin SA, Facanha AR, Bressan-Smith RE. Developing Crop-Specific Irrigation Management Strategies Considering Effects of Drought on Carbon Metabolism in Plants, Water Quality, Soil and Managing Irrigation of Crops, Dr. Teang Shui Lee (Ed.), InTech, doi:10.5772/30419. Available from: http://www.intechopen. com/books/water-quality-soil-and-managing-irrigation-of-crops/developingcrop-specific-irrigation-management-strategies-considering-effects-of-droughton-carbon-m.

79. Bray EA. Plant responses to water deficit. Trends Plant Sci. 1997;2:48–54.

80. Millar AA, Kunst L. Very-long-chain fatty acid biosynthesis is controlled through the expression and specificity of the condensing enzyme. Plant J. 1997;12:121–31.

81. Denic V, Weissman JS. A molecular caliper mechanism for determining very long-chain fatty acid length. Cell. 2007;130:663–77.

82. Bartels D, Nelson D. Approaches to improve stress tolerance using molecular genetic. Plant Cell Environ. 1994;17:659–67.

83. Kosma KD, Bourdenx B, Bernard A, Parsons EP, Lü S, Joubès J, et al. The impact of water deficiency on leaf cuticle lipids of Arabidopsis. Plant Physiol. 2009;151:1918–29.

84. Cameron KD, Teece MA, Smart LB. Increased accumulation of cuticular wax and expression of lipid transfer protein in response to periodic drying events in leaves of tree tobacco. Plant Physiol. 2006;140:176–83.

85. Samdur MY, Manivel P, Jain VK, Chikani BM, Gor HG, Desai S, et al. Genotypic differences and water-deficit induced enhancement in epicuticular wax load in peanut. Crop Sci. 2003;43:1294–99.

86. Jenks MA, Andersen L, Teusink RS, Williams MH. Leaf cuticular waxes of potted rose cultivars as affected by plant development, drought and paclobutrazol treatments. Physiol Plant. 2001;112:62–70.

87. Bondada BR, Oosterhuis DM, Murphy JB, Kim KS. Effect of water stress on the epicuticular wax composition and ultrastructure of cotton (Gossypium hirsutum L.) leaf, bract, and boll. Environ Exp Bot. 1996;36:61–9.

88. Islam MA, Du H, Ning J, Ye H, Xiong L. Characterization of Glossy1- homologous genes in rice involved in leaf wax accumulation and drought resistance. Plant Mol Biol. 2009;70:443–56.

89. Torres-Franklin ML, Repellin A, Huynh VB, d'Arcy-Lameta A, ZuilyFodil Y, Pham-Thi AT. Omega-3 fatty acid desaturase (FAD3, FAD7, FAD8) gene expression and linolenic acid content in cowpea leaves submitted to drought and after rehydration. Environ Exp Bot. 2009;65:162–69.

90. Ng CKY, Carr K, McAinsh MR, Powell B, Hetherington AM. Drought-induced guard cell signal transduction involves sphingosine-1-phosphate. Nature. 2001;410:596–98.

91. Norberg P, Mansson JE, Liljenberg C. Characterisation of glucosylceramide from plasma membranes of plant root cells. Biochem Biophys Acta. 1991;1066:257–60.

92. Spiegel S, Milstein S. Sphingosine-1-phosphate: signalling inside and out. FEBS Lett. 2000;476:55–7.

93. Monteiro De Paula F, Pham Thi AT, Zuily-Fodil Y, Ferrari- Iliou R, Vieira Da Silva J, Mazliak P. Effect of water stress on the biosynthesis and degradation of polyunsaturated lipid molecular species in leaves of Vigna unguiculata. Plant Physiol Biochem. 1993;31:707–15.

94. Sahsah Y, Campos P, Gareil M, Zuily-Fodil FY, Pham-Thi AT. Enzymatic degradation of polar lipids in Vigna unguiculata leaves and influence of drought stress. Physiol Plant. 1998;104:577–86.

95. El Maarouf H, d'Arcy-Lameta A, Gareil M, Zuily-Fodil Y, Pham-Thi AT. Cloning and expression under drought of cDNAs coding for two PI-PLCs in cowpea leaves. Plant Physiol Biochem. 2001;39:167–72.

96. Bargmann BOR, Laxalt AM, ter Riet B, van Schooten B, Merquiol E, Testerink C, et al. Multiple PLDs required for high salinity and water deficit tolerance in plants. Plant Cell Physiol. 2009;50:78–89.

97. Hong Y, Pan X, Welti R, Wang X. Phospholipase Da3 is

involved in the hyperosmotic response in Arabidopsis. Plant Cell. 2008;20:803–16.

98. Wang CR, Yang AF, Yue GD, Gao Q, Yin HY, Zhang JR. Enhanced expression of phospholipase C1 (ZmPLC1) improves drought tolerance in transgenic maize. Planta. 2008;227:1127–40.

99. Wang X. Regulatory functions of phospholipase D and phosphatidic acid in plant growth, development, and stress responses. Plant Physiol. 2005;139:566–73.

100. Zhang M, Barg R, Yin M, Gueta-Dahan Y, Leikin-Frenkel A, Salts Y, et al. Modulated fatty acid desaturation via overexpression of two distinct omega-3 desaturases differentially alters tolerance to various abiotic stresses in transgenic tobacco cells and plants. Plant J. 2005;44:361–71.

101. Hong Y, Zhang W, Wang X. Phospholipase D and phosphatidic acid signalling in plant response to drought and salinity. Plant Cell Environ. 2010;33(4):627–35.

102. Mace ES, Jordan DR. Integrating sorghum whole genome sequence information with a compendium of sorghum QTL studies reveals uneven distribution of QTL and of gene-rich regions with significant implications for crop improvement. Theor App Gen. 2011;123(1):169–91.

103. Barr HD, Weatherley PE. A re-examination of the relative turgidity technique for estimating water deficit in leaves. Aust J Biol Sci. 1962;15:413–28.

104. Mastrorilli M, Katerji N, Rana G. Water efficiency and stress on grain sorghum at different reproductive stages. Agr Water Manag. 1995;28:23–34.

105. Mortazavi A, Williams BA, McCue K, Schaeffer L, Wold B. Mapping and quantifying mammalian transcriptomes by RNA-Seq. Nat Methods. 2008;5:621–28.

106. Benjamini Y, Yekutieli D. The control of false discovery rate in multiple testing dependency. Ann Stat. 2001;29:1165–88.

107. Du Z, Zhou X, Ling Y, Zhang Z, Su Z. agriGO: a GO analysis toolkit for the agricultural community. Nucleic Acids Res. 2010;38:W64–70.

108. Trapnell C, Pachter L, Salzberg SL. TopHat: discovering splice junctions with RNA-Seq. Bioinformatics 2009; 25:1105-11.

CHAPTER 4

INHERITANCE OF RESISTANCE TO SORGHUM SHOOT FLY, ATHERIGONA SOCCATA IN SORGHUM, SORGHUM BICOLOR (L.) MOENCH

Riyazaddin Mohammed, Ashok Kumar Are, Rajendra Sudhakar Munghate, Ramaiah Bhavanasi, Kavi Kishor B. Polavarapu and Hari Chand Sharma

Keywords: Sorghum, *Atherigona soccata*, resistance, combining ability, heritability, general combining ability, specific combining ability

INTRODUCTION

Sorghum [*Sorghum bicolor* (L.) Moench] is an important food and fodder crop of dryland agriculture. It has a wide range of adaptability to various agro-ecological conditions in the semi-arid tropics (SAT). It is a staple food crop for millions of poor people living in the SAT regions of Asia and Africa (Ashok Kumar et al., 2011). It is cultivated on marginal, fragile, and drought-prone environments in the semi-arid tropics. Sorghum is also grown for forage, and is fed to animals as a green chop, silage or hay. Sorghum grain is one of the major ingredients in poultry and cattle feed in USA, China and Australia (Bramel-Cox et al., 1995). It is also emerging as an important bio-fuel crop. India is the largest sorghum grower in the world with an average productivity of 854.4 kg ha^{-1} (FAO, 2014). In the dry land agriculture, the yield and quality of sorghum produced is affected by a wide array of biotic and abiotic constraints. Several insect pests damage sorghum from seedling stage to maturity. Around 150 insect pests attack sorghum (Jotwani et al., 1980; Sharma, 1993), of which sorghum shoot fly, *Atherigona soccata* is a serious pest that reduces sorghum production in the semi-arid tropics. Due to shoot fly damage, a loss of 80–90% of grain, and 68% of fodder yield was recorded in India (Balikai and Bhagwat, 2009; Kahate et al., 2014). Identifying sorghum genotypes with stable shoot fly resistance is highly important as it will help to reduce the cost of cultivation and stabilize the yields.

Shoot fly, *A. soccata* attacks sorghum at 7-30 days after seedling emergence (DAE). It lays white elongated cigar shaped eggs on the lower surface of the leaf, parallel to leaf midrib (Padmaja et al., 2010). Maggot emerges from the egg in 2 days, reaches the central whorl, cuts the central leaf, and starts feeding on the decaying leaf tissue of central whorl. As a result, the central whorl dries off resulting in a typical deadheart (Deeming, 1971; Dhillon et al., 2006a). It completes its life cycle in 17–21 days (Sharma et al., 2003). Many approaches have been used to minimize the losses caused by shoot fly, including agronomic practices, natural enemies, synthetic insecticides, and host plant resistance (Sharma, 1993; Kumar et al., 2008), but implementation of all these practices is not always feasible.

Host plant resistance (HPR) is one of the most economic and practical means for controlling shoot fly damage because it does not involve any extra cost to the farmers or require application skills in pest control techniques (Sharma, 1985; Dhillon et al., 2006a). It is also compatible with other methods of pest control. The negative effects of resistant genotypes on insect populations are continuous and cumulative over time. Reduction in pest abundance through HPR can also enhance the effectiveness of natural enemies, and reduces the need to apply pesticides (Sharma, 1993).

In view of serious economic losses due to shoot fly damage, improvement of sorghum for shoot fly resistance is one of the major goals in sorghum improvement programs. Although several improved varieties and hybrids have been developed

Citation: Mohammed R, Are AK, Munghate RS, Bhavanasi R, Polavarapu KKB and Sharma HC (2016) Inheritance of Resistance to Sorghum Shoot Fly, Atherigona soccata in Sorghum, Sorghum bicolor (L.) Moench. Front. Plant Sci. 7:543. doi: 10.3389/fpls.2016.00543.

and released, the yield gains at farmers› level are minimal (Sharma et al., 2003). Therefore, there is a need to combine shoot fly resistance with high grain yield to increase the productivity of this crop. Because of the poor understanding of inheritance of shoot fly resistance, sorghum improvement for resistance to this pest have not been very effective (Doggett et al., 1970; Riyazaddin et al., 2015a). Combining ability studies is needed to identify better combiners and develop superior hybrids that would be helpful in understanding the nature of gene action and inheritance of quantitative traits (Singh and Chaudhary, 1985; Goyal and Kumar, 1991; Mohammed et al., 2015).

Diallel crosses have been widely used in genetic research to investigate the inheritance of important agronomical and morphological traits. Studies on combining ability estimates are useful to understand the nature of genetic variance, and to predict the relative performance of different lines in hybrid combinations. Information on the nature and magnitude of gene action is important in understanding the genetic potential of a population, and deciding the breeding procedure to be adopted in a given population (Prabhakar and Raut, 2010). Several researchers worked on developing sorghum genotypes with resistance to shoot fly, but the genetic gains were quite low (Ashok Kumar et al., 2011). Therefore, the present studies were carried out to understand the nature of gene action of shoot fly resistance and morphological traits, and their inheritance to study the general and specific combining abilities of parents and crosses, respectively, to establish appropriate strategies for developing sorghum genotypes with improved resistance to shoot fly, *A. soccata*.

MATERIALS AND METHODS

Experimental Material

The experiments were conducted at the International Crops research Institute for the Semi-Arid Tropics (ICRISAT), Patancheru-502 324, Telangana, India. Based on performance of the genotypes in the field (Riyazaddin et al., 2015a,b), 10 sorghum genotypes exhibiting high levels of resistance (ICSV 25019, PS 35805, IS 2123, IS 2146, and IS 18551) or susceptibility (CSV 15 and Swarna) to shoot fly; and ICSV 700, M 35-1, and Phule Anuradha with moderate levels of shoot fly resistance (Annexure I in Supplementary Material) which are genetically diverse were selected for the crossing program. These genotypes were used for crossing in a full diallel fashion i.e., crossing in all possible combinations including reciprocals, to test the hybrid vigor of crosses and combining abilities, and the reciprocal effects of parental genotypes. Crossing of 10 selected parents in a full diallel fashion generated 45 direct crosses, and 45 reciprocal crosses (90 F_1's).

The test material consisting of 10 parents and 90 F_1's was sown in three replications in a randomized complete block design (RCBD). One set of replicated trial was raised under protected conditions during the 2013 rainy and postrainy seasons. The test genotypes were sown in the field in 2.0 m row length, row to row spacing of 75 cm and with a spacing of 10 cm in-between the plants within a row. A basal dose of ammonium phosphate @100 kg/ha was applied to the field. Parents were sown in two rows each, and the F_1's in a single row. The test material was tested under high shoot fly pressure using interlard fishmeal technique (Soto, 1974; Sharma et al., 1992). Thinning was carried out at 7 days after seedling emergence. Earthing up and topdressing with urea (@100 kg/ha) were carried out at 30 DAE. Hand weeding was carried out whenever necessary. Furrow irrigation was given to the experimental material during the 2013 postrainy season.

OBSERVATIONS

Shoot Fly Damage Parameters

Data were recorded on plants with shoot fly eggs and number of shoot fly eggs per plant at 14 DAE, and shoot fly deadhearts at 21 DAE, and expressed as the percentage of plants with shoot fly eggs, number of shoot fly eggs per plant, and percentage of plants with deadhearts. Overall resistance score was recorded on a 1-9 scale before harvesting (1 = plants with < 10% deadhearts and uniform tillers and harvestable panicles, and 9 = plants with >80% deadhearts, and a few or no productive tillers) (Sharma et al., 1992).

Morphological Characteristics

Leaf glossiness was evaluated visually on a 1–5 scale (1 = highly glossy, and 5 = non-glossy) at 10–12 DAE (fifth leaf stage), when the expression of this trait is most apparent in the early morning hours (Sharma and Nwanze, 1997). Leafsheath pigmentation was visually scored on a 1-3 rating scale (1 = leafsheath with deep pink pigmentation, and 3 = green leafsheath and no pink pigmentation) at 7 DAE (Dhillon et al., 2006b). Seedling vigor was recorded at 10 DAE on 1–3 scale (1 = high seedling vigor, and 3 = poor seedling vigor) (Sharma et al., 1992). Waxy bloom was visually scored on 1–3 scale (1 = slightly waxy, and 3 = completely waxy) at the flag leaf stage of the crop. Plant color was evaluated visually on a 1–2 scale (1 = non-tan; red colored plant, and 2 = tan; colorless plant).

The density of trichomes on both the surfaces of leaf was recorded at 12 DAE by taking a 2.5 cm^2 portion from the center of the fifth leaf (Maiti and Bidinger, 1979). The leaf samples were taken from three plants at random, de-stained overnight with 2:1 of acetic acid: alcohol. The de-stained leaf is stored in 90% lactic acid, and observed at 10× magnification under a stereomicroscope. Trichomes on the leaf surface were counted and expressed as numbers of trichomes in a 10× microscopic field (density per microscopic field).

Statistical Analysis

The data were subjected to analysis of variance (ANOVA) using GenStat® 13th version (GenStat, 2010). Significance of the differences between the genotypes was judged by *F*-test, while the genotypic means were compared by least significant difference (LSD) at $P \leq 0.05$. Diallel analysis of parents and F_1's was carried out according to the Griffing's method 1 and model 1 (Griffing, 1956), which partitions the total variation into the general combining ability (*gca*) effects that provide the genetic nature of the parents, and the specific combining ability (*sca*) effects that provide information about the performance of hybrids, using Windowstat software (Indostat Services, 2004). The coefficient of variation at phenotypic and genotypic levels was estimated using the formula adopted by Johnson et al. (1955) and predictability ratio using Baker (1978).

RESULTS

Expression of Resistance to Shoot Fly, *A. soccata* by the F_1 Hybrids (10 × 10 Diallel) in Comparison with the Parents

The variance ratio for 10 parents, 45 F_1's along with 45 reciprocal crosses for all the traits studied were significant at $P \leq 0.01$, with few exceptions. Variance ratio of plants with shoot fly eggs (%), and number of shoot fly eggs/plant were significant only in the postrainy season (Table 1).

Table 1: Expression of resistance to sorghum shoot fly, *A. soccata* **in sorghum (ICRISAT,** Patancheru, 2013–14)

Pedigree	Plants with shoot fly eggs (%)		Number of shoot fly eggs/plant		Shoot fly deadhearts (%)		ORS	
	2013 R	2013 PR	2013 R	2013 PR	2013 R	2013 PR	2013 R	2013 PR
PARENTS								
ICSV 700	92.24	16.55	2.33	1.13	66.44	7.48	6.33	6.67
Phule Anuradha	96.29	10.02	2.15	1.00	75.92	8.06	5.00	6.33
M 35-1	91.41	18.25	1.70	1.06	83.72	17.77	8.33	4.00
CSV 15	100.00	44.49	3.57	1.22	83.33	40.40	8.00	6.33
ICSV 25019	69.73	16.53	1.31	1.05	21.45	6.39	5.00	4.33
PS 35805	96.19	10.21	1.81	1.06	44.62	2.17	4.33	4.00
IS 2123	81.33	10.44	2.27	1.00	60.07	4.30	6.00	4.00
IS 2146	94.80	17.72	2.24	1.13	54.59	5.32	6.00	5.33
IS 18551	89.78	6.30	2.12	1.33	60.21	3.96	6.00	4.00
Swarna	93.60	84.57	1.72	1.38	76.45	53.05	8.67	8.33
DIRECT CROSSES								
ICSV 700 X Phule Anuradha	100.00	9.39	1.71	0.67	88.64	6.67	5.00	7.00
ICSV 700 X M 35-1	100.00	36.57	1.94	1.00	89.56	32.87	5.67	6.00
ICSV 700 X CSV 15	93.94	30.81	1.92	1.08	80.68	26.64	7.00	7.67
ICSV 700 X ICSV 25019	86.90	22.62	2.17	0.89	73.81	7.54	5.67	5.33
ICSV 700 X PS 35805	80.91	27.78	2.15	1.33	70.91	37.04	4.67	5.67
ICSV 700 X IS 2123	87.88	35.91	3.09	1.00	85.61	15.28	6.00	6.33
ICSV 700 X IS 2146	93.94	33.43	1.65	1.39	74.68	10.17	6.67	5.00
ICSV 700 X IS 18551	100.00	35.81	2.10	1.00	83.33	17.22	6.00	7.00
ICSV 700 X Swarna	93.33	56.78	2.10	1.29	93.33	34.30	7.67	7.00
Phule Anuradha X M 35-1	90.48	3.03	1.56	0.33	90.48	3.03	6.67	6.33
Phule Anuradha X CSV 15	88.89	53.70	2.17	1.00	92.59	35.19	6.00	6.00
Phule Anuradha X ICSV 25019	100.00	47.13	1.26	1.00	96.97	25.05	6.00	6.00
Phule Anuradha X PS 35805	100.00	30.71	1.42	1.00	94.44	19.20	6.33	5.00
Phule Anuradha X IS 2123	100.00	28.69	2.15	1.17	85.86	20.36	5.67	4.67
Phule Anuradha X IS 2146	88.33	54.98	1.93	1.00	92.46	41.29	5.33	6.00
Phule Anuradha X IS 18551	100.00	15.58	2.24	1.00	77.78	14.29	5.67	6.33
Phule Anuradha X Swarna	100.00	65.02	2.21	1.47	94.44	51.86	7.00	6.00
M 35-1 X CSV 15	100.00	48.68	2.30	1.17	91.67	52.38	6.67	5.67
M 35-1 X ICSV 25019	91.67	4.17	2.24	0.33	91.38	12.50	4.67	6.00
M 35-1 X PS 35805	100.00	10.37	1.87	0.67	76.67	6.67	6.33	4.00
M 35-1 X IS 2123	86.77	14.54	2.15	1.00	66.27	17.50	6.00	5.67
M 35-1 X IS 2146	95.83	24.44	2.29	0.83	82.37	24.44	6.00	5.67
M 35-1 X IS 18551	96.08	30.00	3.26	1.83	72.16	28.89	6.00	5.00
M 35-1 X Swarna	91.53	29.44	2.04	1.33	72.75	6.67	5.33	6.67
CSV 15 X ICSV 25019	94.44	70.91	1.80	2.03	94.44	55.45	7.00	6.67
CSV 15 X PS 35805	100.00	83.82	1.54	1.07	93.89	37.88	6.00	7.33
CSV 15 X IS 2123	100.00	21.09	1.76	1.42	85.19	14.42	5.00	5.00
CSV 15 X IS 2146	100.00	72.89	2.16	1.29	89.56	63.36	6.00	5.67
CSV 15 X IS 18551	96.30	60.15	2.14	1.23	97.44	42.12	6.00	7.00
CSV 15 X Swarna	93.33	58.18	1.71	1.56	86.67	68.48	7.67	7.00
ICSV 25019 X PS 35805	85.19	16.98	1.84	1.00	44.44	8.89	5.67	6.33
ICSV 25019 X IS 2123	90.48	2.08	2.47	0.33	61.69	20.83	5.00	5.67
ICSV 25019 X IS 2146	100.00	21.59	2.56	1.00	83.33	11.36	4.67	4.33
ICSV 25019 X IS 18551	95.83	7.54	2.19	0.67	95.83	20.71	5.00	5.00
ICSV 25019 X Swarna	86.77	53.52	1.89	1.27	79.37	46.48	7.33	7.67
PS 35805 X IS 2123	90.48	15.76	1.83	0.67	69.05	12.73	4.33	5.67
PS 35805 X IS 2146	90.11	18.52	2.08	0.83	68.42	14.81	5.00	7.00
PS 35805 X IS 18551	87.83	8.10	2.49	0.67	67.20	5.90	4.33	5.67
PS 35805 X Swarna	97.44	49.66	2.06	1.28	66.80	28.28	6.67	7.00
IS 2123 X IS 2146	66.67	12.63	1.33	1.00	66.67	7.87	6.33	6.67
IS 2123 X IS 18551	91.67	15.74	2.18	0.67	77.22	11.57	5.67	6.00
IS 2123 X Swarna	100.00	51.67	1.85	1.41	92.59	24.17	4.33	5.67
IS 2146 X IS 18551	100.00	6.49	2.09	0.67	87.88	2.56	6.33	4.67
IS 2146 X Swarna	93.64	66.32	1.89	1.22	87.58	51.91	5.67	5.33
IS 18551 X Swarna	74.81	61.34	3.44	1.32	85.93	46.41	6.00	5.33
RECIPROCAL CROSSES								
Phule Anuradha X ICSV 700	100.00	26.85	2.21	1.17	78.33	9.72	5.00	5.33
M 35-1 X ICSV 700	100.00	23.65	1.42	1.08	89.26	12.22	5.33	7.00
M 35-1 X Phule Anuradha	100.00	11.11	1.92	0.44	76.07	0.00	6.33	5.67
CSV 15 X ICSV 700	100.00	91.67	2.27	1.19	91.90	45.83	5.33	6.00
CSV 15 X Phule Anuradha	100.00	37.04	1.39	0.88	93.33	39.15	6.00	6.33
CSV 15 X M 35-1	96.97	55.39	2.05	1.15	96.97	61.45	5.67	6.00
ICSV 25019 X ICSV 700	93.33	13.69	2.33	0.67	76.67	8.93	5.00	6.33
ICSV 25019 X Phule Anuradha	84.85	20.00	1.56	1.11	49.90	13.33	5.67	6.00
ICSV 25019 X M 35-1	100.00	27.36	2.14	1.07	91.53	9.26	5.67	5.33
ICSV 25019 X CSV 15	100.00	99.29	2.97	1.56	85.79	46.19	6.33	7.67
PS 35805 X ICSV 700	100.00	16.30	1.88	1.17	56.88	5.56	5.33	6.00
PS 35805 X Phule Anuradha	83.33	4.17	3.08	0.33	53.17	0.00	5.67	6.00
PS 35805 X M 35-1	100.00	33.86	2.23	1.00	89.74	17.99	6.00	6.00
PS 35805 X CSV 15	88.89	20.63	2.14	0.83	77.78	30.16	7.00	7.67
PS 35805 X ICSV 25019	100.00	25.71	1.89	1.00	75.79	0.00	6.33	6.00
IS 2123 X ICSV 700	96.97	23.93	1.82	1.17	63.54	12.26	6.67	6.33
IS 2123 X Phule Anuradha	100.00	28.10	2.29	1.33	80.16	28.69	7.00	4.67
IS 2123 X M 35-1	97.62	42.06	2.33	1.00	84.92	12.96	5.33	6.33

PS 35805 X CSV 15	88.89	20.63	2.14	0.83	77.78	30.16	7.00	7.67
PS 35805 X ICSV 25019	100.00	25.71	1.89	1.00	75.79	0.00	6.33	6.00
IS 2123 X ICSV 700	96.97	23.93	1.82	1.17	63.54	12.26	6.67	6.33
IS 2123 X Phule Anuradha	100.00	28.10	2.29	1.33	80.16	28.69	7.00	4.67
IS 2123 X M 35-1	97.62	42.06	2.33	1.00	84.92	12.96	5.33	6.33
IS 2123 X CSV 15	100.00	62.64	2.80	1.14	95.24	47.19	6.67	6.00
IS 2123 X ICSV 25019	95.83	25.16	2.19	1.00	64.96	16.23	3.67	5.33
IS 2123 X PS 35805	100.00	15.15	1.67	0.67	82.15	15.15	5.00	6.00
IS 2146 X ICSV 700	100.00	16.50	3.47	1.00	93.33	0.00	6.33	5.00
IS 2146 X Phule Anuradha	93.94	12.96	1.87	0.67	90.91	6.48	5.67	5.67
IS 2146 X M 35-1	87.04	16.62	1.25	1.00	57.83	18.22	6.33	5.67
IS 2146 X CSV 15	100.00	59.43	2.27	1.13	84.26	49.20	5.00	4.67
IS 2146 X ICSV 25019	100.00	13.22	1.38	1.17	86.61	12.96	3.50	5.00
IS 2146 X PS 35805	96.97	24.36	2.29	1.00	90.24	11.79	5.33	4.67
IS 2146 X IS 2123	100.00	9.71	2.27	1.33	73.61	16.85	6.00	4.33
IS 18551 X ICSV 700	100.00	24.16	1.64	1.25	69.44	10.94	6.67	6.67
IS 18551 X Phule Anuradha	90.28	15.00	1.73	0.67	60.00	6.67	6.33	5.33
IS 18551 X M 35-1	75.76	30.28	2.00	1.33	63.64	3.33	6.67	6.67
IS 18551 X CSV 15	100.00	33.27	1.72	1.23	91.67	26.88	6.67	4.67
IS 18551 X ICSV 25019	91.67	9.09	1.59	0.44	83.06	24.58	5.00	5.33
IS 18551 X PS 35805	100.00	25.57	2.35	1.07	78.70	13.26	5.00	5.67
IS 18551 X IS 2123	82.22	10.36	2.22	1.17	72.22	5.59	6.00	4.67
IS 18551 X IS 2146	96.97	23.61	2.19	1.53	74.46	13.06	5.33	6.00
Swarna X ICSV 700	100.00	89.17	2.23	1.31	98.04	58.33	4.67	6.33
Swarna X Phule Anuradha	100.00	60.71	1.78	1.12	90.00	44.05	6.33	6.00
Swarna X M 35-1	100.00	65.29	1.49	0.94	96.97	32.42	6.00	6.00
Swarna X CSV 15	100.00	81.96	1.95	1.26	100.00	65.75	8.00	6.67
Swarna X ICSV 25019	100.00	57.41	2.01	1.54	96.97	52.78	7.33	7.33
Swarna X PS 35805	100.00	63.64	2.25	1.29	91.56	25.76	6.00	7.33
Swarna X IS 2123	100.00	63.26	2.37	1.63	74.44	37.88	7.00	6.00
Swarna X IS 2146	100.00	43.59	2.17	1.29	100.00	48.29	6.33	6.33
Swarna X IS 18551	100.00	67.33	2.43	1.47	100.00	47.48	6.00	6.33
Mean	94.59	33.80	2.08	1.08	79.99	23.57	5.92	5.88
SE ±	6.35	10.76	0.43	0.26	10.09	10.38	0.54	0.61
Vr (99, 198)	1.25^{NS}	4.79**	1.14^{NS}	1.45**	2.06**	3.05**	3.06**	2.31**
LSD (P 0.05)	17.72	30.00	1.19	0.72	28.14	28.93	1.51	1.71

*** F-test significant at P = 0.01; R, rainy season; PR, postrainy season; NS, non-significant F value; ORS, Overall resistance score: 1–9 ranking with 1 = plants with uniform tillers and harvestable panicles; 9 = plants with a few or no productive tillers.*

Very high levels of oviposition (2–4 eggs per plant) were recorded during the rainy season as compared to the postrainy season (1–2 eggs/plant) (Table 1). During the rainy season, the genotypes ICSV 25019 (69.73%) and IS 2123 (81.33%), the direct crosses IS 2123 X IS 2146 (66.67%), and IS 18551 X Swarna (74.81%), and the reciprocal cross IS 18551 X M 35-1 (75.76%) exhibited lower oviposition as compared to that of susceptible check Swarna with 93.60% of plants with shoot fly eggs. In the postrainy season, almost all the crosses exhibited lower percentage of plants with shoot fly eggs than the susceptible check Swarna, with few exceptions. Higher ovipostion was observed in the susceptible genotypes CSV 15 and Swarna and in crosses where these genotypes were involved as either of the parents, with 2-3 shoot fly eggs/plant across the seasons.

The percentage of plants with shoot fly deadhearts in the parents varied from 21.45–83.72% during rainy season, and 3.96–53.05% in the postrainy season (Table 1); and for crosses, the deadheart percentage ranged from 44.44 to 100% in the rainy season, and 0.00–68.48% during the postrainy season. The genotypes ICSV 700, Phule Anuradha, ICSV 25019, PS 35805, IS 2123, and IS 2146 exhibited lower shoot fly damage than CSV 15 and Swarna. M 35-1 showed a susceptible reaction in the rainy season and exhibited resistant nature in the postrainy season. Thirty-three crosses exhibited resistance to shoot fly across seasons, and PS 35805 and IS 18551 were involved as one of the parents in most of these crosses.

M 35-1 X Swarna, PS 35805 X Swarna, IS 18551 X ICSV 700, IS 18551 X Phule Anuradha, and ICSV 700 X IS 2146 exhibited antibiosis resistance mechanism i.e., lower numbers of plants with shoot fly deadhearts than the plants with shoot fly eggs (Table 1). Twenty-four crosses showed antibiosis as a mechanism of resistance to shoot fly in the rainy season. The direct crosses ICSV 700 X IS 2123, ICSV 700 X IS 18551, CSV 15 X PS 35805, CSV 15 X IS 18551, PS 35805 X Swarna, IS 2123 X Swarna, and reciprocal crosses CSV 15 X ICSV 700, ICSV 25019 X CSV 15, IS 2123 X M 35-1, IS 18551 X M 35-1, Swarna X M 35-1, Swarna X PS 35805, and Swarna X IS 2123 exhibited lower shoot fly deadhearts than the plants with shoot fly eggs, indicating antibiosis as mechanism of resistance to shoot fly during the postrainy season.

Morphological Traits

The parents ICSV 700, Phule Anuradha, IS 2123, IS 2146 and IS 18551, and the 29 crosses were glossy with high plant vigor, and had leafsheath pigmentation and high trichome density on the abaxial and adaxial leaf surfaces (Table 2). The hybrids ICSV 700 X M 35-1, ICSV 700 X PS 35805, ICSV 700 X IS 2146, Phule Anuradha X M 35-1, M 35-1 X Phule Anuradha, M 35-1 X IS 2146, M 35-1 X IS 18551, IS 2146 X Phule Anuradha, and IS 18551 X Swarna expressed the leaf glossiness, leafsheath pigmentation, high vigor and high trichome density only in the rainy season; whereas the crosses ICSV 700 X CSV 15, M 35-1 X PS 35805, PS 35805 X IS 2123, PS 35805 X IS 18551, IS 2123 X M 35-1, IS 2123 X IS 2146 expressed these traits only in the postrainy season.

Table 2: Morphological characteristics of sorghum (parents and F_1 **crosses) evaluated for resistance to sorghum shoot fly,** *A. soccata* **(ICRISAT,** Patancheru, 2013–14)

Pedigree	Leaf glossy score		Plant vigor score		Leafsheath pigmentation		Trichome density on abaxial leaf surface		Trichome density on adaxial leaf surface		Waxy bloom		Plant color	
	2013 R	2013 PR	2013 R	2013 PR	2013 R	2013 PR	2013 R	2013 PR	2013 R	2013 PR	2013 R	2013 PR	2013 R	2013 PR
PARENTS														
ICSV 700	2.67	1.00	1.33	1.00	1.67	1.00	139.00	89.56	162.67	109.22	1.33	2.00	2.00	2.00
Phule Anuradha	1.50	2.67	1.00	1.33	1.50	2.00	73.33	62.56	98.22	64.00	1.67	2.00	1.00	1.00
M 35-1	3.67	3.00	2.67	1.33	1.67	2.00	83.78	70.67	85.44	57.67	2.00	2.00	1.00	1.00
CSV 15	3.33	5.00	2.00	3.00	1.67	3.00	0.00	0.00	0.00	0.00	3.00	2.67	2.00	2.00
ICSV 25019	2.33	3.33	1.00	2.33	1.00	2.00	45.56	47.56	64.78	45.33	3.00	3.00	2.00	2.00
PS 35805	4.00	3.33	2.33	2.33	1.67	2.00	64.78	61.22	74.11	72.89	3.00	3.00	2.00	2.00
IS 2123	3.67	1.00	2.33	1.00	2.33	2.00	58.78	51.44	60.11	59.28	1.00	2.00	1.00	1.00
IS 2146	2.00	1.33	1.33	1.00	1.67	1.67	106.80	84.28	120.67	97.61	1.00	2.00	1.00	1.00
IS 18551	3.33	1.00	2.00	1.00	1.67	1.33	84.11	84.56	97.56	101.56	1.00	1.00	1.00	1.00
Swarna	3.67	5.00	2.00	2.33	2.00	2.00	33.89	7.11	32.22	13.33	3.00	3.00	1.00	1.00
DIRECT CROSSES														
ICSV 700 X Phule Anuradha	3.67	2.33	2.00	1.67	1.67	1.67	59.67	59.06	92.33	72.11	1.00	2.00	1.00	1.00
ICSV 700 X M 35-1	3.00	3.33	1.33	2.33	2.00	2.00	66.44	83.11	108.67	104.78	1.00	2.00	1.00	1.00
ICSV 700 X CSV 15	3.67	3.33	2.33	2.00	1.33	2.00	51.67	54.78	89.00	101.28	1.67	3.00	2.00	2.00
ICSV 700 X ICSV 25019	2.67	2.67	1.67	1.33	1.00	2.00	68.33	75.89	112.22	102.67	1.33	2.00	2.00	2.00
ICSV 700 X PS 35805	2.67	4.00	1.33	2.33	1.00	2.33	49.56	52.33	74.78	63.56	1.33	2.33	2.00	2.00
ICSV 700 X IS 2123	3.33	3.00	1.67	2.00	1.67	2.00	49.22	64.11	78.89	87.78	1.00	2.00	1.00	1.00
ICSV 700 X IS 2146	2.67	3.00	1.00	2.33	1.33	2.00	54.11	71.44	93.00	74.00	1.33	1.67	1.00	1.00
ICSV 700 X IS 18551	2.33	2.33	1.33	1.33	1.33	1.67	65.11	90.78	102.67	96.44	1.00	1.33	1.00	1.00
ICSV 700 X Swarna	3.00	4.67	1.50	2.33	2.00	2.00	67.11	58.11	104.78	74.78	1.67	2.33	1.00	1.00
Phule Anuradha X M 35-1	2.67	3.33	1.33	2.33	1.67	2.67	51.33	58.56	68.11	72.22	1.00	2.33	1.00	1.00
Phule Anuradha X CSV 15	4.00	4.00	2.00	1.67	1.67	2.67	34.22	36.50	48.00	38.94	1.67	2.67	1.00	1.00
Phule Anuradha X ICSV 25019	3.33	3.33	1.33	2.00	1.33	2.33	52.11	56.00	69.56	56.67	1.33	2.33	1.00	1.00
ICSV 700 X PS 35805	2.67	4.00	1.33	2.33	1.00	2.33	49.56	52.33	74.78	63.56	1.33	2.33	2.00	2.00
ICSV 700 X IS 2123	3.33	3.00	1.67	2.00	1.67	2.00	49.22	64.11	78.89	87.78	1.00	2.00	1.00	1.00
ICSV 700 X IS 2146	2.67	3.00	1.00	2.33	1.33	2.00	54.11	71.44	93.00	74.00	1.33	1.67	1.00	1.00
ICSV 700 X IS 18551	2.33	2.33	1.33	1.33	1.33	1.67	65.11	90.78	102.67	96.44	1.00	1.33	1.00	1.00
ICSV 700 X Swarna	3.00	4.67	1.50	2.33	2.00	2.00	67.11	58.11	104.78	74.78	1.67	2.33	1.00	1.00
Phule Anuradha X M 35-1	2.67	3.33	1.33	2.33	1.67	2.67	51.33	58.56	68.11	72.22	1.00	2.33	1.00	1.00
Phule Anuradha X CSV 15	4.00	4.00	2.00	1.67	1.67	2.67	34.22	36.50	48.00	38.94	1.67	2.67	1.00	1.00
Phule Anuradha X ICSV 25019	3.33	3.33	1.33	2.00	1.33	2.33	52.11	56.00	69.56	56.67	1.33	2.33	1.00	1.00
Phule Anuradha X PS 35805	3.00	3.67	1.67	2.00	2.00	2.33	56.00	53.11	68.22	64.56	1.33	2.00	1.00	1.00
Phule Anuradha X IS 2123	3.00	3.00	2.00	2.00	1.33	2.00	45.00	67.11	61.56	66.22	1.00	1.67	1.00	1.00
Phule Anuradha X IS 2146	3.33	3.67	2.00	1.67	1.67	2.33	81.22	89.56	98.22	88.22	1.67	1.67	1.00	1.00
Phule Anuradha X IS 18551	2.00	3.33	1.00	1.33	1.33	2.00	75.00	81.44	95.22	98.00	1.67	1.00	1.00	1.00
Phule Anuradha X Swarna	4.67	4.67	2.33	2.00	2.00	2.67	27.28	9.17	35.22	18.17	2.33	2.67	1.00	1.00
M 35-1 X CSV 15	3.67	4.33	2.00	2.67	1.33	2.33	2.00	26.56	6.11	32.44	2.00	2.33	1.00	1.00
M 35-1 X ICSV 25019	4.00	3.33	2.67	2.33	1.33	2.00	44.00	40.33	55.33	49.78	1.67	2.33	1.00	1.00
M 35-1 X PS 35805	2.67	3.33	1.67	2.00	1.33	2.33	42.22	55.61	54.22	73.78	1.33	3.00	1.00	1.00
M 35-1 X IS 2123	2.33	3.33	1.33	1.67	1.33	2.00	47.56	36.89	58.11	51.67	1.00	2.00	1.00	1.00
M 35-1 X IS 2146	2.67	3.67	1.67	1.67	1.33	2.33	64.56	71.78	94.56	88.00	1.00	2.00	1.00	1.00
M 35-1 X IS 18551	2.67	3.00	1.33	2.67	1.67	2.00	60.67	94.83	85.78	103.06	1.00	1.33	1.00	1.00
M 35-1 X Swarna	1.67	3.33	1.33	2.33	1.33	2.00	47.67	58.17	72.44	80.89	1.67	2.00	1.00	1.33
CSV 15 X ICSV 25019	4.33	5.00	2.00	2.00	2.00	2.33	0.00	0.00	0.78	0.00	3.00	2.67	2.00	2.00
CSV 15 X PS 35805	4.33	4.00	2.33	2.33	2.67	2.33	0.00	0.00	0.00	0.00	2.67	3.00	2.00	2.00
CSV 15 X IS 2123	4.33	4.67	2.00	2.00	1.67	2.67	6.44	3.89	12.22	7.78	2.33	2.33	1.00	1.00
CSV 15 X IS 2146	4.00	4.00	2.33	2.00	1.67	3.00	34.89	25.00	49.67	44.22	2.33	2.00	1.00	1.00
CSV 15 X IS 18551	4.67	4.00	2.67	2.33	1.33	2.00	36.44	21.00	54.44	28.83	1.67	2.00	1.00	1.00
CSV 15 X Swarna	4.33	4.67	2.67	2.67	1.33	3.00	3.78	0.00	6.22	0.00	2.67	3.00	1.00	1.00
ICSV 25019 X PS 35805	3.00	3.33	1.00	2.67	1.33	2.33	37.89	68.56	57.00	59.61	3.00	3.00	2.00	2.00
ICSV 25019 X IS 2123	3.33	3.67	2.00	2.67	1.00	2.67	44.89	47.22	58.33	61.39	1.33	2.33	1.00	1.00
ICSV 25019 X IS 2146	3.67	4.00	1.67	2.00	1.33	2.00	75.67	80.78	100.44	96.78	1.33	2.00	1.00	1.00
ICSV 25019 X IS 18551	2.33	3.67	1.00	2.00	1.33	2.00	67.00	92.44	94.11	110.11	1.00	2.00	1.00	1.00
ICSV 25019 X Swarna	4.33	5.00	2.67	2.67	2.00	2.00	17.89	8.44	29.22	13.89	3.00	3.00	1.00	1.00
PS 35805 X IS 2123	2.33	2.67	1.33	2.00	1.33	2.00	44.78	52.67	58.56	64.44	1.00	2.00	1.00	1.00
PS 35805 X IS 2146	2.33	2.67	1.67	1.67	1.33	2.33	69.33	43.78	87.67	70.33	1.00	2.33	1.00	1.00
PS 35805 X IS 18551	3.33	3.00	2.00	1.33	2.00	2.00	62.22	55.00	89.22	83.67	1.00	2.00	1.00	1.00
PS 35805 X Swarna	3.33	4.67	2.00	2.67	2.33	2.33	28.67	11.67	44.67	9.89	3.00	3.00	1.00	1.00
IS 2123 X IS 2146	3.33	3.33	1.33	2.33	1.67	2.00	46.44	77.33	70.67	81.67	1.00	2.00	1.00	1.00
IS 2123 X IS 18551	2.00	2.33	1.00	1.33	1.00	2.00	55.22	69.39	85.11	79.78	1.00	1.00	1.00	1.00
IS 2123 X Swarna	3.33	3.67	2.33	2.00	1.33	2.00	24.11	30.33	43.00	32.44	1.67	2.33	1.00	1.00
IS 2146 X IS 18551	1.67	2.00	1.00	1.00	1.33	2.00	88.33	104.33	124.67	115.00	1.33	1.00	1.00	1.00
IS 2146 X Swarna	3.67	4.33	1.67	2.33	1.33	2.00	83.11	46.06	112.11	56.00	2.00	2.67	1.00	1.00
IS 18551 X Swarna	3.33	4.67	2.00	2.33	1.33	2.33	59.56	56.89	87.00	84.11	1.67	2.00	1.00	1.00
RECIPROCAL CROSSES														
Phule Anuradha X ICSV 700	3.00	2.67	1.33	2.00	1.33	2.00	72.78	74.22	105.00	93.00	1.33	2.00	1.00	1.00
M 35-1 X ICSV 700	3.67	3.33	2.33	2.33	1.67	2.00	48.78	64.78	77.78	82.44	1.33	2.00	1.00	1.00
M 35-1 X Phule Anuradha	2.33	4.00	1.33	2.00	1.00	2.33	53.89	41.78	69.22	46.67	1.00	2.00	1.00	1.00
CSV 15 X ICSV 700	4.00	5.00	2.33	2.00	1.67	2.33	20.56	4.33	39.67	13.17	1.67	2.33	2.00	2.00
CSV 15 X Phule Anuradha	3.67	4.33	1.67	2.00	2.67	2.33	3.22	0.00	4.11	0.00	1.67	2.33	1.00	1.00
CSV 15 X M 35-1	3.67	5.00	2.00	2.33	1.33	2.67	5.89	0.00	8.33	0.00	1.67	2.33	1.00	1.00
ICSV 25019 X ICSV 700	2.67	2.67	1.33	1.33	1.67	2.00	58.56	70.22	79.89	76.00	1.67	2.33	2.00	2.00
ICSV 25019 X Phule Anuradha	3.00	3.67	2.00	1.67	1.67	2.33	40.67	72.39	51.89	68.06	1.33	2.33	1.00	1.00
ICSV 25019 X M 35-1	3.67	3.33	2.00	2.33	1.00	2.00	48.78	52.17	57.89	53.61	2.00	2.00	1.00	1.00
ICSV 25019 X CSV 15	3.33	4.67	1.67	2.33	1.67	2.67	4.33	0.00	3.67	0.00	3.00	3.00	2.00	2.00
PS 35805 X ICSV 700	3.00	3.67	1.67	2.00	2.00	2.00	45.22	62.39	79.44	71.94	1.67	2.33	2.00	2.00
PS 35805 X Phule Anuradha	1.67	3.00	1.00	1.67	1.67	2.00	48.00	58.67	65.78	60.11	1.67	2.00	1.00	1.00
PS 35805 X M 35-1	2.33	4.00	1.33	2.33	1.33	2.33	41.00	42.89	60.78	54.61	1.67	2.33	1.00	1.00
PS 35805 X CSV 15	2.67	4.33	1.00	2.00	1.83	2.67	5.78	29.94	12.22	36.39	2.67	3.00	1.67	1.67
PS 35805 X ICSV 25019	2.33	3.67	1.67	2.00	1.67	2.00	42.22	66.61	59.00	64.00	3.00	3.00	2.00	2.00
IS 2123 X ICSV 700	1.67	2.33	1.33	1.33	1.33	1.67	45.78	68.89	71.00	70.33	1.00	2.00	1.00	1.00
IS 2123 X Phule Anuradha	3.67	4.00	1.67	2.33	1.33	2.33	46.83	45.17	59.22	47.78	1.00	2.00	1.00	1.00
IS 2123 X M 35-1	3.67	3.33	1.67	1.67	1.33	2.00	40.11	50.33	60.33	59.00	1.00	2.00	1.00	1.00
IS 2123 X CSV 15	4.67	5.00	2.67	2.33	2.00	2.33	1.00	6.72	7.78	5.44	2.00	2.33	1.00	1.00
IS 2123 X ICSV 25019	2.67	3.33	1.33	1.33	1.00	2.00	31.89	41.39	47.67	51.83	1.00	2.33	1.00	1.00
IS 2123 X PS 35805	3.00	3.00	1.33	1.67	1.33	1.67	63.56	63.44	79.44	60.44	1.33	2.00	1.00	1.00
IS 2146 X ICSV 700	3.33	2.67	1.33	1.33	1.33	2.00	79.33	60.22	110.78	70.89	1.00	2.00	1.00	1.00
IS 2146 X Phule Anuradha	2.33	3.67	1.67	2.00	2.00	2.33	76.00	87.33	109.56	86.78	1.67	2.00	1.00	1.00
IS 2146 X M 35-1	2.33	3.00	1.00	1.33	1.67	2.00	81.39	88.56	85.50	99.44	1.00	2.00	1.00	1.00
IS 2146 X CSV 15	4.00	4.33	2.33	2.33	1.67	2.33	35.89	4.33	53.78	11.33	2.00	2.33	1.00	1.00
IS 2146 X ICSV 25019	2.50	3.33	1.00	2.00	1.00	1.67	95.78	78.11	113.00	94.22	1.00	2.00	1.00	1.00
IS 2146 X PS 35805	2.67	2.67	1.67	1.33	1.33	2.00	91.78	63.56	118.67	66.67	1.33	2.00	1.00	1.00
IS 2146 X IS 2123	2.00	2.00	1.00	1.00	1.67	1.67	90.00	91.33	106.00	110.78	1.00	1.67	1.00	1.00
IS 18551 X ICSV 700	2.67	2.00	1.00	1.00	1.00	1.33	77.89	60.44	101.56	78.78	1.00	1.67	1.00	1.00
IS 18551 X Phule Anuradha	3.00	3.00	1.67	2.00	1.33	2.00	61.78	58.33	95.00	76.89	1.00	1.67	1.00	1.00
IS 18551 X M 35-1	2.67	3.33	1.00	1.67	1.00	2.00	57.33	69.33	83.22	94.22	1.00	1.33	1.00	1.00
IS 18551 X CSV 15	3.67	4.33	2.33	2.33	1.67	2.00	39.00	38.11	67.33	59.39	1.67	1.67	1.00	1.00
IS 18551 X ICSV 25019	3.33	3.00	1.67	1.67	1.67	2.00	81.78	73.00	107.33	77.56	1.00	1.67	1.00	1.00
IS 18551 X PS 35805	2.00	2.33	1.33	2.00	1.00	2.00	64.56	86.22	99.00	96.78	1.00	1.67	1.00	1.00
IS 18551 X IS 2123	2.33	2.00	1.33	1.33	1.00	2.00	70.11	75.00	99.11	93.78	1.00	1.67	1.00	1.00
IS 18551 X IS 2146	2.33	1.67	1.33	1.00	1.67	2.00	89.33	94.44	112.11	107.11	1.00	1.67	1.00	1.00
Swarna X ICSV 700	4.33	4.67	2.33	2.00	1.67	2.00	66.89	58.94	108.89	74.67	2.33	2.00	1.00	1.00
Swarna X Phule Anuradha	3.67	4.33	2.33	2.00	1.67	2.00	16.11	20.89	26.67	32.89	2.33	2.33	1.00	1.00
Swarna X M 35-1	4.33	5.00	2.00	2.67	2.33	2.33	21.89	22.33	26.89	17.89	2.33	2.00	1.00	1.00
Swarna X CSV 15	4.33	5.00	2.67	2.67	2.33	2.33	4.78	0.00	9.78	0.00	2.67	3.00	1.00	1.00
Swarna X ICSV 25019	3.67	4.67	2.67	3.00	1.67	2.00	13.22	2.11	15.89	3.22	3.00	3.00	1.00	1.00
Swarna X PS 35805	4.33	4.67	2.67	2.00	1.33	2.67	16.00	5.44	19.33	5.39	3.00	3.00	1.00	1.00

Swarna X IS 2123	4.67	4.67	2.67	3.00	1.67	2.00	16.89	21.56	29.56	19.44	2.00	2.67	1.00	1.00
Swarna X IS 2146	4.67	4.33	1.67	2.33	1.33	2.00	67.22	55.28	92.78	73.67	2.67	3.00	1.00	1.00
Swarna X IS 18551	3.67	4.67	2.33	2.33	2.00	2.00	50.22	78.22	89.78	97.44	2.33	2.33	1.00	1.00
Mean	3.17	3.52	1.74	1.96	1.55	2.12	49.70	50.90	68.70	60.45	1.70	2.19	1.20	1.16
SE ±	0.57	0.35	0.40	0.34	0.36	0.24	7.69	10.22	9.06	9.92	0.24	0.23	0.03	0.05
Vr (99, 198)	2.04**	8.03**	1.70**	2.18**	1.08^{NS}	1.84**	12.46**	7.75**	14.81**	11.12**	8.07**	4.74**	118.09**	58.48**
LSD (P 0.05)	1.58	0.97	1.12	0.96	0.99	0.68	21.45	28.50	25.26	27.65	0.68	0.65	0.09	0.13

** *F probability significant at P = 0.01; NS, non-significant F-value; R, rainy season; PR, postrainy season.*

Combining Ability Analysis

Analysis of Variance for Combining Ability

The estimation of mean sum of squares (ANOVA) for general combining ability (GCA) of parents, and specific combining ability (SCA) of the hybrids showed significant ($P \leq 0.01$) general combining ability effects for all the traits studied in the postrainy season, and for most of the traits in the rainy season (Table 3). The mean sum of squares due to SCA were significant for most of the traits during the rainy and postrainy seasons, with the exception of leaf glossy score and plants with shoot fly eggs during the rainy season, number of shoot fly eggs/plant and leafsheath pigmentation across seasons, and waxy bloom in the postrainy season. The mean sum of squares for the reciprocal crosses were significant for trichome density on abaxial and adaxial leaf surfaces across seasons; overall resistance score during the rainy season, and plants with shoot fly eggs, and leaf glossy score during the postrainy season, suggesting influence of the cytoplasmic factors in expression of the traits associated with resistance to shoot fly in sorghum.

Table 3: Analysis of variance (ANOVA) showing mean sum of squares of general, specific, and reciprocal combining abilities of $F_1(10 \times 10)$ diallel (ICRISAT, Patancheru, 2013–14)

Source	GCA		SCA		Reciprocal		Error	
	2013 R	2013 PR	2013 R	2013 PR	2013 R	2013 PR	2013 R	2013 PR
Plants with shoot fly eggs (%)	62.20	2707.91**	42.10	524.18**	56.02	256.12**	40.36	110.81
Number of shoot fly eggs/plant	0.20	0.33**	0.19	0.09	0.23	0.06	0.18	0.07
Shoot fly deadhearts (%)	673.87**	2410.48**	190.33**	146.67	136.37	93.05	101.81	107.64
Overall resistance score	3.14**	3.23**	0.89**	0.77**	0.448*	0.51	0.29	0.38
Leaf glossy score	2.89**	7.34**	0.43	0.49**	0.42	0.17*	0.32	0.12
Plant vigor score	1.13**	1.33**	0.24*	0.17*	0.13	0.13	0.16	0.12
Leafsheath pigmentation	0.34	0.65**	0.11	0.06	0.12	0.05	0.13	0.06
Trichome density on abaxial leaf surface	6126.24**	6846.39**	292.89**	236.67**	104.26**	173.84**	59.16	104.42
Trichome density on adaxial leaf surface	10789.18**	9028.91**	346.52**	314.57**	166.63**	285.76**	81.99	98.31
Plant color	0.82**	0.80**	0.12**	0.13**	0.00	0.00	0.00	0.00
Waxy bloom	3.91**	2.22**	0.22**	0.07	0.05	0.05	0.06	0.05

*, ** *F probability significant at P= 0.05 and 0.01, respectively; GCA, general combining ability; SCA, specific combining ability; R, rainy season; PR, postrainy season.*

Estimates of *gca, sca,* and Reciprocal Effects of Parents and Hybrids

General Combining Ability (gca) Effects

gca effects of shoot fly resistance traits

The general combining ability (*gca*) effects for shoot fly deadhearts ranged from –8.13 (PS 35805) to 9.80 (CSV 15) in the rainy season, and from –8.80 (PS 35805) to 20.86 (CSV 15) in the postrainy season (Table4). The genotypes Phule Anuradha, ICSV 25019, PS 35805, IS 2123, and IS 18551 exhibited negative and significant *gca* effects, with a few exceptions. M 35-1 and Phule Anuradha exhibited positive but non-significant *gca* effects in the rainy season but showed negative *gca* effects in the postrainy season. CSV 15 and Swarna exhibited significant positive *gca* effects across seasons. Similar pattern was observed in case of overall resistance score across seasons.

Table 4: Estimates of general combining ability effects of shoot fly resistance and morphological traits of parents (10 × 10 diallel) across seasons (ICRISAT, Patancheru, 2013–14)

Traits	ICSV 700	Phule Anuradha	M 35-1	CSV 15	ICSV 25019	PS 35805	IS 2123	IS 2146	IS 18551	Swarna
Plants with shoot fly eggs (%)	(–0.71)	(–5.25)*	(–6.11)**	(20.65)**	(–4.56)*	(–6.91)**	(–7.54)**	(–4.89)*	(–7.50)**	(22.83)**
Number of shoot fly eggs/plant	(0.02)	(–0.16)**	(–0.09)	(0.16)**	(–0.07)	(–0.13)*	(–0.02)	(0.01)	(0.02)	(0.26)**
Shoot fly deadhearts (%)	–0.45 (–5.25)*	1.88 (–4.51)*	2.4 (–4.15)	9.80** (20.86)**	–6.22** (–3.28)	–8.13** (–8.80)**	–4.91* (–6.26)**	–0.32 (–2.81)	–2.07 (–6.10)**	8.03** (20.30)**
ORS	–0.06 (0.38)**	–0.04 (–0.03)	0.25* (–0.2)	0.58** (0.43)**	–0.45** (–0.08)	–0.44** (–0.03)	–0.24* (–0.43)**	–0.23* (–0.47)**	–0.09 (–0.32)*	0.71** (0.75)**
Leaf glossy score	–0.14 (–0.53)**	–0.22 (–0.05)	–0.11 (0.05)	0.73** (0.98)**	–0.03 (0.13)	–0.22 (–0.05)	–0.02 (–0.45)**	–0.30* (–0.47)**	–0.34** (–0.68)**	0.66** (1.07)**
Plant vigor score	–0.15 (–0.21)**	–0.13 (–0.11)	–0.01 (0.1)	0.41** (0.34)**	–0.08 (0.14)	–0.08 (0.07)	–0.01 (–0.16)*	–0.24** (–0.28)**	–0.17* (–0.31)**	0.45** (0.44)**
Leafsheath pigmentation	(–0.27)**	(0.09)	(0.04)	(0.38)**	(–0.01)	(0.06)	(–0.07)	(–0.06)	(–0.22)**	(0.06)
Trichome density on abaxial leaf surface	16.52** (14.75)**	2.66 (2.82)	–0.07 (4.06)	–35.24** (–38.35)**	–3.93* (0.24)	–2.81 (–1.09)	–5.36** (–0.12)	26.18** (18.19)**	16.76** (22.51)**	–14.72** (–23.01)**
Trichome density on adaxial leaf surface	29.09** (20.90)**	2.27 (0.32)	–3.48 (3.54)	–45.04** (–41.49)**	–6.56** (–3.95)	–4.86* (–2.85)	–8.36** (–1.92)	30.03** (21.07)**	24.69** (28.75)**	–17.76** (–24.38)**
Waxy bloom	–0.32** (–0.12*)	–0.20** (–0.14)**	–0.24** (–0.12)*	0.58** (0.35)**	0.33** (0.28)**	0.28** (0.31)**	–0.44** (–0.17)**	–0.29** (–0.19)**	–0.45** (–0.64)**	0.75** (0.43)**
Plant color	0.24** (0.24)**	–0.16** (–0.16)**	–0.16** (–0.14)**	0.23** (0.22)**	0.24** (0.24)**	0.23** (0.22)**	–0.16** (–0.16)**	–0.16** (–0.16)**	–0.16** (–0.16)**	–0.16** (–0.14)**

*, ** *t-test significant at P = 0.05 and 0.01, respectively; ORS, overall resistance score; The values inside the parentheses are for postrainy season and outside the parentheses for rainy season.*

The genotypes CSV 15 and Swarna showed significant and positive *gca* effects for percentage plants with shoot fly eggs and number of shoot fly eggs/plant in the postrainy season (Table 4). Phule Anuradha, M 35-1, ICSV 25019, PS 35805, IS 2123, IS 2146, and IS 18551 exhibited significant negative *gca* effects for plants with shoot fly eggs (%), whereas Phule Anuradha and PS 35805 exhibited significant negative *gca*effects for number of shoot fly eggs/plant in the postrainy season.

gca effects of morphological traits

The *gca* effects of leaf glossy score ranged from –0.34 (IS 18551) to 0.73 (CSV 15), and for plant vigor score from –0.24 (IS 2146) to 0.45 (Swarna) in the rainy season. In the postrainy season the *gca* effects for leaf glossy score ranged from –0.68 (IS 18551) to 1.07 (Swarna), for plant vigor score from –0.31 (IS 18551) to 0.44 (Swarna) and for leafsheath pigmentation from –0.27 (ICSV 700) to 0.38 (CSV 15) in the postrainy season (Table 4). IS 2146 and IS 18551 exhibited significant and negative *gca* effects, and CSV 15 and Swarna exhibited significant and positive *gca* effects for leaf glossy score and for plant vigor score across seasons. ICSV 700 and IS 2123 exhibited significant negative *gca* effects in the postrainy season, both for leaf glossy score and plant vigor score. ICSV 700 and IS 18551 exhibited significant negative and CSV 15 significant positive *gca* effects for leafsheath pigmentation in the postrainy season.

The general combining ability effects for trichome density on the abaxial leaf surface ranged from –35.24 (CSV 15) to 16.52 (ICSV 700) in the rainy season; and from –38.35 (CSV 15) to 22.51 (IS 18551) in the postrainy season (Table 4). The *gca* effects of trichome density on the adaxial leaf surface ranged from –45.04 (CSV 15) to 30.03 (IS 2146) in the rainy season, and from –41.49 (CSV 15) to 28.75 (IS 18551) in the postrainy season. CSV 15, ICSV 25019, IS 2123, PS 35805, and Swarna exhibited significant negative *gca*effects, while ICSV 700, IS 2146, and IS 18551 exhibited significant positive *gca* effects in the rainy season and CSV 15 and Swarna showed significant negative, and ICSV 700, IS 2146, and IS 18551 significant positive *gca* effects for trichome density on abaxial and adaxial leaf surfaces in the postrainy season.

The *gca* effects of waxy bloom ranged from -0.45 (IS 18551) to 0.75 (Swarna) in the rainy season, and from –0.64 (IS 18551) to 0.43 (Swarna) in the postrainy season (Table 4). ICSV 700, Phule Anuradha, M 35-1, IS 2123, IS 2146, and IS 18551 exhibited significant and negative *gca* effects; while CSV 15, ICSV 25019, PS 35805, and Swarna exhibited significant and positive *gca* effects across seasons. The general combining ability effect of plant color ranged from –0.16 to 0.24 in the rainy season, and –0.14 to 0.24 in the postrainy season.

Specific Combining Ability (*sca*) Effects

sca Effects of Shoot Fly Resistance Traits

The *sca* effects of plants with shoot fly eggs (%) ranged from –13.04 to 37.56 in the postrainy season (Table 5). The range of *sca* effects for shoot fly deadhearts (%) varied from –12.42 to 17.74 in the rainy season, and the overall resistance score from –1.16 to 1.15 in the rainy season, and from –0.80 to 1.22 in the postrainy season.

Table 5: Estimates of specific combining ability effects of shoot fly resistance traits of F_1 **crosses (10 × 10** diallel) of sorghum, across seasons (ICRISAT, Patancheru, 2013-14)

Pedigree	Plants with shoot fly eggs (%)	Shoot fly deadhearts (%)	ORS	
	2013 PR	2013 R	2013 R	2013 PR
ICSV 700 X Phule Anuradha	-7.39	2.06	-0.83*	-0.07
ICSV 700 X M 35-1	5.46	7.48	-0.61	0.43
ICSV 700 X CSV 15	9.82	-3.05	-0.28	0.13
ICSV 700 X ICSV 25019	-8.03	1.91	-0.09	-0.35
ICSV 700 X PS 35805	-1.82	-7.51	-0.43	-0.40
ICSV 700 X IS 2123	6.69	-0.05	0.71*	0.50
ICSV 700 X IS 2146	-0.92	4.78	0.86*	-0.80*
ICSV 700 X IS 18551	6.71	-1.08	0.56	0.88*
ICSV 700 X Swarna	19.38**	8.12	-0.41	-0.35
Phule Anuradha X M 35-1	-13.04	-0.99	0.37	0.35
Phule Anuradha X CSV 15	-1.50	1.30	-0.46	-0.12
Phule Anuradha X ICSV 25019	11.93	-2.22	0.40	0.23
Phule Anuradha X PS 35805	-1.87	0.07	0.56	-0.32
Phule Anuradha X IS 2123	9.71	6.05	0.69	-0.75
Phule Anuradha X IS 2146	12.64	10.14	-0.15	0.45
Phule Anuradha X IS 18551	-3.43	-10.91	0.21	0.30
Phule Anuradha X Swarna	13.82*	2.33	0.07	-0.60
M 35-1 X CSV 15	6.02	2.14	-0.58	-0.28
M 35-1 X ICSV 25019	-5.02	15.29*	-0.55	0.07
M 35-1 X PS 35805	3.66	8.96	0.44	-0.65
M 35-1 X IS 2123	10.48	-1.88	-0.26	0.75
M 35-1 X IS 2146	0.06	-11.96	0.23	0.45
M 35-1 X IS 18551	12.27	-12.42	0.26	0.47
M 35-1 X Swarna	-0.83	-5.55	-1.21**	-0.10

M 35-1 X CSV 15	6.02	2.14	-0.58	-0.28
M 35-1 X ICSV 25019	-5.02	15.29*	-0.55	0.07
M 35-1 X PS 35805	3.66	8.96	0.44	-0.65
M 35-1 X IS 2123	10.48	-1.88	-0.26	0.75
M 35-1 X IS 2146	0.06	-11.96	0.23	0.45
M 35-1 X IS 18551	12.27	-12.42	0.26	0.47
M 35-1 X Swarna	-0.83	-5.55	-1.21**	-0.10
CSV 15 X ICSV 25019	37.56**	6.55	0.61	0.93*
CSV 15 X PS 35805	7.01	4.18	0.44	1.22**
CSV 15 X IS 2123	-2.72	5.34	-0.43	-0.38
CSV 15 X IS 2146	18.93**	-2.56	-0.77*	-0.68
CSV 15 X IS 18551	2.09	6.84	-0.08	-0.17
CSV 15 X Swarna	-4.88	-4.48	0.62	-0.23
ICSV 25019 X PS 35805	1.37	-5.52	0.96**	0.40
ICSV 25019 X IS 2123	-5.73	-5.53	-0.90*	0.13
ICSV 25019 X IS 2146	-4.60	11.52	-1.16**	-0.67
ICSV 25019 X IS 18551	-11.08	17.74**	-0.39	-0.32
ICSV 25019 X Swarna	5.75	6.37	1.15**	0.95*
PS 35805 X IS 2123	-1.57	8.66	-0.58	0.42
PS 35805 X IS 2146	1.76	7.80	-0.09	0.45
PS 35805 X IS 18551	-0.24	3.17	-0.73*	0.13
PS 35805 X Swarna	9.26	-0.70	0.14	0.57
IS 2123 X IS 2146	-7.87	-4.62	0.71*	0.52
IS 2123 X IS 18551	-3.39	1.72	0.24	0.20
IS 2123 X Swarna	10.70	0.41	-0.73*	-0.37
IS 2146 X IS 18551	-4.04	3.57	0.23	0.23
IS 2146 X Swarna	5.55	6.09	-0.40	-0.33
IS 18551 X Swarna	17.53*	7.02	-0.55	-0.48

*, ** t-test significant at P = 0.05 and 0.01, respectively; ORS, overall resistance score; R, rainy season; PR, postrainy season

ICSV 700 X Swarna, Phule Anuradha X Swarna, CSV 15 X ICSV 25019, CSV 15 X IS 2146, and IS 18551 X Swarna exhibited significant positive *sca* effects for percentage plants with shoot fly eggs in the postrainy season. M 35-1 X ICSV 25019 and ICSV 25019 X IS 18551 exhibited significant positive *sca* effects for shoot fly deadhearts (%) in the rainy season (Table 5). ICSV 700 X IS 2123, ICSV 700 X IS 2146, ICSV 25019 X PS 35805, ICSV 25019 X Swarna, and IS 2123 X IS 2146 in the rainy season; and ICSV 700 X IS 18551, CSV 15 X ICSV 25019, CSV 15 X PS 35805, and ICSV 25019 X Swarna in the postrainy season exhibited significant positive *sca* effects for overall resistance score. ICSV 700 X Phule Anuradha, M 35-1 X Swarna, CSV 15 X IS 2146, ICSV 25019 X IS 2123, ICSV 25019 X IS 2146, PS 35805 X IS 18551, and IS 2123 X Swarna in the rainy season; and ICSV 700 X IS 2146 in the postrainy season exhibited significant negative *sca* effects for overall resistance score.

sca Effects of Morphological Traits

The *sca* effects of leaf glossy score ranged from –0.73 to 0.90 (in the postrainy season), for plant vigor score from –0.52 to 0.68 and –0.55 to 0.48, for trichome density on the abaxial leaf surface –25.68 to 15.47 and –22.85 to 18.53, for trichome density on the adaxial leaf surface from –25.92 to 26.80 and –29.98 to 25.95, and for waxy bloom from –0.55 to 0.72 (in the rainy season only) respectively, in the rainy and postrainy seasons (Table 6).

Table 6: Estimates of specific combining ability effects of morphological traits of F_1crosses (10 × 10 diallel) of sorghum across seasons (ICRISAT, Patancheru, 2013-14)

Pedigree	Leaf glossy score	Plant vigor score		Trichome density on abaxial leaf surface		Trichome density on adaxial leaf surface		Waxy bloom
	2013 PR	2013 R	2013 PR	2013 R	2013 PR	2013 R	2013 PR	2013 R
ICSV 700 X Phule Anuradha	-0.43	0.20	0.20	-2.70	-1.84	-1.38	0.89	0.02
ICSV 700 X M 35-1	0.30	0.25	0.48*	-8.57	4.22	-1.09	8.72	0.05
ICSV 700 X CSV 15	0.20	0.33	-0.09	5.10	2.24	11.59	17.36**	-0.26
ICSV 700 X ICSV 25019	-0.45*	-0.02	-0.55*	1.11	7.16	4.85	11.93	-0.18
ICSV 700 X PS 35805	0.90**	-0.02	0.35	-16.06**	-7.21	-15.78**	-10.75	-0.13
ICSV 700 X IS 2123	0.13	-0.08	0.08	-13.41**	0.96	-14.49*	-0.38	0.09
ICSV 700 X IS 2146	0.32	-0.18	0.36	-25.68**	-18.01**	-25.92**	-29.98**	0.10
ICSV 700 X IS 18551	-0.13	-0.25	-0.27	-11.53*	-12.56	-20.37**	-22.50**	0.10
ICSV 700 X Swarna	0.62**	-0.13	-0.02	15.47**	15.88*	26.80**	17.75**	-0.10
Phule Anuradha X M 35-1	0.15	-0.28	0.21	0.29	-7.61	1.19	-4.86	-0.23
Phule Anuradha X CSV 15	-0.28	-0.19	-0.35	1.58	2.88	0.14	0.20	-0.38*
Phule Anuradha X ICSV 25019	-0.10	0.13	-0.15	-2.08	10.24	-3.69	5.54	-0.46**
Phule Anuradha X PS 35805	-0.08	-0.21	-0.09	2.43	3.26	0.93	4.42	-0.25
Phule Anuradha X IS 2123	0.48*	0.23	0.48*	-1.13	2.54	-2.22	-1.84	-0.03
Phule Anuradha X IS 2146	0.67**	0.46	0.26	0.04	6.54	2.87	5.67	0.49**
Phule Anuradha X IS 18551	0.38	-0.11	0.13	-0.77	-6.34	-0.55	-2.07	0.32*
Phule Anuradha X Swarna	-0.03	0.27	-0.29	-15.98**	-15.68*	-22.27**	-10.86	0.12
M 35-1 X CSV 15	0.12	-0.14	0.10	-10.49*	-3.34	-12.94*	-6.28	-0.18
M 35-1 X ICSV 25019	-0.37	0.68*	0.13	0.67	-8.95	-2.05	-8.35	0.07
M 35-1 X PS 35805	0.15	-0.16	0.03	-5.22	-4.63	-2.83	3.05	-0.21
M 35-1 X IS 2123	0.22	-0.23	-0.24	-0.45	-11.24	2.40	-6.74	0.00
M 35-1 X IS 2146	0.23	-0.16	-0.29	-2.86	7.02	-5.20	8.66	-0.15

Phule Anuradha X IS 18551	0.38	-0.11	0.13	-0.77	-6.34	-0.55	-2.07	0.32*
Phule Anuradha X Swarna	-0.03	0.27	-0.29	-15.98**	-15.68*	-22.27**	-10.86	0.12
M 35-1 X CSV 15	0.12	-0.14	0.10	-10.49*	-3.34	-12.94*	-6.28	-0.18
M 35-1 X ICSV 25019	-0.37	0.68*	0.13	0.67	-8.95	-2.05	-8.35	0.07
M 35-1 X PS 35805	0.15	-0.16	0.03	-5.22	-4.63	-2.83	3.05	-0.21
M 35-1 X IS 2123	0.22	-0.23	-0.24	-0.45	-11.24	2.40	-6.74	0.00
M 35-1 X IS 2146	0.23	-0.16	-0.29	-2.86	7.02	-5.20	8.66	-0.15
M 35-1 X IS 18551	0.28	-0.39	0.41	-7.39	4.60	-5.41	5.89	0.02
M 35-1 X Swarna	-0.47*	-0.52*	0.00	-0.16	8.30	2.24	9.77	-0.18
CSV 15 X ICSV 25019	0.20	-0.24	-0.27	-8.40	-12.79	-14.88*	-15.02*	0.42**
CSV 15 X PS 35805	-0.28	-0.41	-0.20	-8.80	3.50	-12.68*	2.09	0.14
CSV 15 X IS 2123	0.78**	0.19	0.03	-5.41	-7.13	-5.31	-10.43	0.35*
CSV 15 X IS 2146	0.13	0.43	0.15	-5.29	-16.08*	-1.97	-12.25	0.20
CSV 15 X IS 18551	0.35	0.53*	0.35	6.46	-5.52	12.54*	-3.61	-0.13
CSV 15 X Swarna	-0.73**	0.07	-0.07	4.51	10.45	2.12	5.42	-0.33*
ICSV 25019 X PS 35805	-0.10	-0.26	0.16	-2.93	18.53**	0.73	8.15	0.72**
ICSV 25019 X IS 2123	0.30	0.01	0.06	-2.06	-6.71	-0.78	2.03	-0.40*
ICSV 25019 X IS 2146	0.48*	-0.09	0.18	13.72**	10.12	14.57*	17.93**	-0.55**
ICSV 25019 X IS 18551	0.37	-0.16	0.05	11.83*	9.07	13.89*	8.58	-0.55**
ICSV 25019 X Swarna	0.12	0.55*	0.30	-15.52**	-22.85**	-21.82**	-23.58**	0.25
PS 35805 X IS 2123	-0.18	-0.33	-0.04	12.61*	8.36	13.54*	6.77	-0.35*
PS 35805 X IS 2146	-0.33	0.24	-0.25	7.46	-14.33*	9.32	-10.17	-0.49**
PS 35805 X IS 18551	-0.12	0.18	-0.05	-0.29	-1.72	5.63	3.87	-0.49**
PS 35805 X Swarna	0.13	0.22	-0.14	-9.86	-18.25**	-14.07*	-25.58**	0.30
IS 2123 X IS 2146	0.07	-0.33	0.15	-2.33	15.36*	-2.03	16.63*	0.05
IS 2123 X IS 18551	-0.22	-0.39	-0.15	1.54	-1.10	7.08	-0.51	0.22
IS 2123 X Swarna	0.03	0.32	0.26	-9.15	-1.83	-6.31	-8.21	-0.15
IS 2146 X IS 18551	-0.53*	-0.16	-0.37	-3.84	7.79	-5.03	0.79	0.24
IS 2146 X Swarna	0.22	-0.28	0.21	13.97**	4.59	21.47**	7.69	0.20
IS 18551 X Swarna	0.77**	0.15	0.25	3.12	17.15*	12.76*	25.95**	0.04

*, ** *t-test significant at P = 0.05 and 0.01, respectively; R, rainy season; PR, postrainy season.*

ICSV 700 X PS 35805, ICSV 700 X Swarna, Phule Anuradha X IS 2123, Phule Anuradha X IS 2146, CSV 15 X IS 2123, ICSV 25019 X IS 2146, and IS 18551 X Swarna exhibited significant positive *sca* effects for leaf glossy score in the postrainy season (Table 6). ICSV 700 X ICSV 25019, M 35-1 X Swarna, CSV 15 X Swarna, and IS 2146 X IS 18551 exhibited significant negative *sca* effects for leaf glossy score in the postrainy season. M 35-1 X ICSV 25019, CSV 15 X IS 18551, and ICSV 25019 X Swarna in the rainy season; and ICSV 700 X M 35-1, Phule Anuradha X IS 2123 in the postrainy season exhibited significant positive*sca* effects for plant vigor score. M 35-1 X Swarna in the rainy season, and ICSV 700 X ICSV 25019 in the postrainy season exhibited negative *sca* effects for plant vigor score.

ICSV 700 X PS 35805, ICSV 700 X IS 2123, ICSV 700 X IS 2146, ICSV 700 X IS 18551, Phule Anuradha X Swarna, M 35-1 X CSV 15, ICSV 25019 X Swarna in the rainy season; and ICSV 700 X IS 2146, Phule Anuradha X Swarna, CSV 15 X IS 2146, ICSV 25019 X Swarna, PS 35805 X IS 2146, and PS 35805 X Swarna in the postrainy season showed significant negative *sca* effects for trichome density on the abaxial leaf surface (Table 6). ICSV 700 X Swarna across seasons, and ICSV 25019 X IS 2146, ICSV 25019 X IS 18551, PS 35805 X IS 2123, and IS 2146 X Swarna in the rainy season, and ICSV 25019 X PS 35805, IS 2123 X IS 2146, and IS 18551 X Swarna in the postrainy season exhibited significant positive *sca* effects for trichome density on the abaxial leaf surface.

The crosses ICSV 700 X PS 35805, ICSV 700 X IS 2123, ICSV 700 X IS 2146, ICSV 700 X IS 18551, Phule Anuradha X Swarna, M 35-1 X CSV 15, CSV 15 X ICSV 25019, CSV 15 X PS 35805, ICSV 25019 X Swarna, and PS 35805 X Swarna in the rainy season; and ICSV 700 X IS 2146, ICSV 700 X IS 18551, CSV 15 X ICSV 25019, ICSV 25019 X Swarna, and PS 35805 X Swarna in the postrainy season showed significant negative *sca* effects for trichome density on the adaxial leaf surface (Table 6). ICSV 700 X Swarna, CSV 15 X IS 18551, ICSV 25019 X IS 2146, ICSV 25019 X IS 18551, PS 35805 X IS 2123, IS 2146 X Swarna, and IS 18551 X Swarna in the rainy season; and ICSV 700 X CSV 15, ICSV 700 X Swarna, ICSV 25019 X IS 2146, IS 2123 X IS 2146 and IS 18551 X Swarna in the postrainy season exhibited significant positive *sca* effects for trichome density on the adaxial leaf surface.

The *sca* effects of Phule Anuradha X CSV 15, Phule Anuradha X ICSV 25019, CSV 15 X Swarna, ICSV 25019 X IS 2123, ICSV 25019 X IS 2146, ICSV 25019 X IS 18551, PS 35805 X IS 2123, PS 35805 X IS 2146, and PS 35805 X IS 18551 in the rainy season exhibited significant negative *sca* effects for waxy bloom trait. Phule Anuradha X IS 2146, Phule Anuradha X IS 18551, CSV 15 X ICSV 25019, CSV 15 X IS 2123, ICSV 25019 X PS 35805 in the rainy season exhibited significant positive *sca* effects for waxy bloom.

Reciprocal Combining Ability Effects

The crosses CSV 15 X ICSV 700, ICSV 25019 X CSV 15, IS 2123 X M 35-1, IS 2123 X CSV 15, Swarna X ICSV 700, Swarna X M 35-1 exhibited significant negative reciprocal effects; while the crosses PS 35805 X CSV 15, IS 2146 X Phule Anuradha showed significant positive reciprocal effects for plants with shoot fly eggs (%) in the postrainy season (Table 7).

Table 7: Estimates of reciprocal combining ability effects of reciprocal crosses (10 × 10 diallel) of sorghum across seasons (ICRISAT, Patancheru, 2013-14)

Pedigree	Plants with shoot fly eggs (%)	ORS	Leaf glossy score	Trichome density on abaxial leaf surface		Trichome density on adaxial leaf surface	
	2013 PR	2013 R	2013 PR	2013 R	2013 PR	2013 R	2013 PR
Phule Anuradha X ICSV 700	-8.73	-	-0.17	-6.55	-7.58	-6.33	-10.44
M 35-1 X ICSV 700	6.46	0.17	0.00	8.85	9.17	15.40*	11.17
M 35-1 X Phule Anuradha	-4.04	0.17	-0.33	-1.28	8.39	-0.57	12.78
CSV 15 X ICSV 700	-30.43**	0.80*	-0.83**	15.6**	25.22**	24.70**	44.05**
CSV 15 X Phule Anuradha	8.33	-	-0.17	15.5**	18.25**	21.90**	19.47**
CSV 15 X M 35-1	-3.36	0.50	-0.33	-1.93	13.28*	-1.10	16.22*
ICSV 25019 X ICSV 700	4.47	0.33	0.00	4.90	2.84	16.20**	13.33*
ICSV 25019 X Phule Anuradha	13.57	0.17	-0.17	5.72	-8.20	8.85	-5.70
ICSV 25019 X M 35-1	-11.60	-0.50	0.00	-2.37	-5.92	-1.27	-1.92
ICSV 25019 X CSV 15	-14.18*	0.33	0.17	-2.17	0.00	-1.45	0.00
PS 35805 X ICSV 700	5.74	-0.33	0.17	2.15	-5.03	-2.35	-4.19
PS 35805 X Phule Anuradha	13.27	0.33	0.33	4.02	-2.78	1.22	2.22
PS 35805 X M 35-1	-11.75	0.17	-0.33	0.60	6.36	-3.27	9.58
PS 35805 X CSV 15	31.59**	-0.50	-0.17	-2.88	-14.97*	-6.10	-18.19**
PS 35805 X ICSV 25019	-4.37	-0.33	-0.17	-2.17	-0.03	-0.98	-2.20
IS 2123 X ICSV 700	5.99	-0.33	0.33	1.72	-2.39	3.97	8.72
IS 2123 X Phule Anuradha	0.30	-0.67	-0.50*	-0.93	10.97	1.15	9.22
IS 2123 X M 35-1	-13.77*	0.33	0.00	3.72	-6.72	-1.12	-3.67
IS 2123 X CSV 15	-20.77**	-0.80*	-0.17	2.72	-1.42	2.22	1.17
IS 2123 X ICSV 25019	-11.54	0.67	0.17	6.52	2.92	5.33	4.78
IS 2123 X PS 35805	0.30	-0.33	-0.17	-9.40	-5.39	-10.43	2.00
IS 2146 X ICSV 700	8.47	0.17	0.17	-12.60*	5.61	-8.88	1.56
IS 2146 X Phule Anuradha	21.01**	-0.17	0.00	2.62	-8.89	-5.67	0.72
IS 2146 X M 35-1	3.91	-0.17	0.33	-8.42	-8.39	4.53	-5.72
IS 2146 X CSV 15	6.73	0.50	-0.17	-0.52	10.33	-2.06	16.45*
IS 2146 X ICSV 25019	4.19	0.58	0.33	-10.10*	1.33	-6.30	1.28
IS 2146 X PS 35805	-2.92	-0.17	0.00	-11.20*	-9.89	-15.50*	1.83
IS 2146 X IS 2123	1.46	0.17	0.67**	-21.80**	-7.00	-17.70**	-14.56*
IS 18551 X ICSV 700	5.83	-0.33	0.17	-6.38	15.17*	0.57	8.83
IS 18551 X Phule Anuradha	0.29	-0.33	0.17	6.62	11.56	0.13	10.56
IS 18551 X M 35-1	-0.14	-0.33	-0.17	1.67	12.75	1.28	4.42
IS 18551 X CSV 15	13.44	-0.33	-0.17	-1.28	-8.56	-6.45	-15.28*
IS 18551 X ICSV 25019	-0.78	-	0.33	-7.40	9.72	-6.62	16.28*
IS 18551 X PS 35805	-8.74	-0.33	0.33	-1.17	-15.61*	-4.90	-6.56
IS 18551 X IS 2123	2.69	-0.17	0.17	-7.43	-2.81	-7.00	-7.00
IS 18551 X IS 2146	-8.57	0.50	0.17	-0.50	4.94	6.28	3.95
Swarna X ICSV 700	-16.19*	1.5**	0.00	0.10	-0.42	-2.05	0.06
Swarna X Phule Anuradha	2.16	0.33	0.17	5.57	-5.86	4.30	-7.36
Swarna X M 35-1	-17.92*	-0.33	-0.83**	12.90*	17.92**	22.8**	31.50**
Swarna X CSV 15	-11.89	-0.17	-0.17	-0.52	0.00	-1.78	0.00
Swarna X ICSV 25019	-1.95	-	0.17	2.33	3.17	6.65	5.33
Swarna X PS 35805	-6.99	0.33	0.00	6.32	3.11	12.70*	2.25
Swarna X IS 2123	-5.80	-1.30**	-0.50*	3.63	4.39	6.73	6.50
Swarna X IS 2146	11.37	-0.33	0.00	7.93	-4.61	9.67	-8.83
Swarna X IS 18551	-3.00	-	0.00	4.67	-10.67	-1.42	-6.67

*, ** *t*-test significant at $P = 0.05$ and 0.01, respectively; ORS, overall resistance score.

IS 2123 X CSV 15 and Swarna X IS 2123 exhibited significant negative reciprocal effects; and CSV 15 X ICSV 700 and Swarna X ICSV 700 exhibited significant positive reciprocal effects for overall resistance score during the rainy season. CSV 15 X ICSV 700, IS 2123 X Phule Anuradha, Swarna X M 35-1, and Swarna X IS 2123 exhibited significant negative reciprocal effects, and IS 2146 X IS 2123 exhibited positive reciprocal effects for leaf glossy score in the postrainy season.

IS 2146 X ICSV 700, IS 2146 X ICSV 25019, IS 2146 X PS 35805, IS 2146 X IS 2123 in the rainy season, and PS 35805 X CSV 15, IS 18551 X ICSV 25019 in the postrainy season exhibited significant negative reciprocal effects for trichome density on the abaxial leaf surface (Table 7). CSV 15 X ICSV 700, CSV 15 X Phule Anuradha, and Swarna X M 35-1 across seasons, and CSV 15 X M 35-1, and IS 18551 X IS 2123 in the postrainy season exhibited significant positive reciprocal effects for trichome density on the abaxial leaf surface.

IS 2146 X PS 35805 in the rainy season, IS 2146 X IS 2123 across seasons, and IS 18551 X CSV 15 in the postrainy season exhibited significant negative reciprocal effects for trichome density on the adaxial leaf surface (Table 7). M 35-1 X ICSV 700 and Swarna X PS 35805 in the rainy season, and CSV 15 X ICSV 700, CSV 15 X Phule Anuradha, ICSV 25019 X ICSV 700, Swarna X M 35-1 across seasons, and CSV 15 X M 35-1, IS 2146 X CSV 15 and IS 18551 X ICSV 25019 in the postrainy season exhibited significant positive reciprocal effects for trichome density on the adaxial leaf surface.

Combining Ability Estimates and Genetic Parameters of Shoot Fly Resistance and Morphological Traits

Shoot fly deadhearts (%) in the rainy season, and plants with shoot fly eggs (%) in the postrainy season showed higher variance due to specific combining ability (σ^2s) than the variance due to general combining ability (σ^2g), indicating the predominance of dominance gene action in controlling the expression of resistance to sorghum shoot fly, *A. soccata* (Table 8). This was confirmed by higher dominance variance than the additive variance for these traits. The other traits that had σ^2g and σ^2s on par with each other exhibited both additive and non-additive type of gene action.

Table 8: Estimates of various genetic parameters for different shoot fly resistance and morphological traits of sorghum across seasons (ICRISAT, Patancheru, 2013-14)

Traits	Plants with shoot fly eggs (%)	Number of shoot fly eggs/plant	Shoot fly deadhearts (%)	ORS	Leaf glossy score	Plant vigor score	Leafsheath pigmentation	Trichome density on abaxial leaf surface	Trichome density on adaxial leaf surface	Waxy bloom	Plant color
σ^2g	(129.86)	(0.01)	28.6 (115.14)	0.14 (0.14)	0.13 (0.36)	0.05 (0.06)	(0.03)	303.35 (337.10)	535.36 (446.53)	0.19 (0.11)	0.04 (0.04)
σ^2s	(413.37)	(0.03)	88.51	0.59 (0.39)	(0.37)	0.08 (0.05)	-	233.73 (132.26)	264.53 (216.25)	0.16	0.12 (0.13)
σ^2r	(72.66)	-	-	0.08	(0.03)	-	-	22.55 (34.71)	42.32 (93.72)	-	-
σ^2e	(110.81)	(0.07)	101.81 (107.64)	0.29 (0.38)	0.32 (0.12)	0.16 (0.12)	(0.06)	59.16 (104.42)	81.99 (98.31)	0.06 (0.05)	-
σ^2a	(259.71)	(0.03)	57.21 (230.28)	0.29 (0.29)	0.26 (0.72)	0.10 (0.12)	(0.06)	606.71 (674.20)	1070.72 (893.06)	0.39 (0.22)	0.08 (0.08)
σ^2d	(413.37)	(0.03)	88.51	0.59 (0.39)	(0.37)	0.08 (0.05)	-	233.73 (132.26)	264.53 (216.25)	0.16	0.12 (0.13)
σ^2p	(856.54)	(0.11)	264.81 (369.66)	1.25 (1.12)	0.74 (1.24)	0.32 (0.3)	(0.12)	922.15 (945.58)	1459.56 (1301.35)	0.60 (0.28)	0.21 (0.21)
h_{ns^2}	(0.30)	(0.23)	0.22 (0.62)	0.23 (0.26)	0.35 (0.58)	0.30 (0.41)	(0.51)	0.66 (0.71)	0.73 (0.69)	0.65 (0.77)	0.40 (0.39)
h_{b^2}	(0.79)	(0.46)	0.55 (0.73)	0.70 (0.61)	0.50 (0.88)	0.55 (0.59)	(0.54)	0.91 (0.85)	0.92 (0.85)	0.91 (0.82)	0.99 (0.99)
GCA/SCA ratio	(0.31)	(0.51)	0.32	0.24 (0.37)	(0.97)	0.59 (1.13)	-	1.30 (2.55)	2.02 (2.07)	1.22	0.33 (0.32)
Predictability ratio	(0.39)	(0.51)	0.39	0.32 (0.42)	(0.66)	0.54 (0.69)	-	0.72 (0.84)	0.80 (0.81)	0.71	0.40 (0.39)

σ^2g, gca variance; σ^2s, sca variance; σ^2r, reciprocal variance; σ^2e, environmental/error variance; σ^2a, additive variance; σ^2d, dominance variance; σ^2p, phenotypic variance; h_{ns^2}, narrow sense heritability; h_{b^2}, broad sense heritability; GCA, general combining ability; SCA, specific combining ability; the values outside the parentheses are for rainy season, whereas inside the parentheses are for postrainy season.

Variance due to general combining ability (σ^2g) was higher than the variance due to specific combining ability (σ^2s) for trichome density on abaxial and adaxial leaf surfaces across seasons; indicating the predominance of additive gene action in controlling the expression of these traits (Table 8). Trichome density on abaxial and adaxial leaf surfaces showed higher additive variance (σ^2a) than the dominance variance (σ^2d) across seasons. Leaf glossy score and plant vigor score in the postrainy season exhibited higher additive variance than the dominance variance. Overall resistance score exhibited high dominance variance to the additive variance across seasons. Plant vigor score and trichome density on the abaxial and adaxial leaf surfaces across seasons, and leaf glossy score in the postrainy season exhibited high GCA/SCA ratio, indicating the additive type of gene action controlling the expression of these traits. These traits also had high predictability ratios.

Heritability estimates for narrowsense (h_{ns^2}) and broadsense (h_{b^2}) heritability of all the traits ranged from 0.22 to 0.73 (Table 8). Plants with shoot fly deadhearts and overall resistance score in the rainy season, and plants with shoot fly eggs, number of shoot fly eggs/plant, and overall resistance score in the postrainy season exhibited low ($\leq$0.30) narrowsense heritability. The other traits exhibited moderate to high narrowsense heritability. All the shoot fly resistance and morphological traits across seasons exhibited high broadsense heritability values.

DISCUSSION

The significant *F*-values for all the traits studied indicated the diverse nature of the genotypes used in this study. The *per se* performance of the parents and hybrids against shoot fly, *A. soccata* indicated the existence of genetic potential for shoot fly resistance in the material used. The evaluation of parents and their hybrids during the rainy and postrainy seasons indicated variation in expression of shoot fly resistance across seasons; with greater susceptibility in the rainy season (Dhillon et al., 2006c). This was largely because of available favorable environmental conditions suitable for shoot fly multiplication in the rainy season. The performance of hybrids was season specific indicating the influence of environmental factors in the expression of shoot fly resistance (Riyazaddin et al., 2015a,b). High G X E interactions for shoot fly deadhearts percentage, has been reported earlier (Padmaja et al., 2010; Aruna et al., 2011). Most of the crosses and their reciprocals showed resistance to shoot fly, suggesting that shoot fly resistance can be easily transferred into the progenies. Most of the crosses exhibited oviposition non-preference and tolerance mechanisms of resistance, which are the major components of resistance in sorghum to shoot fly (Doggett et al., 1970; Raina et al., 1981; Sharma and Nwanze, 1997; Kamatar et al., 2003; Dhillon et al., 2005, 2006b; Sivakumar et al., 2008). F_1 hybrids based on shoot fly-resistant parents exhibited leaf glossiness, high plant vigor, leafsheath pigmentation, and high trichome density across seasons, suggesting that both the parents should be resistant to shoot fly. The genotypes ICSV 700, M 35-1, ICSV 25019, PS 35805, IS 2123, IS 2146, and IS 18551 showed resistance to shoot fly across seasons, and hence, these can be effectively utilized in developing shoot fly resistant sorghums.

Estimates of combining ability will be helpful in selecting the desirable parents for developing shoot fly-resistant hybrids. Significant mean sum of squares due to GCA for all characters, and SCA for most of the traits indicated contribution of both additive and non-additive types of gene action for resistance to shoot fly in sorghum (Borikar and Chopde, 1982; Riyazaddin et al., 2015b). Significant reciprocal mean sum of squares for plants with shoot fly eggs, overall resistance score and trichome density indicated the cytoplasmic influence in inheritance of these traits. Hence, care should be taken while selecting the parents for these traits for use in sorghum improvement. Leaf glossy score, plant vigor and trichome density had greater σ^2g, with high GCA/SCA and predictability ratios, suggesting predominance of additive gene action in inheritance of these traits. Plants with shoot fly deadhearts, plants with shoot fly eggs, and ORS exhibited higher SCA variance, with lower GCA/SCA and predictability ratios, suggesting the predominance of non-additive (dominance) gene action in inheritance of these traits. Hence, heterosis breeding could be used for improvement of sorghum for these traits.

Predominance of different types of gene action, and their heritability differs with the shoot fly population pressure (Rana et al., 1981; Borikar and Chopde, 1982). Almost all the traits across seasons exhibited high broadsense heritabilities, indicating high inheritance of shoot fly resistance. Genotypes with significant negative *gca* effects were good combiners for shoot fly resistance. Genotypes with negative *gca* effects for plants with shoot fly eggs, number of shoot fly eggs/plant, shoot fly deadhearts, leaf glossy score, plant vigor score and leafsheath pigmentation, significant positive *gca* effects for trichome density can be selected and effectively utilized in the breeding program (Sharma et al., 1977; Hallali et al., 1982; Dhillon et al., 2006b). The crosses with negative *sca* effects for leaf glossy and plant vigor scores can be utilized in hybrid breeding process. Significant reciprocal combining ability of some of the crosses indicted influence of maternal factors in inheritance of plants with shoot fly eggs, leaf glossy score and trichome density. This information should be taken into consideration to select genotypes for use as male or female parents.

CONCLUSIONS

The present studies indicated that the parents involved in the crossing program should possess the shoot fly resistance genes for breeding high yielding shoot fly-resistant sorghums. Both the additive and non-additive type of gene action controlled expression of resistance to shoot fly, *A. soccata*. Most of the traits exhibited high narrow and broadsense heritabilities, indicating high inheritance of the shoot fly resistance traits. Breeding for shoot fly resistance exhibited season specificity, and utmost care should be taken while selecting the parental lines. The presence of dominance gene action for shoot fly traits indicated heterosis breeding as ideal for improving shoot fly resistance in sorghum genotypes. The predominance of additive nature of gene action for leaf glossy score, plant vigor, leafsheath pigmentation and trichome density indicated simple selection techniques and recombination breeding with pedigree as methods of selection for improving these traits. The variation in expression of shoot fly resistance across seasons was due to the involvement of non-additive genetic components of resistant traits. Crosses with significant positive or negative *sca* effects for shoot fly resistance suggested that hybridization is necessary to increase the levels of shoot fly resistance. Parents involved in the crosses with significant specific combining abilities can be utilized directly in hybrid breeding process. The genotypes with good general combining ability for shoot fly resistance, and high grain yield can be used in developing shoot fly resistant cultivars for sustainable crop production.

AUTHOR CONTRIBUTIONS

HS: Planned, supervised the research, and contributed in preparing the manuscript. RM: Executed the experiments, carried out the statistical analysis and prepared the manuscript. RM, RB: Scientific Officers, helped in carrying out the field experiments. AA: Provided financial support for field experiments and helped in preparing the manuscript. KP: Supervised the research and helped in preparing the manuscript.

CONFLICT OF INTEREST STATEMENT

The authors declare that the research was conducted in the absence of any commercial or financial relationships that could be construed as a potential conflict of interest.

ACKNOWLEDGMENTS

We are thankful to the sorghum entomology and breeding staff for their help in carrying out the field experiments and Bill & Melinda Gates Foundation for financial support through HOPE-Sorghum and Millets Project. We also thank Dr. Rajeev Gupta for reviewing the manuscript and providing constructive suggestions.

REFERENCES

1. Aruna, C., Bhagwat, V. R., Vittal, S., Hussain, T., Ghorade, R. B., Khandalkar, H. G., et al. (2011). Genotype X Environment interactions for shoot fly resistance in sorghum [*Sorghum bicolor* (L.) Moench]: response of recombinant inbred lines. *Crop Prot.* 30, 623–630. doi: 10.1016/j.cropro.2011.02.007
2. Ashok Kumar, A., Reddy, B. V. S., Sharma, H. C., Hash, C. T., Srinivasa Rao, P., Ramaiah, B., et al. (2011). Recent advances in sorghum genetic enhancement research at ICRISAT. *Am. J. Plant Sci.* 2, 589–600. doi: 10.4236/ajps.2011.24070
3. Baker, R. J. (1978). Issues in diallel analysis. *Crop Sci.* 18, 533–536. doi: 10.2135/cropsci1978.0011183X0 01800040001x
4. Balikai, R. A., and Bhagwat, V. R. (2009). Evaluation of integrated pest management components for the management of shoot fly, shoot bug and aphid in rabi sorghum. *Karnataka J. Agric. Sci.* 22, 532–534.
5. Borikar, S. T., and Chopde, P. R. (1982). Stabililty parameters for shoot fly resistance in sorghum. *Indian J. Genet.* 42, 155–158.
6. Bramel-Cox, P. J., Kumar, K. A., Hancock, J. D., and Andrews, D. J. (1995). "Sorghum and millets for forage and feed," in *Sorghum and Millets Chemistry and Technology*, ed D. A. V. Dendy (St. Paul, MN: American Association Cereal Chemists, Inc.), 325–364.
7. Deeming, J. C. (1971). Some species of *Atherigona rondani* (Diptera: Muscidae) from northern Nigeria, with special reference to those injurious to cereal crops. *Bull Entomol. Res.* 61, 133–190. doi: 10.1017/S0007485300057527
8. Dhillon, M. K., Sharma, H. C., Naresh, J. S., Singh, R., and Pampapathy, G. (2006c). Influence of cytoplasmic male sterility on expression of different mechanisms of resistance in sorghum to *Atherigona soccata* (Diptera: Muscidae). *J. Econ. Entomol.* 99, 1452–1461. doi: 10.1603/0022-0493-99.4.1452
9. Dhillon, M. K., Sharma, H. C., Ram, S., and Naresh, J. S. (2005). Mechanisms of resistance to shoot fly, *Atherigona soccata* in sorghum.*Euphytica* 144, 301–312. doi: 10.1007/s10681-005-7400-4
10. Dhillon, M. K., Sharma, H. C., Reddy, B. V. S., Ram, S., and Naresh, J. S. (2006a). Inheritance of Resistance to Sorghum Shoot Fly,*Atherigona soccata*. *Crop Sci.* 46, 1377–1383. doi: 10.2135/cropsci2005.06-0123
11. Dhillon, M. K., Sharma, H. C., Singh, R., and Naresh, J. S. (2006b). Influence of cytoplasmic male-sterility on expression of physico-chemical traits associated with resistance to sorghum shoot fly, *Atherigona*

soccata (Rondani). *SABRAO J. Breed. Genet.* 38, 105–122.

12. Doggett, H., Starks, K. J., and Eberhart, S. A. (1970). Breeding for resistance to the sorghum shoot fly. *Crop Sci.* 10, 528–531. doi: 10.2135/cropsci1970.0011183X001000050023x

13. FAO (2014). *Crops Primary Equivalent.* Retrieved on 2nd March, 2015. Available online at: www.faostat.fao.org

14. GenStat (2010). *Introduction to GenStat for Windows Genstat, 13th Edn.* Hemel Hempstead: Lawes Agricultural Trust, Rothamsted Experimental Station.

15. Goyal, S. N., and Kumar, S. (1991). Combining ability for yield component and oil content in sesame. *Indian J. Genet. Plant Breed.* 51, 311–314.

16. Griffing, B. (1956). Concept of general and specific combining ability in relation to diallel crossing systems. *Australian J. Biol. Sci.* 9, 463–493.

17. Hallali, M. S., Gowda, B. T. S., Kulkarni, K. A., and Goud, J. V. (1982). Inheritance of resistance to shoot fly [*Atherigona soccata* (Rond.)] in sorghum [*Sorghum bicolor* (L.) Moench]. *SABRAO J. Breed. Genet.* 14, 165–170.

18. Indostat Services (2004). *Windostat.* Hyderabad: Indostat Services.

19. Johnson, H. W., Robinson, H. F., and Comstock, R. E. (1955). Estimates of genetic and environmental variability in soybean. *Agron. J.*47, 314–318. doi: 10.2134/agronj1955.00021962004700070009x

20. Jotwani, M. G., Teetes, G. L., and Young, W. R. (1980). "Elements of integrated control of sorghum pests," in *FAO Plant Production and Protection Paper.* Rome: FAO.

21. Kahate, N. S., Raut, S. M., Ulemale, P. H., and Bhogave, A. F. (2014). Management of Sorghum Shoot Fly. *Popular Kheti.* 2, 72–74.

22. Kamatar, M. Y., Salimath, P. M., Kumar, R. L. R., and Rao, T. S. (2003). Heterosis for biochemical traits governing resistance to shoot fly in sorghum [*Sorghum bicolor* (L.) Moench]. *Indian J. Genet.* 63, 124–127.

23. Kumar, A. A., Reddy, B. V. S., Sharma, H. C., and Ramaiah, B. (2008). Shoot fly (*Atherigona soccata*) resistance in improved grain sorghum hybrids. *E J. SAT Agric. Res.* 6, 1–4.

24. Maiti, R. K., and Bidinger, F. R. (1979). Simple approaches to the identification of shoot fly tolerance in sorghum. *Indian J. Plant Prot.* 7, 135–140.

25. Mohammed, R., Are, A. K., Bhavanasi, R., Munghate, R. S., Kavi Kishor, P. B., and Sharma, H.C. (2015). Quantitative genetic analysis of agronomic and morphological traits in sorghum, *Sorghum bicolor*. *Front. Plant Sci.* 6:945. doi: 10.3389/fpls.2015.00945

26. Padmaja, P. G., Madhusudhana, R., and Seetharama, N. (2010). *Sorghum Shoot Fly.* Hyderabad: Directorate of Sorghum Research.

27. Prabhakar and Raut, M. S. (2010). Exploitation of heterosis using diverse parental lines in Rabi Sorghum. *Elec. J. Plant Breed.* 1, 680–684.

28. Raina, A. K., Thindwa, H. Z., Othieno, S. M., and Corkhill, R. T. (1981). Resistance in sorghum to sorghum shoot fly: larval development and adult longevity and fecundity on selected cultivars. *Insect Sci. Appl.* 2, 99–103.

29. Rana, B. S., Jotwani, M. G., and Rao, N. G. P. (1981). Inheritance of host plant resistance to sorghum shoot fly. *Insect Sci. Appl.* 2, 105–100.

30. Riyazaddin, M. D., Kavi Kishor P. B., Are, A. K., Reddy, B. V. S., Rajendra, S. M., and Sharma, H. C. (2015a). Mechanisms and diversity of resistance to sorghum shoot fly, *Atherigona soccata*. *Plant Breed.* 134, 423–436. doi: 10.1111/pbr.12276

31. Riyazaddin, M. D., Munghate, R. S., Are, A. K., Kavi Kishor P. B., Reddy, B. V. S., and Sharma, H. C. (2015b). Components of resistance to sorghum shoot fly, *Atherigona soccata*. *Euphytica* 207, 419–438. doi: 10.1007/s10681-015-1566-1

32. Sharma, G. C., Jotwani, M. G., Rana, B. S., and Rao, N. G. P. (1977). Resistance to the sorghum shoot fly, *Atherigona soccata* (Rondani) and its genetic analysis. *J. Entomol. Res.* 1, 1–12.

33. Sharma, H. C., and Nwanze, K. F. (1997). *Mechanisms of Resistance to Insects and their Usefulness in Sorghum Improvement.*Information Bulletin no. 55, Patancheru: International Crops Research Institute for the Semi-Arid Tropics (ICRISAT).

34. Sharma, H. C., Taneja, S. L., Kameswara Rao, N., and Prasada Rao, K. E. (2003). *Evaluation of Sorghum Germplasm for Resistance to Insect Pests.* Information Bulletin no. 63, Patancheru: International Crops Research Institute for the Semi-Arid Tropics (ICRISAT).

35. Sharma, H. C., Taneja, S. L., Leuschner, K., and Nwanze, K. F. (1992). *Techniques to Screen Sorghums for Resistance to Insect Pests.*Information Bulletin no. 32. Patancheru: International crops research institute for the Semi-Arid tropics (ICRISAT).

36. Sharma, H. C. (1985). Future strategies for pest control in sorghum in India. *Trop. Pest Manag.* 31, 167–185. doi: 10.1080/09670878509370977

37. Sharma, H. C. (1993). Host-Plant Resistance to insects in sorghum and its role in integrated pest management. *Crop Prot.* 12, 11–34. doi: 10.1016/0261-2194(93)90015-B

38. Singh, R. K., and Chaudhary, B. D. (1985). *Biometrical methods in quantitative genetics.* New Dehli: Kalyani Publishers, 102–127.

39. Sivakumar, C., Sharma, H. C., Lakshmi Narasu, M., and Pampapathy, G. (2008). Mechanisms and diversity of resistance to shoot fly,*Atherigona soccata* in *Sorghum bicolor*. *Indian J. Plant Prot.* 36, 249–256.

40. Soto, P. E. (1974). Ovipositional preference and antibiosis in relation to resistance to sorghum shoot fly. *J. Econ. Entomol.* 67, 265–267. doi: 10.1093/jee/67.2.265

CHAPTER 5

EVALUATION OF SORGHUM [SORGHUM BICOLOR (L.)] REFERENCE GENES IN VARIOUS TISSUES AND UNDER ABIOTIC STRESS CONDITIONS FOR QUANTITATIVE REAL-TIME PCR DATA NORMALIZATION

Palakolanu Sudhakar Reddy, Dumbala Srinivas Reddy, Kaliamoorthy Sivasakthi, Pooja Bhatnagar-Mathur, Vincent Vadez and Kiran K. Sharma

Keywords: qPCR, RefFinder, *Sorghum bicolor*, gene expression stability, reference gene, normalization

INTRODUCTION

Sorghum is the fifth most important cereal crop that is widespread in the semi-arid regions of the world with an annual production of 65.5 Mt (FAO, 2011), possesses strong drought-tolerance traits and a high forage value. Sorghum seeds are utilized for both human and animal feed, and cultivars are processed for syrup, sugar, and alcohol. Additionally, sorghum cultivars have great economic potential as they can be valorized for second-generation biofuels to produce environment-friendly energy (Vermerris, 2011;Calviño and Messing, 2012). Sorghum is related to many C4 plants including maize, sugarcane, foxtail millet and switchgrass, which tend to have larger polyploid genomes containing repetitive sequences (Price et al., 2005). In contrast, sorghum is a diploid with a small genome (750 Mbp) and possesses an extraordinary germplasm diversity that greatly aids gene discovery and analysis through comparative and functional genomics, making it highly useful as model cereal for structural and functional genomic studies aimed at improving agronomically important traits (Paterson et al., 2009).

Coupled with species phenotypes, functional genomics can provide important insights into complex biological processes including abiotic stress responses. Functional genomic studies are now being used to identify the roles of regulatory and structural genes. The importance of sorghum functional genomics has increased following the sequencing of its genome (Paterson et al., 2009). Despite a wide range of available experimental approaches for exploring gene expression at the transcriptional level, e.g., northern blotting, ribonuclease protection assays, reverse transcription PCR (RT-PCR), quantitative real-time PCR (qPCR) and DNA microarrays (Valasek and Repa, 2005). Among them, qPCR is the most efficient for quantification of gene expression levels due to its simplicity, sensitivity, accuracy and cost (Wong and Medrano, 2005); however, despite these advantages, the utility of this method is often limited due to the absence of reliable reference genes for qPCR data normalization.

Classically, most of the reference genes fall in the category of housekeeping genes, which have major roles in the maintenance of basic cellular metabolism irrespective of physiological conditions (Bustin, 2002). These reference genes have been widely used in qPCR assays in molecular and functional genomic studies as internal calibrators (Vandesompele et al., 2002; Eisenberg and Levanon, 2013); however, recent studies have shown that these reference genes are often not stably expressed under various experimental

Citation: Sudhakar Reddy P, Srinivas Reddy D, Sivasakthi K, Bhatnagar-Mathur P, Vadez V and Sharma KK (2016) Evaluation of Sorghum [Sorghum bicolor (L.)] Reference Genes in Various Tissues and under Abiotic Stress Conditions for Quantitative Real-Time PCR Data Normalization. Front. Plant Sci. 7:529. doi: 10.3389/fpls.2016.00529.

conditions (Thellin et al., 1999; Zhu et al., 2013; Gimeno et al., 2014). Consequently, the selection and validation of suitable reference genes that are uniformly and stably expressed across experimental conditions have become imperative (Jian et al., 2008; Li et al., 2012). Various statistical tools, e.g., geNorm, NormFinder, BestKeeper and RefFinder, have been developed to determine reference gene suitability for qPCR data normalization (Vandesompele et al., 2002; Andersen et al., 2004; Pfaffl et al., 2004). Stable reference gene validation has been performed in many plant species, including model species and crop plants such as *Arabidopsis thaliana* (Czechowski et al., 2005; Remans et al., 2008),*Brassica napus* (Yang et al., 2014a; Machado et al., 2015), *Hordium vulgare* (Zmienko et al., 2015),*Fagopyrum tataricum* (Demidenko et al., 2011), *Cajanus cajan* (Sinha et al., 2015), *Oryza sativum* (Jain et al., 2006; Li et al., 2010; Ji et al., 2014), *Arachis hypogaea* (Reddy et al., 2013), *Pennisetum glaucum*(Saha and Blumwald, 2014; Reddy et al., 2015b), *Solanum tuberosum* (Nicot et al., 2005; Mariot et al., 2015), *Glycine max* (Jian et al., 2008; Nakayama et al., 2014), *Saccharum officinarum* (Iskandar et al., 2004; Ling et al., 2015), *Lycopersicon esculentum* (Expósito-Rodríguez et al., 2008), *Triticum aestivum*(Paolacci et al., 2009), and *Zea mays* (Manoli et al., 2012; Lin et al., 2014).

Until now, sorghum qPCR assays were performed using a single reference gene to quantify target gene expression in response to various experimental conditions. Traditional reference genes including *18S RNA, Actin, EIF4α, Tubulin*, and *Ubiquitin* have been used as calibrators for quantifying sorghum gene expression in response to abiotic stresses and during plant growth and development (Yang et al., 2004;Jain et al., 2008; Cook et al., 2010; Shen et al., 2010; Wang et al., 2010; Dugas et al., 2011; Ishikawa et al., 2011; Koegel et al., 2013; Li et al., 2013; Gelli et al., 2014; Shakoor et al., 2014; Yang et al., 2014b; Yin et al., 2014; Kebrom and Mullet, 2015; Li et al., 2015; Walder et al., 2015). Nevertheless, the reference genes used in these studies were randomly selected from various sources without any experimental validation. Moreover, single gene quantification qPCR assays are well known to frequently exhibit variability in gene expression under various experimental conditions (Dheda et al., 2005). The Minimum Information for Publication of Quantitative Real-Time PCR Experiments (MIQE) guidelines developed for the proper selection (Bustin et al., 2009) and validation of stable candidate reference genes for qPCR experiments highly recommend averaging data from more than 2 reference genes (Bustin et al., 2010). To date, only one study has evaluated reference genes in virus-infected *Sorghum bicolor* tissues (Zhang et al., 2013), and to the best of our knowledge, there are no systematic studies regarding the selection of suitable stable sorghum reference genes in various tissues and abiotic stress conditions.

This study tested 15 potential candidate reference genes, *Acyl Carrier Protein 2 (ACP2), ADP-Ribosylation Factor (ADPRF), alpha Tubulin (α-TUB), beta Tubulin (β-TUB), Eukaryotic Initiation Factor 4A-1 (EIF4a), Elongation Factor P (EF-P), Glyceraldehyde 3-phosphate Dehydrogenase (GAPDH), Malonyl CoA-Acyl Carrier Protein (MACP), Malate Dehydrogenase (MDH), Cyclophilin/ Peptidylprolyl Isomerase (CYP), Protein Phosphatase 2C (PP2C), 6-Phosphogluconate Dehydrogenase (6PGDH), S-adenosyl methionine Decarboxylase (SAMDC), Serine/ threonine-Protein Phosphatase (PP2A), and Ubiquitin Protein 1, Isoform c (UBC1)*. The stabilities of these reference genes were analyzed in sets of samples from treatments with various abiotic stresses (salt, cold, heat, water stress, and ABA stress) and tissues (seedlings, leaves, panicles, roots and mature seeds). Additionally, to verify and support the identification of the best-ranked candidate reference genes, *SbHSF5* and *SbHSF13*gene expression levels were assayed. The stable reference genes identified in this study will be helpful in future qPCR-based molecular and functional studies in sorghum and related species.

MATERIALS AND METHODS

Plant Materials and Abiotic Stress Treatments

Sorghum bicolor genotypes (Parbani Moti, Phule Vasudha (PVS), S35, M35, and BTx623) seeds were collected from the sorghum breeding unit of the International Crops Research Institute for the Semi-Arid Tropics (ICRISAT) in Patancheru, India. Seeds were sown in pots filled with soil mixture (3:2:1 clay:sand:manure) in a glasshouse. Various abiotic stress treatments (salt, cold, heat, water stress, and ABA stresses) were performed and various tissues (seedlings, leaves, panicles, roots, and mature seeds) were sampled according to (Reddy et al., 2015a). Leaf tissues were collected from sorghum cultivars including Parbani Moti, Phule Vasudha, S35, M35, and BTx623 were grown under control conditions and under progressive drought where at normalized transpiration ratio (NTR) of reached at 0.1 (10% of soil moisture remaining in the pot) (Vadez and Sinclair, 2001) then leaf samples were collected. Tissue samples were collected from three individual seedlings for three independent biological replicates, immediately snap frozen in liquid nitrogen and stored at −80°C until RNA extraction.

Total RNA Isolation and cDNA Synthesis

Total RNA was extracted from tissues using the Plant RNA mini spin kit (MACHEREY-NAGEL GmbH and Co. KG, Neumann-Neander-Straße 6-8, Duren, Germany) with the in-column DNase I treatment according to the manufacturer's instructions. Sample quantity and purity were determined using a NanoVue Plus Spectrophotometer (GE Healthcare). RNA samples with an OD_{260}/OD_{280} absorbance ratio between 1.9 and 2.2 were used for further analysis, and RNA integrity was assessed on 1% denaturing formaldehyde agarose gel. For each sample, 1 μg of total RNA was reverse transcribed using the SuperScript First-Strand Synthesis System for RT-PCR (Invitrogen) and oligo (dT) primers according to manufacturer›s instructions. The reverse transcribed cDNAs were then diluted 1:12 with nuclease-free water and used for qPCR analysis.

Candidate Reference Gene Selection and qPCR Primer Design

Fifteen candidate reference genes were identified from the literature and the sequences of their corresponding homologs were extracted from the NCBI and the Phytozome databases (Table 1). The selected candidate reference genes included *GAPDH, UBC1, MACP, ACP2, α-TUB, EF-P, 6PGDH, SAMDC, CYP, MDH, ADPRF, β-TUB, PP2A, EIF4α,* and *PP2C* that were earlier reported to be stably expressed in monocots such as rice, pearl millet, barley, foxtail millet, wheat, and maize (Jain et al., 2006;Paolacci et al., 2009; Manoli et al., 2012; Kumar et al., 2013; Reddy et al., 2015b; Zmienko et al., 2015). Primers were designed using the Primer3 software (Untergasser et al., 2012) with the following parameters: 59–62°C annealing temperature, 20–22 bp primer length, 45–55% GC contents, and 90–150 bp amplicon length (Table 1). Pooled and diluted cDNA samples were used for qPCR and 2% agarose gel electrophoresis was used to check primer pair specificity prior to sequencing. The sequences amplified by each primer combination were compared with GenBank sequences using the BLASTN and BLASTX algorithms to verify amplicon specificity.

Table 1: Comprehensive details of the reference genes used for the normalization in Sorghum

S.No.	Gene symbol	Gene name	Accession number	Cellular function	Primers (F/R) (5''–3'')	Amplicon length (bp)	Tm (°C)	PCR efficiency	Regression coefficient (R2)
1	*GAPDH*	*Glyceraldehyde 3-phosphate-Dehydrogenase*	XM_002439118	Glycolysis and gluconeogenesis	AAGGCCGGCATTGCTTTGAAT ACATGTGGCAGATCAGGTCGA	107	84.1	1.00	0.998
2	*UBC1*	*Ubiquitin Protein 1 Isoform C*	XM_002452660	Protein degradation	AGAGGCTCATCTTCGCTGGG CTCAGTTAACACGGCCACCAC	120	88.1	0.98	0.994
3	*MACP*	*Malonyl CoA-Acyl Carrier protein*	XM_002465363	Fatty acid biosynthesis	GCATTGAGAACATCGGGGCTT ATGAGTGGAAACTTCGTTCCA	139	82.8	0.98	0.997
4	*ACP2*	*Acyl Carrier Protein-2*	XM_002462703	Fatty acid and Polyketides biosynthesis	ACGAACTTGTTGCGGCAGAAG GAACAAGAAGGGATGCGCTGG	110	84.3	1.00	0.996
5	*α-TUB*	*Alpha –Tubulin*	XM_002466490	Cytoskeleton structure protein	GAGGGTGAGTTCTCTGAGGCC CTCCCTCATCTCCATCCTCGC	103	85.2	1.00	0.997
6	*EF-P*	*Elongation Factor P*	XM_002463651	Peptide bond synthesis	TGAAGCGGGTGAGAAGATTGT AGCCAAATCATACTCGCCCA	114	80.3	1.00	0.999
7	*6PGDH*	*6-Phosphogluconate dehydrogenase*	XM_002449451	Involved in the pentose phosphate pathway	CACACGGAATGGACCAAGCTG AGAGAAATCACCCAGAGCAGCA	91	85.3	0.97	0.997
8	*SAMDC*	*S-adenosylmethionine decarboxylase*	XM_002446695	Synthesis of polyamines	GTGGCGGACTCCTCATCTACC CCAGTTTGCGGTCCTTCACAG	132	85.1	0.98	0.996
9	*CYP*	*Peptidylprolyl Isomerase*	XM_002453800	*Cis*-trans isomerization of prolineimidic peptide bonds	GTATCTGTGCTCGCCGTCTCT TTCACCCAACTCCTCAACCCC	108	81.1	1.01	0.989
10	*MDH*	*Malate dehydrogenase*	XM_002467034	Citric acid cycle and gluconeogenesis	TGCAGTGGTGGTGAATGGAA GCGTCTTCTCTTCCGACAGC	103	82.5	1.01	0.972
11	*ADP-RF*	*ADP-Ribosylation Factor*	XM_002441244	Regulators of vesicular traffic and actin remodeling	GTCTGTCGGATGTGGGGATGT CACAGCACACAGTCGGACATG	136	84.5	0.96	0.998
12	*β-TUB*	*Beta Tubulin*	XM_002439758	Cytoskeleton structure protein	GCGTGTGAGTCATCCGTTCAC CCGCCTCAGTGAACTCCATCT	98	84.3	1.03	0.999
13	*PP2A*	*Serine/threonine-Protein Phosphatase*	XM_002453490	Control the specific dephosphorylation	AACCCGCAAAACCCCAGACTA TACAGGTCGGGCTCATGGAAC	138	82.7	0.97	0.991
14	*EIF4α*	*Eukaryotic Initiation Factor 4A*	XM_002451491	Eukaryotic translation	CAACTTTGTCACCCGCGATGA TCCAGAAACCTTAGCAGCCCA	144	84.8	1.02	0.993
15	*PP2C*	*Protein Phosphatase 2C*	XM_002468507	Signal transduction	GTAACCCTTCCCGCGAAATCC GCGCTACACTTTGCTGCTTTT	140	85.4	0.94	0.992

Details of candidate reference genes, NCBI accession number, their putative functions, primer sequences, product size and amplicon characteristics, i.e., melting temperature, PCR efficiency and regression coefficient values.

Quantitative Real-Time RT-PCR (qPCR)

The qPCR reactions were performed using a Realplex Real-Time PCR system (Eppendorf, Germany) and SYBR Green mix (Bioline) in 96 well optical reaction plates (Axygen, USA) sealed with ultra-clear sealing film (Platemax). The reactions were performed in a 10 µl total volume containing 5 µl of 2x SensiMix SYBR No ROX mix (Bioline), 400 nM of each primer, 1.0 µl of diluted cDNA and nuclease-free water. The reaction conditions were 95°C for 2 min, followed by 40 cycles of 15 s at 95°C and 30 s at 62°C with fluorescent signal recording. After amplification, melt curves were generated for each reaction to ensure specific amplification. All qPCR reactions, including the non-template control, were performed in biological and technical triplicates. The mean values obtained from the nine values (triplicates of each biological triplicate) were used to calculate the final quantification cycle values (Cq).

Data Analysis

The quantitative cycle (Cq) values were recorded using the RT-PCR system default settings in which the baseline was automatically corrected and threshold values were estimated using the noise band mode. Statistical analysis (mean and CV) of the Cq values was performed using a Microsoft Excel 2010 spreadsheet. PCR efficiencies (*E*) for candidate genes were evaluated using the dilution series method and pooled cDNA samples. The 12-fold diluted pooled cDNA sample was used for 2-fold serial dilutions. Five serially diluted cDNA samples were used as templates for the construction of standard curves for each primer pair using the above PCR composition and conditions. Standard curves were generated using linear regression based on the quantitative cycle (Cq) values for the dilution series. The correlation coefficients (R^2) and slope values were obtained from the standard curves, and the PCR amplification efficiencies (*E*) were calculated according to the following equation: $E = (10^{-1/slope}-1)$.

Statistical Tools for Normalization

The genEX Professional software geNorm and NormFinder algorithms (MultiD Analyses AB, Sweden) were used to identify and analyze the stably expressed gene (s) in diverse experimental samples. The raw Cq values for each gene were corrected according to their PCR efficiencies and then converted into relative quantities. The mean values for the biological replicates were used as the input data for the geNorm and NormFinder analyses. The geNorm program calculates the candidate gene expression stability (M value) using pairwise comparisons and stepwise exclusion (Vandesompele et al., 2002). Pairwise variation analysis was performed to identify the optimal number of reference genes required for normalization in each sample set using geNorm and qBase plus software (v: 2.4; Biogazelle, Belgium) (Hellemans et al., 2007). NormFinder also measures candidate reference gene expression stability using stepwise exclusion by estimating the intra- and inter-group variation. A low stability value indicates low combined variation and high expression stability (Andersen et al., 2004). RefFinder is a web-based tool (http://www.leonxie.com/referencegene.php) that integrates the four major computational programs currently available [geNorm (Vandesompele et al., 2002), NormFinder (Andersen et al., 2004), BestKeeper (Pfaffl et al., 2004), and the comparative ΔCt method (Silver et al., 2006)], and calculates the geometric mean for comprehensive ranking.

SbHSF5 and *SbHSF13* Genes Expression

Sorghum heat shock transcription factors *SbHSF5* and *SbHSF13* were selected as target genes for validating stabilities of the most and least stable reference genes by quantifying the gene expression levels in various experimental samples. Sample collections and experiments were performed as described above. *SbHSF5* and *SbHSF13* gene expression levels were normalized using the two most stable candidate reference genes, *PP2A* and *EIF4α*, as well as the two least stable reference genes in the all sample group, *UBC1* and *GAPDH*, individually and in combination. For reference gene validations, root tissues of variety Parbani Moti, and leaf tissues of five sorghum varieties under water stress conditions were used for estimating relative expression levels of *SbHSF5* and *SbHSF13* using REST software (Pfaffl et al., 2002).

RESULTS

Primer Specificity and PCR Efficiency Calculations

The candidate reference genes selected for this study represent various functional classes and gene families. Primer pairs for 15 candidate reference genes were used for qPCR amplification of sorghum cDNA and yielded single PCR products of the expected sizes (Figure 1A), as well as single melting curve peaks (Figure 1B). The amplification products were sequenced, and verification using the NCBI database BLASTN algorithm revealed 100% matches, demonstrating the gene specificity of the qPCR primer pairs. While the amplification efficiencies (*E*) of the candidate reference genes ranged from 0.94 (*PP2C*) to 1.03 (β-*TUB*), the regression coefficient values (R^2) ranged from 0.972 (*UBC1*) to 0.999 (*FE-P* and β-*TUB*) (Table 1).

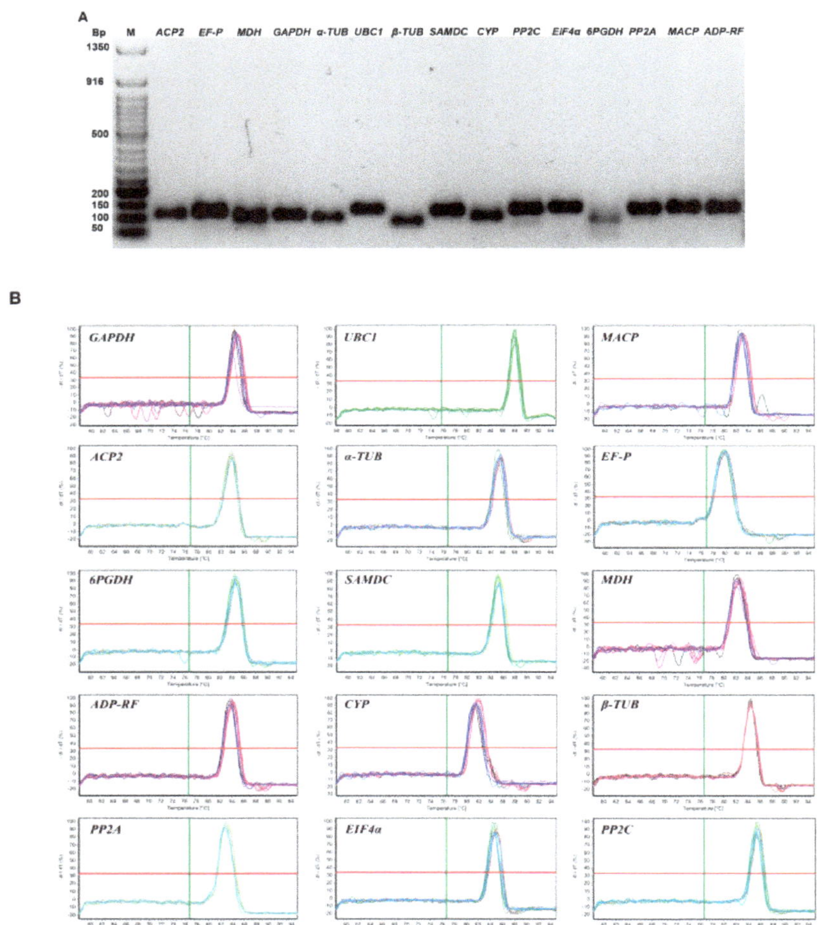

Figure 1: Specificity of primer pairs for qPCR amplification. (A) Agarose gel (2.4%) electrophoresis showing PCR products of the expected sizes for 15 candidate genes. M: 50 bp DNA marker (NEB). (B) Dissociation curves of 15 candidate reference genes under various experimental conditions, each showing a single peak.

Sorghum Reference Gene Expression Analysis

A total of 15 candidate reference genes were selected for qPCR normalization. A SYBR Green-based qPCR assay was used for transcriptional profiling of the 15 candidate reference genes in 10 samples, including five from abiotic stress conditions (water stress, salt, ABA, cold and heat shock) and five different tissues (leaves, roots, panicles, mature seeds and seedlings). Candidate reference gene transcript levels were determined using the quantification cycle (Cq) values, and these genes varied in their abundance (Figure2). The mean candidate gene Cq values ranged from 17.70 to 28.92, with most falling between 22 and 25.*GAPDH* exhibited the lowest mean Cq value (17.70), indicating that *GAPDH* was the most highly expressed, while *PP2C, ACP2*, α-*TUB, 6PGDH, MDH, ELFP, EIF4*α, and *PP2A* were moderately expressed (mean Cq 22.31–24.14), whereas β-*TUB* (mean Cq 26.69) and *UBC1* (mean Cq 28.91) were expressed at low levels. Across the tested samples, the *GAPDH* gene showed the least variation in expression levels (coefficient of variation 2.59%), while α-*TUB* was the most variable (13.53%). The degree of variation in reference gene expression across the tested samples is shown in Figure 2. These results indicated that none of the reference gene expression levels were constant and that they varied from one assay to another.

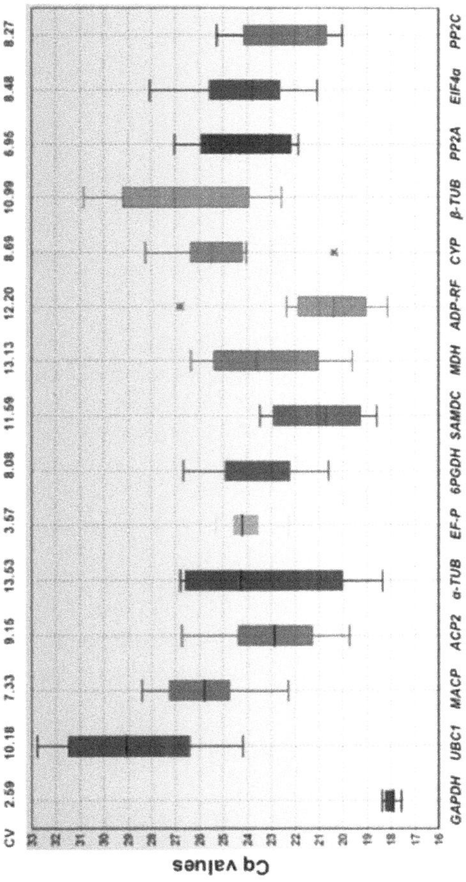

Figure 2: Expression levels of 15 candidate reference genes in all experimental samples displaying the Ct distribution for each candidate reference gene in all tested samples. Whiskers represent the maximum and minimum Ct values, and the line across the box indicates the median value, while the asterisks marks outliers. The coefficient of variance (CV) for each gene is given as a percentage. The x-axis represents the genes and the y-axis represents the Cq values.

Reference Gene Ranking and Expression Stability Analysis

The stability values for each reference gene or combination of reference genes may vary from one experimental set to another. The orders of candidate gene stability ranking for the three sample sets were determined separately to identify the most stable reference genes using three different statistical tools, geNorm, NormFinder and RefFinder.

GeNorm Analysis

GeNorm software was used to determine the expression stability of the 15 candidate sorghum reference genes; for this analysis, lower M values indicate greater gene expression stability. The software recommends a cutoff M value of 1.5 to identify sets of reference genes that are stably expressed. Figure 3displays the expression stability rankings for the tested candidate genes; the lowest M value was calculated for the *PP2A* and *EIF4*α pair (M = 0.76) and corresponded to the most stable expression in the all sample group, whereas the M values for *GAPDH* and *UBC1* were considerably higher than for the remaining candidate genes (Figure 3). For the abiotic stress treatments, *PP2A* and *CYP* were the most stable reference genes, while *UBC1* and β-*TUB* were the least stable (Figure 3). In the tissue group,*SAMDC* and *PP2A* were the most stable candidate genes while *CYP* and *GAPDH* were the least stable (Figure 3). The optimal numbers of reference genes required for gene expression normalization for the various sample groups were determined using geNorm. As shown in Figure 4, the V2/3 values of all three-sample sets were below the threshold value of 0.15 (0.144 for all pooled samples, 0.112 for the abiotic treatments, and 0.14 for the various tissue sample sets), indicating that two reference genes were sufficient for sorghum gene expression data normalization.

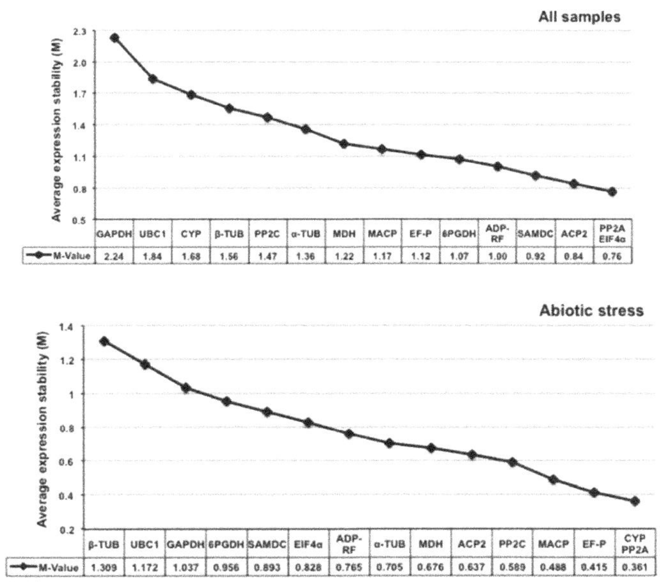

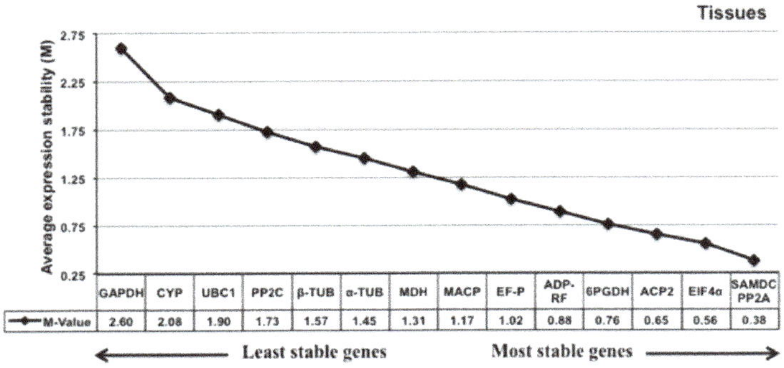

Figure 3: geNorm expression stability and ranking of the 15 candidate reference genes in various sample sets. The cutoff M value was set at 1.5; a lower M value indicates greater stability and the largest value indicates the least stable reference gene. The direction of the arrow indicates the most and least stable reference genes. The most stable genes are listed on the right and the least stable genes are listed on the left.

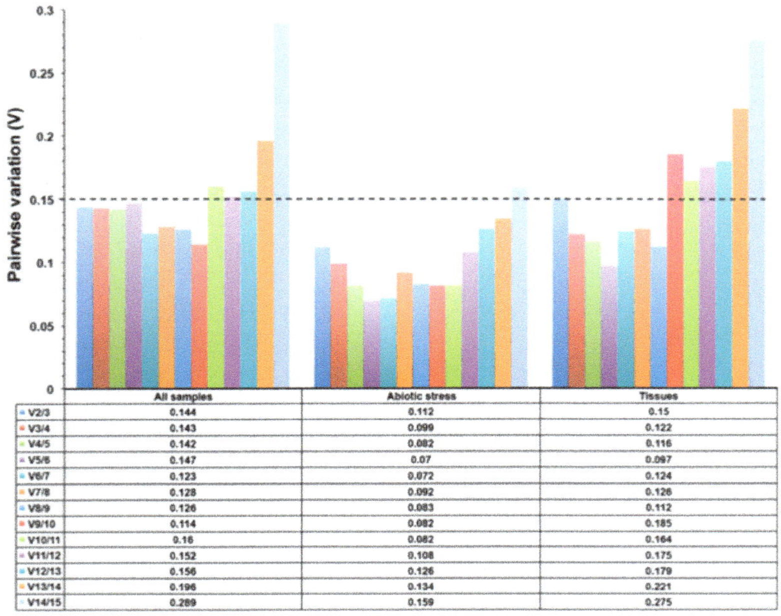

Figure 4: Optimal number of reference genes required for accurate normalization in all three experimental groups using the geNorm tool. The pairwise variation (Vn/Vn+1) was calculated by geNorm tool to determine the minimum number of reference genes for accurate normalization in each experimental set. The cutoff value was 0.15, below which additional reference genes are not necessary for gene expression normalization. The dotted line indicates the optimal number of reference genes.

NormFinder Analysis

NormFinder rankings differed slightly from the geNorm rankings (Figure 5). Both tools ranked *PP2A* and*EIF4α* as being the most stable genes in all the sample sets, while *GAPDH* and *UBC1* were the least stable genes (Figure 5); in contrast, α-*TUB* and *PP2A* emerged as the most stably expressed genes in the abiotic stress sample set, while geNorm ranked α-*TUB* eighth (Figure 5). For the tissue sample set, *EIF4α* and*PP2A* reference genes occupied the top positions, while geNorm ranked *EIF4α* third (Figure 5). *GAPDH*and *UBC1* were also the least stable reference genes in the abiotic stress and tissue sample sets (Figure 5).

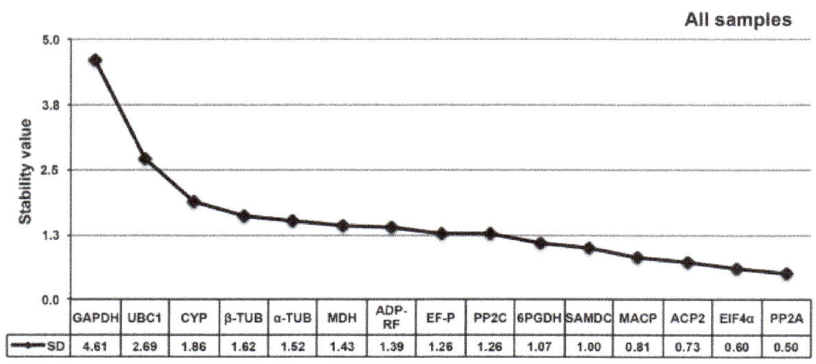

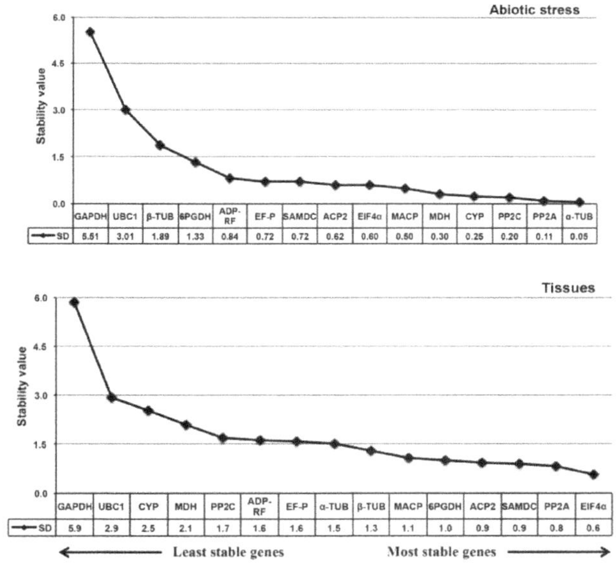

Figure 5: NormFinder expression stability values and candidate reference gene ranking in *Sorghum bicolor* **samples**. Lower values indicate greater stability and larger values indicate the least stable reference genes. The direction of the arrow indicates the most and least stable reference genes. The most stable genes are listed on the right and the least stable are listed on the left.

RefFinder Analysis

The comprehensive ranking of the candidate reference genes for all three experimental sample sets using RefFinder was highly consistent with the findings of geNorm and NormFinder. RefFinder analysis revealed that *PP2A* and *EIF4α* were the most stable in all the sample set (Table 2). Similarly, *PP2A* and*CYP* were highly stable under abiotic stress conditions, while *EIF4α* and *PP2A* were the two most stable genes for the tissue sample set (Table 2). Comprehensive ranking revealed that *GAPDH* was the least stable gene in the all sample and tissue sample sets, while β-*TUB* was the least stable in the abiotic sample set. *UBC1* was the second most stable gene in all three-sample sets (Table 2). Comprehensive ranking for the individual abiotic stress treatments revealed a pattern similar to that of the abiotic stress sample set in which *PP2A* was second for heat, salt and water stresses. Similarly, *EF-P* remained in the third position for the ABA and clod stresses (Supplementary Table 1).

Table 2: The stability ranking of 15 candidate *Sorghum bicolor* **reference genes in three experimental sample sets calculated by delta Ct (DCT), geNorm (GN), NormFinder (NF), BestKeeper (BK) and RefFinder (RF) algorithms**

Rank	All Samples					Abiotic stress					Tissues				
	DCT	GN	NF	BK	RF	DCT	GN	NF	BK	RF	DCT	GN	NF	BK	RF
1	*PP2A*	*PP2A*	*PP2A*	*EF-P*	*PP2A*	*PP2A*	*CYP*	*PP2A*	*GAPDH*	*PP2A*	*EIF4α*	*SAMDC*	*EIF4A*	*EF-P*	*EIF4α*
2	*ACP2*	*EIF4α*	*EIF4α*	*MACP*	*EIF4α*	*CYP*	*PP2A*	*CYP*	*EF-P*	*CYP*	*SAMDC*	*PP2A*	*PP2A*	*PP2C*	*PP2A*
3	*EIF4α*	*ACP2*	*ACP2*	*PP2C*	*ACP2*	*PP2C*	*EF-P*	*α-TUB*	*PP2A*	*EF-P*	*PP2A*	*EIF4α*	*SAMP*	*CYP*	*SAMDC*
4	*MACP*	*SAMDC*	*MACP*	*6PGDH*	*MACP*	*EF-P*	*MACP*	*MDH*	*CYP*	*PP2C*	*ACP2*	*ACP2*	*ACP2*	*6PGDH*	*ACP2*
5	*SAMDC*	*ADP-RF*	*SAMDC*	*CYP*	*EF-P*	*MDH*	*PP2C*	*PP2C*	*EIF4α*	*α-TUB*	*6PGDH*	*6PGDH*	*6PGDH*	*MACP*	*6PGDH*
6	*6PGDH*	*6PGDH*	*6PGDH*	*EIF4α*	*6PGDH*	*α-TUB*	*ACP2*	*EF-P*	*MACP*	*MACP*	*MACP*	*ADP-RF*	*MACP*	*ACP2*	*EF-P*
7	*EF-P*	*EF-P*	*PP2C*	*PP2A*	*SAMDC*	*MACP*	*MDH*	*MACP*	*α-TUB*	*MDH*	*ADP-RF*	*EF-P*	*β-TUB*	*EIF4α*	*MACP*
8	*ADP-RF*	*MACP*	*EF-P*	*ACP2*	*PP2C*	*ACP2*	*α-TUB*	*ACP2*	*ACP2*	*GAPDH*	*EF-P*	*MACP*	*A-TUB*	*UBC1*	*PP2C*
9	*MDH*	*MDH*	*ADP-RF*	*ADP-RF*	*ADP-RF*	*EIF4α*	*ADP-RF*	*EIF4α*	*MDH*	*ACP2*	*β-TUB*	*MDH*	*ELFP*	*PP2A*	*ADP-RF*
10	*PP2C*	*α-TUB*	*MDH*	*SAMDC*	*MDH*	*ADP-RF*	*EIF4α*	*SAMDC*	*ADP-RF*	*EIF4α*	*α-TUB*	*α-TUB*	*ADPRF*	*SAMDC*	*CYP*
11	*α-TUB*	*PP2C*	*α-TUB*	*MDH*	*CYP*	*SAMDC*	*SAMDC*	*ADP-RF*	*PP2C*	*ADP-RF*	*MDH*	*B-TUB*	*PP2C*	*β-TUB*	*β-TUB*
12	*β-TUB*	*β-TUB*	*β-TUB*	*UBC1*	*α-TUB*	*6PGDH*	*6PGDH*	*6PGDH*	*SAMDC*	*SAMDC*	*PP2C*	*PP2C*	*MDH*	*ADP-RF*	*α-TUB*
13	*CYP*	*CYP*	*CYP*	*α-TUB*	*β-TUB*	*GAPDH*	*GAPDH*	*GAPDH*	*6PGDH*	*6PGDH*	*CYP*	*UBC1*	*CYP*	*MDH*	*MDH*
14	*UBC1*	*UBC1*	*UBC1*	*β-TUB*	*UBC1*	*UBC1*	*UBC1*	*UBC1*	*UBC1*	*UBC1*	*UBC1*	*CYP*	*UBC1*	*α-TUB*	*UBC1*
15	*GAPDH*	*GAPDH*	*GAPDH*	*GAPDH*	*GAPDH*	*β-TUB*	*β-TUB*	*β-TUB*	*β-TUB*	*β-TUB*	*GAPDH*	*GAPDH*	*GAPDH*	*GAPDH*	*GAPDH*

Validation of the Best and Least Ranked Sorghum Reference Genes

To validate the selected candidate reference genes, the expression patterns of *SbHSF5* and *SbHSF13* were assayed in root tissues, water-stressed samples and different varieties (Figures 6A–C). In sorghum,*SbHSF5* and *SbHSF13* genes have been shown to express in different tissues and induced by abiotic stress treatments (Nagaraju et al., 2015). In this study, the two most stable reference genes from all the sample set, *PP2A* and *EIF4α*, were used for normalization: in roots under normal growth conditions, *SbHSF5* and*SbHSF13* expression were not significantly different. In contrast, the *SbHSF5* and *SbHSF13* expression patterns were very different when *GAPDH* and *UBC1*, the least stable genes from all the sample set, were used for normalization. Normalization with *UBC1* indicated a 2- to 3-fold increase in expression, while normalization with *GAPDH* and *GAPDH+UBC1* indicated downregulation of both genes in root tissues under normal conditions (Figure 6A). To further validate the selected reference genes, the transcript levels were quantified in five sorghum varieties under water stress conditions (Figure 6C). For water stress samples, *SbHSF5* and *SbHSF13* expression levels increased1 to 12-fold under water stress conditions when normalized with the two most stable reference genes from all the sample set (*PP2A* and *EIF4α*; Figures 6B,C). As expected, normalization with the least stable genes (*GAPDH* and *UBC1*), either alone or in combination with *UBC1* resulted in discrepancies for expression of both *SbHSF5* and *SbHSF13* gene expression (Figures 6B,C).

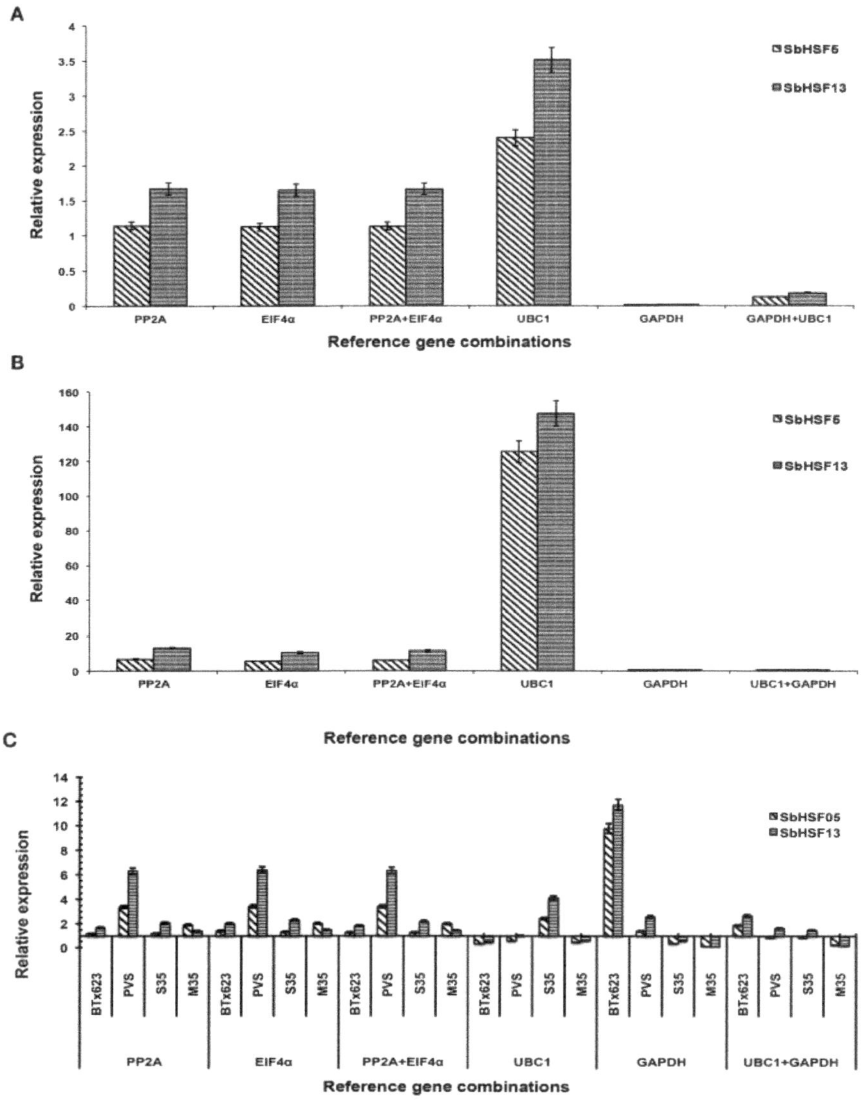

Figure 6: Relative expression of *SbHSF5* **and** *SbHSF13* **genes in root tissues (A) leaves during water stress treatment (B) selected varieties (C) using different reference genes selected by the RefFinder tool**. *SbHSF5* and *SbHSF13* expression levels were normalized in a single and combined manner with either a stable or unstable reference genes. All samples were analyzed in triplicate, in three independent experiments.

DISCUSSION

Accurate gene expression primarily relies on the stable expression of reference genes under diverse experimental conditions; however, thus far, no single reference gene has been shown to be stably expressed across experimental conditions (Olsvik et al., 2008; Cruz et al., 2009). The stability value of the most stable reference gene or the best combination of genes may vary from one experimental setup to another, meaning that the reference genes used for normalization should be validated under certain experimental conditions by statistically and experimentally. In the present study, 15 candidate reference genes were evaluated for expression stability among diverse experimental samples including various tissues and abiotic stress conditions. Expression analysis revealed that all 15 candidate reference genes varied significantly across the 10 tested samples; therefore, reference genes for optimal normalization were chosen from a set of candidate genes for each experiment either alone or in combination. To determine the comprehensive ranking of candidate reference genes in a sample set, we also used RefFinder, which considers the results of the four algorithms (geNorm, NormFinder, BestKeeper and the ΔCq method) together. Based on their stability rankings, various candidate genes have been proposed for their suitability as reference genes depending on the experimental conditions, implying that the reference genes used for abiotic stress studies in sorghum might not be suitable for gene expression analysis across genotypes and species.

In all the sample set, all three algorithms identified two reference genes, *PP2A* and *EIF4α*, as the two most stable genes. The pairwise variation calculated by geNorm also indicated that the use of two genes would be sufficient for normalization (Figure 4). Based on these results, it is inferred that *PP2A* and *EIF4α* would be appropriate for qPCR data normalization for the sorghum all sample set. For the abiotic stress treatment sample set, the pairwise variation indicated that the top ranking two genes (*PP2A* and *CYP*) could be used for normalization (Figure 4). Similarly, *PP2A* and *EIF4α* would be sufficient for the normalization of the tissue sample set data (Figure 4). *PP2A* was identified as one of the two most stable genes in all three studied sample sets

(Table 2). This result is in agreement with previous studies of maturing *B. napus* embryos (Chen et al., 2010), multiple stress conditions (Wang et al., 2014), development and under various environmental conditions in *A. thaliana* (Czechowski et al., 2005) and*Caragana intermedia* (Zhu et al., 2013), as well as for virus-infected *Nicotiana benthamiana* (Liu et al., 2012) and *Z. mays* (Zhang et al., 2013). Furthermore, *PP2A* was the most stably expressed gene among various Striga life stages (Fernández-Aparicio et al., 2013). *EIF4α* was second most stably expressed gene in the all sample and tissue sample sets (Table 2). Similar results were observed in *Carica papaya*, where*EIF4α* expression was stable under most of the experimental conditions tested (Zhu et al., 2012). *EIF4α* was also reported as being a stably expressed gene in sexual and apomictic *Brachiaria brizantha* (Silveira et al., 2009), and in sexual and apomictic accessions of *Cenchrus ciliaris* (Simon et al., 2013). In the present study, *CYP* was the third most stable gene; it was stably expressed under abiotic stress treatments and in the tissue sample set (Table 2). The results obtained in the present study are in agreement with our previous studies in peanuts, where *CYP* was the most stably expressed gene during vegetative stages and under abiotic stress conditions (Reddy et al., 2013) and was also stably expressed in various tissues of *G. max* (Jian et al., 2008), *Vicia faba* (Gutierrez et al., 2011) and salt-stressed *S. tuberosusm* (Nicot et al., 2005). The *CYP* gene was moderately stable in developing and germinating *G. max* seeds (Li et al., 2012) and in bananas under various experimental conditions (Chen et al., 2011). In contrast, our recent study in*Cicer arietinum* indicated that *CYP* was the least stably expressed gene under various experimental conditions (Reddy et al., 2016), consistent with a previous study of grapevine berry development (Reid et al., 2006). In conclusion, *PP2A, EIF4α*, and *CYP* are recommended as suitable reference genes for the qPCR-based normalization of gene expression in sorghum under abiotic stress and in various tissues.

To further validate the candidate reference genes selected from the RefFinder comprehensive ranking, the expression levels of two sorghum heat shock transcription factors, *SbHSF5* and *SbHSF13*, were assessed. Plant heat shock transcription factors play major roles in higher plants under various abiotic stresses. In sorghum, *SbHSF5* and *SbHSF13* are expressed under abiotic stress conditions: *SbHSF5* is moderately expressed under cold stress and upregulated under drought stress, and *SbHSF13* is moderately expressed under cold and heat stress and upregulated under drought and high salinity stress (Nagaraju et al., 2015). In the present study, *SbHSF5* and *SbHSF13* expression levels were upregulated under water stress conditions and no significant changes were observed in root tissue gene expression levels when normalized to the most stable reference genes, *PP2A* and *EIF4α* (Figures 6A,B). While, *SbHSF5* and*SbHSF13* genes significantly upregulated under water stress conditions in leaf tissues of all five sorghum genotypes with relatively higher expression when normalized with the sable reference genes (*PP2A* and*EIF4α*), the normalization was obscured when least stable reference gene (s) (*UBC1* and *GAPDH*) were used (Figures 6B,C). The upregulation of *SbHSF5* and *SbHSF13* under this study is in agreement with another recent study in sorghum (Nagaraju et al., 2015) confirming the stability validation of the selected candidate reference genes and demonstrated the disadvantages of using unstable reference genes for normalization.

CONCLUSION

This study reports on the selection and validation of stable reference genes for qPCR-based gene expression studies in sorghum in response to various abiotic stresses and in different tissues. Three major statistical tools, geNorm, NormFinder and RefFinder, were used to analyze the suitability of reference genes. Some slight differences were observed between the statistical tools with regards to the selected stable reference genes. Overall, *PP2A* and *EIF4α* are the most stable candidate reference genes, and *UBC1*and *GAPDH* were found to be least stable. The use of two reference genes would be optimal for the normalization of sorghum gene expression levels in various tissues and abiotic stress conditions. The most and least stable reference genes were further validated for their suitability as reference genes by normalizing *SbHSF5* and *SbHSF13* expression levels under various experimental conditions across five genotypes. This work will benefit future studies of gene expression in related experiments using different tissues and abiotic stress conditions in *S. bicolor* and related crops.

AUTHOR CONTRIBUTIONS

PS, KS, and VV designed the experiments. PS, DS, SK, and PB performed the experiments and data analysis. PS, DS and KS drafted the manuscript.

FUNDING

Financial support to PS through the INSPIRE Faculty Award (IFA11-LSPA-06) and the Young Scientist Scheme SB/YS/LS-12/2013 provided by the Department of Science and Technology, Government of India, New Delhi is gratefully acknowledged. This work was performed as part of the CGIAR Research Program on Dryland Cereals.

CONFLICT OF INTEREST STATEMENT

The authors declare that the research was conducted in the absence of any commercial or financial relationships that could be construed as a potential conflict of interest.

ACKNOWLEDGMENTS

PS acknowledges INSPIRE Faculty Award (IFA11-LSPA-06) and the Young Scientist Scheme (SB/YS/LS-12/2013) from the Department of Science and Technology, Government of India, New Delhi. This work was performed as part of the CGIAR Research Program on Dryland Cereals. ICRISAT is a member of the CGIAR Consortium.

REFERENCES

1. Andersen, C. L., Jensen, J. L., and Ørntoft, T. F. (2004). Normalization of real-time quantitative reverse transcription-PCR data: a model-based variance estimation approach to identify genes suited for normalization, applied to bladder and colon cancer data sets.*Cancer Res.* 64, 5245–5250. doi: 10.1158/0008-5472.CAN-04-0496

2. Bustin, S. A. (2002). Quantification of mRNA using real-time reverse transcription PCR (RT-PCR): trends and problems. *J. Mol. Endocrinol.* 29, 23–39. doi: 10.1677/jme.0.0290023

3. Bustin, S. A., Beaulieu, J. F., Huggett, J., Jaggi, R., Kibenge, F. S. B., Olsvik, P. A., et al. (2010). MIQE precis: practical implementation of minimum standard guidelines for fluorescence-based quantitative real-time PCR experiments. *BMC Mol. Biol.* 11:74. doi: 10.1186/1471-2199-11-74

4. Bustin, S. A., Benes, V., Garson, J. A., Hellemans, J., Huggett, J., Kubista, M., et al. (2009). The MIQE guidelines: minimum information for publication of quantitative real-time PCR experiments. *Clin. Chem.* 55, 611–622. doi: 10.1373/clinchem.2008.112797

5. Calviño, M., and Messing, J. (2012). Sweet sorghum as a model system for bioenergy crops. *Curr. Opin. Biotechnol.* 23, 323–329. doi: 10.1016/j.copbio.2011.12.002

6. Chen, L., Zhong, H. Y., Kuang, J. F., Li, J. G., Lu, W. J., and Chen, J. Y. (2011). Validation of reference genes for RT-qPCR studies of gene expression in banana fruit under different experimental conditions. *Planta* 234, 377–390. doi: 10.1007/s00425-011-1410-3

7. Chen, X., Truksa, M., Shah, S., and Weselake, R. J. (2010). A survey of quantitative real-time polymerase chain reaction internal reference genes for expression studies in *Brassica napus*. *Anal. Biochem.* 405, 138–140. doi: 10.1016/j.ab.2010.05.032

8. Cook, D., Rimando, A. M., Clemente, T. E., Schröder, J., Dayan, F. E., Nanayakkara, N. P., et al. (2010). Alkylresorcinol synthases expressed in *Sorghum bicolor* root hairs play an essential role in the biosynthesis of the allelopathic benzoquinone sorgoleone. *Plant Cell* 22, 867–887. doi: 10.1105/tpc.109.072397

9. Cruz, F., Kalaoun, S., Nobile, P., Colombo, C., Almeida, J., Barros, L. M. G., et al. (2009). Evaluation of coffee reference genes for relative expression studies by quantitative real-time RT-PCR. *Mol. Breed.* 23, 607–616. doi: 10.1007/s11032-009-9259-x

10. Czechowski, T., Stitt, M., Altmann, T., Udvardi, M. K., and Scheible, W. R. (2005). Genome-wide identification and testing of superior reference genes for transcript normalization in *Arabidopsis*. *Plant Physiol.* 139, 5–17. doi: 10.1104/pp.105.063743

11. Demidenko, N. V., Logacheva, M. D., and Penin, A. A. (2011). Selection and validation of reference genes for quantitative real-time PCR in buckwheat (*Fagopyrum esculentum*) based on transcriptome sequence data. *PLoS ONE* 6:e19434. doi: 10.1371/journal.pone.0019434

12. Dheda, K., Huggett, J. F., Chang, J. S., Kim, L. U., Bustin, S. A., Johnson, M. A., et al. (2005). The implications of using an inappropriate reference gene for real-time reverse transcription PCR data normalization. *Anal. Biochem.* 344, 141–143. doi: 10.1016/j.ab.2005.05.022

13. Dugas, D. V., Monaco, M. K., Olsen, A., Klein, R. R., Kumari, S., Ware, D., et al. (2011). Functional annotation of the transcriptome of*Sorghum bicolor* in response to osmotic stress and abscisic acid. *BMC Genomics* 12:514. doi: 10.1186/1471-2164-12-514

14. Eisenberg, E., and Levanon, E. Y. (2013). Human housekeeping genes, revisited. *Trends Genet.* 29, 569–574. doi: 10.1016/j.tig.2013.05.010

15. Expósito-Rodríguez, M., Borges, A. A., Borges-Perez, A., and Pérez, J. A. (2008). Selection of internal control genes for quantitative real-time RT-PCR studies during tomato development process. *BMC Plant Biol.* 8:131. doi: 10.1186/1471-2229-8-131

16. FAO. (2011). *FAOSTAT Database.* Available online at: http://faostat.fao.org

17. Fernández-Aparicio, M., Huang, K., Wafula, E. K., Honaas, L. A., Wickett, N. J., Timko, M. P., et al. (2013). Application of qRT-PCR and RNA-Seq analysis for the identification of housekeeping genes useful for normalization of gene expression values during *Striga hermonthica* development. *Mol. Biol. Rep.* 40, 3395–3407. doi: 10.1007/s11033-012-2417-y

18. Gelli, M., Duo, Y. C., Konda, A. R., Zhang, C., Holding, D., and Dweikat, I. (2014). Identification of differentially expressed genes between sorghum genotypes with contrasting nitrogen stress tolerance by genome-wide transcriptional profiling. *BMC Genomics* 15:179. doi: 10.1186/1471-2164-15-179

19. Gimeno, J., Eattock, N., Van Deynze, A., and Blumwald, E. (2014). Selection and validation of reference genes for gene expression analysis in switchgrass (*Panicum virgatum*) using quantitative real-time RT-PCR. *PLoS ONE* 9:e91474. doi: 10.1371/journal.pone.0091474

20. Gutierrez, N., Gimenez, M. J., Palomino, C., and Avila, C. M. (2011). Assessment of candidate reference genes for expression studies in*Vicia faba* L. by real-time quantitative PCR. *Mol. Breed.* 28, 13–24. doi: 10.1007/s11032-010-9456-7

21. Hellemans, J., Mortier, G., De Paepe, A., Speleman, F., and Vandesompele, J. (2007). qBase relative quantification framework and software for management and automated analysis of real-time quantitative PCR data. *Genome Biol.* 8:r19. doi: 10.1186/gb-2007-8-2-r19

22. Ishikawa, R., Aoki, M., Kurotani, K., Yokoi, S., Shinomura, T., Takano, M., et al. (2011). Phytochrome B regulates Heading date 1 (Hd1)-mediated expression of rice florigen Hd3a and critical day length in rice. *Mol. Genet. Genomics* 285, 461–470. doi: 10.1007/s00438-011-0621-4

23. Iskandar, H. M., Simpson, R. S., Casu, R. E., Bonnett, G. D., Maclean, D. J., and Manners, J. M. (2004). Comparison of reference genes for quantitative real-time polymerase chain reaction analysis of gene expression. *Plant Mol. Biol. Rep.* 22, 325–337. doi: 10.1007/BF02772676

24. Jain, M., Chourey, P. S., Li, Q. B., and Pring, D. R. (2008). Expression of cell wall invertase and several other genes of sugar metabolism in relation to seed development in sorghum (*Sorghum bicolor*). *J. Plant Physiol.* 165, 331–344. doi: 10.1016/j.jplph.2006.12.003

25. Jain, M., Nijhawan, A., Tyagi, A. K., and Khurana, J. P. (2006). Validation of housekeeping genes as internal control for studying gene expression in rice by quantitative

real-time PCR. *Biochem. Biophys. Res. Commun.* 345, 646–651. doi: 10.1016/j.bbrc.2006.04.140

26. Ji, Y., Tu, P., Wang, K., Gao, F., Yang, W., Zhu, Y., et al. (2014). Defining reference genes for quantitative real-time PCR analysis of anther development in rice. *Acta Biochim. Biophys. Sin. (Shanghai).* 46, 305–312. doi: 10.1093/abbs/gmu002

27. Jian, B., Liu, B., Bi, Y. R., Hou, W. S., Wu, C. X., and Han, T. F. (2008). Validation of internal control for gene expression study in soybean by quantitative real-time PCR. *BMC Mol. Biol.* 9:59. doi: 10.1186/1471-2199-9-59

28. Kebrom, T. H., and Mullet, J. E. (2015). Photosynthetic leaf area modulates tiller bud outgrowth in sorghum. *Plant Cell Environ.* 38, 1471–1478. doi: 10.1111/pce.12500

29. Koegel, S., Ait Lahmidi, N., Arnould, C., Chatagnier, O., Walder, F., Ineichen, K., et al. (2013). The family of ammonium transporters (AMT) in *Sorghum bicolor*: two AMT members are induced locally, but not systemically in roots colonized by arbuscular mycorrhizal fungi. *New Phytol.* 198, 853–865. doi: 10.1111/nph.12199

30. Kumar, K., Muthamilarasan, M., and Prasad, M. (2013). Reference genes for quantitative real-time PCR analysis in the model plant foxtail millet (*Setaria italica* L.) subjected to abiotic stress conditions. *Plant Cell Tissue Organ. Cult.* 115, 13–22. doi: 10.1007/s11240-013-0335-x

31. Li, J. Q., Wang, L. H., Zhan, Q. W., Liu, Y. L., Fu, B. S., and Wang, C. M. (2013). Sorghum bmr6 mutant analysis demonstrates that a shared MYB1 transcription factor binding site in the promoter links the expression of genes in related pathways. *Funct. Integr. Genomics* 13, 445–453. doi: 10.1007/s10142-013-0335-2

32. Li, J., Wang, L., Zhang, Q., and Liu, Y. (2015). Map-based cloning and expression analysis of BMR-6 in sorghum. *J. Genet.* 94, 445–452. doi: 10.1007/s12041-015-0550-9

33. Li, Q., Fan, C. M., Zhang, X. M., and Fu, Y. F. (2012). Validation of reference genes for real-time quantitative PCR normalization in soybean developmental and germinating seeds. *Plant Cell Rep.* 31, 1789–1798. doi: 10.1007/s00299-012-1282-4

34. Li, Q. F., Sun, S. S. M., Yuan, D. Y., Yu, H. X., Gu, M. H., and Liu, Q. Q. (2010). Validation of candidate reference genes for the accurate normalization of real-time quantitative rT-PCR data in rice during seed development. *Plant Mol. Biol. Rep.* 28, 49–57. doi: 10.1007/s11105-009-0124-1

35. Lin, F., Jiang, L., Liu, Y. H., Lv, Y. D., Dai, H. X., and Zhao, H. (2014). Genome-wide identification of housekeeping genes in maize. *Plant Mol. Biol.* 86, 543–554. doi: 10.1007/s11103-014-0246-1

36. Ling, H., Wu, Q., Guo, J., Xu, L., and Que, Y. (2015). Comprehensive selection of reference genes for gene expression normalization in sugarcane by real time quantitative RT-PCR. *PLoS ONE* 10:e97469. doi: 10.1371/journal.pone.0097469

37. Liu, D. S., Shi, L. D., Han, C. G., Yu, J. L., Li, D. W., and Zhang, Y. L. (2012). Validation of reference genes for gene expression studies in virus-infected *Nicotiana benthamiana* using quantitative real-time PCR. *PLoS ONE* 7:e46451. doi: 10.1371/journal.pone.0046451

38. Machado, R. D., Christoff, A. P., Loss-Morais, G., Margis-Pinheiro, M., Margis, R., and Körbes, A. P. (2015). Comprehensive selection of reference genes for quantitative gene expression analysis during seed development in *Brassica napus*. *Plant Cell Rep.* 34, 1139–1149. doi: 10.1007/s00299-015-1773-1

39. Manoli, A., Sturaro, A., Trevisan, S., Quaggiotti, S., and Nonis, A. (2012). Evaluation of candidate reference genes for qPCR in maize. *J. Plant Physiol.* 169, 807–815. doi: 10.1016/j.jplph.2012.01.019

40. Mariot, R. F., De Oliveira, L. A., Voorhuijzen, M. M., Staats, M., Hutten, R. C., Van Dijk, J. P., et al. (2015). Selection of reference genes for transcriptional analysis of edible tubers of potato (*Solanum tuberosum* L.). *PLoS ONE* 10:e0120854. doi: 10.1371/journal.pone.0120854

41. Nagaraju, M., Reddy, P. S., Kumar, S. A., Srivastava, R. K., Kavi Kishor, P. B., and Rao, D. M. (2015). Genome wide scanning and characterization of *Sorghum bicolor* L. heat shock transcription factors. *Curr. Genom.* 16, 279–291. doi: 10.2174/1389202916666150313230812

42. Nakayama, T. J., Rodrigues, F. A., Neumaier, N., Marcelino-Guimarães, F. C., Farias, J. R., De Oliveira, M. C., et al. (2014). Reference genes for quantitative real-time polymerase chain reaction studies in soybean plants under hypoxic conditions. *Genet. Mol. Res.* 13, 860–871. doi: 10.4238/2014.February.13.4

43. Nicot, N., Hausman, J. F., Hoffmann, L., and Evers, D. (2005). Housekeeping gene selection for real-time RT-PCR normalization in potato during biotic and abiotic stress. *J. Exp. Bot.* 56, 2907–2914. doi: 10.1093/jxb/eri285

44. Olsvik, P. A., Softeland, L., and Lie, K. K. (2008). Selection of reference genes for qRT-PCR examination of wild populations of Atlantic cod *Gadus morhua*. *BMC Res. Notes* 1:47. doi: 10.1186/1756-0500-1-47

45. Paolacci, A. R., Tanzarella, O. A., Porceddu, E., and Ciaffi, M. (2009). Identification and validation of reference genes for quantitative RT-PCR normalization in wheat. *BMC Mol. Biol.* 10:11. doi: 10.1186/1471-2199-10-11

46. Paterson, A. H., Bowers, J. E., Bruggmann, R., Dubchak, I., Grimwood, J., Gundlach, H., et al. (2009). The *Sorghum bicolor* genome and the diversification of grasses. *Nature* 457, 551–556. doi: 10.1038/nature07723

47. Pfaffl, M. W., Horgan, G. W., and Dempfle, L. (2002). Relative expression software tool (REST (c)) for group-wise comparison and statistical analysis of relative expression results in real-time PCR. *Nucleic Acids Res.* 30:e36. doi: 10.1093/nar/30.9.e36

48. Pfaffl, M. W., Tichopad, A., Prgomet, C., and Neuvians, T. P. (2004). Determination of stable housekeeping genes, differentially regulated target genes and sample integrity: BestKeeper - excel-based tool using pair-wise correlations. *Biotechnol. Lett.* 26, 509–515. doi: 10.1023/B:BILE.0000019559.84305.47

49. Price, H. J., Dillon, S. L., Hodnett, G., Rooney, W. L., Ross, L., and Johnston, J. S. (2005). Genome evolution in the genus Sorghum (Poaceae). *Ann. Bot.* 95, 219–227. doi: 10.1093/aob/mci015

50. Reddy, D. S., Bhatnagar-Mathur, P., Cindhuri, K. S., and Sharma, K. K. (2013). Evaluation and validation of reference genes for normalization of quantitative real-time PCR based gene expression studies in peanut. *PLoS ONE* 8:e78555. doi: 10.1371/journal.pone.0078555

51. Reddy, D. S., Bhatnagar-Mathur, P., Reddy, P. S., Sri Cindhuri, K., Sivaji Ganesh, A., and Sharma, K. K. (2016). Identification and validation of reference genes and their impact on normalized gene expression studies across cultivated and wild cicer species. *PLoS ONE*11:e0148451. doi: 10.1371/journal.pone.0148451

52. Reddy, P. S., Rao, T. S. R., Sharma, K. K., and Vadez, V. (2015a). Genome-wide identification and characterization of the aquaporin gene family in *Sorghum bicolor* (L.). *Plant Gene* 1, 18–28. doi: 10.1016/j.plgene.2014.12.002

53. Reddy, P. S., Reddy, D. S., Bhatnagar-Mathur, P., Sharma, K. K., and Vadez, V. (2015b). Cloning and validation of reference genes for normalization of gene expression studies in pearl millet [*Pennisetum glaucum* (L.) R. Br.] by quantitative real-time PCR. *Plant Gene* 1, 35–42. doi: 10.1016/j.plgene.2015.02.001

54. Reid, K. E., Olsson, N., Schlosser, J., Peng, F., and Lund, S. T. (2006). An optimized grapevine RNA isolation procedure and statistical determination of reference genes for real-time RT-PCR during berry development. *BMC Plant Biol.* 6:27. doi: 10.1186/1471-2229-6-27

55. Remans, T., Smeets, K., Opdenakker, K., Mathijsen, D., Vangronsveld, J., and Cuypers, A. (2008). Normalisation of real-time RT-PCR gene expression measurements in *Arabidopsis thaliana* exposed to increased metal concentrations. *Planta* 227, 1343–1349. doi: 10.1007/s00425-008-0706-4

56. Saha, P., and Blumwald, E. (2014). Assessing reference genes for accurate transcript normalization using quantitative real-time PCR in pearl millet [*Pennisetum glaucum* (L.) R. Br.]. *PLoS ONE* 9:e106308. doi: 10.1371/journal.pone.0106308

57. Shakoor, N., Nair, R., Crasta, O., Morris, G., Feltus, A., and Kresovich, S. (2014). A *Sorghum bicolor* expression atlas reveals dynamic genotype-specific expression profiles for vegetative tissues of grain, sweet and bioenergy sorghums. *BMC Plant Biol.* 14:35. doi: 10.1186/1471-2229-14-35

58. Shen, C., Bai, Y., Wang, S., Zhang, S., Wu, Y., Chen, M., et al. (2010). Expression profile of PIN, AUX/LAX and PGP auxin transporter gene families in *Sorghum bicolor* under phytohormone and abiotic stress. *FEBS J.* 277, 2954–2969. doi: 10.1111/j.1742-4658.2010.07706.x

59. Silveira, E. D., Alves-Ferreira, M., Guimaraes, L. A., Da Silva, F. R., and Carneiro, V. T. (2009). Selection of reference genes for quantitative real-time PCR expression studies in the apomictic and sexual grass *Brachiaria brizantha*. *BMC Plant Biol.* 9:84. doi: 10.1186/1471-2229-9-84

60. Silver, N., Best, S., Jiang, J., and Thein, S. L. (2006). Selection of housekeeping genes for gene expression studies in human reticulocytes using real-time PCR. *BMC Mol. Biol.* 7:33. doi: 10.1186/1471-2199-7-33

61. Simon, B., Conner, J. A., and Ozias-Akins, P. (2013). Selection and validation of reference genes for gene expression analysis in apomictic and sexual *Cenchrus ciliaris*. *BMC Res. Notes* 6:397. doi: 10.1186/1756-0500-6-397

62. Sinha, P., Singh, V. K., Suryanarayana, V., Krishnamurthy, L., Saxena, R. K., and Varshney, R. K. (2015). Evaluation and validation of housekeeping genes as reference for gene expression studies in pigeonpea (*Cajanus cajan*) under drought stress conditions. *PLoS ONE*10:e0122847. doi: 10.1371/journal.pone.0122847

63. Thellin, O., Zorzi, W., Lakaye, B., De Borman, B., Coumans, B., Hennen, G., et al. (1999). Housekeeping genes as internal standards: use and limits. *J. Biotechnol.* 75, 291–295. doi: 10.1016/S0168-1656(99)00163-7

64. Untergasser, A., Cutcutache, I., Koressaar, T., Ye, J., Faircloth, B. C., Remm, M., et al. (2012). Primer3-new capabilities and interfaces.*Nucleic Acids Res.* 40, e115. doi: 10.1093/nar/gks596

65. Vadez, V., and Sinclair, T. R. (2001). Leaf ureide degradation and N(2) fixation tolerance to water deficit in soybean. *J. Exp. Bot.* 52, 153–159. doi: 10.1093/jexbot/52.354.153

66. Valasek, M. A., and Repa, J. J. (2005). The power of real-time PCR. *Adv. Physiol. Educ.* 29, 151–159. doi: 10.1152/advan.00019.2005

67. Vandesompele, J., De Preter, K., Pattyn, F., Poppe, B., Van Roy, N., De Paepe, A., et al. (2002). Accurate normalization of real-time quantitative RT-PCR data by geometric averaging of multiple internal control genes. *Genome Biol.* 3:RESEARCH0034. doi: 10.1186/gb-2002-3-7-research0034

68. Vermerris, W. (2011). Survey of genomics approaches to improve bioenergy traits in maize, sorghum and sugarcane. *J. Integr. Plant Biol.*53, 105–119. doi: 10.1111/j.1744-7909.2010.01020.x

69. Walder, F., Brulé, D., Koegel, S., Wiemken, A., Boller, T., and Courty, P. E. (2015). Plant phosphorus acquisition in a common mycorrhizal network: regulation of phosphate transporter genes of the Pht1 family in sorghum and flax. *New Phytol.* 205, 1632–1645. doi: 10.1111/nph.13292

70. Wang, S. K., Bai, Y. H., Shen, C. J., Wu, Y. R., Zhang, S. N., Jiang, D. A., et al. (2010). Auxin-related gene families in abiotic stress response in *Sorghum bicolor*. *Funct. Integr. Genomics* 10, 533–546. doi: 10.1007/s10142-010-0174-3

71. Wang, Z., Chen, Y., Fang, H. D., Shi, H. F., Chen, K. P., Zhang, Z. Y., et al. (2014). Selection of reference genes for quantitative reverse-transcription polymerase chain reaction normalization in *Brassica napus* under various stress conditions. *Mol. Genet. Genomics* 289, 1023–1035. doi: 10.1007/s00438-014-0853-1

72. Wong, M. L., and Medrano, J. F. (2005). Real-time PCR for mRNA quantitation. *BioTechniques* 39, 75–85. doi: 10.2144/05391RV01

73. Yang, H. L., Liu, J., Huang, S. M., Guo, T. T., Deng, L. B., and Hua, W. (2014a). Selection and evaluation of novel reference genes for quantitative reverse transcription PCR (qRT-PCR) based on genome and transcriptome data in *Brassica napus* L. *Gene* 538, 113–122. doi: 10.1016/j.gene.2013.12.057

74. Yang, S., Murphy, R. L., Morishige, D. T., Klein, P. E., Rooney, W. L., and Mullet, J. E. (2014b). Sorghum phytochrome B inhibits flowering in long days by activating expression of SbPRR37 and SbGHD7, repressors of SbEHD1, SbCN8 and SbCN12. *PLoS ONE*9:e105352. doi: 10.1371/journal.pone.0105352

75. Yang, X., Scheffler, B. E., and Weston, L. A. (2004). SOR1, a gene associated with bioherbicide production in sorghum root hairs. *J. Exp. Bot.* 55, 2251–2259. doi: 10.1093/jxb/erh252

76. Yin, L., Wang, S., Liu, P., Wang, W., Cao, D., Deng, X., et al. (2014). Silicon-mediated changes in polyamine and 1-aminocyclopropane-1-carboxylic acid are involved in silicon-induced drought resistance in *Sorghum bicolor* L. *Plant Physiol. Biochem.* 80, 268–277. doi: 10.1016/j.plaphy.2014.04.014

77. Zhang, K., Niu, S. F., Di, D. P., Shi, L. D., Liu, D. S., Cao, X. L., et al. (2013). Selection of reference genes for gene expression studies in virus-infected monocots using quantitative real-time PCR. *J. Biotechnol.* 168, 7–14. doi: 10.1016/j.jbiotec.2013.08.008

78. Zhu, J. F., Zhang, L. F., Li, W. F., Han, S. Y., Yang, W. H., and Qi, L. W. (2013). Reference gene selection for quantitative real-time PCR normalization in *Caragana intermedia* under different abiotic stress conditions. *PLoS ONE* 8:e53196. doi: 10.1371/journal.pone.0053196

79. Zhu, X. Y., Li, X. P., Chen, W. X., Chen, J. Y., Lu, W. J., Chen, L., et al. (2012). Evaluation of new reference genes in papaya for accurate transcript normalization under different experimental conditions. *PLoS ONE* 7:e44405. doi: 10.1371/journal.pone.0044405

80. Zmienko, A., Samelak-Czajka, A., Goralski, M., Sobieszczuk-Nowicka, E., Kozlowski, P., and Figlerowicz, M. (2015). Selection of reference genes for qPCR- and ddPCR-based analyses of gene expression in Senescing Barley leaves. *PLoS ONE* 10:e0118226. doi: 10.1371/journal.pone.0118226

CHAPTER 6

TRANSCRIPTOME CHARACTERIZATION AND FUNCTIONAL MARKER DEVELOPMENT IN SORGHUM SUDANENSE

Jieqin Li, Lihua Wang, Qiuwen Zhan, Yanlong Liu, Xiaocui Yang

INTRODUCTION

Sorghum is a diverse genus consisting of cultivated and wild species, many of which are diploid, possessing considerable genetic and morphological diversity [1]. *Sorghum bicolor* (*S.bicolor*), commonly called sorghum, is an important cultivated species in the *Sorghum* genus and is used as food, forage and biofuel [2]. *Sorghum sudanense* (*S. sudanense*), commonly called sudangrass, is used as forage for ruminants. It was introduced into the United States in 1909 and rapidly became popular as forage [3]. Now, it is widely cultivated in Russia, Eastern Europe and South Asia [4]. Hybrids of these two species exhibit favorable forage yields, improved quality and disease resistance and therefore, are widely planted in China and the United States [5].

S. bicolor and *S. sudanense* exhibit big differences in morphology [6]. It is expected that genome-wide differences exist between these two species as well, but few studies have been focus on this. In addition, little genetic mechanism research on hybrids between *S. sudanense*and *S. bicolor* has been conducted due to lack of genomic information of *S. sudanense*. Therefore, development of genomic resources for *S. sudanense* is the basis of molecular biology studies on *S. sudanense* and hybrids of *S. bicolor* and *S. sudanense*. RNA-sequencing (RNA-Seq) is a powerful tool for transcription profiling, providing a rapid access to a collection of expressed sequences (transcriptome). RNA-Seq offers novel opportunities in functional genomics and for developing molecular markers in non-model plants [7]. Up to now, RNA-Seq has been successfully applied in different domains of life from yeast to plants [8,9,10].

In this study, the leaf transcriptomes of two *Sorghum* species, *S. bicolor* and *S. sudanense*, were characterized. The main objective was to compare the transcriptomes of the two*Sorghum* species, and to develop DNA markers for genetic mapping of the genes for traits specific to *S. sudanense*. In addition, substitution rates of orthologous transcript pairs were estimated to calculate functional sequence diversity.

Citation: Li J, Wang L, Zhan Q, Liu Y, Yang X (2016) Transcriptome Characterization and Functional Marker Development in Sorghum Sudanense. PLoS ONE 11(5): e0154947. doi:10.1371/journal.pone.0154947.

MATERIALS AND METHODS

Plant materials

Sorghum Tx623B and sudangrass S722 were grown and maintained in a greenhouse under natural daylight condition. When the plants grew to the seven-leaf stage, all the visible leaves of three biological replicates were taken, frozen in liquid N_2 and stored at -80°C until further use. Each replicate involved three plants.

RNA extraction and Illumina sequencing

Total RNA was extracted using an RNA Prep Pure Plant kit (LabKit Co., China) and analyzed for quality and quantity by an Agilent 2100 Bioanalyzer (Agilent Technologies, Santa

Clara, CA, USA). Equal quantities of total RNA from the three biological replicates were pooled before used for cDNA library construction and sequencing. cDNA libraries were constructed using the Illumina RNAseq kit (Illumina, USA). Solexa adapters were then ligated to the ends of the cDNA fragments for Solexa sequencing. High-throughput sequencing was performed using Illumina Hiseq 2500 (Illumina, USA). Reads were 125 bases in length and generated from each end of the DNA fragments in paired-end sequencing.

Gene annotation and differential expression analysis

The adaptor sequences and low-quality sequence reads were removed from the data sets. Raw sequences were transformed into clean tags after data processing. These clean tags were then mapped to the reference genome sequence (http://phytozome.net/, *Sorghum bicolor* v2.1). GO (Gene Ontology, http://www.geneontology.org/) and COG (Clusters of Orthologous Groups of proteins, http://www.ncbi.nlm.nih.gov/COG/) analyses were subsequently performed.

The expression level of each gene was measured as the normalized number of matched clean tags. The normalization method of reads per kilobase per million mapped (RPKM) was used in this study [11]. The IDEG6 software was used to test the significance of gene expression. The false discovery rate (FDR) method was used to tune the threshold *P* value. An FDR <0.01 and an absolute $\log_2$ ratio ≥1 were used to judge the significance of differences in gene expression.

SNP and indel detection

All S722 reads were used to mine SNP markers by aligned to the sorghum reference genome using TopHat [12]. SNP loci were detected using SAMtools between the library and the reference genome [13]. For the nucleotides to be counted as a SNP, the following parameters were required: (1) The sequencing depth was more than 5X, but less than 100X; (2) the quality value of variant calling was larger than 20%; and (3) the distance between SNP loci was longer than 5 bases.

Indel loci were detected using GATK by aligning S722 reads to the reference genome using STAR. The parameters for indel loci detection were as follows: (1) The quality value was larger than 20; and (2) the distance between detected indel loci was longer than 35 bases.

Indel marker design and genetic map construction

All indel sequences (500 bases) were extracted from the sorghum reference genome. Then, indel markers were designed using Primer 3.0 [14]. The parameters were as follows: (1) T_m value was 55–60°C; and (2) PCR target product was 150–350 bases. A hundred indel markers (more than 3 bases of difference) were synthesized by the Springen Biotechnology Company (Nanjing, China). A hundred indel markers (10 markers for each chromosome) were screened for polymorphism between Tx623B and S722 by size detection using 8% non-denaturing polyacrylamide-based electrophoresis. To prove the value of these indel markers, a set of 92 F_2 individuals derived from a cross between sorghum Tx623A and sudangrass Sa was used as the mapping population. Then, 15 indel markers were used to construct the genetic map. The genetic map was constructed using Windows QTL IciMapping version 4.0 [15]. The Kosambi map function was used to calculate the genetic distances.

Analysis of non-synonymous and synonymous mutations

All reads were aligned to the reference genome for coding frame prediction using ClustalW2. The ratio of non-synonymous to synonymous mutations (Ka/Ks) was calculated for these coding reads using PAML_yn00 program. The related parameters for PAML_yn00 program were verbose = 0, icode = 0, weighting = 1, commonf3x4 = 0 and ndata = 1.

RESULTS

Transcriptome sequencing output and aligning to the reference genome

High quality sequencing data is the base of transcriptome analysis. RNA-Seq of the two cDNA libraries, Tx623B and S722, was performed. The Illumina sequencing generated 48,155,337 sequence reads and 9.62 Gb of sequence data. Both GC contents of the sequences of the two libraries were larger than 50% (58.26% and 58.15% for Tx623B and S722, respectively), and CycleQ20 was 100%. All these values indicated that the accuracy and quality of the sequence data were high enough for further analysis.

All reads were aligned to the *S. bicolor* genome which contained 33,472 gene models. The results showed that most reads of the two libraries (88.79% and 83.88% for Tx623B and S722, respectively) matched to the *S. bicolor* genes (Table 1). The majority of the mapped reads were "perfectly mapped reads" without any base-pair mismatch or indel. Totally, 21,572 and 20,752 genes were identified in the reads of the Tx623B and S722 libraries, respectively. The results showed that more than 60% putative genes in the reference genome were expressed in the seven-leaf stage of the two *Sorghum* species.

Table 1: RNA-sequencing results for sorghum Tx623B and sudangrass S722

	Tx623B	S722
Total reads	24,257,222	23,898,115
Total base	4.85 Gb	4.77 Gb
GC (%)	58.26	58.15
CycleQ20 (%)	92.90	93.03
Mapped reads	43,077,834 (88.79%)	40,089,957 (83.88%)
Perfectly mapped reads	26,780,019 (62.17%)	39,373,636 (53.32%)
Mismatched reads	5,881,880 (13.66%)	8,354,470 (20.84%)
Indel reads	8,825,105 (20.49%)	7,949,694 (19.83%)
Indel and mismatched reads	1,590,830 (3.6%)	2,408,455 (6.01%)
Identities	98.60%	98.21%

doi:10.1371/journal.pone.0154947.t001

Differentially expressed gene analysis and new gene analysis

A total of 2,397 differentially expressed genes (DEGs) were detected by RNA-Seq between the two libraries, including 849 up-regulated ones and 1,548 down-regulated ones. The results showed that there were wide differences in gene expression between the two *Sorghum*species. And most of the DEGs were down-regulated in *Sorghum sudanense*. To get annotations, the DEGs of the two libraries were mapped using the GO database. After GO analysis, the 1,619 unigenes were divided into three groups: molecular function, cellular component and biological process (Fig 1). Then, gene set enrichment analysis was applied to unveil that common unigenes and DEG unigenes mainly differed in protein transcription factor activity, rhythmic process, viral reproduction and growth. These results explained the difference in growth rate between *S. bicolor* and *S. sudanense* in the seven-leaf stage.

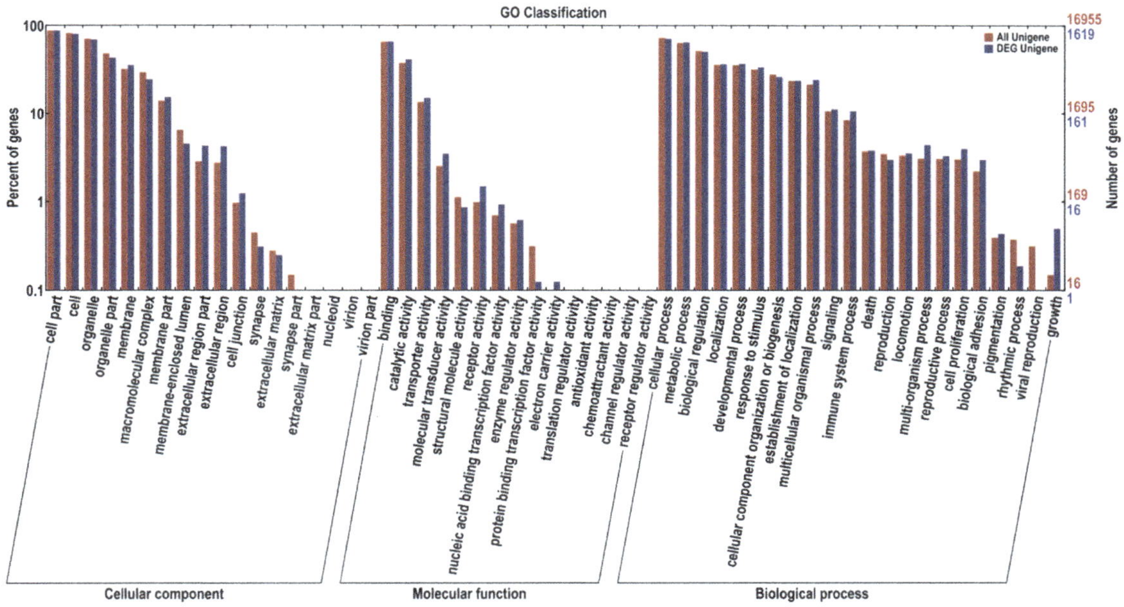

Figure 1: Functional annotation of sequences based on Gene Ontology (GO) categorization.

A total of 1,470 new genes were identified in S722. These new genes, of different lengths, were annotated using different databases. Totally, 102 new genes were categorized into 25 functional groups according to the COG classification (Table 2). The group'replication, recombination and repair' contained the highest number of new genes, followed successively by the groups of 'general function prediction only', 'transcription', 'cell cycle control, cell division, chromosome partitioning' and 'signal transduction mechanism' (Fig 2). The percentage of annotated new genes was 60.6%. The results indicated that further analysis of the unannotated new genes may shed light on the genetic basis for the morphological differences between sorghum and sudangrass.

Table 2: The 102 new genes annotated using different annotation databases

Anno_Database	Annotated_Number	300≤Length<1000	Length≥1000
COG_Annotation	102	23	78
GO_Annotation	425	125	299
nr_Annotation	876	287	584
All_Annotated	891	290	596

doi:10.1371/journal.pone.0154947.t002

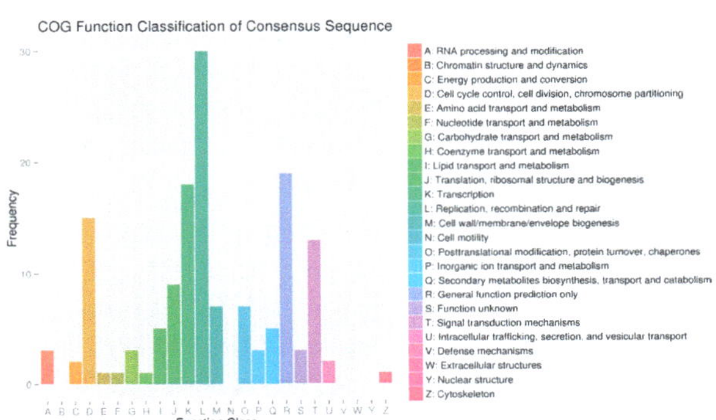

Figure 2: Functional classification of the new genes using COG.

SNP and indel discovery and validation

SNP and indel loci are important information to develop markers for molecular research of sorghum, sudangrass and hybrids of sorghum and sudangrass. A total of 44,495 SNPs were discovered by aligning S722 reads to the reference genome. The highest and lowest number of SNPs was found on chromosome 1 and 5, respectively (Table 3) with a mean of 4,415. This indicated that every 15.7 Kb of the sudangrass genome had 1 SNP between *S.bicolor* and *S. sudanense*, which was good enough for developing markers. Of these SNPs, 33,751 were located in genes, while 10,744 were in intergenic regions. It was found that more than 70% of the SNPs could be developed into functional markers. Further analysis showed that 61.37% of the SNPs were transition, and the value was almost the same among different chromosomes.

Table 3: SNPs identified in *Sorghum sudanense* S722

Chromosome	SNP_number	SNP_number_gene	SNP_number_intergenic	Transition (%)	Transversion (%)
Chromosome_1	6,965	5,545	1,420	61.84	38.16
Chromosome_2	5,528	4,335	1,193	61.03	38.97
Chromosome_3	5,438	4,308	1,130	59.07	40.93
Chromosome_4	5,401	4,203	1,198	61.78	38.22
Chromosome_5	2,965	2,010	955	60.20	39.80
Chromosome_6	4,132	3,140	992	61.79	38.21
Chromosome_7	2,830	2,052	778	62.30	37.70
Chromosome_8	3,287	2,350	937	61.15	38.85
Chromosome_9	3,649	2,668	981	61.61	38.39
Chromosome_10	3,957	2,958	999	63.58	36.42
Super_10	343	182	161	60.79	39.21
Total	44,495	33,751	10,744	61.37	38.63

doi:10.1371/journal.pone.0154947.t003

Indel markers are ubiquitous, co-dominant and locus-specific [16]. They are a type of useful marker in molecular genetics. In this research, 16,928 indel loci were identified by aligning S722 reads to the sorghum reference genome. On average, one indel was found in every 41.3 Kb of the sudangrass genome. Of these indels, most were only one-base different between *S. bicolor* and *S. sudanense*; 68.1% were 1- to 3-base different (Fig 3A). The highest and lowest number of indel was found on chromosome 1 and 5, respectively (Fig 3B).

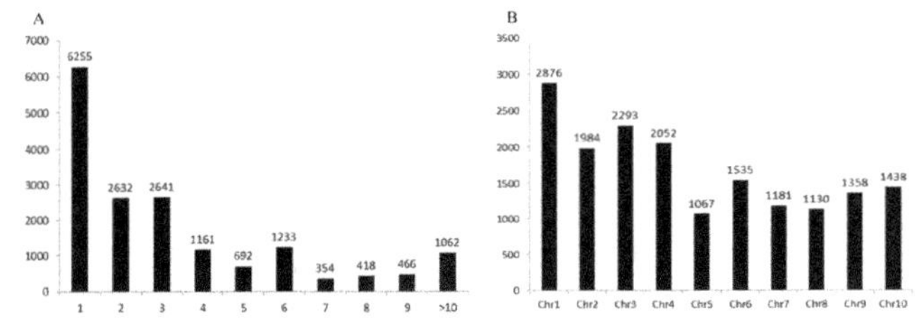

Figure 3: The distribution of indel loci in sorghum (A: the distribution of base differences, X axis is base difference, and Y axis is number; B: the distribution of indel in different chromosomes, X axis is chromosome, and Y axis is number).

SNP and indel validation and genetic map construction

To validate the SNPs, 30 SNP loci were randomly selected to be sequenced using the Sanger sequencing method. The results confirmed that all the SNP loci should be reliable.

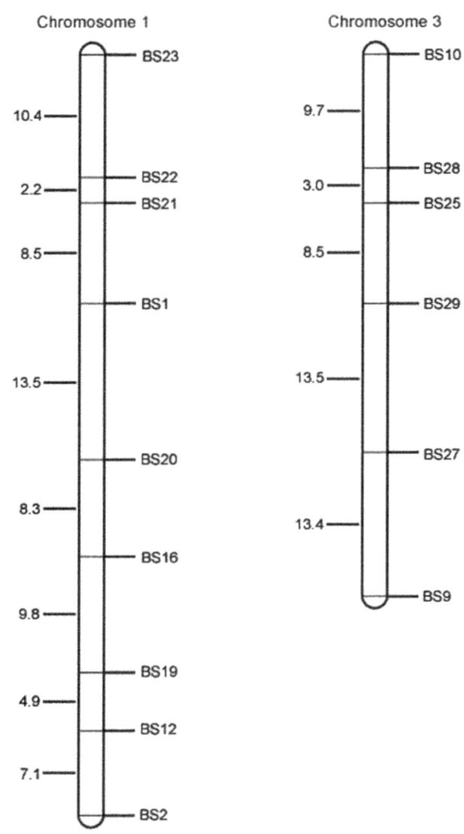

Figure 4: The genetic map of the F_2 **population derived from the cross of sorghum Tx623A and** sudangrass Sa.

A total of 5,344 indel primers were designed for the 16,928 indel loci using Primer 3.0, 100 of which (more than 3 bases of difference) were synthesized and screened between Tx623B and S722. It turned out that 5 did not amplify PCR product. Ninety-five indel markers showed polymorphism between the two accessions. The percentage of polymorphism was 82% between Tx623A and Sa, further proving the universality of the 100 indel markers. Then, 15 out of the 100 indel markers were selected to construct the genetic map of the F_2population derived from the cross of Tx623A and Sa (Fig 4). All the results indicated that the SNPs and indels could provide important information for marker mining and applications in molecular genetic research and molecular marker-assisted breeding.

Synonymous and non-synonymous mutation analysis

To detect functional sequence diversity, putative exon sequences were compared between S722 and the reference genome. Comparison between synonymous base substitutions (Ks) and non-synonymous base substitutions (Ka) can tell sequences are influenced by purifying or diversifying selection. The average Ka/Ks value was 0.39 for orthologous pairs. About 95% of the orthologous pairs exhibited a Ka/Ks<1.0, indicating that the genes were influenced by purifying selection (Fig 5). And 5% of the orthologous pairs showed a Ka/Ks>1.0, indicating that the genes were influenced by diversifying selection. Functional annotation was performed for the 15 pairs under diversifying selection. The largest numbers of annotated pairs were of catalytic activity and DNA/RNA binding molecular functions.

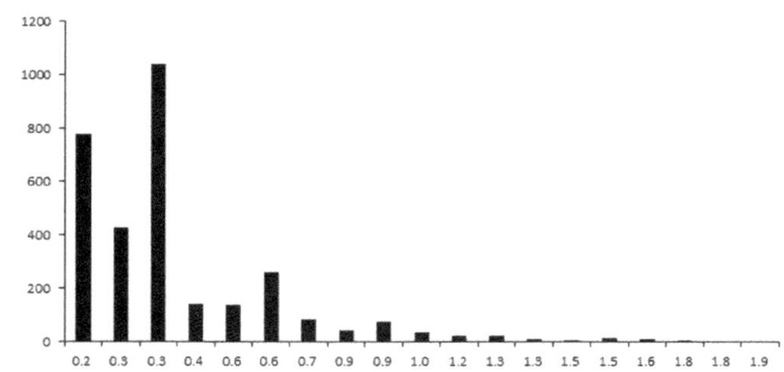

Figure 5: A histogram of Ka/Ks distribution.

DISCUSSION

Transcriptome sequencing provides a new tool for genomic studies on model or non-model organisms [10,17]. In this study, the transcriptome of sudangrass, which is an important forage in temperate regions, was characterized. Compared to sorghum, sudangrass has a higher forage yield and a higher growth rate [5]. So, there is a need to understand the genetic mechanisms underlying the phenotypic differences between *S. bicolor* and *S. sudanense* so that *S. sudanense* genes can be utilized to develop related breeding strategies. Learning the expression profiles of *S. bicolor* and *S. sudanense* and the sequence polymorphisms provides an opportunity to understand the genetic basis of phenotypic differences. In this research, some up-regulated transcriptional factors involved in plant growth and development were also identified. For example, the up-regulated *Sorghum_newGene_2588* in sudangrass was annotated as a Myb-related transcriptional factor involved in plant growth and development [18].

It is controversial about whether *S. bicolor* and *S. sudanense* are actually of a same species. Snowden treated sudangrass as *S. sudanense*, a species different from sorghum in spikelet, anthotaxy and plant traits [19]. De Wet suggested that sudangrass be placed as a subspecies,*drummondii* (steud), of *S. bicolor* (L.) Moench [20]. The results from this study showed that the GC contents of sorghum Tx623B and sudangrass S722 were just slightly different. The percentages of the mapped reads aligned to the reference genome also differed only slightly between the two transcriptomes. All these data indicated a high genomic similarity between the two transcriptomes. Although the results of this research could not support that sudangrass should be placed as a subspecies of *S. sorghum*, they provided a clue for further studies on the place of sudangrass in taxonomy.

The polymorphism of SSR markers developed from the sorghum genome was low (19.48%) between sorghum and sudangrass [21]. Therefore, it is important to develop high diversity markers for genetic research on sudangrass and hybrids of sorghum and sudangrass. As indel markers within genes are functional markers, they are superior to random DNA markers, such as RAPD, SSR, and AFLP, because functional markers are completely linked with trait locus alleles [22]. A total of 16,928 indel loci between S722 and the reference genome were detected, and 5,344 indel markers were developed. The percentage of polymorphism was 95% in the 100 selected indel markers between Tx623B and S722; the percentage of polymorphism was 82% between Tx623B and Sa. It was indicated that the indel loci were reliable and the indel markers were diverse between sorghum and sudangrass. To further verify the value of these indel markers, they were used to construct the genetic maps of chromosomes 1 and 3 derived from a cross between sorghum Tx623A and sudangrass Sa. The results demonstrated that these indel markers developed will be useful for QTL mapping and molecular-assisted breeding in sorghum, sudangrass, and hybrids of sorghum and sudangrass.

SNP detection is an important part of molecular genetic research because SNP loci can be exploited to construct high-density genetic maps and genome-wide association studies [8,23,24]. A total of 44,495 SNP loci between S722 and sorghum genome were detected in this work. The frequencies of transition and transversion were comparable to those observed in other plant species [25, 26, 27]. The density of SNP loci was 2.6 times that of indel loci. Therefore, SNP loci are an important supplement to indel markers in marker development for construction of high-density genetic maps or gene mapping. Additionally, it was found that 24.3% of the SNP loci were genic. These genic SNP loci will be more appropriate for developing phenotype-associated markers.

The ratio of Ka to Ks provides a measure of the selection pressure to which a gene pair is subjected and has been used to tell whether a gene is under positive/diversifying selection or under negative/purifying selection. Our results are consistent with the findings by other researchers that most orthologous genes are under purifying selection. It was also suggested that positive selection pressure results in the maintenance of divergence between the two species, which is also observed in other organisms [25, 28].

CONCLUSION

In this study, the transcriptome of *Sorghum sudanense* was sequenced and characterized using the next-generation sequencing platform. A total of 2,397 DEGs were detected by RNA-Seq between the two libraries, including 849 up-regulated ones and 1,548 down-regulated ones. A total of 44,495 SNPs were discovered by aligning S722 reads to the sorghum reference genome. A total of 16,928 indel loci were identified between the two genomes. A total of 5,344 indel markers were designed, 15 of which were selected to construct the genetic map derived from the cross of Tx623A and Sa. This work provides a set of functional markers and a resource for mapping of and molecular genetic research on sudangrass and other *Sorghum*species.

ACKNOWLEDGMENTS

This work was supported by grants from the National Natural Science Foundation of China (31301383), the Key-construction Subject Plan of Anhui Province (No: WanJiaoMiKe[2014]28), the young talent project of Anhui province (gxyqZD2016217), the key project of natural science research of Anhui provincial education department (KJ2016A177), the Key-construction Subject Plan of Anhui Science and Technology University (AKZDXK2015A03).

AUTHOR CONTRIBUTIONS

Conceived and designed the experiments: QWZ. Performed the experiments: LHW JQL YLL. Analyzed the data: JQL LHW. Contributed reagents/materials/analysis tools: XCY. Wrote the paper: JQL.

REFERENCES

1. Menz MA, Klein RR, Mullet JE, Obert JA, Unruh NC, Klein PE. A high-density genetic map of Sorghum bicolor (L.) Moench based on 2926 AFLP, RFLP and SSR markers. Plant Mol Biol.2002; 48: 483–499. pmid:11999830
2. Carpita NC, McCann MC. Maize and sorghum: genetic resources for bioenergy grasses. Trends Plant Sci. 2008; 13: 415–420. doi: 10.1016/j.tplants.2008.06.002. pmid:18650120
3. Hacker JB. Sorghum drummondii (Steud) Millsp.& Chase.In:Mannetje LT, Jones RM. eds. Plant resources of South-East Asia 4 Forages. Wageningen: Pudoc Scientific Publishers; 1992.pp.206–208.
4. Beck PA, Hutchison S, Gunter SA, Losi TC, Stewart CB, Capps PK, et al. Chemical composition and in situ dry matter and fiber disappearance of sorghum x Sudangrass hybrids. J Anim Sci.2007; 85: 545–555. pmid:17235037 doi: 10.2527/jas.2006-292
5. Zhan QW, Qian ZQ. Heterosis utilization of hybrid between sorghum [Sorghum bicolor (L.) Moench] and Sudangrass [Sorghum sudanense (Piper) Stapf]. Acta Agron Sin. 2004; 30: 73–77. doi: 10.1046/j.1439-037x.2003.00001.x
6. Zhan QW, Zhang TZ, Wang BH, Li JQ. Diversity comparison and phylogenetic relationships of S. bicolor and S. sudanense revealed by SSR markers. Plant Sci. 2008; 174: 9–16. doi: 10.1016/j.plantsci.2007.09.007
7. Chen XR, Xing YP, Li YP, Tong YH, Xu JY. RNA-Seq reveals infection-related gene expression changes in Phytophthora capsici. PLoS One.2013; 8: e74588. doi: 10.1371/journal.pone.0074588. pmid:24019970
8. Novaes E, Drost DR, Farmerie WG, Pappas GJ, Grattapaglia D, Sederoff RR, et al. High-throughput gene and SNP discovery in Eucalyptus grandis, an uncharacterized genome. BMC Genomics. 2008; 9: 312.

doi: 10.1186/1471-2164-9-312. pmid:18590545

9. Kim MY, Lee S, Van K, Kim TH, Jeong SC, Choi IY, et al. Whole-genome sequencing and intensive analysis of the undomesticated soybean (Glycine sojaSieb. and Zucc.) genome. Proc Natl Acad Sci U S A. 2010; 107: 22032–22037. doi: 10.1073/pnas.1009526107. pmid:21131573
10. Zheng W, Ma L, Zhao J, Li Z, Sun F, Lu X. Comparative transcriptome analysis of two rice varieties in response to rice stripe virus and small brown plant hoppers during early interaction. PLoS One. 2013; 8: e82126. doi: 10.1371/journal.pone.0082126. pmid:24358146
11. Wagner G, Kin K, Lynch V. Measurement of mRNA abundance using RNA-seq data: RPKM measure is inconsistent among samples. Theory in Biosciences. 2012; 131: 281–285. doi: 10.1007/s12064-012-0162-3. pmid:22872506
12. Kim D, Pertea G, Trapnell C, Pimentel H, Kelley R, Salzberg SL. TopHat2: accurate alignment of transcriptomes in the presence of insertions, deletions and gene fusions. Genome Biol. 2013; 14: R36. doi: 10.1186/gb-2013-14-4-r36. pmid:23618408
13. Li H, Handsaker B, Wysoker A, Fennell T, Ruan J, Homer N, et al. The Sequence Alignment/Map format and SAMtools. Bioinformatics. 2009; 25: 2078–2079. doi: 10.1093/bioinformatics/btp352. pmid:19505943
14. Untergasser A, Cutcutache I, Koressaar T, Ye J, Faircloth BC, et al. Primer3—new capabilities and interfaces. Nucleic Acids Res, 2012; 40: e115. pmid:22730293 doi: 10.1093/nar/gks596
15. Meng L, Li H, Zhang L, Wang J. QTL IciMapping: Integrated software for genetic linkage map construction and quantitative trait locus mapping in biparental populations. The Crop Journal. 2015; 3: 269–283. doi: 10.1016/j.cj.2015.01.001
16. Tan S, Cheng JW, Zhang L, Qin C, Nong DG, Li WP, et al. Construction of an interspecific genetic map based on InDel and SSR for mapping the QTLs affecting the initiation of flower primordia in pepper (Capsicum spp.). PLoS One. 2015; 10: e119389. doi: 10.1371/journal.pone.0119389
17. Muriira NG, Xu W, Muchugi A, Xu J, Liu A. De novo sequencing and assembly analysis of transcriptome in the Sodom apple (Calotropis gigantea). BMC Genomics. 2015; 16: 723. doi: 10.1186/s12864-015-1908-3. pmid:26395839
18. Jin H, Martin C. Multifunctionality and diversity within the plant MYB-gene family. Plant Mol Biol. 1999; 41: 577–585. pmid:10645718
19. Snowden JD. The Cultivated Races of Sorghum: Adlard& Son, Ltd.; 1936. pp. 274.
20. De Wet JMJ, Huckabay JP. The origin of Sorghum bicolor. II.Distribution and domestication. Evolution. 1967; 21: 787–802. doi: 10.2307/2406774
21. Liu YL, Wang LH, Li JQ, Zhan QW, Zhang Q, Li JF, et al. QTL mapping of forage yield and forage yield component traits in Sorghum bicolor x S. sudanense. Genet Mol Res. 2015; 14: 3854–3861. doi: 10.4238/2015.April.22.14. pmid:25966155
22. Liu J, Qu J, Yang C, Tang D, Li J, Lan H, et al. Development of genome-wide insertion and deletion markers for maize, based on next-generation sequencing data. BMC Genomics. 2015; 16: 601. doi: 10.1186/s12864-015-1797-5. pmid:26269146
23. Guo M, Rupe MA, Danilevskaya ON, Yang X, Hu Z. Genome-wide mRNA profiling reveals heterochronic allelic variation and a new imprinted gene in hybrid maize endosperm. Plant J. 2003; 36: 30–44. pmid:12974809 doi: 10.1046/j.1365-313x.2003.01852.x
24. Duran C, Appleby N, Clark T, Wood D, Imelfort M, Batley J, et al. AutoSNPdb: an annotated single nucleotide polymorphism database for crop plants. Nucleic Acids Res. 2009; 37: 951–953. doi: 10.1093/nar/gkn650
25. Jhanwar S, Priya P, Garg R, Parida SK, Tyagi AK, Jain M. Transcriptome sequencing of wild chickpea as a rich resource for marker development. Plant Biotechnol J. 2012; 10: 690–702. doi: 10.1111/j.1467-7652.2012.00712.x. pmid:22672127
26. Wu X, Ren C, Joshi T, Vuong T, Xu D, Nguyen HT. SNP discovery by high-throughput sequencing in soybean. BMC Genomics. 2010; 11: 469. doi: 10.1186/1471-2164-11-469. pmid:20701770
27. Milano I, Babbucci M, Panitz F, Ogden R, Nielsen RO, Taylor MI, et al. Novel tools for conservation genomics: comparing two high-throughput approaches for SNP discovery in the transcriptome of the European hake. PLoS One. 2011; 6: e28008. doi: 10.1371/journal.pone.0028008. pmid:22132191
28. Schlueter JA, Dixon P, Granger C, Grant D, Clark L, Doyle JJ, et al. Mining EST databases to resolve evolutionary events in major crop species. Genome. 2004; 47: 868–876. pmid:15499401 doi: 10.1139/g04-047

CHAPTER 7

NEW INSIGHTS INTO THE POST-TRANSLATIONAL MODIFICATION OF MULTIPLE PHOSPHOENOLPYRUVATE CARBOXYLASE ISOENZYMES BY PHOSPHORYLATION AND MONOUBIQUITINATION DURING SORGHUM SEED DEVELOPMENT AND GERMINATION

Isabel Ruiz-Ballesta, Guillermo Baena, Jacinto Gandullo, Liqun Wang, Yi-Min She, William Charles Plaxton and Cristina Echevarría

Keywords: Germination, monoubiquitination, phosphorylation, phosphoenolpyruvate carboxylase, post-translational modifications, seed development, seeds, Sorghum bicolor, tissue-specific gene expression

INTRODUCTION

Phosphoenolpyruvate (PEP) carboxylase (PEPC; EC 4.1.1.31) is an important, tighly regulated, cytosolic metabolic enzyme that plays a pivotal photosynthetic role in primary CO_2 fixation by C_4 and Crassulacean acid metabolism plants, but also has a variety of additional functions, including roles in seed development and germination (O'Leary *et al.*, 2011*a*). It catalyses the irreversible β-carboxylation of PEP to form oxaloacetate (OAA) and orthophosphate (P_i) (Chollet *et al.*, 2006). The*plant-type PEPC* (*PTPC*) genes encode closely related 100–110kDa polypeptides that typically exist as Class-1 PEPC tetramers that are post-translationally regulated by a combination of several allosteric effectors, including Glc-6-P (activation) and L-malate (inhibition), reversible phosphorylation (Nimmo, 2003; O'Leary *et al.*, 2011*a*), and monoubiquitination (Uhrig *et al.*, 2008, Shane *et al.*, 2013; Ruiz-Ballesta *et al.*, 2014). The metabolite control is integrated with the enzyme's reversible phosphorylation at a highly conserved serine residue located near the N-terminus of its PTPC subunits (Chollet *et al.*, 1996; Echevarría and Vidal, 2003; O'Leary *et al.*, 2011*a*). Phosphorylation activates the enzyme by making it less sensitive to feedback inhibition by malate. Phosphorylation is catalysed by a PTPC-kinase (more commonly referred to as PEPC-kinase; PPCK), whereas dephosphorylation is catalysed by a protein phosphatase type-2A (Nimmo, 2000; Izui *et al.*, 2004; O'Leary *et al.*, 2011*a*). Both controls are interactive and thus allow a fine modulation of the enzyme's kinetic properties as required by the physiological context (Echevarría and Vidal, 2003). Regulatory PTPC monoubiquitination has been demonstrated in the endosperm of germinating castor oil seeds (COS) (Uhrig *et al.*, 2008), in developing proteoid roots of P-deficient harsh hakea (Shane *et al.*, 2013), and in lily pollen (Igawa *et al.*, 2010). A recent study by our group has shown monoubiquitination of SbPPC3 in sorghum, a cereal with starch-storing seeds (Ruiz-Ballesta *et al.*, 2014). Monoubiquitination is inhibitory as it results in increased K_m (PEP) values and enhanced sensitivity to allosteric inhibitors (Uhrig *et al.*, 2008; Shane*et al.*, 2013; Ruiz-Ballesta *et al.*, 2014). In COS, PTPC monoubiquitination and phosphorylation appear to be mutually exclusive, while PTPC polypeptides are alternatively phosphorylated or monoubiquitinated in developing proteoid

Citation: Isabel Ruiz-Ballesta, Guillermo Baena, Jacinto Gandullo, Liqun Wang, Yi-Min She, William Charles Plaxton and Cristina Echevarría, New insights into the post-translational modification of multiple phosphoenolpyruvate carboxylase isoenzymes by phosphorylation and monoubiquitination during sorghum seed development and germination, doi: 10.1093/jxb/erw186.

roots of harsh hakea (Shane *et al.*, 2013). In addition, the elimination of photosynthate supply caused by depodding in COS triggers *in vivo* dephosphorylation and subsequent partial monoubiquitination of PTPC polypeptides (Uhrig *et al.*, 2008; O'Leary *et al.*, 2011*b*). Previous studies have indicated that PTPC polypeptides become *in vivo* phosphorylated in de-embryonated sorghum seed during imbibition (i.e. on the basis of a significant increase in the enzyme's IC_{50} for L-malate and *in vivo* ^{32}P incorporation) (Nhiri *et al.*, 2000). In other starch-storing cereal seeds such as barley and wheat, the phosphorylation of both PTPC polypeptides during germination has also been confirmed by immunoprecipitation, followed by SDS-PAGE and autoradiography of the *in vivo* ^{32}P-labeled PEPC (Osuna *et al.*, 1999; Feria *et al.*, 2008). As far as we know, no previous study has focused on how these post-translational modifications (PTMs) are combined in the different subunits and isoenzymes of a PEPC's protein family. Deciphering how monoubiquitination and phosphorylation influence each PTPC isoenzyme during the life cycle of sorghum seeds was a major objective of the current study.

Bacterial-type PEPC (*BTPC*) genes encode larger 116–118kDa polypeptides exhibiting low (<40%) sequence identity with PTPCs. BTPC lacks the conserved N-terminal seryl phosphorylation motif characteristic of PTPCs, and in vascular plants it appears to only exist as catalytic and regulatory subunits of novel Class-2 PEPC heteromeric complexes composed of tightly associated PTPC and BTPC subunits (Igawa *et al.*, 2010; O'Leary *et al.*, 2011*a*; Park *et al.*, 2012). Using western blot analysis and specific anti-BTPC-IgGs in crude extract of barley seeds, Class-2 PEPC complexes have not been identified (Feria *et al.*, 2008). This is an interesting finding as the BTPC and Class-2 PEPC appear to play an important role during COS development (Uhrig *et al.*, 2008). Assessing, by MS/MS and qRT-PCR, the possible occurrence of BTPC (SbPPC6) and/or Class-2 PEPC during the course of the life cycle of sorghum seeds was also an important objective of this work.

Sorghum (*Sorghum bicolor* L. Moench) is the world's fifth most important cereal crop (www.fao.org) and provides food, feed, fiber, fuel, and chemical/biofuel feedstocks across a range of environments and production systems (Kresovich *et al.*, 2005). Sorghum encodes five *PTPC*genes (Paterson *et al.*, 2009): *Sb02g021090* and *Sb04g008720* encode housekeeping and inducible SbPPC2 and SbPPC3, respectively (Lepiniec *et al.*, 1993), whilst *Sb07g014960* and *Sb03g035090* encode PTPC isoenzymes SbPPC4 and SbPPC5 but these have unknown functions. All of these are C_3-type PTPCs. *Sb10g021330* encodes C_4 photosynthetic SbPPC1 subunits that are highly expressed in sorghum leaves. In addition,*Sb03g008410* (*SbPPC6*) encodes a single BTPC (Table 1).

Table 1: The sorghum *PEPC* gene family. Names used in this work for the different *PEPC*genes, the equivalence with other names, and assigned functions described by Lepiniec et al. (1993, 1994), and orthologs established by Paterson et al. (2009). Sb, Sorghum bicolor; Sv, Sorghum vulgare; Zm, Zea mays; Os Oryza sativa

Name of gene	Accession number[a]	Named by Lepiniec *et al.* (1993) and Ruiz-Ballesta *et al.* (2014)	Known function	Orthologs (Paterson *et al.*, 2009)
SbPPC1	Sb10g021330	CP46	C_4 PTPC, photosynthetic C_4 gene	X63756 (SvPEPC1); NM001111948 (ZmPEPC1)
SbPPC2	Sb02g021090	CP28	C_3 PTPC, housekeeping gene	X59925 (SvPEPC4); NM001111968 (ZmPEPC2)
SbPPC3	Sb04g008720	CP21	C_3 PTPC, root-inducible gene	X65137 (SvPEPC3); NM001112033 (ZmPEPC3
SbPPC4	Sb07g014960	-	C_3 PTPC, unknown so far	Os08g0366000
SbPPC5	Sb03g035090	-	C_3 PTPC, unknown so far	Os01g0758300
SbPPC6	Sb03g008410	-	BTPC, as Class-2 PEPC	Os01g0110700

[a] NCBI database

In this study, the biochemical and immunological properties of PTPCs during development and germination of starch-storing sorghum seeds were further characterized. We also elucidated the transcript abundance of*PTPC* genes, and determined subunit composition, relative abundance, and PTMs of each PTPC isoenzyme during germination. As far as we know, this represents the most detailed study to date that examines both phosphorylation and monoubiquitination of a PTPC protein family in any plant species. Furthermore, we show that enhanced PTPC monoubiquitination always accompanies radical emergence during seed germination, and that both processes are accelerated in parallel under aerobic conditions or elevated temperature.

MATERIALS AND METHODS

Plant material

Mature sorghum [*Sorghum bicolor* (L.) Moench, var. PR87G57 and var. PR88Y20, Pioneer Hi-Bred, Spain] seeds were sterilized and germinated for up to 96h at 25 or 35 °C as previously reported (Nhiri *et al.*, 2000). For immunological studies or to analyze PEPC activity, the aleurone layers, starchy endosperm (aleurone-endosperm), and embryos were carefully dissected at the indicated times post-imbibition. Sterilized sorghum seeds were also submerged in distilled water (hypoxic condition) or placed on moistened filter paper (normoxic condition) at 25 °C for 48 and 72h. Harvested tissues were frozen in liquid nitrogen and stored at –80 °C until used. Developing seeds were obtained from sorghum plants cultivated in a greenhouse under a 14h day (25 °C)/10-h night (18 °C) cycle. Seeds were harvested at various stages of development, ground to a powder under liquid nitrogen, lyophilized, and stored at 4 °C for later analyses.

Antibodies

Polyclonal antibodies against native C_4-photosynthetic SbPPC1 from sorghum leaves (anti-C_4 PTPC) were prepared as described in Pacquit *et al.* (1995).

Anti-C19 and anti-N24 were raised, respectively, against synthetic peptides corresponding to the C-terminal [(Y) EDTLILTMK GIAAGMQNTG] and the dephosphorylated N-terminal [ERHHSIDAQLRALAPGKVSEE24(YG)] ends of SbPPC1 (i.e., sorghum C_4-photosynthetic PTPC) as previously described (Ruiz-Ballesta *et al.*, 2014). Anti-COS PTPC was raised against native Class-1 PEPC purified from the endosperm of developing castor oil seeds as described in Tripodi *et al.* (2005).

Immunohistochemistry

Seeds germinated at 24h post-imbibition were fixed in 4% (w/v) paraformaldehyde and in 0.25% (v/v) glutaraldehyde dehydrated in a graded series of aqueous ethanol solutions, and embedded in Paraplast Plus (Panreac) as described by González *et al.* (1998). Sections (15 µm thick) were prepared using a microtome (model RM2165, Leica) and placed on poly-L-Lys-coated microscope slides. After deparaffinizing in xylol and rehydrating in decreasing concentrations of ethanol, sections were blocked for 30min in TBS buffer containing 3% (w/v) BSA. Then, 500 µl of the same solution containing affinity-purified anti-C_4 PTPC (3 µg of protein) (Pacquit *et al.*, 1995) or pre-immune serum was placed on the samples and incubated overnight at 4 °C. Unbound antibodies were removed by three 10-min washes in TBS. Tissue sections were then incubated with alkaline phosphatase-conjugated goat anti-(rabbit IgG)-IgG (Bio-Rad) for 2h at 37 °C. The reaction of alkaline phosphatase was developed colorimetrically with a BCIP/NBT Liquid Substrate System (Sigma).

Protein extraction

Frozen powder (0.4g) from whole developing or germinated sorghum seeds was ground in a chilled mortar with sand in 1ml of an ice-cold buffer that contained 100mM Tris-HCl (pH 7.5), 10mM $MgCl_2$, 1mM EDTA, 20% (v/v) glycerol, 14mM β-mercaptoethanol, 1mM phenylmethylsulfonyl fluoride, 10 µg ml^{-1} chymostatin, 10 µM leupeptin, 50mM KF, 1mM Na_2MoO_4, 1mM $Na_3VO_4{}^-$, and 50nM microcystin-LR. Homogenates were centrifuged at 17000 *g* for 7min and the supernatants (0.5ml) desalted through 3-ml Sephadex G-25 spin columns equilibrated in extraction buffer (lacking PVPP and PVP) prior to enzymatic analysis.

Electrophoresis and immunoblotting

SDS-PAGE, non-denaturing PAGE and in-gel PEPC activity staining, immunoblotting, and visualization of antigenic polypeptides using an alkaline-phosphatase-conjugated secondary antibody with chromogenic detection were performed as previously described (Rivoal *et al.*, 2002;Tripodi *et al.*, 2005). Alternatively, immunoreactive polypeptides were detected using a chemiluminescence system (SuperSignal West Pico Rabbit IgGs; Thermo Scientific) according to the manufacturer's instructions in a Fujifilm LAS3000 mini-system. The immunoreactive polypeptides were quantified via analysis of the scanned blots using Multi-Gauge V3.0 software (Fujifilm).

Determination of PEPC activity and protein concentration

PEPC activity was assayed at 340nm using a Molecular Devices Spectramax Kinetics Microplate reader. Optimized assay conditions were: 50mM HEPES-KOH (pH 8.0), 2.5mM PEP, 5mM $NaHCO_3$, 5mM $MgCl_2$, 0.15mM NADH, 10% (v/v) glycerol, 1mM DTT, and 5 units ml^{-1} porcine heart malate dehydrogenase (0.2ml final volume). One unit of activity is defined as the amount of PEPC catalysing the production of 1 µmol of oxaloacetate min^{-1} at 25 °C. Protein concentrations were determined by the Coomassie Blue G-250 dye-binding method using bovine γ-globulin as the standard (Tripodi *et al.*, 2005).

Immunoprecipitation and immunoaffinity chromatography

Immunoprecipitation of PTPC polypeptides from clarified seed extracts was performed using anti-C_4 PTPC as previously described (Osuna *et al.*, 1999).

For immunoaffinity chromatography, frozen whole 48-h germinated sorghum seeds (5g) were homogenized (1:2.5; w/v) using a Polytron PT-3100 homogenizer in ice-cold buffer A that contained 100mM HEPES-KOH (pH 7.5), 1mM EDTA, 1mM EGTA, 15% (v/v) glycerol, 5mM $MgCl_2$,100mM KCl, 10mM NaCl, 25mM NaF, 1mM $Na_3VO_4{}^-$, 0.1% (w/v) polyvinylpolypyrrolidone, 0.1% (v/v) Triton X-100, 2mM 2,2′-dipyridyl disulfide, and 10 µl ml^{-1} ProteCEASE-100 (G-BioSciences). Homogenates were centrifuged at 4 °C for 10min at 48500 *g*, and the supernatant fluid re-centrifuged for 5min and filtered through a layer of Miracloth (Calbiochem). Clarified extracts were pre-cleared at 25 °C by eluting at 0.5ml min^{-1} through an AminoLink column (1×4cm) that had been pre-saturated with 1M Tris-HCl (pH 7.4). Unbound proteins were immediately absorbed at 1ml min^{-1} onto an anti-COS PTPC immunoaffinity column (1×2cm; prepared as described by Uhrig *et al.*, 2008) that had been pre-equilibrated with PBS. Flow-through fractions were collected by eluting the column at 1ml min^{-1} with PBS until A_{280} decreased to baseline, and bound proteins were eluted at 0.5ml min^{-1} with 100mM Gly-HCl (pH 2.8), and immediately neutralized (1-ml fractions collected into 0.1ml of unbuffered 1M Tris). A_{280} absorbing fractions were concentrated to ~1mg ml^{-1} and analyzed for their polypeptide composition by SDS-PAGE and immunoblotting.

Mass spectrometry

In the initial MALDI-TOF MS analysis, PTPCs at the different stages of seed development and germination were extracted and purified by immunoprecipitation from the clarified extracts, followed by SDS-PAGE of the solubilized immunoprecipitates (Osuna *et al.*, 1999). Protein-staining bands of interest were manually excised from micro-preparative gels using biopsy punches, then reduced, alkylated, and digested with trypsin according to Sechi and Chait (1998). Briefly, gel bands were washed twice with water, shrunk for 15min with 100% acetonitrile, and dried in a Savant SpeedVac for 30min. Then the samples were reduced with 10mM dithioerythritol in 25mM ammonium bicarbonate (NH_4HCO_3) for 30min at 56 °C and subsequently alkylated with 55mM iodoacetamide in 25mM NH_4HCO_3 for 15min in the dark. Finally, samples were digested with 12.5ng µl^{-1} sequencing grade trypsin (Roche Molecular Biochemicals) in 25mM NH_4HCO_3 (pH 7.6) overnight at 37 °C. After digestion, 1 µl of each sample was spotted onto a MALDI plate and allowed to air-dry at room temperature. Then, 0.4 µl of 3mg ml^{-1} α-cyano-4-hydroxy-cinnamic acid as the matrix (Sigma) in 50% acetonitrile was added onto the spots and they were again allowed to air-dry at room temperature. Sample analyses were performed on a 4800 Plus Proteomics Analyzer MALDI-TOF/TOF mass spectrometer (AB Sciex, Toronto, Canada). The instrument was operated in the positive reflector mode with an accelerating voltage of 20 000V. Mass spectra were calibrated internally using the peptides of trypsin autolysis. Mass spectrometric measurements yielded a list of monoisotopic masses of the observed peptides at a signal-to-noise ratio greater than 10, which were used for peptide mapping.

Peptides of interest were further analyzed by MS/MS in order to assess their sequences. Suitable precursor ions were selected from the MS spectra and broken down using collision induced dissociation (CID) with atmospheric air as the collision gas, at a collision energy of 1kV and in the positive ion reflector mode. The precursor ions were selected within the mass windows of ±4Da. The non-redundant NCBI database (downloaded on 24/2/2011; 13135398 sequences) and in-house databases with the sorghum entries from NCBInr (71862 sequences) and the sorghum PEPC sequence only were used for protein identification, and the mass spectrometric data were searched using the MASCOT 2.3 server through the Global Protein Server v3.6 (AB Sciex). The search parameters were set up to no limitation in taxonomy, trypsin as the cleavage enzyme, carbamidomethylation on cysteine as the fixed modification, and the oxidation of methionine and ubiquitination (Gly-Gly adducts) of lysine, phosphorylation of serine or threonine as variable modifications. Mass tolerances of single MS and MS/MS were set up to 80ppm and 0.3Da, respectively. For trypsin digestion, one missed cleavage was allowed. In such analyses, the probability scores greater than the score fixed by mascot were considered to be significant at a *P*-value of <0.05. *De novo*sequencing of the MS/MS spectra of the peptides was performed using the DeNovo tool software (AB Sciex), the retrieved peptide sequences were manually checked and validated. A subsequent homology search of these sequences was obtained by BLAST (www.ncbi.nlm.nih.gov/BLAST).

PTPC PTMs were analyzed by a high-accuracy and sensitivity LC MS/MS. Coomassie blue-stained polypeptides were excised from SDS-PAGE gels, in-gel reduced, alkylated, and digested with trypsin using standard protocols (Uhrig *et al.*, 2008). Tryptic peptides were extracted using acetonitrile/0.1% trifluoroacetic acid (TFA) (v/v, 60:40), and dried using a CentriVap refrigerated centrifugal concentrator (Labconco Corp.). Peptides were reconstituted in 4 µl of 0.1% (v/v) formic acid (FA) and identified using two instrumentation methods: nanoAcquity ultra performance LC systems (Waters, Milford, MA) coupled with a TripleTOF 5600+ quadrupole time-of-flight mass spectrometer (AB Sciex, Concord, ON, Canada) at low-energy CID; and an Orbitrap Fusion Tribrid MS system (Thermo Fisher Scientific Inc.Watham, MA) in higher energy collision induced dissociation (HCD) mode. The sample was trapped by a 2G-V/MT Trap symmetry C18 column (5 µm particles, 180 µm id × 20mm length) at a flow rate of 5 µl min^{-1}, and separated on a BEH130 C18 analytical column (1.7 µm particles, 100 µm id ×100mm length) at 300 nl min^{-1} for 60min. The mobile phase was set up to a linear gradient from 5–30% solvent B (0.1% FA in acetonitrile), followed by 85% solvent B over 10min for peptide elution. TripleTOF MS and MS/MS scans were acquired with a high resolution of 30.000, and the *m*/*z* regions of the MS survey scan and MS/MS were selected from *m*/*z* 350–1600 and *m*/*z* 100–1250, respectively, and by data-dependent scanning of the top twenty ions at multiply charged states of 2+, 3+ and 4+. Dynamic exclusion was set to a period of time of 30s. For comparison purposes, similar analyses were performed on the gel-separated proteins using the Orbitrap Fusion Tribrid LC-MS system under high-accuracy MS survey scan at a resolution of 60.000 followed by sequential HCD scans at the twenty top ions. MS/MS data from the two instruments were searched against the viridiplantae (green plants) protein sequences in the NCBI database using the Mascot Server (version 2.4.0, Matrix Science, London, UK). The search parameters were restricted to tryptic peptides at a maximum of two missed cleavages. Cysteine carbamidomethylation was designated as a fixed modification, and deamidation of asparagine and glutamine, oxidation of methionine, phosphorylation of serine/threonine/tyrosine, ubiquitination of lysine were considered as variable modifications. Mass tolerances were set up to 30ppm for TripleTOF 5600 MS ions and 0.05Da for MS/MS fragment ions, 20ppm for Orbitrap MS and 0.8Da for the MS/MS fragments. Peptide assignments were filtered by an ion score cut off at 15, and the identified MS/MS spectra were also verified manually.

RNA Extraction and cDNA synthesis

Total RNA was isolated from 100mg of frozen, powdered half-embryonated germinating seeds or whole developing seeds using the IQeasy™ Plant RNA Extraction (Intron Biotechnology). Extracted nucleic acids were DNase treated to eliminate genomic DNA. RNA was quantified using a NanoDrop 2000 spectrophotometer (Thermo Scientific). Reverse-transcription reactions were performed using 1 µg of purified total RNA, 1 µl ImProm-IITM Reverse Transcriptase (Promega) and a reaction buffer containing 0.5mM dNTP, 6mM $MgCl_2$, 20 units recombinant RNasin® ribonuclease inhibitor, and 0.5 µg oligo(dt)15.

qRT-PCR

Quantitative polymerase chain reaction (qRT-PCR) was performed in a final volume of 20 µl consisting of 1 µl of the cDNA, 10 µl of SensiFAST SYBR No-ROX kit (Roche) and 15 µM of the gene-specific primers pairs as follows: *SbPPC3* (forward 5′–TGTTGAACAGTTTCTGGAACCTCTT-3′, reverse 5′-GCTTCA CAAGGGCAAGCCCAAAG-3′), *SbPPC2* (forward 5′-CCGCCT CGCAACACCTGAAACA-3′, reverse 5′-ACCGGGAGGTGGAA CCGTGT-3′),*SbPPC4* (forward 5′-TGAGCTTCGGGCACAAGC AGATG-3′, reverse 5′-GCTCCAAAGGCTCTAAGAACTGCT C-3′), and *18 S* rRNA (forward 5′-GGGGAAACTTACCA GGTCCA-3′, reverse 5′-GGATGGCTCCGCATAGCTA-3′). PCR was conducted on the MiniOpticon™ Real-Time PCR Detection System (Biorad), and the threshold cycles (Ct) were determined using Opticon Monitor™ analysis software for all treatments. To normalize the obtained values, 18S ribosomal RNA was used as internal control in each sample. This gene displays a steady RNA level across the experimental conditions.

Statistical analysis

Statistical analysis was performed using SigmaStat (Systat Software Inc., San José, CA, USA). Data were analyzed using the Student's t-test or with the Mann-Whitney U test. Means were considered to be significantly different at $P<0.05$.

RESULTS AND DISCUSSION

PEPC activity, and PTPC integrity and cellular localization

PEPC was characterized throughout the development, maturation, and germination of sorghum seeds, focusing on the enzyme's activity, integrity, subunit composition, and immunolocalization. Stages of sorghum seed development and germination are described; among these, stage IV of development marks the beginning of seed desiccation, whilst stage II of germination marks the protrusion of the radicle and the completion of germination (Bewley *et al.*, 2013). During seed development, PEPC activity of desalted extracts increased by about 3-fold to reach a maximum level at stage II–III and then decreased (Fig. 1). During germination, PEPC activity in the embryo increased to a maximum at 72 to 96h post-imbibition, while in the aleurone/endosperm it did not change significantly and was 10–15-fold lower than that of embryos (Fig. 1).

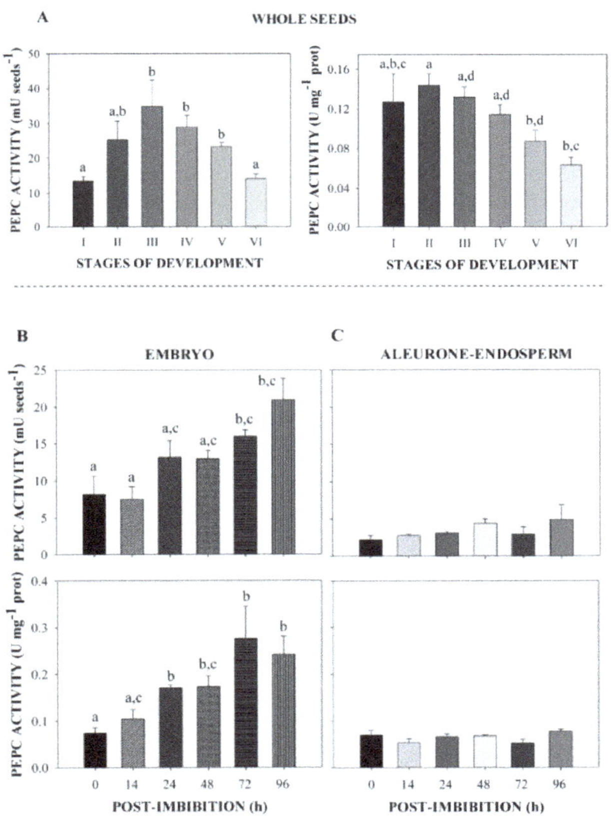

Figure 1: Time course of PEPC activity in developing and germinating sorghum seeds. PEPC activity was assayed at optimal pH (8.0), 2.5mM PEP, and 30 °C in clarified extracts from: (A) whole seeds during development; and (B) embryos or (C) aleurone-endosperm of germinating seeds. (A) PEPC activity is expressed on a per seed basis (mU seeds^{-1}) or on a protein basis (U mg^{-1} protein). Results are means ±SE of three independent experiments. Mean values that are not significantly different ($P<0.05$, Student's t-test) are indicated with the same letters. Data for aleurone-endosperm are not significantly different.

In germinating COS, the p110 and p107 PTPC polypeptides appear to be *in vivo* truncated by 19 amino acids such that their N-terminal phosphorylation domain is absent (Uhrig *et al.*, 2008). In contrast, non-monoubiquitinated p107 and monoubiquitinated p110 PTPCs from germinating sorghum seeds at 48h post-imbibition were shown to be intact and to therefore contain the phosphorylation motif (Ruiz-Ballesta *et al.*, 2014). In this study, we demonstrate the integrity of PTPC polypeptides throughout the life cycle of sorghum seeds since the anti-C19 and anti-N24 both cross-reacted with the p110 and p107. PTPC was also immunolocalized in germinating sorghum seed tissues characterized by high metabolic activities (epicotyl, radicle, aleurone layer, scutellum, scutellum's epithelium) (Fig. 2). Immunohistochemistry studies have localized PTPC in the protein bodies of wheat grains, where it was suggested to contribute to amino acid and protein biosynthesis during grain development (Araus *et al.*, 1993). PTPC has also been localized in tissues of developing and germinating wheat grains characterized by high metabolic activity (González *et al.*, 1998). It is noteworthy that PEPC was detected as an abundant protein in the aleurone layer using immuno-localization (Fig. 2E). Consistent with this result, the role of PEPC in malate production by the barley aleurone layer has been described by Macnicol and Raymond, (1998).

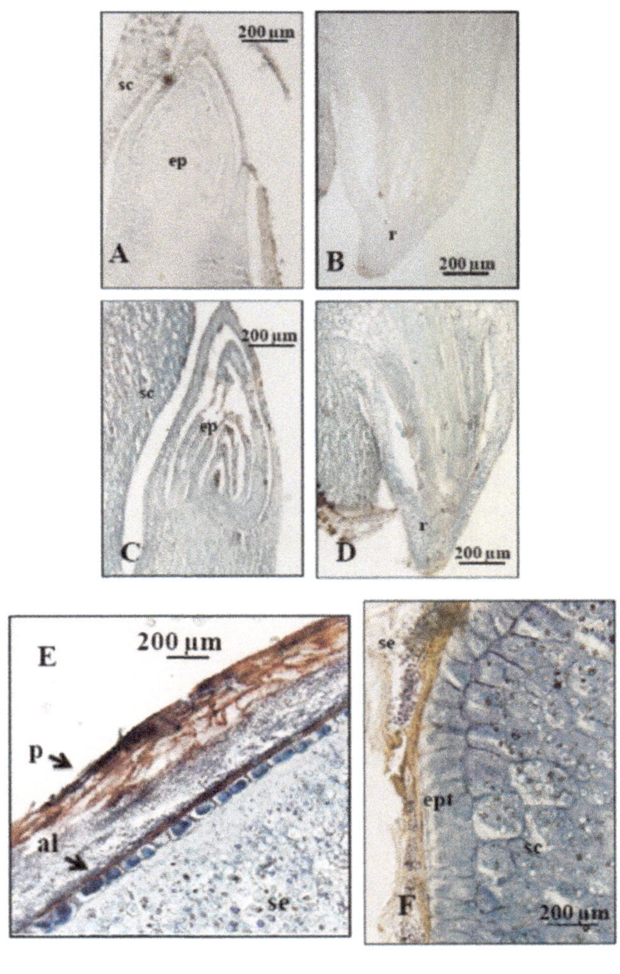

Figure 2: Immunological localization of PTPC in germinating sorghum seeds. Grains were imbibed for 24h. Sections (15 µm) were probed with: (A, B), preimmune serum, or (C–F) polyclonal anti-C_4PTPC (SbPPC1). Abbreviations: al, aleurone layer; ep, epicotyl; ept, epithelium; sc, scutellum; se, starchy endosperm; r, radicle; p, pericarp.

Collectively, these results indicated the stability and consistent functional relevance of PTPC throughout the life cycle of sorghum seeds.

Expression of PEPC genes in developing and germinating seeds

Sorghum PEPC is encoded by a small multigene family consisting of six genes: *SbPPC2*, *SbPPC3*, *SbPPC4*, *SbPPC5* (all C_3 PTPCs), and *SbPPC1* (C_4photosynthetic PTPC) encode closely related PTPCs, whilst *SbPPC6* encodes the distantly related BTPC (Table 1) (Paterson *et al.*, 2009). In this present study, we used qRT-PCR to analyze the pattern of their transcript abundance. *SbPPC2, SbPPC3*, and *SbPPC4* transcripts were present in developing and germinating sorghum seeds (Fig. 3). Transcripts of*SbPPC2*, the housekeeping PTPC (Lepiniec *et al.*, 1994), were detected during seed development and its abundance increased slightly during the maturation phase (Fig. 3A).

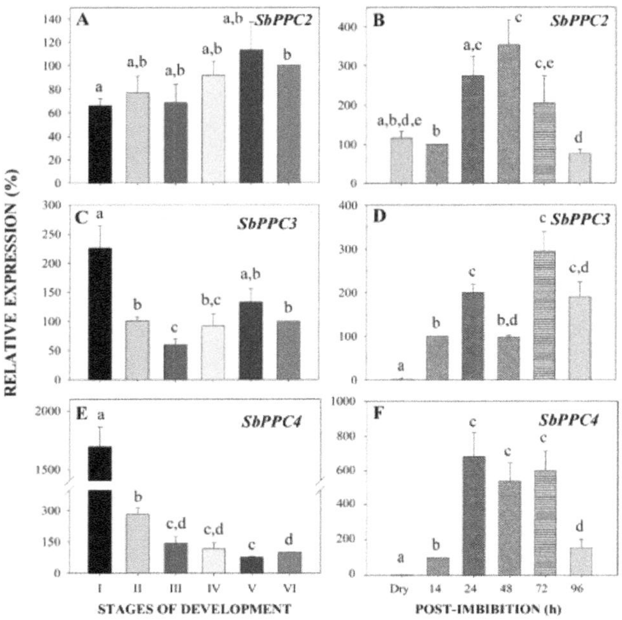

Figure 3: qRT-PCR analysis of *PTPC*transcript levels in developing and germinating sorghum seeds. To normalize the values, 18S RNA was used as internal control in each sample. Data were normalized to stage VI for developing seeds (A, C, E), or 14h post-imbibition germinating seeds (B, D, F). Results are means ±SE of at least three independent experiments. Mean values that are not significantly different ($P<0.05$, Student's *t*-test) are indicated with identical letters.

In the germinating seeds, *SbPPC2* showed maximal transcript abundance at 24–48h post-imbibition and then decreased (Fig. 3B). By contrast, *SbPPC3* transcript levels were maximum at the beginning of seed development, during the period of cellularization (Bewley *et al.*, 2013); this was followed by a decrease of *SbPPC3* transcripts until stage III and then an increase during the desiccation period with a peak at stage V (Fig. 3C). *SbPPC3* transcripts were also detected during germination with two important peaks at 24 and 72h post-imbibition (Fig. 3D). Finally, *SbPPC4* transcripts were quite abundant at the beginning of development, suggesting a possible specific role for SbPPC4 during the phase of cellularization (Fig. 3E). Cellularization is the massive production of new cells. Malate is a ubiquitous molecule that provides carbon skeletons and reducing power and is needed for the synthesis of storage end-products by developing seeds. *SbPPC4* transcripts were also very abundant at 24–72h post-imbibition in germinating seeds (Fig. 3F). By contrast, transcripts of *SbPPC5* and *SbPPC6* (*BTPC*) were very low during sorghum seed development and germination and, as expected, no *SbPPC1*transcripts were detected (results not shown). These results show a novel pattern of the steady-state of transcript abundance of *PTPC* isogenes during the life cycle of sorghum seeds, with an important presence of*SbPPC3* and also an important contribution of the *SbPPC4* isogene to the cellularization stage of development (stage I) and during germination.

Different PTPC isoenzymes co-exist in developing and germinating seeds

Peptide mass fingerprinting by MALDI-TOF MS of the p110 and p107 polypeptides immunoprecipitated using anti-C_4 PTPC revealed that different PTPC isoenzymes co-exist in developing and germinating sorghum seeds (Table 2; asterisks denote confirmation of the identification via LC MS/MS sequencing of tryptic peptides). SbPPC3 was the most abundant during the course of the life cycle of the seeds in both p107 and p110. SbPPC3 was highly abundant in the embryo (Table 2) and less abundant in the aleurone/endosperm of the germinating seed; however, SbPPC3 was the only isoenzyme detected in this tissue (Table 2). The housekeeping SbPPC2 (Lepiniec *et al.*, 1993, 1994) was present in early development only as a p107 (i.e. probably non-monoubiquitinated) subunit (Table 2, stages I and III) and was identified as a p110 and p107 subunit in stage VI (Table 2). SbPPC2 was also present as a monoubiquitinated p110 subunit in the embryo of dry and 48-h post-imbibition seeds, but was not identified in the aleurone (Table 2). Finally, the *SbPPC4* isogene was shown to be translated into the corresponding p107 PTPC subunits in early development (Table 2, stages I and III), in agreement with the transcript profiling results of Fig. 3E. In addition, SbPPC4 was also detected in p110, indicating the probable monoubiquitination of this PTPC isoenzyme. Although the precise physiological function of SbPPC4 remains unclear, our results support a role for this isoenzyme during the early stages of seed development. During the phase of cellularization SbPPC4 is hypothesized to generate the C-skeletons needed as precursors for synthesis of new components such as amino acids/proteins, and membrane or storage lipids. Finally, SbPPC5 was detected only as p110 polypeptides in stage I (Table 2). SbPPC6 polypeptides were never detected by MS/MS analysis, in agreement with the absence of *SbPPC6* expression during sorghum seed development and germination. Nor was a high molecular mass Class-2 PEPC hetero-octamer evident when clarified extracts from different stages of development and germination were subjected to non-denaturing PAGE followed by in-gel PEPC activity staining. However, tetrameric and dimeric Class-1 PEPC oligomers were observed, with the dimeric form being the most abundant *in vitro*. This dimeric form was not altered by de-ubiquitination of the enzyme after incubation with 2 μM of the catalytic subunit of ubiquitin specific protease-2 (USP_{2c}).

Table 2: Identification of p107 and p110 PTPC isoforms from developing and germinating sorghum seeds. Proteins were identified via MALDI-TOF/TOF MS peptide mass fingerprinting of tryptic peptides derived from immunoprecipitated samples of representative stages. Asterisks indicate identification via MS/MS sequencing of tryptic peptides and BLAST analysis. DPA, days post-anthesis

STAGE	Subunit	Protein name	Accession number[a]	Score	Sequence coverage (%)	N°. of Matching Peptides
I (7–12 DPA)	p110	PEPC SbPPC3*	gi|241931686	248	35	31
		PEPC SbPPC4*	gi|241940554	118	24	22
		PEPC SbPPC5	gi|241930496	98	22	21
	p107	PEPC SbPPC3	gi|241931686	325	45	39
		PEPC SbPPC4	gi|241940554	114	26	21
		PEPC SbPPC2	gi|241925556	107	23	20
III (16–20 DPA)	p110	PEPC SbPPC3	gi|241931686	103	24	21
	p107	PEPC SbPPC3	gi|241931686	206	33	29
		PEPC SbPPC4	gi|241940554	66	19	16
		PEPC SbPPC2	gi|241925556	61	17	15
VI (31–40 DPA)	p110	PEPC SbPPC3	gi|241931686	60	10	13
		PEPC SbPPC2*	gi|241925556	36	9	8
	p107	PEPC SbPPC3	gi|241931686	121	15	18
		PEPC SbPPC2*	gi|241925556	76	8	10
Dry seed Aleurone/endosperm	p110	PEPC SbPPC3	gi|241931686	30	13	9
	p107	PEPC SbPPC3	gi|241931686	72	20	15
Dry seed Embryo	p110	PEPC SbPPC3*	gi|241931686	310	45	39
		PEPC SbPPC2	gi|241925556	91	21	18
	p107	PEPCSb PPC3	gi|241931686	283	42	48
		PEPC SbPPC2	gi|241925556	63	16	14
48 h post-imbibition	p110	PEPC SbPPC3*	gi|241931686	95	24	16
Aleurone/endosperm	p107	PEPC SbPPC3	gi|241931686	67	16	15
48 h post-imbibition	p110	PEPC SbPPC3	gi|241931686	305	44	38
Embryo		PEPC SbPPC2	gi|241925556	54	15	14
	p107	PEPC SbPPC3*	gi|241931686	164	29	38
		PEPC SbPPC2	gi|241925556	55	15	13

[a]NCBI database

A number of important points emerge from these results, as follows. (i) Multiple C_3 PTPC isoenzymes are present during the development and the germination of sorghum seeds. SbPPC3 appears to have an essential role as it is the isoenzyme most consistently present during the course of the life cycle of the seeds , and was the only one detected in the aleurone/endosperm. (ii) The SbPPC4 isozyme appears to play a major role during the very early stage of seed development, known as the phase of cellularization (Bewley *et al.*, 2013). (iii) BTPC is not significantly expressed throughout the life cycle of the seeds, which is an important difference relative to developing COS, a non-green oil seed, in which p118 BTPC polypeptides, and thus Class-2 PEPC complexes are highly expressed (O'Leary *et al.*, 2011*b*).

Post-translational modifications of the different Class-1 PEPC isoenzymes during germination

In the endosperm of germinating COS, p110 PTPC subunits are monoubiquitinated but not phosphorylated (Uhrig *et al.*, 2008; O'Leary *et al.*, 2011*b*). On the other hand, simultaneous p110 monoubiquitination and p107 phosphorylation have been described as an alternative PTM pattern for Class-1 PEPC of immature proteoid (cluster) roots of harsh hakea (Shane *et al.*, 2013). However, the coincident occurrence of both PTMs in sorghum seed p110 and p107 PTPC (i.e. SbPPC3) subunits has only been shown *in vitro* (Ruiz-Ballesta *et al.*, 2014). In this study, we show both PTMs also occur *in vivo* with three of the four PTPC isoenzymes present in germinating sorghum seeds. To this end, PTPC polypeptides of clarified extracts from 48-h post-imbibition germinating seeds were isolated by immunoaffinity chromatography with anti-COS PEPC. Coomassie blue R-250-stained p110 and p107 were excised from SDS gels of the immunopurified PTPC and individually subjected to in-gel tryptic digestion and detailed analysis by LC MS/MS. A Mascot database search of p110 and p107 datasets derived from nano HPLC TripleTOF MS of their tryptic peptides (Table 3) revealed the presence of SbPPC2, SbPPC3, and SbPPC4 PTPCs in p110 and p107, as well as SbPPC5 in p107. Consistent with our earlier results (Ruiz-Ballestaet *al.*, 2014), p107 N-terminal tryptic peptides of SbPPC3 containing non-phosphorylated Ser7 were identified by Orbitrap LC MS/MS. However, the monoubiquitinated SbPPC3 p110 subunits were also determined to be phosphorylated at Ser7 (Table 3). Monoubiquitinated SbPPC2 and SbPPC4 (i.e. p110 subunits), and deubiquitinated SbPPC5 (i.e. p107 subunits) were also shown to be phosphorylated at their conserved N-terminal seryl phosphorylation sites; i.e. Ser13, Ser10, and Ser11 for SbPPC2, SbPPC4, and SbPPC5, respectively (Table 3). To the best of our knowledge, this is the first evidence that *in vivo* phosphorylation of different PTPC isoenzymes can simultaneously occur in any plant tissue. MS/MS also confirmed the monoubiquitination of SbPPC3 p110 subunits at Lys624 (Table 3), as previously reported (Ruiz-Ballesta *et al.*, 2014). In addition, TripleTOF MS/MS analysis of the quadruply charged SbPPC2 p110 peptide ion of m/z 455.7462 (LC retention time 16.61min) and the triply charged ion of m/z 507.9268 (LC retention time 18.01min), corresponding to residues 623–637 (VAKDFGVKLTMFHGR) and 626–637 (DFGVKLTMFHGR), showed that the masses of the C-terminal fragments up to y7 matched the predicted values, but the mass shift of the fragment y8 by 114Da indicated a ubiquitination site localized at Lys-630, where a characteristic mass increment of 114Da indicates a Gly-Gly motif attachment of ubiquitin at the side chain. LC MS/MS measurements of the doubly and triply charged ions of m/z 769.3827 and m/z 513.2607 (LC retention time 23.22min) using a high-resolution Orbitrap Fusion Tribrid instrument identified the ubiquitination of the oxidized peptide at residues 626–637. Under high-resolution HCD fragmentation, the mass shift of 114Da on both the N-terminal *b5* ion and the C-terminal *y8* ion unambiguously confirmed Lys-630 as p110's monoubiquitination site in SbPPC2. This residue precisely aligns with the Lys-624 and Lys-628 (a highly conserved PTPC residue that is immediately proximal to catalytic residues involved in PEP binding) monoubiquitination sites, respectively determined for p110 subunits of sorghum seed SbPPC3 PTPC (Table 3 and Ruiz-Ballesta *et al.*, 2014) and PTPC (RcPPC3) from germinated COS (Uhrig *et al.*, 2008).

Table 3: Identification of PTMs of immunopurified PTPC polypeptides from 48h germinated sorghum seeds using TripleTOF 5600 and Orbitrap Fusion mass spectrometry. pSer; phosphorylated-Ser; Ub, monoubiquitinated

Peptide	m/z(obs.)	Mr(exp.)	Mr(calc.)	ppm	Score	Peptide sequence	Modification sites
p107: PEPC SbPPC2, 107 kDa, NCBI access # gi|242048870, sequence coverage 90%.							
8–19	749.8647	1497.7148	1497.7010	9.2	18	K.MERLS**S**IDAQLR.M	pSer13
11–19	541.7654	1081.5163	1081.5169	-0.5	49	R.LS**S**IDAQLR.M	pSer13
p107: PEPC SbPPC3, 107 kDa, NCBI access # gi|242061132, sequence coverage 87%.							Not detected
p107: PEPC SbPPC4, 107 kDa, NCBI access # gi|242078871, sequence coverage 89%.							
8–16	542.7460	1083.4764	1083.4784	-1.9	34	K.MA**S**IDAQLR.M	pSer10
8–16	550.7449	1099.4753	1099.4733	1.8	34	K.MA**S**IDAQLR.M	pSer10; Met-Ox
p107: PEPC SbPPC5, 107 kDa, NCBI access # gi|242058751, sequence coverage 64%.							
4–17	527.9304	1580.7694	1580.7559	8.5	16	R.NAVDKAT**S**IDAQLR.L	pSer11
4–17	791.3842	1580.7528	1580.7559	2.0	58	R.NAVDKAT**S**IDAQLR.L	pSer11
p110: PEPC SbPPC2, 110kDa, NCBI access # gi|242048870, sequence coverage 94%.							
8–19	749.8642	1497.7128	1497.7010	8.2	22	K.MERLS**S**IDAQLR.M	pSer13
11–19	541.7646	1081.5146	1081.5169	-2.1	43	R.LS**S**IDAQLR.M	pSer13
623–637	455.7462	1818.9535	1818.9563	1.6	110	K.VAKDFGV**K**LTMFHGR.G	UbLys-630
626–637	507.9268	1520.7570	1520.7558	-1.0	60	K.DFGV**K**LTMFHGR.G	UbLys-630
626–637	513.2607	1536.7587	1536.7507	5.5	19	K.DFGV**K**LTMFHGR.G	UbLys-630, Met-OX
626–637	769.3827	1536.7498	1536.7507	1.0	30	K.DFGV**K**LTMFHGR.G	UbLys-630, Met-OX
p110: PEPC SbPPC3, 110kDa, NCBI access # gi|242061132, sequence coverage 93%.							
2–13	765.3659	1528.7173	1528.7147	1.6	19	M.PERHQ**S**IDAQLR.L	pSer7
620–631	780.4022	1558.7888	1558.7827	4.1	90	K.HYGV**K**LTMFHGR.G	UbLys-624
620–631	390.7011	1558.7752	1558.7827	-4.8	42	K.HYGV**K**LTMFHGR.G	UbLys-624
620–631	520.5975	1558.7707	1558.7827	-7.6	71	K.HYGV**K**LTMFHGR.G	UbLys-624
p110: PEPC SbPPC4, 110kDa, NCBI access # gi|242078871, sequence coverage 90%.							
8–16	542.7472	1083.4798	1083.4784	1.3	17	K.MA**S**IDAQLR.M	pSer10
8–16	550.7423	1099.4700	1099.4733	-3.0	37	K.MA**S**IDAQLR.M	pSer10

Collectively, these results establish for the first time several important points. (i) Different PTPC isoenzymes are co-expressed during sorghum seed germination and all of them except the p107 subunits of SbPPC3 are phosphorylated during germination. (ii) Monoubiquitination of the corresponding conserved Lys was confirmed by MS/MS analysis for SbPPC3, and determined in this study for SbPPC2, demonstrating that this PTM occurs with multiple PTPC isoenzymes at a conserved lysine residue within a single plant species. (iii) Monoubiquitination and phosphorylation can simultaneously occur on the same PTPC polypeptide and are therefore not mutually exclusive PTMs, as appears to be the case with PTPCs from developing proteoid roots of harsh hakea (Shane *et al.*, 2013), as well as various green and non-green tissues of the castor oil plant (Uhrig *et al.*, 2008; O'Leary *et al.*, 2011*b*).

Is PTPC monoubiquitination linked to specific stages and/or environmental conditions during sorghum seed germination?

Because of its recent discovery (Uhrig *et al.*, 2008), there are few studies about the influence of environmental factors or biotic/abiotic stresses on PTPC monoubiquitination during seed development and germination. Elimination of photosynthate translocation (a type of abiotic stress) from the shoot to the seed by depodding (fruit excision) of developing COS triggered rapid *in vivo* p107 dephosphorylation, followed by monoubiquitination of 50% of the p107 subunits to form p110 characteristic of germinating COS (O'Leary *et al.*, 2011b). Sorghum is a cereal cultivated in warm environments and seeds germinate well within a wide range of temperature from 25–35 °C, with an optimum of over 30 °C (Kanemasu *et al.*, 1975). This allowed us to correlate the effect of temperature with germination and the pattern of PTPC monoubiquitination (i.e. appearance of p110 polypeptides on PTPC immunoblots). To this end, seeds from two sorghum varieties PR88Y20 and PR87G57, respectively characterized by rapid and slow germination velocities, were imbibed in water for up to 96h at 25 and 35 °C. As defined by Bewley *et al.* (2013), germination begins with water uptake by the seed (imbibition, Phase I) and ends with the emergence of the radicle, which marks the end of Phase II and the beginning of Phase III (Bewley *et al.*, 2013). Phase II is also associated with a slow rate of water uptake, where the little net movement of water into the seed is due mainly to a lowering of the solute potencial ($_{s}$), which drives an increase in volume that results in cracking of the testa. Germination does not include seedling growth. Accordingly, eight stages of germination were established in this study. Among them stage II represents the emergence of the root, and the beginning of the post-germination Phase III (Bewley *et al.*, 2013). Germination rates were calculated by determining the average of the different stages of germination in a pooled of seeds at a given time post-imbibition (0, 24, 48, 72, and 96h). Figure 4 shows germination rate for the two sorghum varieties at 25 and 35 °C. Both varieties had a higher germination velocity at 35 than 25 °C, with the seeds germinating at 35 °C having an average lead of two stages over those incubated at 25 °C (Fig. 4). Moreover, PR88Y20 germinated more rapidly than PR87G57 at both 25 and 35 °C (Fig. 4). In parallel, clarified extracts from the same pool of seeds (the 15–20 seeds in the most advanced stages of germination at each time of imbibition) were analyzed by SDS-PAGE and immunoblotting with anti-C_4 PEPC (SbPPC1) (Fig. 5). PR88Y20 at 25 °C attained maximal PTPC monoubiquitination (i.e. formation of p110) between 24 and 48h post-imbibition (Fig. 5A, arrow), with a p110:p107 ratio of approximately 2. By contrast, maximal PTPC monoubiquitination in PR87G57 appeared to be delayed by up to 48–72h, in accordance with its slower germination rate (Fig. 5A, dashed arrow), with a p110:p107 ratio of 3 at 48h. PTPC monoubiquitination in seeds germinated at 35 °C occurred much earlier (Fig. 5B, var. PR88Y20), starting between 9 and 12h post-imbibition and reaching a maximal p110:p107 ratio (around 1.7) at 18h (Fig. 5B, arrow). These results indicate a strong correlation between the germination and post-germination velocity and the pattern of monoubiquitination of p110 PTPC subunits. Further analysis of the results indicated that, regardless of the time post-imbibition, PTPC monoubiquitination starts around stage II of germination for each variety or temperature and decreases around stage IV–V (Fig. 4). Thus PTPC monoubiquitination is well correlated with Stage II's period of high cell division during which there is massive mobilization of reserves, and corresponding biosynthesis of the new macromolecules needed for development of the growing seedling. In addition, PTPC monoubiquitination decreased around stage IV–V when the coleoptile is about 1cm long and becomes green to initiate photoautotrophic seedling growth.

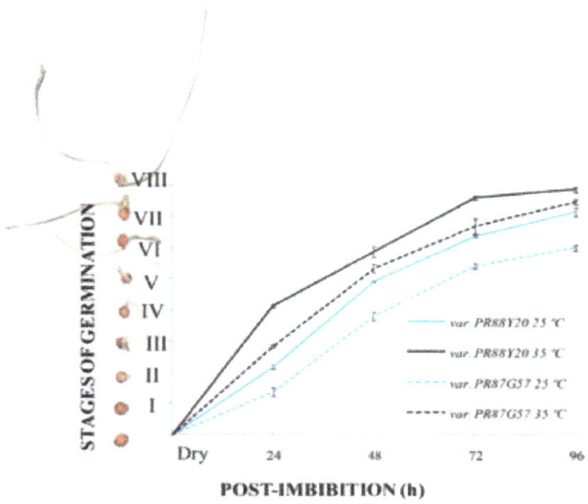

Figure 4: Germination rate of two different sorghum varieties at 25 and 35 °C. PR88Y20 and PR87G57 were germinated from dry seed to 96h post-imbibition. The average stage of germination was calculated as described in the Materials and Methods. Results are means ±SE of at least three independent experiments.

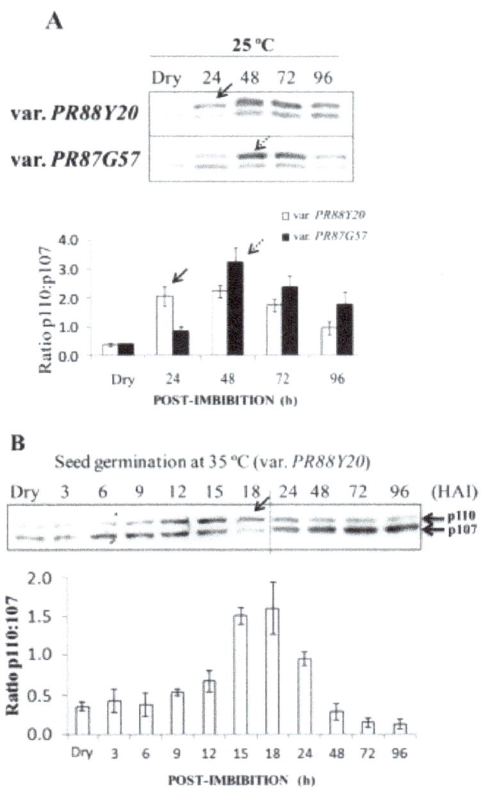

Figure 5: Patterns of PTPC monoubiquitination in two different varieties of sorghum at 25 and 35 °C. (A) Clarified extracts from dry seeds to 96h post-imbibition were subjected to 8% SDS-PAGE (50 µg protein lane^{-1}) and analyzed by immunoblotting using anti-C_4 PTPC SbPPC1. The plot represents the ratio of p110:p107 determined from scanned immunoblots. (A) and (B) show seeds imbibed at 25 or 35 °C, respectively. Results are means ±SE of at least three independent experiments. The arrows indicate the maximum level of monoubiquitination and p110:p107 ratio. HAI, hours after imbibition.

To further analyze this phenomenon, seeds were imbibed with water for periods up to 24h and classified as germinated seeds (stage II) or non-germinated seeds (stage I) depending on the emergence or not of the radicle. PTPC monoubiquitination was far more prevalent in germinating seeds where the emergence of the radicle occurred (Fig. 6A, Germ.). These results confirmed that PTPC monoubiquitination in germinating sorghum seeds coincides with metabolic events associated with the emergence of the radicle (stage II) and early seedling growth.

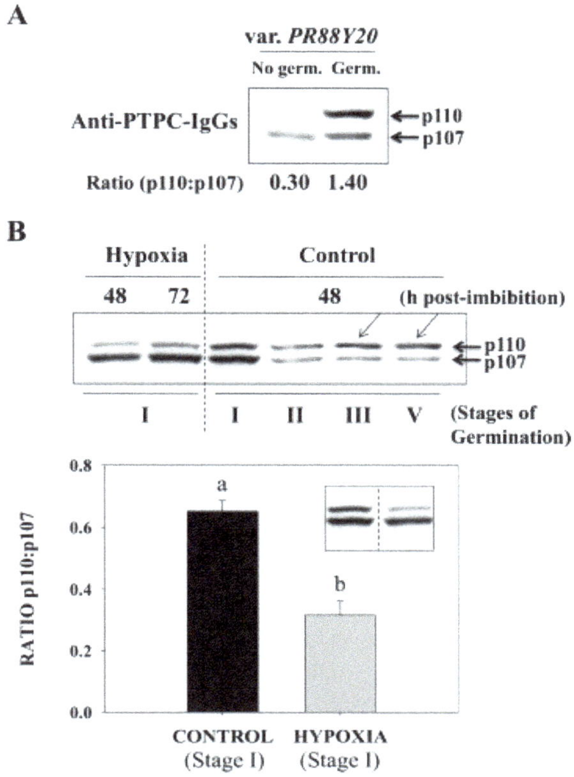

Figure 6: PTPC monoubiquitination occurs in viable seeds that have increased aerobic respiration and reserve mobilization associated with radicle emergence. (A) Seeds (var. PR88Y20) at 24h post-imbibition were separated into two pools: germinated seeds (where the testa was cracked with protrusion of the radicle; Germ) and non-germinated seeds (where the testa was not disrupted, No Germ.). Clarified extracts were subjected to 8% SDS-PAGE and analyzed by immunoblotting with anti-C_4 PTPC. (B) Seeds were imbibed submerged in water (Hypoxia) or on moistened filter paper (Control, normoxic). At 48h post-imbibition control seeds mostly ranged from stage II to IV of germination, although some remained at stage I (Control, stages I, II, III and IV), while hypoxic seeds remained at stage I even at 72h post-imbibition. Results in the plot represent the ratio of p110:p107 in control and hypoxic seeds at stage I, 48h post-imbibition, and represent means ±SE of at least three independent experiments. Mean values that are significantly different (P<0.01, Student's t-test) are indicated with different letters.

As the relative rates of glycolysis and gluconeogenesis and adenylate energy charge are influenced by oxygen availability (Bewley *et al.*, 2013), it was also of interest to study the impact of hypoxia stress on PTPC monoubiquitination during seed germination. To this end, seeds were imbibed on moist filter paper (control, normoxic) or submerged in water (hypoxic). At 48h post-imbibition most of the normoxic seeds reached stage II, a smaller number remained at stage I, whereas others had attained stage III–IV (Fig. 6, control). However, the submerged seeds remained at stage I, at 72h post-imbibition (Fig. 6, Hypoxia). The hypoxia-stressed seeds were not dead since they germinated well after they were transferred to normoxic conditions (data not shown). As expected, PTPC monoubiquitination of normoxic seeds markedly increased at stages II to III, but remained low in hypoxia-stressed seeds, even up to 72h post-imbibition (Fig. 6B). These results establish that enhanced PTPC monoubiquitination is intrinsically linked to the germination of the seed and corresponding enhanced aerobic respiration, cracking of the testa, and the protrusion of the radicle. This makes sense since the ubiquitination machinery is ATP dependent, and abundant ATP probably requires abundant aerobic respiration (Callis, 2014). In addition, even if germination did not occur (i.e. the radicle did not emerge; stage I), PTPC monoubiquitination was enhanced under normoxic relative to hypoxic conditions (Fig. 6B).

These are the first results showing that PTPC monoubiquitination in germinating sorghum seeds: (i) is very low during Phase I; (ii) always increases at stage II (emergence of the radicle); (iii) is maintained during the aerobic period of high activity of cell division and reserve mobilization; and (iv) remains until stage IV–V when coleoptiles are about 1cm long and the formation of the photosynthetic tissues begins.

CONCLUDING REMARKS

Deciphering the occurrence, roles, and PTMs of different PTPC isoenzymes during seed development and germination has been poorly studied. Monoubiquitination was recently discovered to be a novel but relatively common PTM of PTPC subunits of Class-1 PEPCs from diverse tissues and plant species (Uhrig *et al.*, 2008; Igawa *et al.*, 2010; O'Leary *et al.*, 2011b;Ruiz-Ballesta *et al.*, 2014; Shane *et al.*, 2013). However, the occurrence of PTPC monoubiquitination and phosphorylation has been specifically described only in the limited examples of developing and germinating COS (Tripodi *et al.*, 2006; Uhrig *et al.*, 2008; O'Leary *et al.*, 2011*b*), and during proteoid root formation in harsh hakea (Shane *et al.*, 2013). This present study describes the patterns of transcript levels and relative abundance of different PTPC isozymes during sorghum seed development and germination, as well as their PTM by monoubiquitination and phosphorylation during germination. The results reveal new and important points; for example, the high level of *SbPPC4* transcripts during early development, at stage I (Fig. 3). This is the first report in which a possible role of this PTPC isozyme during cellularization of the developing seed (Bewley *et al.*, 2013) is suggested. The absence of *BTPC* transcripts and protein represent an important difference with developing COS, a non-green oil seed in which BTPC is highly expressed and appears to play an important role as a catalytic and regulatory subunit of the unusual Class-2 PEPC hetero-octameric complex (Uhrig *et al.*, 2008; O'Leary *et al.*, 2011*a,b*). *SbPPC2* and *SbPPC3* transcripts and SbPPC2 and SbPPC3 isoenzymes were extensively detected throughout the life cycle of sorghum seeds (Fig. 3 and Table 2). As SbPPC3 expression in roots depends on nitrogen supply (Lepiniec *et al.*, 1993), we propose an important role of this enzyme in providing the precursors needed for amino acid synthesis and transamination reactions in the germinating seed. In addition, SbPPC3 was the only isoenzyme found in the aleurone/endosperm, where it is proposed to play an important role in regulating the production of malic acid, which is either excreted for the acidification of the starchy endosperm (Macnicol and Raymond, 1998), or transported to the growing seedling for the anaplerotic function (Osuna *et al.*, 1996, 1999). Among the immunopurified PTPCs revealed by MS/MS analysis SbPPC2, SbPPC3, and SbPPC4 are monoubiquitinated *in vivo*. These results provide the first evidence that monoubiquitination is not restricted to a single PTPC isoenzyme within a specific plant species or physiological context, but can occur with multiple PTPC isozymes *in vivo*. SbPPC2, SbPPC3, and SbPPC4 were simultaneously monoubiquitinated and phosphorylated (Table 3), showing that both PTMs can occur *in vivo* in the same PTPC polypeptide (Table 3); i.e., they are not mutually exclusive PTMs, as appears to be the case in castor oil plants (Uhrig *et al.*, 2008; O'Leary *et al.*, 2011*b*). Class-1 PEPC phosphorylation at the single highly conserved serine residue close to the N-terminus of the protein uniformly results in the enzyme activation at physiological pH by decreasing and increasing its sensitivity to allosteric inhibitors and activators, respectively (Echevarría *et al.*, 1994). By contrast, PTPC monoubiquitination has an opposite kinetic effect (Uhrig*et al.*, 2008; Shane *et al.*, 2013). In germinating COS and developing proteoid roots of P-deficient harsh hakea, both PTPC PTMs appear to be exclusive, leading to a clear picture about the impact of each PTM on the activity of the enzyme. However, we show in this study that monoubiquitination and phosphorylation can simultaneously occur with three of the four PTPC isoenzymes expressed in sorghum seeds. This suggests that a combination of both PTMs may serve to fine tune the precise kinetic and regulatory properties for each PTPC isoenzyme at each specific stage of development and germination. However, it is at present difficult to establish the precise impact of phosphorylation versus monoubiquitination in each PTPC isoenzyme at each stage of the life cycle of sorghum seed. The use of phosphorylation-site-specific PTPC antibodies (i.e. anti-pSer13 and anti-pSer7 for SbPPC2 and SbPPC3, respectively), which we are presently characterizing, will help us to evaluate the specific pattern of *in vivo* phosphorylation for the two predominant PTPC isoenzymes in sorghum seeds. In this study we have documented the remarkable versatility and complexity of *in vivo* PTMs of different PTPC isoenzymes during the life cycle of sorghum seeds.

This is also the first study where the extent of PEPC monoubiquitination was observed to increase at stage II of germination. This stage corresponds to radicle protrusion and the beginning of the post-germination phase (Phases II and III, respectively, as described by Bewley*et al.*, 2013), and also to initiation of the second respiratory burst due to the cracking of the testa (Bewley *et al.*, 2013).

Future studies are needed in order to clarify the extent of PEPC monoubiquitination in other seeds, its relation with the efficiency of germination, and its possible regulation by metabolites and other elements of the signal transduction pathway that lead to this PTM. It will also be important to assess whether this PTM influences PEPC's stability, protein–protein interactions, and/or subcellular location. It will be equally important to identify how this PTM is co-ordinated with PPCK-mediated phosphorylation-activation of PTPCs at their conserved N-terminal seryl residue. We are currently attempting to disrupt *PPCK* and *SbPPC3*expression in transgenic sorghum in order to study the impact of the absence of PTPC phosphorylation (*PPCK* silenced), or the SbPPC3 PTPC isoenzyme on seed development and germination.

ACKNOWLEDGEMENTS

We gratefully acknowledge Felipe Clemente and Dolores Gutierrez from Unidad de Proteómica, Centro de Genómica y Proteómica, Universidad Complutense de Madrid for helpful comments on the MS analyses shown in Table 1. We thank Dr Jean Vidal for providing important scientific comments and discussion, Dr Rosario Álvarez and Dr José Antonio Monreal for collaborating in the immunohistochemistry detection and oligonucleotide design, respectively. This research was supported by the 'Ministerio de Economía y Competitividad' grant no. AGL2012-35708, by 'La Junta de Andalucía' grant no. P06-CVI-02186 and BIO298, and the 'Plan Propio de la Universidad de Sevilla'. This work was also supported by grants from the Natural Science and Engineering Research Council of Canada and Queens's Research Chairs program to W.C.P., and the 100 Technical Talents Program of the Chinese Academy of Sciences to Y. M. She.

REFERENCES

1. Araus JL Bort J Brown RH Bassett CL Cortadellas N . 1993. Inmunocytochemical localization of phosphoenolpyruvate carboxylase and photosynthesis gas-exchange characteristics in ears of Triticum aestivum Desf. Planta 191, 507–514.
2. Bewley JD Bradford KJ Hilhorst HWM Nonogaki H . 2013. Seeds: physiology of development, germination and dormancy , 3rd edn. New York, NY: Springer.
3. Callis A . 2014. The ubiquitination machinery of the ubiquitin system. The Arabidopsis Book, 06, e0174. doi: 10.1199/tab.0174.
4. Chollet R Vidal J O'Leary M . 1996. Phosphoenolpyruvate carboxylase: a ubiquitous, highly regulated enzyme in plants. Annual Review of Plant Physiology and Plant Molecular Biology 47, 273–298.
5. Echevarría C Vidal J . 2003. The unique phosphoenolpyruvate carboxylase kinase. Plant Physiology and Biochemistry 41, 541–547.
6. Echevarría C Pacquit V Bakrim N Osuna L Delgado B Arrio-Dupont M Vidal J . 1994. The effect of pH on the covalent and metabolic control of C4 phosphoenolpyruvate carboxylase from sorghum leaf. Archive of Biochemistry and Biophysics 313, 425–430.
7. Feria AB Álvarez R Cochereau L Vidal J García-Mauriño S Echevarría C . 2008. Regulation of phosphoenolpyruvate carboxylase phosphorylation by metabolites and ABA during the development and germination of barley seeds. Plant Physiology 148, 761–774.
8. González MC Osuna L Echevarría C Vidal J Cejudo FJ . 1998. Expression and localization of PEP carboxylase in developing and germinating wheat grains. Plant Physiology 116, 1249–1258.
9. Igawa T Fujiwara M Tanaka I Fukao Y Yanagawa Y . 2010. Characterization of bacterial-type phosphoenolpyruvate carboxylase expressed in male gametophyte of higher plants. BMC Plant Biology 10, 200.
10. Izui K Matsumura H Furumoto T Kai Y . 2004. Phosphoenolpyruvate carboxylase: a new era of structural biology. Annual Review of Plant Biology 55, 69–84.
11. Kanemasu ET Bark DL Chin Choy E . 1975. Effect of soil temperature on sorghum emergence. Plant Soil 43, 411–417.
12. Kresovich S Barbazuk B Bedell JA et al. 2005. Toward sequencing the sorghum genome. A U.S. National Science Foundation–Sponsored Workshop Report. Plant Physiology 138, 1898–1902.
13. Lepiniec L Keryer E Philippe H Gadal P Cretin C . 1993. Sorghum phosphoenolpyruvate carboxylase gene family, structure, function and molecular evolution. Plant Molecular Biology 21, 487–502.
14. Lepiniec L Vidal J Chollet R Gadal P Crétin C . 1994. Phosphoenolpyruvate carboxylase: structure, regulation and evolution. Plant Science 99, 111–124.
15. Macnicol PK Raymond P . 1998. Role of phosphoenolpyruvate carboxylase in malate production by the developing barley aleurone layer. Physiolgia Plantarum 103, 132–138.
16. Nhiri M Bakrim N Bakrim N El Hachimi-Messouak Z Echevarría C Vidal J . 2000. Posttranslational regulation of phosphoenolpyruvate carboxylase during germination of sorghum seeds, influence of NaCl and L-malate. Plant Science 151, 29–37.
17. Nimmo HG . 2000. The regulation of phosphoenolpyruvate carboxylase in CAM plants. Trends in Plant Science 5, 75–80.
18. Nimmo HG . 2003. Control of the phosphorylation of phosphoenolpyruvate carboxylase in higher plants. Archives of Biochemistry and Biophysics 414, 189–196.
19. O'Leary B Fedosejevs ET Hill AT Bettridge J Park J Rao SK Leach CA Plaxton WC . 2011 b. Tissue-specific expression and post-translational modifications of plant- and bacterial-type phosphoenolpyruvate carboxylase isozymes of the castor oil plant, Ricinus communis L. Journal of Experimental Botany 62, 5485–5495.
20. O'Leary B Park J Plaxton WC . 2011 a. The remarkable diversity of plant PEPC (phosphoenolpyruvate carboxylase): recent insights into the physiological function and post-translational controls of non-photosynthetic PEPCs. Biochemical Journal 436, 15–34.
21. Osuna L González MC Cejudo FJ Vidal J Echevarría C . 1996. In vivo and in vitro phosphorylation of the phosphoenolpyruvate carboxylase from wheat seeds during germination. Plant Physiology 111, 551–558.
22. Osuna L Pierre JN González MC Álvarez R Cejudo FJ Echevarría C Vidal J . 1999. Evidence for a slow-turnover form of the Ca2+-independent phosphoenolpyruvate carboxylase kinase in the aleurone endosperm tissue of germinating barley seeds. Plant Physiology 119, 511–520.
23. Pacquit V Giglioli N Crétin C Pierre JN Vidal J Echevarría C . 1995. Regulatory phosphorylation of C4 phosphoenolpyruvate carboxylase from sorghum. An immunological study using specific anti-phosphorylation site-antibodies. Photosynthesis Research 43, 283–288.
24. Park J Khuu N Howard ASM Mullen RT Plaxton WC

. 2012. Bacterial- and plant-type phosphoenolpyruvate carboxylase isozymes from developing castor oil seeds interact in vivo and associate with the surface of mitochondria. The Plant Journal 71, 251–262.

25. Paterson AH Bowers JE Bruggmann R et al. 2009. The Sorghum bicolor genome and the diversification of grasses. Nature 457, 551–556.

26. Rivoal J Turpin DH Plaxton WC . 2002. In vitro phosphorylation of phosphoenolpyruvate carboxylase from the green alga Selenastrum minutum . Plant Cell Physiology 43, 785–792.

27. Ruiz-Ballesta I Feria AB Ni H She YM Plaxton WC Echevarría C . 2014. In vivo monoubiquitination of anaplerotic phosphoenolpyruvate carboxylase occurs at Lys624 in germinating sorghum seeds. Journal of Experimental Botany 65, 443–451.

28. Sechi S Chait BT . 1998. Modification of cysteine residues by alkylation. A tool in peptide mapping and protein identification. Analytical Chemistry 70, 5150–5158.

29. Shane MW Fedosejevs ET Plaxton WC . 2013. Reciprocal control of anaplerotic phosphoenolpyruvate carboxylase by in vivo monoubiquitination and phosphorylation in developing proteoid roots of phosphate-deficient Harsh Hakea. Plant Physiology 161, 1634–1644.

30. Tripodi KE Turner WL Gennidakis S Plaxton WC . 2005. In vivo regulatory phosphorylation of novel phosphoenolpyruvate carboxylase isoforms in endosperm of developing castor oil seeds. Plant Physiology 139, 969–978.

31. Uhrig RG She YM Leach CA Plaxton WC . 2008. Regulatory monoubiquitination of phosphoenolpyruvate carboxylase in germinating castor oil seeds. Journal of Biological Chemistry 283, 29650–29657.

CHAPTER 8

BIOCHEMICAL ANALYSES OF SORGHUM VARIETIES REVEAL DIFFERENTIAL RESPONSES TO DROUGHT

Chukwuma C. Ogbaga, Piotr Stepien, Beth C. Dyson, Nicholas J. W. Rattray, David I. Ellis, Royston Goodacre, Giles N. Johnson

INTRODUCTION

Drought is one of the major limitations on food production especially in developing countries in Africa and Asia [1]. There is a growing need to produce crops that can survive extreme environmental conditions such as drought in order to maximise crop production. Detrimental effects of water deficiency can be seen in the roots, shoots and leaves of green plants at morphological, physiological and biochemical levels. In this paper, the emphasis will be on the plant's biochemical responses to drought, in terms of its metabolism and quantification of selected proteins.

Exposure to drought triggers a variety of biochemical responses in plants and these responses include the accumulation of a variety of different compatible solutes and up-regulation of specific drought-induced proteins e.g. heat shock proteins (HSPs) and dehydrins (DHNs). Compatible solutes are osmolytes that protect plants from osmotic stress whilst maintaining cytosolic osmotic balance during extreme conditions in plants [2]. Drought has been reported to induce the accumulation of compatible solutes in various herbaceous plants [3–6]. Compatible solute accumulation has been described as either being a sign of stress or induced as a protective mechanism, which can minimise the effects of stress [2, 7–9]. On the one hand, their accumulation, enhanced by the up-regulation of abscisic acid (ABA), is seen as a plant's signalling mechanism that can potentially mitigate the production of reactive oxygen species (ROS) [8], whilst on the other, their accumulation is thought to contribute to the maintenance of the cytosolic energy balance of a cell [9]. Examples of compatible solutes that have been identified as drought sensitive include nitrogen-containing compounds–e.g. glycine betaine, proline–and sugars–e.g. trehalose, glucose and sucrose [2, 10, 11]. These osmolytes are metabolically expensive to produce and, in the case of amino acids, represent an important pool of nitrogen [12]. Although compatible solutes are thought to maintain cellular homeostasis, there is no consensus yet regarding their role and mechanism of regulation *in vivo* in plants, especially during drought. What is known is that exogenous application of the above amino acids and their derivatives, improves stress tolerance [13]. It has been argued however that accumulation of solutes could be as a consequence of stress rather than an adaptive response [13, 14].

HSPs and DHNs are broadly expressed in grasses under different stress conditions, such as high light, temperature, salinity, oxidative stress, heat and cold, and play crucial roles in maintaining cellular homeostasis, thereby protecting plants from stress [1, 15–17]. HSPs help maintain protein stability, aiding folding and refolding, and conversely are involved in the degradation of damaged proteins under stress conditions. HSPs are found throughout the cell and play very important roles in maintaining normal protein conformation [16]. Although HSPs have been suggested to have protective roles [16, 17], especially in response to heat stress, their functions in other stress conditions, e.g. drought, is less clear [18]. The levels of the chloroplast HSP60 were maintained in drought stressed *Sorghum bicolor* while those of cytosolic Heat Shock Protein HSC 70 were reportedly induced in *Festuca arundinacea* [18, 19]. However, the roles of other HSP families in relation to drought are unclear.

DHNs are Group II of late embryogenesis abundant (LEA) proteins that are hydrophilic, thermostable and accumulate in the cytosol, nucleus, chloroplast and mitochondria [20]. They

Citation: Ogbaga CC, Stepien P, Dyson BC, Rattray NJW, Ellis DI, Goodacre R, et al. (2016) Biochemical Analyses of Sorghum Varieties Reveal Differential Responses to Drought. PLoS ONE 11(5): e0154423. doi:10.1371/journal.pone.0154423.

are characterised as having three conserved motifs, K, Y and S segments, named based on predominance of the amino acids–Lysine, Tyrosine and Serine respectively [21]. Both classes of proteins–HSPs and DHNs–have been shown to be expressed in sorghum and maize in response to drought [15, 22, 23] and have been suggested to have protective roles in both plants [16, 17].

Amongst widely cultivated crops, the C_4 grass *Sorghum bicolor* is one of the most drought tolerant species [24]. Having C_4 biochemistry allows it to maintain photosynthesis at low CO_2concentrations, e.g. when stomata are significantly closed. However, this alone does not explain its drought tolerance–its close relative maize has similar C_4 metabolism but much lower drought tolerance. We showed previously contrasting responses in two sorghum varieties—Samsorg 17 and Samsorg 40 with differing degrees of drought tolerance, specifically in terms of high constitutive sugar content and loss of proteins respectively [19]. We also examined their morphological and physiological responses to drought [19]. We observed that soil in pots containing Samsorg 17 dried slower than those with Samsorg 40 but both varieties lost specific leaf area in response to drought [19]. Samsorg 40 had higher above ground biomass but lost chlorophyll per unit leaf area with increasing drought in contrast to Samsorg 17. Samsorg 17 maintained relative water content, open stomata, net assimilation and photosystem II efficiency longer than with Samsorg 40 [19]. In the present study, biochemical and metabolic responses to drought in these two sorghum cultivars are compared using a non-targeted metabolomic approach to gain a fuller understanding of the metabolic differences between varieties. We focussed on identifying the specific sugars accumulated by Samsorg 17 and the nitrogen sinks for lost proteins in Samsorg 40. The two cultivars are also shown to differ substantially in their expression of HSPs and DHNs and in their accumulation of metabolites. The results are discussed in terms of identifying potential targets for developing better drought tolerance.

RESULTS

Changes in metabolites during drought

Plant pots were allowed to dry to a range of different water contents and then maintained at those levels of dryness for five days, by re-watering pots daily with water lost in the previous 24 h, using an experimental approach described previously [19]. Leaf samples were collected 8h into photoperiod and flash frozen. Ground leaf material was then analysed using FT-IR analysis From this analysis, it can be seen that there are clear differences in chemical composition between cultivars even under control conditions, with differences in the carbohydrate, lipid and amide regions (Fig 1).

Principal component-discriminant function analysis (PC-DFA) was used to define key differences in FT-IR spectra. FT-IR spectra showed discrimination between plant lines under control conditions and with differing degrees of drought (Fig 2A). In well-watered plants the two parent lines were well separated. With increasing water deficit, there is variation between the samples that is reflected mainly in PC-DF2. Drying to 40% resulted in a substantial change in both plants, with this being most evident in Samsorg 40. This change could be described by both PC-DF2 and, to a lesser extent PC-DF1. Further drought resulted in a tendency for changes in PC-DF2 to reverse in Samsorg 40, but less so in Samsorg 17, with further changes being described by PC-DF1. Under the most extreme drought conditions, there was a tendency for the two lines to converge. Examination of the PC-DF loadings indicated that changes in carbohydrates, amides and lipids all contributed to both PC-DF1 and 2 suggesting that these are affected by drought (Fig 2B).

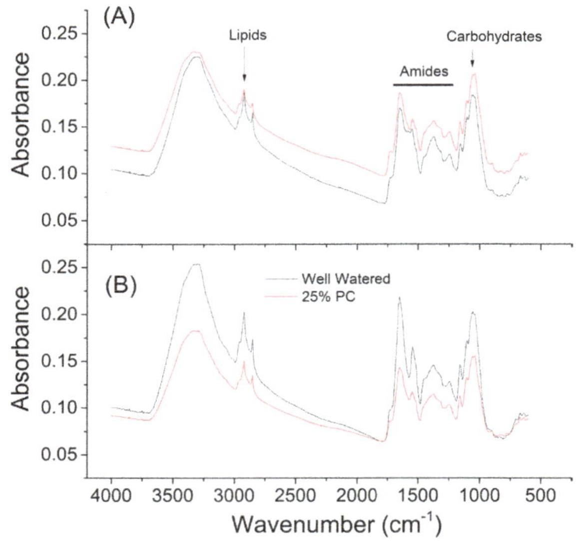

Figure 1: Example FT-IR Spectra.

(A) Samsorg 17. (B) Samsorg 40. Extracts from well-watered plants (black lines) or plants exposed to 25% pot capacity (red lines) for up to 5 days. FT-IR raw data were plotted in Origin Pro 8 (OriginLab, Northampton, MA).

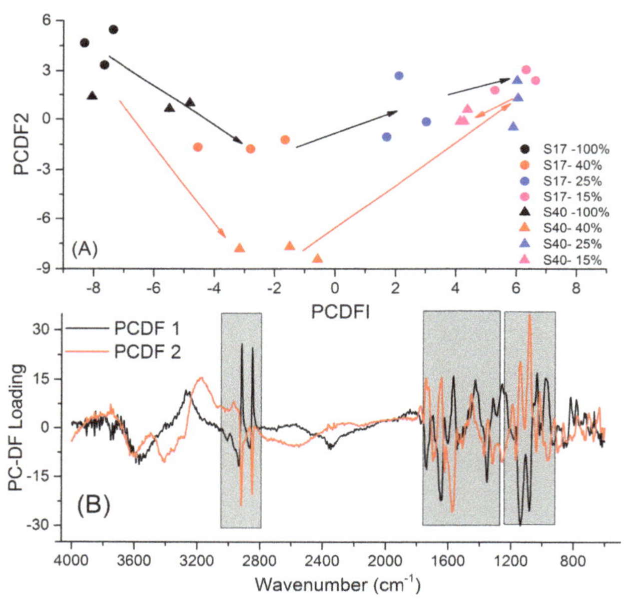

Figure 2: Principal components discriminant function analysis.

(A) Scores. (B) Loadings plots of Samsorg 17 and Samsorg 40 acquired using FT-IR spectroscopy. Acquired FT-IR raw data were converted into principal components using the R statistical package. Principal component discriminant function analysis (PC-DFA) was performed on the principal components of the data. The generated discriminant functions were plotted using Origin Pro 8 (OriginLab, Northampton, MA). The shaded regions (from right to left) represent changes in carbohydrates, amides and lipids. Coding used was 'S17' or 'S40' for Samsorg 17 and Samsorg 40, respectively, and '-X%' refers to percentage SWC.

Changes in glucose and sucrose in response to drought

PC-DFA of FT-IR data indicated the occurrence of changes in signals associated with carbohydrates in response to drought in both sorghum varieties. Total soluble sugars were shown previously to be present at high concentrations constitutively in the tolerant Samsorg 17 whilst being induced in response to drought in Samsorg 40 [19]. Sorghum, especially sweet sorghum, is known to accumulate sucrose [25, 26]. Therefore, to determine whether the differences in sugar content between the two varieties studied here was due to sucrose, this was assayed using a linked enzyme assay, which also allowed us to estimate glucose content. When exposed to mild drought (40% pot capacity), there was a transient decrease in glucose content in both cultivars, with the glucose content recovering at more severe drought levels. This effect was greater in Samsorg 17, where glucose content peaked at a concentration significantly above control levels (Fig 3A). Both cultivars showed a further loss of glucose at the highest level of drought, compared to milder conditions. In Samsorg 17, there was a progressive accumulation of sucrose with increasing levels of drought. In Samsorg 40, there was no significant accumulation of sucrose. Previously, we observed that the former contained high concentrations of soluble sugars constitutively, whilst the latter accumulated sugars under drought conditions [19]. This suggests that sucrose accumulation is not making a dominant contribution to the total sugar pool or that other sugars are changing in parallel with changes in sucrose and glucose. In an attempt therefore to identify further sugars as well as other metabolites that might accumulate under drought in the two varieties, plant material was analysed using non-targeted GC-MS analysis.

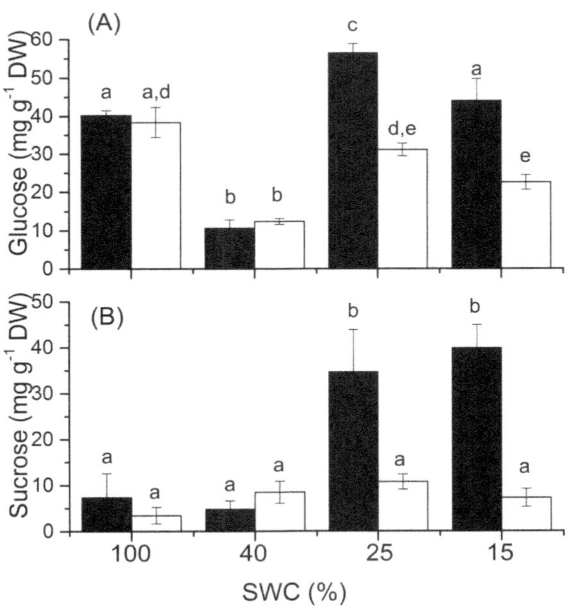

Figure 3: Responses of the soluble sugars.

(A) Glucose. (B) Sucrose to drought. Black bars are plants of Samsorg 17 while open bars are those of Samsorg 40. Data are mean of at least 3 replicates and error bars are ± 1 SE. Different letters represent significant difference; ANOVA, $p < 0.05$.

GC-MS analysis of differences between Samsorg 17 and Samsorg 40

In this non-targeted metabolomic profiling analysis [27] a total of 188 compounds were detected in the two sorghum varieties using GC-MS, of which 142 were at least tentatively identified (S1 Table. Complete list of compounds detected by GC-MS).

GC-MS revealed that the drought tolerant Samsorg 17 contained, in addition to glucose and sucrose, significant amounts of fructose, galactose, lactose, cellobiose and sedoheptulose, relative to Samsorg 40. The former also made more sorbose and a compound putatively identified as trehalose, but similar amounts of arabinose and 1,6-anhydroglucose compared with the latter (Fig 4).

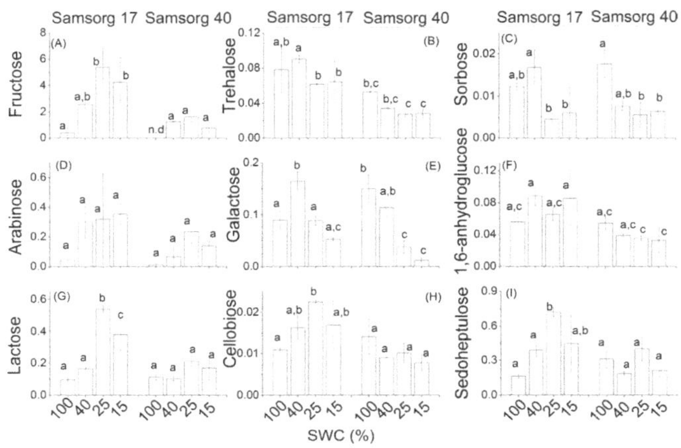

Figure 4: Differences in sugars in the two sorghum varieties.

For each panel, the responses of Samsorg 17 plants are shown first at four levels of drought followed by those of Samsorg 40. Data are mean of at least 3 replicates and error bars are ± 1 SE. Data were normalised to Glycine-d_5. Different letters represent significant difference; n.d, not detected; ANOVA, $p < 0.05$. Sugar alcohols have been suggested to play a role in drought tolerance, by acting as osmoprotectants, hence are thought to induce osmotic adjustments [28, 29]. The sugar alcohols–ribitol, *myo*-inositol and xylitol–were significantly accumulated in Samsorg 17 but not in Samsorg 40 in response to drought (Fig 5).

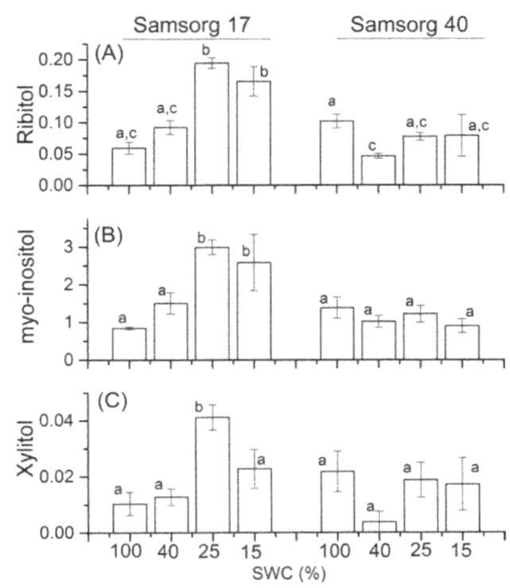

Figure 5: Differences in sugar alcohols in the two sorghum varieties.

For each panel, the responses of Samsorg 17 plants are shown first at four levels of drought followed by those of Samsorg 40. Data are a mean of at least 3 replicates and error bars are ± 1 SE. Data were normalised to Glycine d_5. Different letters represent significant difference; ANOVA, $p < 0.05$.

The osmoprotective role of amino acid accumulation during drought tolerance has been widely studied. It is thought that the accumulation of amino acids can remove reactive oxygen species formed in the cell, thereby improving tolerance [30, 31]. In our study, the less drought tolerant Samsorg 40 accumulated the amino acids proline, phenylalanine, tryptophan, and serine in response to drought, relative to Samsorg 17 (Fig 6).

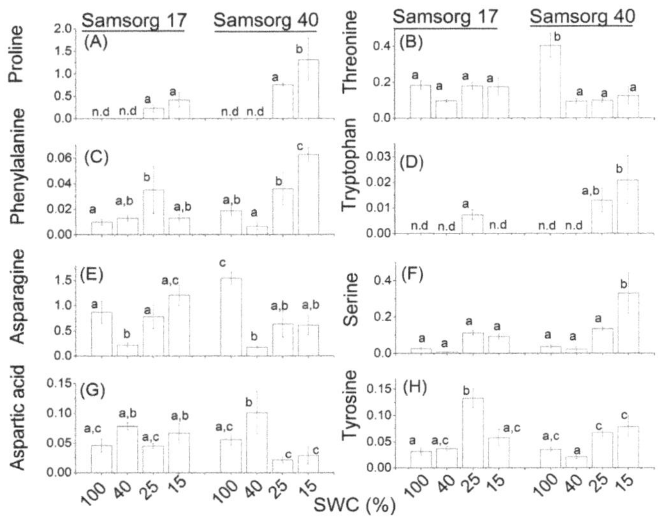

Figure 6: Differences in amino acids in the two sorghum varieties.

For each panel, the responses of Samsorg 17 plants are shown first at four levels of drought followed by those of Samsorg 40. Data are mean of at least 3 replicates and error bars are ± 1 SE. Data were normalised to Glycine d_5. Different letters represent significant difference, n.d; not detected; ANOVA, $p < 0.05$.

Threonine and asparagine were however higher under well watered conditions in Samsorg 40. The only amino acid that was accumulated in higher proportions in Samsorg 17 was tyrosine.

Various organic acids were detected in the sorghum leaves. Ascorbic acid, an important antioxidant, has widely been reported to be induced by stress and is believed to offer some degree of tolerance [32–34]. We found that ascorbic acid did not differ significantly when compared between the two sorghum lines. Amongst the other organic acids detected were propanoic, 2-ketoglutaric, pyruvic, itaconic and gluconic acids. These organic acids also did not differ significantly between the cultivars. The amount of pyroglutamic acid detected was higher under well-watered condition in the less tolerant Samsorg 40 but with drought, dropped to the level seen in Samsorg 17. Tartaric acid was not detected in any treatments in Samsorg 17 but was present in significant amounts in Samsorg 40 dropping in concentration with drought (Fig 7).

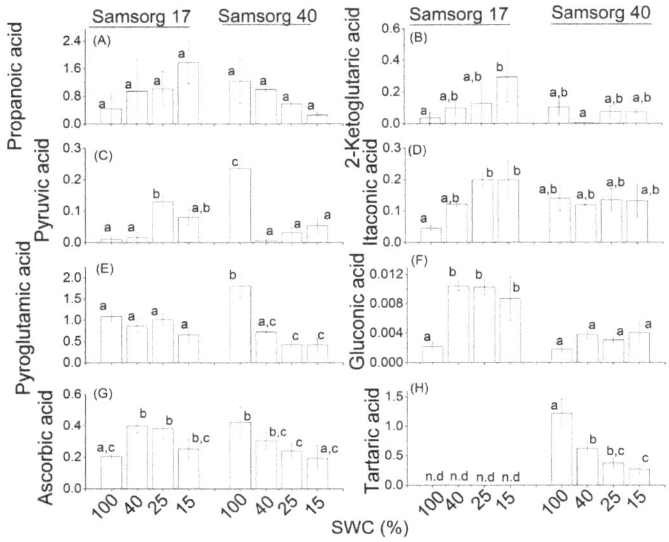

Figure 7: Differences in other organic compounds in the two sorghum varieties.

For each panel, the responses of Samsorg 17 plants are shown first at four levels of drought followed by those of Samsorg 40. Data are mean of at least 3 replicates and error bars are ± 1 SE. Data were normalised to Glycine d_5. Different letters represent significant difference; n.d, not detected; ANOVA, $p < 0.05$).

Changes in Heat Shock Proteins

Previous reports have shown that the chloroplast localised 60-kDa HSP, known as chaperonin Cpn60, acts as the Rubisco binding protein (RBP) involved in the folding and assembly of the newly synthesised Rubisco subunits to produce the native enzyme [15–17, 35]. In the leaves of Samsorg 40, Cpn60 declined in parallel with RbcL degradation [19]. Levels of this peptide were maintained in Samsorg 17 throughout the drought period [19]. In order to determine whether the response of HSP60 was typical of other HSPs, we examined other HSP families—HSP70, HSP90 and HSP101.

Chloroplast localised chHSP70 is thought to be involved in protein import and translocation processes in chloroplasts. It also participates in the repair of damaged PSII proteins, by promoting either synthesis or assembly of new reaction center components [36–38]. chHSP70 was significantly up-regulated from 40% PC in Samsorg 17. In Samsorg 40, a similar up-regulation was observed but this effect was less marked.

Chloroplast localised HSPs were not detected in root tissues (Fig 8). The cytosolic protein, Heat Shock Cognate 70 (HSC70) ensures correct folding of polypeptide chains to proteins [17]. HSC70 was constitutively expressed in leaves of both Samsorg 17 and Samsorg 40, however the level of expression appeared stronger in the former than the latter (Fig 8A). There was no clear indication that this expression was affected in response to drought in these tissues. In roots, there was some indication that HSC70 was induced by drought, especially in Samsorg 40 (Fig 8B). Cytosolic HSP90, a molecular chaperone that only interacts with well-folded proteins was up-regulated in response to drought in leaves and roots of both cultivars, while Samsorg 17 appeared to show stronger expression than Samsorg 40. HSP 101, which belongs to a family of heat shock proteins named HSP 100s (or Clp proteins) and functions in protein disaggregation and degradation, accumulated in leaves during the late stages of drought in both lines, but considerably more so in Samsorg 40 (Fig 8A). A similar pattern was observed in roots of Samsorg 40, whilst in Samsorg 17 a more even expression pattern was seen with differing degrees of drought (Fig 8B).

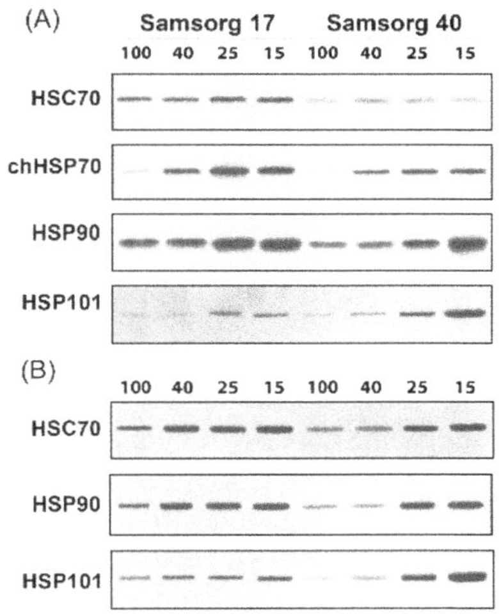

Figure 8: Heat Shock Protein expression.

(A) In the leaf. (B) root of Samsorg 17 and Samsorg 40. Plants were maintained at four different water levels -100, 40, 25 and 15% of pot capacity for 5 days. Equal amounts of whole leaf and root proteins (25 µg) were separated using electrophoresis on Glycine-SDS-PAGE (4–20%), electroblotted onto nitrocellulose membrane, and probed with sets of antibodies raised against HSP chaperone proteins. Blots are representative of at least three independent experiments.

Dehydrin/ Dehydrin-like protein expression

Dehydrins are Group II of Late Embryogenesis Abundant (LEA) proteins. They have a conserved lysine rich amino acid sequence and are thought to contribute to stress tolerance [21, 39]. Samsorg 17 and Samsorg 40 showed contrasting DHN/ DHN-like protein expression when subjected to differing degrees of drought. In leaf tissue, peptides reacting to a DHN antibody ranged in size from 21 to 115 kDa. Amongst these, different expression patterns could be seen. Fewer bands were observed in roots, with bands at approx. 115, 37 and 21 kDa being absent. The band at approx. 37kDa was constitutively expressed in the leaves of both cultivars; however bands appeared more intense in Samsorg 17 than Samsorg 40 (Fig 9A). Other bands showed more or less clear induction in response to drought. Bands in the range 52–69 kDa were observed in both cultivars and were induced in response to drought in both shoots and roots; however these tended to be induced sooner and/or more strongly in Samsorg 17 than in Samsorg 40. In contrast, the leaf only bands at 21 and 115 kDa showed earlier and stronger induction in Samsorg 40. The band at around 28 kDa showed similar expression patterns in roots and shoots of both cultivars, although expression appeared stronger in roots of Samsorg 17 (Fig 9).

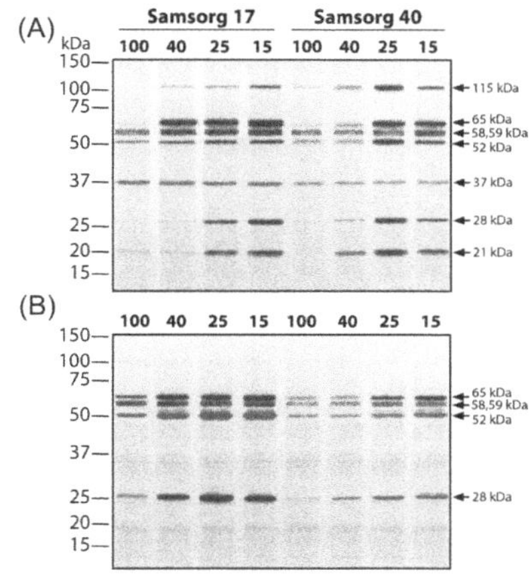

Figure 9: DHN/ DHN-Like protein expressions in *Sorghum bicolor* **L. (var.** Samsorg 17 and Samsorg 40) under drought treatments-100 (Control), 40, 25 and 15% of pot capacity.

Equal amounts of whole leaf (A) and root proteins (B) (25 µg) were separated on Glycine-SDS-PAGE (4–20%), eblotted onto nitrocellulose membrane and probed with set of antibodies raised against the K-segment of DHN. Blots are representative of at least three independent experiments.

DISCUSSION

To increase the drought tolerance of crop species, we need to understand better the traits possessed by drought tolerant plants so that these can be transferred into new varieties. In this study, two varieties of Sorghum–Samsorg 17, a highly drought tolerant cultivar, and Samsorg 40, a less drought tolerant one–were studied. Our previous results have shown that these varieties differ markedly in their physiological responses to drought, with evidence for strongly contrasted biochemical responses to drought in closely related plants [19]. Results presented here, reinforce those earlier results, and show that Samsorg 17 induced higher levels of heat shock proteins (HSPs) and dehydrins (DHNs) but differential accumulation of metabolites compared with Samsorg 40.

FT-IR analyses the infrared absorption spectra of chemical species that absorb in the mid infrared part of the electromagnetic spectrum [40]. Together with PC-DFA loadings, it is possible to determine what types of molecules are being affected by stress in plants. PC-DFA performed showed clear variation between the sorghum varieties. The sorghum samples were clearly different under well-watered and moderate drought conditions but converged at extreme drought. PC-DFA loadings showed that sugars, amides (from proteins) and lipids are affected by drought in the two sorghum cultivars (Fig 2) corroborating the changes identified on the raw spectra (Fig 1).

Previously, we showed that plants of Samsorg 17 contain significantly higher total sugar content than those of Samsorg 40 [19]. It has been shown that sorghum, especially sweet sorghum, accumulates sucrose in its stems [25, 26] making this a strong candidate to explain the high sugar content of Samsorg 17. Both sorghum varieties accumulate high concentrations of glucose in response to drought but Samsorg 17 accumulates sucrose to significantly higher levels than Samsorg 40. However, neither this accumulation of sucrose nor the concentrations of glucose explain our earlier findings, that total soluble sugar levels were constitutively high in Samsorg 17 whilst high levels were induced by drought in Samsorg 40 [19]. The conditions of drought treatment in our earlier study were different–there drought was allowed to develop progressively, whilst here, plants were maintained at a particular level of drought. Nevertheless, it seems likely that the presence of other sugars, explains the differences between the varieties.

To identify other sugars accumulating in the two lines, leaf samples were analysed using GC-MS. PCA revealed a distinct response at lower levels of drought in the two cultivars (S1 Fig) Typically the sugars quantified in this way showed a similar behaviour to sucrose, tending to be present at higher concentrations in Samsorg 17 and being induced to a greater extent in response to drought. Increases in fructose, lactose, cellobiose and sedoheptulose were induced by drought in Samsorg 17 (Fig 4). There were also apparent higher levels of the monosaccharides–galactose, 1, 6-anhydroglucose, sorbose, arabinose, and a disaccharide tentatively identified as trehalose in Samsorg 17 (Fig 4). This is consistent with high sugar content or sugar accumulation playing a crucial role in maintaining leaf function and photosynthesis during drought, as reported earlier [5, 41, 42]. These data do not explain our previous finding that Samsorg 17 stores high constitutive amounts of soluble carbohydrate, suggesting that other sugars or related compounds, not detectable are important in Samsorg 17. Sugar alcohols also increased significantly in Samsorg 17 suggesting that these may have a role as osmoprotectants and may contribute to the stabilization of protein structure against degradation (Fig 5) [11, 28, 29]. Overall, higher accumulation of soluble sugars and sugar alcohols in Samsorg 17 suggest that these may play a role in lowering shoot and root water potential for better water uptake.

In addition to differences in sugar accumulation, Samsorg 17 showed significant up-regulation of gluconic acid in contrast to Samsorg 40. Higher level of gluconic acid suggests that it is protective. Ascorbic acid was not markedly different between the cultivars but was induced to a lesser extent with progressive drought in Samsorg 17 relative to Samsorg 40 (Fig 7). Ascorbic acid has antioxidant properties and is thought to offer some degree of tolerance to plants. The effects of many forms of environmental stress are known to be mediated, at least partially, by an enhanced generation of reactive oxygen species (ROS) in the organelles with highly oxidizing activities and sustained electron flows–chloroplasts, mitochondria and peroxisomes [43, 44]. A cellular system to scavenge ROS, including the low-molecular-weight antioxidants such as ascorbate, may protect membrane lipids and proteins from damage [32–34]. Although ascorbate was not clearly different in the two sorghum lines with drought in this study, these results are consistent with the high, stress-induced levels of ascorbate found in both the mesophyll and bundle sheath chloroplasts isolated from Samsorg 17 compared with Samsorg 40 (Stepien et al, unpublished data). The roles of the other organic acids detected are not clear; however, some of them are speculated to be imported and exuded from the root system which could improve an acquisition of the mineral elements under drought conditions.

It is generally accepted that the uptake of nutrient by crop plants is reduced in dry-soil conditions. Drought provokes nutritional deficiencies, even in well-fertilized environment, as the physiochemical properties of soil may reduce availability of individual nutrients. In addition, any negative effect of drought on the mineralization process of organic matter depending on the soil microorganisms and enzymes activity, affects the amount of inorganic nutrients [45,46]. On the other hand, drought may also increase the allocation of resources and increase production of root exudates, contributing to increased chelation and uptake of nutrients [47,48]. Organic acids are an important class of exudates and their release to the rhizosphere contributes to plant health in several ways, including mobilization of the unavailable forms of essential nutrients. The organic acids detected by GC-MS are believed to be the ones released from the roots of the plants. These acids may solubilise P, Fe, Mn and/or Zn from soil [47].

Previously we observed an accumulation of the amino acid proline and a loss of soluble protein in plants of Samsorg 40 exposed to progressive drought. Here (Fig 6), examining plants exposed to sustained drought, we observed the same accumulation of proline, consistent with a large accumulation of this amino acid, which is widely recognised as playing a role in drought tolerance. This increase in proline is also seen in Samsorg 17, but to a much lower level, again consistent with our previous results. The accumulation of proline in Samsorg 40 is also accompanied by the accumulation of most other amino acids measured (Fig 6).

These metabolic alterations were accompanied by clear changes in the abundance of protein present in the two sorghum lines [19]. Amongst these, changes were noted in the various HSPs and DHNs. These proteins have previously been characterised as being expressed in grasses in response to various stress conditions including light, temperature, salinity, oxidative stress, heat and cold [1, 15–17]. However, little is known about the roles of these proteins in response to drought. What is known is that the maintenance of chloroplast localised chaperonin (Cpn60) and the cytosolic Heat Shock Protein (HSC70) offer some level of drought tolerance [15, 18, 19, 22, 23].

Amongst the HSPs examined, chloroplast localised chHSP70, the constitutively expressed HSC70 and HSP90 all showed either higher apparent expression and/or induction at a lower level of drought in Samsorg 17 than Samsorg 40 (Fig 8). chHSP70 participates in PSII repair process by promoting either synthesis or assembly of new reaction center components [37]. Greater expression of chHSP70 in Samsorg 17 might be responsible, at least partially, for the maintenance of PSII protein complex and PSII electron transport under stress conditions [19]. HSC70 was constitutively expressed in both cultivars, albeit higher in Samsorg 17 (Fig 8). HSC70 is a house-keeping protein that is constitutively expressed in the cytosol [49]. It prevents protein aggregation and hence ensures correct folding of polypeptide chains to proteins [17]. This probably contributed to the maintenance of protein structure in Samsorg 17, which improved the tolerance of the plant. HSP90 is another HSP that was accumulated in higher level in the more tolerant Samsorg 17 relative to Samsorg 40. HSP90 is a molecular chaperone that only interacts with well-folded proteins. It is distinct from many other molecular chaperones in that most of its known substrates are signal transduction proteins such as hormone receptors and signaling kinases. HSP90 only functions when large protein complexes involving co-chaperones such as chHSP70 have induced folding in the chloroplast [16, 17]. The apparent up-regulation or earlier induction of HSP90 in the leaves and roots of Samsorg 17 is consistent with their role being to enhance the drought tolerance and continued survival of these plants under drought stress. In contrast HSP101, a family member of the Clp heat shock proteins, was the only HSP examined where apparent expression was higher in Samsorg 40, with only trace levels of this protein being detectable in leaves of Samsorg 17. Rather than the regular chaperone function of preventing protein aggregation and misfolding, this family functions in protein disaggregation and degradation [16, 17]. The greater induction of this protein in Samsorg 40 is consistent with damage being induced and with damaged proteins having to be degraded. Curiously however, a comparatively higher level of expression was seen in the roots of Samsorg 17, possibly reflecting an earlier reaction to drought in roots.

We found a generally higher induction of DHNs in Samsorg 17 than Samsorg 40 (Fig 9). The functions of dehydrins (DHNs) are still under investigation, however the relationship between DHNs and drought has been examined [50]. They are characterized by a high content of glycine, threonine, asparagine, glutamine, serine, aspartic and glutamic acids, explaining their highly hydrophilic nature [21, 39, 51]. It is believed they are induced by ABA during drought [50, 52]. They are known to be thermostable and hence are thought, like HSPs, to have chaperone-like properties and roles in membrane structure stabilization [20, 53]. The higher expression levels and earlier induction of these proteins in Samsorg 17 is again consistent with their playing a role in maintaining cell function for longer in response to drought.

In conclusion, we have observed that Samsorg 17 and 40 show contrasted responses to drought. In the drought tolerant Samsorg 17, there was a general up-regulation of protective proteins- HSPs, DHNs- sugars and sugar alcohols. In contrast, Samsorg 40, the less tolerant sorghum cultivar, showed a lower accumulation of potentially protective metabolites, but did accumulate significantly more of various amino acids, probably linked to the breakdown of protein previously reported [19]. The up-regulation and higher levels of sugar molecules, in particular high amount of fructose, in Samsorg 17, suggest that they play protective roles in maintaining leaf function and possibly maintaining water content during water deficit. The mobilisation of N compounds in Samsorg 40, may allow movement of N to young tissues for possible recovery. In the future, it would be useful to examine contributions of these different responses, for example using recombinant inbred lines or a segregation population generated from the two sorghum lines.

MATERIALS AND METHODS

Plant material and growth conditions

Seeds of Sorghum (*Sorghum bicolor* L) cv Samsorg 17 (more drought tolerant) and cv Samsorg 40 (less drought tolerant) were provided by the Institute of Agricultural Research, Zaria, Nigeria. The seeds were grown in a growth room as described previously using John Innes No. 3 compost [19]. To estimate relative soil water content (SWC), % maximum pot capacity, fully saturated soil was weighed and thereafter oven dried at 105°C to constant weight. The weight difference between the water-saturated and oven-dried soil was taken as the weight of water needed to bring pots to pot capacity and lower water levels (% pot capacity) estimated accordingly. Plants were allowed to establish for 21 days under well watered conditions before the onset of drought treatments. For droughted plants, watering was stopped and pots allowed to dry until the soil SWC fell to a particular level. Pots were then maintained at different water levels by re-watering daily with water lost in the previous 24 hours, for 5 days before samples were harvested. Samples for metabolomic analysis were harvested by flash freezing leaf material collected 8 h into photoperiod. Samples of leaves and roots for protein analysis were harvested in the same way. For root analysis, prior to sample collection, roots were washed with distilled water to remove debris. 250mg of whole-root soluble proteins were ground into a fine powder in liquid nitrogen and extracted in ice-cold 150mM Tris-Hcl buffer as described previously [19].

Fourier transform infrared (FT-IR) spectroscopy

Fully expanded leaf and shoot tissues from different soil water contents were excised 8 h into the photoperiod with a pair of scissors, flash frozen in liquid nitrogen and lyophilised (freeze-dried) with a ScanVac Coolsafe freeze dryer (Vacuubrand, Wertheim, Germany) for 48 hr. FT-IR spectroscopy was performed on 30 mg DW of tissues homogenised, transferred to 2 mL microcentrifuge tubes and extracted in 600 μL of distilled water. 5 μL of homogenate was loaded onto the wells of a silicon 96 target plate (Bruker, MA, USA) and dried at 60°C. The Si plate was then placed in a Bruker Equinox-55 spectrometer and raw FT-IR data recorded in the absorbance mode of wavenumber 4000–600 cm^{-1} with a resolution of ~4 cm^{-1}.

After data generation principal components analysis (PCA) was used to analyse the data and this was performed in the R software environment (available at http://www.r-project.org/). PCA scores and loadings plots were generated in Origin Pro 8 (OriginLab, Northampton, MA). Principal component discriminant function analysis (PC-DFA), a supervised multivariate technique was performed on the principal components (PCs) of samples harvested at different soil water contents. This was performed to determine the spectral regions of compounds (from PC-DFA loadings plots) that change with water deficit. PC-DFA is used to discriminate between samples based on their principal components with knowledge of which spectra were replicates; in this case 8 groups were used for the 2 plants × the 4 SWC conditions. For clarity, the spectral regions of major compound classes are found for lipids at ~ 3050–2800 cm^{-1}, proteins at ~ 1750–1250 cm^{-1} and carbohydrates at ~ 1250–900 cm^{-1} [54–56].

Gas Chromatography-Mass Spectrometry (GC-MS)

Fully expanded leaf and shoot tissues from plants exposed to different soil water contents were excised 8 h into the photoperiod with a pair of scissors, flash frozen in liquid nitrogen and lyophilised (freeze-dried) with ScanVac Coolsafe freeze dryer (Vacuubrand) for 48 h. 30 mg dry weight (DW) of tissues were extracted in methanol-chloroform-water mixture as described in Lisec, Schauer (57) and Allwood, Ellis (55). Polar phase extracts were dried using a speed vacuum concentration with 100 μL of an internal standard containing 200 μg/mL of succinic-$d_4$acid, lysine-d_4 and glycine-d_5 (Sigma-Aldrich, St Louis, Missouri, USA). This was followed by a two-step derivatisation process according to Fiehn, Kopka [58], Lisec, Schauer [57] and Allwood, Ellis [55], which involves first the addition of methoxyamine hydrochloride dissolved in pure pyridine (Sigma-Aldrich) and second, treatment with a silylating compound *N*-methyl-*N*-(trimethylsilyl)trifluoroacetamide (MSTFA reagent; Macherey-Nagel, Duren, Germany). Subsequently, 20 μL of a retention index (RI) mixture was also added containing 3 mg/mL of docosane, nonadecane and 3 μL/mL decane, dodecane, pentadecane. Samples were then analysed using a Leco Pegasus III (4D) GC x GC/MS in GC/MS mode (Leco Corp., St Joseph, MO) in conjunction with a Gerstel MPS-2 autosampler (Gerstel, Baltimore, MD) and an Agilent 6890N gas chromatograph (Agilent Technologies, Santa Clara, CA, USA). Within the system, an electron ionisation (EI) source was employed for ion formation/fragmentation and mass detection was conducted using a time-of-flight (TOF) mass analyser. Quality control (QC) samples were obtained by mixing equal volumes of the upper polar phase of all the sample extracts and were analysed after every 5th sample. The retention time and/or retention index of metabolites identified were compared and matched with those from pure chemical standards using in-house libraries from the Manchester Metabolomics Database [55]. The GC-MS data was aligned using an in-house LOESS alignment algorithm and missing values using a KNN infill within Matlab. The PCA (using PCs 1 & 2) was plotted with Simca-P version 11 (Umetrics AB, Umea, Sweden).

SDS-PAGE and Immunoblot analysis

Protein extraction and chloroplast isolation were performed as described previously [1919]. Protein samples for SDS-PAGE were incubated in 50 mM Tris-HCl (pH 6.8), containing 2% (w/v) SDS, 50 mM DTT, 1 mM PMSF, 5 mM EDTA, 10% (w/v) glycerol and 0.05% (w/v) bromophenol blue at 95°C for 3 min. Total protein from Sorghum leaf/root (25 μg) or chloroplast (15 μg) was separated on 4–20% Precise Tris-Glycine SDS-PAGE (Thermo Scientific, Waltham, MA, USA) in Mini-PROTEAN 3 cell (Bio-Rad, CA, USA) at 35 mA per gel. For immunoblot analyses, proteins were transferred to Hybond-ECL nitrocellulose membrane (Amersham, Little Chalfront, UK) by electroblotting in western transfer buffer (25 mM Tris-HCl, pH 8.3, 192 mM glycine and 20% (v/v) methanol) in Mini Trans-Blot cell (Bio-Rad; 60 min at 90 V). Blots were checked by Ponceau S staining to ensure even transfer. Membranes were blocked with 5% (w/v) non-fat milk powder in TBS-T (20 mM Tris-HCl, pH 7.6, 125 mM NaCl and 0.5% [v/v] Tween-20) and then incubated overnight at 4°C with primary antibodies against: Dehydrins (Agrisera, Vannas, Sweden). Heat Shock Proteins–HSC/HSP70 (Enzo, Exeter, UK), chHSP70, HSP90-2 and HSP101 (Agrisera, Vannas, Sweden). Dehydrins (DHNs) are Group II of Late Embryogenesis Abundant proteins with conserved motif in C-terminal amino acid sequence termed K-Segment. Anti-DHN antibody raised against this sequence recognises a number of polypeptides in the range 9–200 kDa. Binding of the primary antibody was detected using appropriate horseradish peroxidase (HRP)-conjugated secondary antibodies (1:25000 polyclonal goat anti-rabbit IgG; Sigma) and ECL detection reagent (Amersham) according to the manufacturer›s instructions.

Sucrose and Glucose Assays

100 mg fresh weight (FW) of sorghum samples were weighed and extracted in 1 mL of distilled water. 0.1 mL of sorghum extracts from different soil water contents (% pot capacity) were assayed for sucrose and glucose using Sucrose/D-Glucose assay kit (catalogue number K-SUCGL; Megazyme, Ltd., Bray, Ireland). Assays were performed according to the manufacturer's instructions. The principle of the kit involves the hydrolysis of sucrose by the enzyme β-fructosidase to D-glucose and D-fructose. Free D-glucose catalysed by glucose oxidase and peroxidase in the presence of *p*-hydroxybenzoic acid and 4-aminoantipyrine was then determined based on the concentration of a red coloured quinoneimine dye. FW results obtained were subsequently converted to DW using the formula: %DW = %FW / [1-(RWC/100)], where DW is dry weight and RWC is relative water content.

Statistics

A two-way ANOVA (analysis of variance) with a Tukey post-hoc test was conducted using SPSS Statistics 20 statistical software (IBM, Portsmouth, UK), to compare the responses of the two sorghum lines to water deficit. Different letters represent significant differences at $p < 0.05$.

ACKNOWLEDGMENTS

We would like to acknowledge Matthew Miller for help with FT-IR data analysis. BCD was supported by a BBSRC grant (BB/J004103/1).

AUTHOR CONTRIBUTIONS

Conceived and designed the experiments: CCO PS GNJ. Performed the experiments: CCO PS. Analyzed the data: CCO. Contributed reagents/materials/analysis tools: BCD NJWR DIE RG. Wrote the paper: CCO GNJ.

REFERENCES

1. Ngara R, Ndimba R, Borch-Jensen J, Jensen O, Ndima B. Identification and profiling of salinity stress-responsive proteins in Sorghum bicolor seedlings. J Proteomics. 2012;75:4139–50. doi: 10.1016/j.jprot.2012.05.038. pmid:22652490
2. Kavi Kishor P, Sangam S, Amrutha R, Sri Laxmi P, Naidu K, Rao S, et al. Regulation of proline biosynthesis, degradation, uptake and transport in higher plants: Its implications in plant growth and abiotic stress tolerance. Curr Sci. 2005;88:424–38.
3. Nakayama H, Yoshida K, Ono H, Murooka Y, Shinmyo A. Ectoine, the Compatible Solute of Halomonas elongata, Confers Hyperosmotic Tolerance in Cultured Tobacco Cells. Plant Physiol. 2000;122(4):1239–48. doi: 10.1104/pp.122.4.1239. pmid:10759521
4. Akashi K, Miyake C, Yokota A. Citrulline, a novel compatible solute in drought-tolerant wild watermelon leaves, is an efficient hydroxyl radical scavenger. FEBS Lett 2001;508:438–42. pmid:11728468 doi: 10.1016/

s0014-5793(01)03123-4

5. Garg AK, Kim J-K, Owens TG, Ranwala AP, Choi YD, Kochian LV, et al. Trehalose accumulation in rice plants confers high tolerance levels to different abiotic stresses. Proc Natl Acad Sci USA. 2002;99(25):15898–903. doi: 10.1073/pnas.252637799. pmid:12456878
6. An Y, Liang Z. Drought tolerance of Periploca sepium during seed germination: antioxidant defense and compatible solutes accumulation. Acta Physiol Plant. 2013;35(3):959–67. doi: 10.1007/s11738-012-1139-z.
7. Chen Z, Cuin TA, Zhou M, Twomey A, Naidu BP, Shabala S. Compatible solute accumulation and stress-mitigating effects in barley genotypes contrasting in their salt tolerance. J Exp Bot. 2007;58(15–16):4245–55. doi: 10.1093/jxb/erm284. pmid:18182428
8. Chaves MM, Flexas J, Pinheiro C. Photosynthesis under drought and salt stress: regulation mechanisms from whole plant to cell. Ann Bot. 2009;103(4):551–60. doi: 10.1093/aob/mcn125. pmid:18662937
9. Kavi Kishor PB, Sreenivasulu N. Is proline accumulation per se correlated with stress tolerance or is proline homoeostasis a more critical issue? Plant Cell Environ 2013;37:300–11. doi: 10.1111/pce.12157. pmid:23790054
10. Serraj R, Sinclair T. Osmolyte accumulation: can it really help increase crop yield under drought conditions. Plant Cell Environ. 2002;25:333–41. pmid:11841674 doi: 10.1046/j.1365-3040.2002.00754.x
11. Yobi A, Wone B, Xu W, Alexander D, Guo L, Ryals J, et al. Comparative metabolic profiling between desiccation-sensitve and desiccation-tolernat species of Selaginella reveals insights into the resurrection trait. Plant J. 2012;72:983–99. doi: 10.1111/tpj.12008. pmid:23061970
12. Westhoff P, Gowik U. Evolution of C4 Photosynthesis—Looking for the Master Switch. Plant Physiol. 2010;154(2):598–601. doi: 10.1104/pp.110.161729. pmid:20921192
13. Ashraf M, Foolad MR. Roles of glycine betaine and proline in improving plant abiotic stress resistance. Environ Exp Bot. 2007;59(2):206–16. doi: http://dx.doi.org/10.1016/j.envexpbot.2005.12.006.
14. Chen THH, Murata N. Glycinebetaine protects plants against abiotic stress: mechanisms and biotechnological applications. Plant Cell Environ. 2011;34(1):1–20. doi: 10.1111/j.1365-3040.2010.02232.x. pmid:20946588
15. Jagtap V. Comparative effect of water, heat and light stresses on photosynthetic reactions in Sorghum bicolor (L.) Moench. J Exp Bot. 1998;49:1715. doi: 10.1093/jexbot/49.327.1715
16. Wang W, Vinocur B, Shoseyov O, Altman A. Role of plant heat-shock proteins and molecular chaperoness in the abiotic stress response Trends Plant Sci 2004;9:244–52. pmid:15130550 doi: 10.1016/j.tplants.2004.03.006
17. Xu Y, Zhan C, Huang B. Heat Shock Proteins in Association with Heat Tolerance in Grasses. Int J Proteomics. 2011;2011:1–11. doi: 10.1155/2011/529648
18. Jiang Y, Huang B. Protein Alterations in Tall Fescue in Response to Drought Stress and Abscisic Acid. Crop Sci 2002;42:202–7. pmid:11756275 doi: 10.2135/cropsci2002.0202
19. Ogbaga CC, Stepien P, Johnson GN. Sorghum (Sorghum bicolor) varieties adopt strongly contrasting strategies in response to drought. Physiol Plant. 2014;152:389–401. doi: 10.1111/ppl.12196. pmid:24666264
20. Borovskii G, Stupnikova I, Antipina A, Vladimirova S, Voinikov V. Accumulation of dehydrin-like proteins in the mitochondria of cereals in response to cold, freezing, drought and ABA treatment. BMC Plant Biol. 2002;2:1–7. pmid:12057012
21. Rorat T. Plant dehydrins—tissue location, structure and function. Cell Mol Biol Lett 2006;11:536–56. pmid:16983453 doi: 10.2478/s11658-006-0044-0
22. Mohammadkhani N, Heidari R. Effects of Drought Stress on Soluble Proteins in two Maize Varieties. Turk J Biol. 2008;32:23–30.
23. Wood AJ, Goldsbrough PB. Characterization and expression of dehydrins in water-stressed Sorghum bicolor. Physiol Plant. 1997;99:144–52. doi: 10.1111/j.1399-3054.1997.tb03442.x.
24. Paterson AH, Bowers JE, Bruggmann R, Dubchak I, Grimwood J, Gundlach H, et al. The Sorghum bicolor genome and the diversification of grasses. Nature. 2009;457:551–6. doi: 10.1038/nature07723. pmid:19189423
25. Hoffmann-Thoma G, Hinkel K, Nicolay P, Willenbrink J. Sucrose accumulation in sweet sorghum stem internodes in relation to growth. Physiol Plant. 1996;97(2):277–84. doi: 10.1034/j.1399-3054.1996.970210.x.
26. Tarpley L, Vietor D. Compartmentation of sucrose during radial transfer in mature sorghum culm. BMC Plant Biol. 2007;7(1):33. doi: 10.1186/1471-2229-7-33.
27. Dunn WB, Ellis DI. Metabolomics: current analytical platforms and methodologies. Trends Anal Chem. 2005;24(4):285–94. doi: 10.1016/j.trac.2004.11.021
28. Moing A. Sugar alcohols as carbohydrate reserves in some higher plants. Dev Crop Sci. Volume 26: Elsevier; 2000. p. 337–58. doi: 10.1016/s0378-519x(00)80017-3
29. dos Reis S, Lima M, Baptista de Souza C. Recent Molecular Advances on Downstream Plant Responses to Abiotic Stress. Int J Mol Sci 2012;13:8628–47. doi: 10.3390/ijms13078628. pmid:22942725
30. Good AG, Zaplachinski ST. The effects of drought stress on free amino acid accumulation and protein synthesis in Brassica napus. Physiol Plant. 1994;90(1):9–14. doi: 10.1111/j.1399-3054.1994.tb02185.x.
31. Montesinos-Pereira D, Barrameda-Medina Y, Romero L, Ruiz JM, Sánchez-Rodríguez E. Genotype differences in the metabolism of proline and polyamines under moderate drought in tomato plants. Plant Biol (Stuttg). 2014. doi: 10.1111/plb.12178.
32. Shalata A, Neumann PM. Exogenous ascorbic acid (vitamin C) increases resistance to salt stress and reduces lipid peroxidation. J Exp Bot. 2001;52(364):2207–11. doi: 10.1093/jexbot/52.364.2207. pmid:11604460
33. Wang Z, Xiao Y, Chen W, Tang K, Zhang L. Increased

Vitamin C Content Accompanied by an Enhanced Recycling Pathway Confers Oxidative Stress Tolerance in Arabidopsis. J Integr Plant Biol. 2010;52(4):400–9. doi: 10.1111/j.1744-7909.2010.00921.x. pmid:20377702

34. Farooq M, Irfan M, Aziz T, Ahmad I, Cheema SA. Seed Priming with Ascorbic Acid Improves Drought Resistance of Wheat. J Agron Crop Sci. 2013;199(1):12–22. doi: 10.1111/j.1439-037X.2012.00521.x.

35. Demirevska-Kepova K, Simova-Stoilova L, Kjurkchiev S. Barley Leaf Rubisco, Rubisco binding protein and Rubisco Activase and their protein/protein interactions. Bulg J Plant Physiol. 1999;25:31–44. doi: 10.1007/s10535-005-0045-2

36. Medueno F, Napier J, Gray J. The chloroplast chaperonin is also necessary for the reconstitution of CF1 core peptides of ATP synthase. Plant Cell. 1993;5:1865–76. pmid:12271059

37. Schroda M, Vallon O, Wollman F-A, Beck CF. A Chloroplast-Targeted Heat Shock Protein 70 (HSP70) Contributes to the Photoprotection and Repair of Photosystem II during and after Photoinhibition. The Plant Cell. 1999;11:1165–78. doi: 10.1105/tpc.11.6.1165. pmid:10368186

38. Cao Q, Sha H, Gu Y, Tao X, Huang C, Zhang Y. Cloning, characterization and expression analysis of HSP70 gene from sweet potato. Afr J Microbiol Res. 2012;6:6325–32. doi: 10.5897/ajmr11.1403

39. Hanin M, Brini F, Ebel C, Toda Y, Takeda S, Masmoudi K. Plant dehydrins and stress tolerance: versatile proteins for complex mechanisms. Plant Signaling Behav. 2011;6:1503–9. doi: 10.4161/psb.6.10.17088

40. Griffiths P, de Haseth J. Fourier Transform Infrared Spectrometry: Wiley-Blackwell; 2007.

41. Massacci A, Battistelli A, Loreto F. Effect of Drought Stress on Photosynthetic Characteristics, Growth and Sugar Accumulation of Field-Grown Sweet Sorghum. Funct Plant Biol 1996;23(3):331–40. doi: 10.1071/PP9960331.

42. Merewitz EB, Du H, Yu W, Liu Y, Gianfagna T, Huang B. Elevated cytokinin content in ipt transgenic creeping bentgrass promotes drought tolerance through regulating metabolite accumulation. J Exp Bot. 2011;63(3):1315–28. doi: 10.1093/jxb/err372. pmid:22131157

43. Mittler R. Oxidative stress, antioxidants and stress tolerance. Trends Plant Sci. 2002;7(9):405–10. doi: http://dx.doi.org/10.1016/S1360-1385(02)02312-9. pmid:12234732

44. Suzuki N, Koussevitzky S, Mittler R, Miller G. ROS and redox signalling in the response of plants to abiotic stress. Plant Cell Environ. 2012;35(2):259–70. doi: 10.1111/j.1365-3040.2011.02336.x. pmid:21486305

45. Amtmann A, Blatt MR. Regulation of macronutrient transport. New Phytol. 2009;181(1):35–52. doi: 10.1111/j.1469-8137.2008.02666.x. pmid:19076716

46. Rouphael Y, Cardarelli M, Schwarz D, Franken P, Colla G. Effects of Drought on Nutrient Uptake and Assimilation in Vegetable Crops. In: Aroca R, editor. Plant Responses to Drought Stress: Springer Berlin Heidelberg; 2012. p. 171–95.

47. Henry A, Doucette W, Norton J, Bugbee B. Changes in crested wheatgrass root exudation caused by flood, drought and nutrient stress. J Environ Qual. 2007;36:904–12. pmid:17485723 doi: 10.2134/jeq2006.0425sc

48. Kohli A, Narciso JO, Miro B, Raorane M. Root proteases: reinforced links between nitrogen uptake and mobilization and drought tolerance. Physiol Plant. 2012;145(1):165–79. doi: 10.1111/j.1399-3054.2012.01573.x. pmid:22242864

49. Sung DY, Guy CL. Physiological and Molecular Assessment of Altered Expression of Hsc70-1 in Arabidopsis. Evidence for Pleiotropic Consequences. Plant Physiol. 2003;132:979–87. doi: 10.1104/pp.102.019398. pmid:12805626

50. Lopez CG, Banowetz GM, Peterson CJ, Kronstad W. Dehydrin expression and drought tolerance in seven wheat cultivars Crop Sci. 2003;43:577–82. doi: 10.2135/cropsci2003.0577

51. Allagulova CR, Gimalov FR, Shakirova FM, Vakhitov VA. The Plant Dehydrins: Structure and Putative Functions. Biochemistry (Moscow). 2003;68(9):945–51. doi: 10.1023/a:1026077825584.

52. Schroeder JI, Kwak JM, Allen GJ. Guard cell abscisic acid signalling and engineering drought hardiness in plants. Nature. 2001;410:327–30. pmid:11268200

53. Demirevska K, Simova-Stoilova L, Vassileva V, Vaseva I, Grigorova B, Feller U. Drought-Induced leaf protein alterations in sensitive and tolerant wheat varieties. Gen Appl Plant Physiol. 2008;34:79–102.

54. Correia I, Nunes A, Barros AS, Delgadillo I. Protein profile and malt activity during sorghum germination. J Sci Food Agric. 2008;88:2598–605. doi: 10.1002/jsfa.3348.

55. Allwood JW, Ellis DI, Goodacre R. Metabolomic technologies and their application to the study of plants and plant–host interactions. Physiol Plant. 2008;132(2):117–35. doi: 10.1111/j.1399-3054.2007.01001.x. pmid:18251855

56. Ellis DI, Goodacre R. Metabolic fingerprinting in disease diagnosis: biomedical applications of infrared and Raman spectroscopy. Analyst (Cambridge, UK). 2006;131(8):875–85. doi: 10.1039/b602376m

57. Lisec J, Schauer N, Kopka J, Willmitzer L, Fernie AR. Gas chromatography mass spectrometry-based metabolite profiling in plants. Nat Protoc. 2006;1(1):387–96. pmid:17406261 doi: 10.1038/nprot.2006.59

58. Fiehn O, Kopka J, Trethewey R, Willmitzer L. Identification of Uncommon Plant Metabolites Based on Calculation of Elemental Compositions Using Gas Chromatography and Quadrupole Mass Spectrometry. Anal Chem. 2000;72:3573–80. pmid:10952545 doi: 10.1021/ac991142i

CHAPTER 9

EFFECTS OF GENOTYPE AND GROWTH TEMPERATURE ON THE CONTENTS OF TANNIN, PHYTATE AND IN VITRO IRON AVAILABILITY OF SORGHUM GRAINS

Gangcheng Wu, Stuart K. Johnson, Janet F. Bornman, Sarita J. Bennett, Vijaya Singh, Azra Simic, Zhongxiang Fang

INTRODUCTION

Sorghum (*Sorghum bicolor* (L.) Moench) is the fifth leading cereal crop in the world after rice, wheat, barley and maize. It provides a staple food for people living in resource-poor households in many parts of the developing world, particularly in semi-arid areas such as sub-Saharan Africa [1]. Various foods are made from sorghum grain, with porridge being one of the most important. However, the presence of anti-nutritional factors, such as phytate and tannins, has significant negative effects on the nutritional quality of sorghum. For example, both phytate and tannins can bind to minerals, such as iron (Fe), and reduce its digestibility and bioavailability to cause iron deficiency in humans [2–4]. Low iron bioavailability may cause some physical problems such as tiredness and shortness of breath [5].

Phytate and tannin contents in sorghum grains vary with genotype [6]. Sorghum genotypes with the *B1-B2-* gene contain high levels of tannin, and tannin levels can vary about two fold between genotypes which can have important consequences for the bioavailability of iron in the grain [3, 7]. However, little information is available on tannin, phytate and mineral variability in the sorghum genotypes from the Australian sorghum breeding program, and whether the levels of these compounds are influenced by temperature during plant growth.

Temperature is an important factor controlling the growth and development of the sorghum plant. According to published climate models, the global mean surface temperature will rise by 1.8 to 4.0°C by the year 2100, which means crops including sorghum will likely be grown under higher temperatures in the future [8]. Only very limited research has been reported to date on the effect of temperature during plant growth on the chemical composition of sorghum grain. For example, the free phenolic compounds in white pericarp sorghum increased under a warm and moist environment [9]. However, well controlled studies on the effect of genotype and temperature and their interaction during plant growth on sorghum grain production and composition are limited. Li et al. [10] and Nguyen et al. [11] planted diverse sorghum genotypes in temperature-controlled growth chambers and found seed-set of all genotypes, plus the activities of starch biosynthetic enzymes were reduced under a high temperature regime. To date, no published information is available on the effects of genotype and temperature during plant growth on sorghum grain tannin, phytate and mineral levels, or the associated iron availability.

The objective of this study was to determine the effects of two different temperature regimes during plant production on grain phytate, tannin and minerals levels and *in vitro* iron availability in six different sorghum genotypes from the Australian sorghum breeding program.

MATERIALS AND METHODS

Grain growth experimental details

Six sorghum genotypes with three different pericarp colours (red, brown and white) and two breeding lines (hybrid and inbred) were used in the study (Table 1). CHH1, CHH2 and AQL33/QL36 all originated from Australia. Ai4, PI563516 and IS8525 originated from China, Mali and Ethiopia, respectively. They were supplied by the Australia sorghum pre-breeding program which is a partnership between the University of Queensland, The Queensland Department of Agriculture and Fisheries and the Grains research and

Citation: Wu G, Johnson SK, Bornman JF, Bennett SJ, Singh V, Simic A, et al. (2016) Effects of Genotype and Growth Temperature on the Contents of Tannin, Phytate and In Vitro Iron Availability of Sorghum Grains. PLoS ONE 11(2): e0148712. doi:10.1371/journal.pone.0148712.

Development Corporation. The sorghum cultivars were grown in the Controlled Environment Facility, Queensland Bioscience Precinct, University of Queensland, St Lucia, Brisbane, Australia in 2013. Two growth chambers were used to grow three replicate plants of each genotype under two different temperature regimens: optimum temperature (OT) of 32°C maximum, 21°C minimum; and high temperature (HT) of 38°C maximum, 21°C minimum. The experimental details of the temperature growth trial within the Controlled Environment Facility have been reported in detail previously [11]. After harvest, the grain was air-dried to 10% moisture content and manually cleaned after which it was vacuum-packed and kept at 4°C in the dark until analysis.

Table 1. Breeding line and grain pericarp/testa colour of sorghum genotypes used for temperature trials

Sample number	Genotype	Inbred line/hybrid	Grain pericarp/testa colour
1	CHH2	Commercial hybrid	Red
2	CHH1	Commercial hybrid	Red
3	AQL33/QL36	Hybrid	Red
4	Ai4	Inbred	Red
5	PI563516	Inbred	White
6	IS8525	Inbred	Brown

doi:10.1371/journal.pone.0148712.t001

Sample preparation

Flour.

Sorghum whole grain samples were ground for 5 min using a grain mill CEMOTEC 1090 (Foss Tecator, Höganäs, Sweden), to enable 100% of the flour to pass through a 500 µm sieve. The flour was vacuum-packed and stored at 4°C in the dark until analysis and porridge preparation.

Porridge.

The sorghum whole grain flours were made into a thick porridge with three replications by the method described by Kruger, Taylor and Oelofse [7]. The porridge was freeze-dried with a rotational vacuum concentrator 2–18 CDplus freeze drier (Christ, Osterote, Lower Saxony, Germany) at -30°C and 0.37 mbar for 20 h. The dried porridge samples were passed 100% through a 500 µm sieve to break down agglomerates before analysis. Ground samples were vacuum packed and stored at 4°C until analysis. Each sample was prepared in triplicate.

Analytical methods

Condensed tannin content.

Condensed tannin content of the raw flours and dried porridge was determined by the vanillin HCl assay as described by Price, Van Scoyoc and Butler [12]. Ten ml acidified methanol (1% HCl in methanol) was added to 1 g of sorghum flour or dried porridge in triplicate, and the mixture was extracted for 20 min with constant shaking at room temperature. The extracted sample was centrifuged at 3,220 g for 10 min at room temperature and the supernatant collected for analysis. The residue was re-extracted twice with 5 ml acidified methanol. Vanillin HCl reagent (equal volumes of 2% vanillin in methanol and 8% concentrated HCL in methanol) was prepared daily. To 1 ml of extract was added 5 ml of vanillin HCl reagent. The solution was mixed and incubated for 20 min at room temperature. The absorption of the resulting solution was measured at 500 nm on a UV-1800 spectrophotometer (Shimadzu, Canby, USA). Catechin was used as the standard for the calibration curve. Condensed tannin content was expressed as g catechin equivalents (CE) Kg^{-1} sample (dry basis (db)).

Phytate content.

Phytate content of the raw flours and dried porridge was measured using the spectrophotometric method of Haug and Lantzsch [13]. In brief, the sample solution of ground sorghum (about 0.5 g) was extracted from both raw flour and dried porridge with 20 ml of 0.2 M HCl for 90 min in a shaking water bath at 25°C. The solution was centrifuged at 3,220 g for 10 min after extraction at room temperature, and the supernatant collected. The residue was washed twice with 10 ml 0.2 M HCl, centrifugation was repeated between each washing and the resulting supernatants combined. The combined supernatants were made up to 50 ml with 0.2 M HCl. Each sample was extracted in duplicate.

An 0.5 ml aliquot of the supernatant or a standard phytate solution (phytate phosphorus concentrations from 3–30 µg/ml) was pipetted into capped test tubes. A 2 ml ferric solution (0.2 g $FeNH_4(SO_4)_2$ 12 H_2O in 1L 0.2 M HCl) was added to each tube and mixed thoroughly. The capped tubes were heated in a boiling water bath for 30 min, after which they were cooled in ice water to room temperature. Two ml of 2,2′-bipyridyl solution (10 g 2,2′-bipyridyl and 10 ml thioglycollic acid in 1 L distilled water) was added. After mixing thoroughly, the absorption of the solutions was measured at 519 nm on a UV-1800 spectrophotometer (Shimadzu, Canby, USA). Absorption must be read within 0.5–1 min because reactions between bipyridine and iron phytate can change the colour with time. Distilled water was used as a blank. Phytate phosphorus was used as the standard. Phytate content was expressed as g Kg^{-1} sample (db).

Mineral analysis.

Mineral content of the raw flours was measured using a Vista Pro inductively coupled plasma optical emission spectrometer (ICP-OES) (Varian, Palo Alto, USA) at the National Measurement Institute, Perth, Western Australia. Sorghum flour (1 g) was digested with 3 mL of concentrated HNO_3 plus 3 mL of concentrated HCl in a DEENA automated digestion block (Thomas Cain, Nebraska, USA) at 95°C for 2 hours.

After digestion, the sample tubes were made up to 40 ml with distilled water, and solutions were left to settle. Samples were diluted 5-fold with distilled water before they were analysed using the ICP-OES. Appropriate emission wavelengths that had higher sensitivity and lower interferences were chosen to analyse elements of Fe, Ca, P and Zn. The wavelengths used to measure iron, calcium, phosphorus and zinc were 238.204nm, 327.395nm, 117.434nm and 213.857nm, respectively. All results were expressed as g Kg-1 sample (db).

In vitro iron availability.

The dialysis method as described by Luten et al [14] was used to estimate the *in vitro* iron availability of raw flour and dried porridge samples. Two equal amounts of samples (around 1 g) were weighted, and each sample was mixed with 5 ml distilled water, and the pH adjusted to 2 with 6 M HCl. For simulated gastric digestion, 0.15 ml of freshly prepared pepsin solution (4 g pepsin, 2,500 units/mg, from Toxfree (Bedford St, Gillman SA)) dissolved in 25 ml 0.1 M HCl was added to each sample and the mixture incubated at 37°C for 2 h in a shaking water bath. Titratable acidity was determined in one sample of the gastric digest in which a freshly prepared pancreatin-bile extract mixture (400 mg pancreatin from Toxfree (Gillman, SA, Australia) and 2.5 g bovine bile extract from Sigma (St. Louis, MO, USA) in 100 ml 0.1 M $NaHCO_3$) was added, and the pH adjusted to 7.5 with 0.2 M NaOH. Titratable acidity of the samples was defined as the amount of 0.2 M NaOH required to reach the pH 7.5. In the simulated intestinal stage, segments of dialysis tubing with a molecular mass cut off at 10 kDa (Thermo Scientific, IL, USA) containing 25 ml of sodium bicarbonate (equivalent in moles to the NaOH used to measure the titratable acidity), were placed in tubers with the gastric digest and incubated at 37°C for 30 min, after which 5 ml pancreatin-bile solution were added into tubers and incubated for a further 2 h. The iron content in the dialysis bag was determined as described in Section 2.3.4.

Statistical analysis

The analysis of tannin, phytate, minerals and *in vitro* iron availability were repeated in triplicate and the results were expressed as the mean ± standard deviation (SD). The data were statistically analysed using SPSS V20 (SPSS Inc., an IBM Company, Chicago, IL, USA). The main effects of genotype (G) and temperature (T) and their interaction were investigated by two-way ANOVA. Tukey post hoc tests were used to separate individual sample means. Pearson correlation was used to investigate the associations between in vitro iron availability and each of condensed tannin and phytate levels. $P \leq 0.05$ indicated significant difference.

RESULTS AND DISCUSSION

Grain physical characteristics and seed set

The pollen germination and seed-set percentage, which are positively related to sorghum yields, of all genotypes under both temperature treatments have been reported previously in detail [11]. HT significantly reduced pollen germination and seed-set percentage of all genotypes, and genotypes differed significantly in pollen viability percentage (17–63%) and seed-set percentage (7–65%) under HT [11], indicating the sorghum production yields were reduced by high temperature, the extent depending on the genotypes. The physical characteristics including hardness, weight and diameter of different sorghum genotype grains with two temperature treatments were recorded. Apart from CHH1, HT increased the grain weight and diameter of all other sorghum genotypes, indicating that most genotypes might be suitable to grow under predicted future higher temperatures. The seed set of these samples were also measured and reported in the work of Nguyen et al. [11], which showed that these selected genotypes have relatively high percentage of seed set under HT, compared to other genotypes. Singh et al. [15] demonstrated that sorghum seed set percentage had a positive correlation with grain yield. Therefore, the sorghum genotypes used in this study might have good tolerance to increased temperature compared to other genotypes in terms of yield.

Tannin

The influence of genotype and temperature during plant growth on the grain tannin content is presented in Table 2. Temperature, genotype and their interaction had a significant effect on tannin content of the uncooked whole grains ($P \leq 0.05$). There were wide variations in the tannin content ranging from 2.16 to 42.01 g CE Kg^{-1}, with genotype of AQL33/QL36 and CHH1 having the lowest ($P \leq 0.05$) and IS8525 having the highest ($P \leq 0.05$) content, respectively, under OT. However, under HT, the tannin levels in the six sorghum whole grain genotypes were significantly lower than under OT and ranged from 0.79 to 32.42 g CE Kg^{-1}, with PI563516 and IS8525 having the lowest and highest levels ($P \leq 0.05$), respectively (Table 2). This implies that tannin biosynthesis in the sorghum grains might be inhibited under the higher growing temperature in the study. It has been reported that tannin is polymerized from other polyphenols (catechin and epicatechin), but the mechanism still remains speculative [16, 17].

Table 2: Influence of genotype and temperature during plant growth under optimal temperature (OT) and high temperature (HT) on tannin content (g catechin equivalents Kg^{-1} db) in sorghum whole grain flour (uncooked) and porridge

Genotype	Whole grain flour (OT)	Whole grain flour (HT)	Porridge (OT)	Porridge (HT)
CHH2	2.73±0.11 ^{bC}	1.49±0.07 ^{cB}	1.91±0.12 ^{dB}	0.34±0.04 ^{aA}
CHH1	2.29±0.03 ^{aD}	1.67±0.06 ^{cC}	1.14±0.18 ^{bB}	0.68±0.05 ^{aA}
AQL33/QL36	2.16±0.09 ^{aD}	1.00±0.05 ^{bC}	0.84±0.05 ^{aB}	0.49±0.02 ^{aA}
Al4	4.13±0.13 ^{cD}	3.40±0.27 ^{dC}	1.46±0.04 ^{cB}	0.38±0.03 ^{aA}
PI563516	3.02±0.43 ^{bC}	0.79±0.03 ^{aB}	0.72±0.05 ^{aB}	0.46±0.07 ^{aA}
IS8525	42.01±2.22 ^{dD}	32.42±0.66 ^{eC}	5.08±0.07 ^{eB}	4.02±0.03 ^{bA}

$^{a, b, c, d, e}$ Values with different letters in the same column are significantly different ($P \leq 0.05$).
$^{A, B, C, D}$ Values with different letters in the same row are significantly different ($P \leq 0.05$)

doi:10.1371/journal.pone.0148712.t002

It has previously been reported that sorghum genotypes with a pigmented testa contained significantly higher levels of tannin than those without a pigmented testa [18]. In the present study, the tannin content of IS8525 was about 10–15 times higher than the others, suggesting that IS8525 may have both dominant B_1 and B_2 genes, which control the presence or absence of a pigmented testa. Interestingly, although high temperature could significantly reduce the tannin content in all genotypes, the tannin content was still mainly depended on genotype (Table 2).

Sorghum grains are usually cooked prior to human consumption, so understanding changes in tannins during the cooking process is important. The results indicate that cooking into a thick porridge significantly reduced the tannin content in all the sorghum genotypes ($P \leq 0.05$), irrespective of the growing temperature (Table 2). The tannin content in the sorghum porridge ranged from 0.72 to 5.08 g CE Kg^{-1} under OT, and from 0.38 to 4.02 g CE Kg^{-1} under HT. The tannin contents in the porridge were still the highest in the genotype IS8525, irrespective of growing temperature, and were the lowest in PI563516 and CHH2 ($P \leq 0.05$) for the OT, whereas at HT porridges of all of the genotypes except IS8525 had similar levels ($P > 0.05$). After cooking, there was a large reduction in tannin content (about 87%) in IS8525, while in the other five genotypes it was reduced by between 30 and 76%. These results are in agreement with those of Dlamini, Taylor and Rooney [19] and Wedad et al. [20], who showed that tannin content in sorghum grain was reduced through cooking by 36%–80%. This might be the phenolic hydroxyl groups of tannin reacting with food compounds (protein and minerals such as iron) to form insoluble complexes during the heating process [21]. Since the measurable tannin content was lower after cooking, this may help to reduce the undesirable astringent and bitter taste caused by tannins in the grain.

However, tannin has been commonly referred to as an anti-nutritional factor, because of its high affinity to proteins and tendency to complex with divalent metal ions, thus decreasing their availability [22, 23]. However, these negative effects can be reduced or eliminated by decreasing the tannin content in sorghum grains through cooking. In the present experiment, the tannin content was also significantly reduced by growing under the higher temperatures (Table 2). This suggests that sorghum could contain lower levels of tannin under the expected higher future temperatures, which would increase its nutritional value as human food or animal feed. Tannin has also been identified as the major contributor of the antioxidant activity of some genotypes of sorghum grains and tannins may have gastro-protective and cholesterol-lowering properties [24]. For future production of high tannin sorghum for manufacture into high antioxidant foods, the genotype IS8525 could be a good candidate since it contains by far the highest levels of tannin in both OT and HT conditions in the present study (Table 2).

Phytate content

The phytate content of the whole grain flours among the six sorghum genotypes varied between 10.60 g Kg^{-1} (CHH2) and 17.44 (IS8525) g Kg^{-1} under OT (Table 3). There was a significant effect of genotype, temperature and their interaction on phytate content in the uncooked flours ($P \leq 0.05$). The phytate values reported here are in the range previously reported for sorghum (2.11–17.62 g Kg^{-1}) [20]. HT had diverse impacts on phytate of sorghum grain. For three hybrid lines with the same red colour (Table 1), the phytate in CHH1 and CHH2 increased 32.24% and 26.13% respectively but showed no significant changes in AQL33/QL36 when growth temperature was increased. The white inbred line: PI563516 had higher levels of phytate while the other two (brown and white) showed no significant changes under HT. These results suggest that the phytate content in sorghum grain was not affected or increased by HT, depending on the genotype. The effect of temperature during plant growth on the phytate content in sorghum grains has not been reported previously. In an experiment on other crops, including maize, however, Horvatic and Balint [25] found that lower temperatures and increased irrigation decreased the phytate levels, but they did not separate the influence of temperature and irrigation.

Table 3: Influence of genotype and temperature during plant growth under optimal temperature (OT) and high temperature (HT) on the phytate content (g Kg^{-1}, db) of sorghum whole grain flour (uncooked) and porridge

Genotype	Whole grain flour (OT)	Whole grain flour (HT)	Porridge (OT)	Porridge (HT)
CHH2	10.60 ± 0.39^{aA}	13.73 ± 0.13^{aB}	10.61 ± 1.13^{aA}	12.98 ± 0.50^{aB}
CHH1	12.25 ± 0.49^{bA}	16.20 ± 0.74^{cB}	12.37 ± 0.15^{bA}	16.74 ± 0.47^{cB}
AQL33/QL36	15.80 ± 0.23^{dA}	15.02 ± 0.82^{bA}	14.89 ± 0.21^{dA}	14.37 ± 0.17^{bA}
Ai4	15.56 ± 0.23^{dA}	15.92 ± 0.82^{bcA}	15.53 ± 0.52^{dA}	15.70 ± 1.62^{bcA}
PI563516	14.07 ± 0.10^{cB}	12.96 ± 0.94^{aA}	13.68 ± 0.92^{cB}	12.87 ± 0.63^{aA}
IS8525	17.44 ± 0.40^{eA}	16.39 ± 0.41^{cA}	16.47 ± 0.89^{eA}	16.24 ± 0.13^{cA}

a, b, c, d, e Values with different letters in the same column are significantly different ($P \leq 0.05$).
A, B Values with different letters in the same row are significantly different ($P \leq 0.05$).

doi:10.1371/journal.pone.0148712.t003

The phytate level in the porridge was not significantly different to that in the raw whole grain flour for all sorghum genotypes ($P > 0.05$) within each growth temperature (Table 3). Endogenous phytases that are present in sorghum grains have the potential to break down phytate. However, they may have been deactivated by heat during the porridge preparation [26]. Previous research found that phytate in sorghum seeds was reduced under the traditional porridge preparation method, probably because of longer cooking time [20] than that used in our research. In the present study, cooking into a simple porridge did not reduce the phytate content. Phytate chelates metal ions, such as iron, to form insoluble complexes in the gastrointestinal tract, and these irons cannot be digested [27]. These results might indicate that cooking process cannot improve the mineral bioavailability.

Mineral content

The iron content of the whole grain flour, which ranged from 14.07 to 47.83 mg/kg with PI563516 having the lowest and CHH1 having the highest level ($P \leq 0.05$) under OT (Table 4), was significantly affected by genotype and temperature ($P \leq 0.05$). It is interesting to note that the iron content in the hybrid lines (36.9×10^{-3}–47.83×10^{-3} g kg^{-1}) was significantly higher than that in the inbred lines (14.07×10^{-3}–37.43×10^{-3} g kg^{-1}) ($P \leq 0.05$). A significant difference has previously been reported between the sorghum genotypes in iron acquisition and storage during plant growth and in particular among breeding lines [7]. Under HT the iron content in the whole grain flours varied between 19.57×10^{-3} and 54.53×10^{-3} g kg^{-1}, with the level in CHH1 and PI563516 remaining the lowest and highest respectively ($P \leq 0.05$) (Table 4). Temperature was found to have a marked effect on the iron content in some of the sorghum flours ($P \leq 0.05$). For example, the iron content in CHH1 and IS8525 was significantly higher ($P \leq 0.05$) in HT than OT, whereas that of AQL33/QL36 was lower under HT ($P \leq 0.05$).

Table 4: Influence of genotype and growing temperature during plant growth under optimal temperature (OT) and high temperature (HT) on mineral (Fe, Ca, P, Zn) contents (g kg^{-1} db) in sorghum wholegrain flour (uncooked)

Genotype	Fe		Ca		P		Zn	
	OT	HT	OT	HT	OT	HT	OT	HT
CHH2	$36.90\times10^{-3}\pm0.10\times10^{-3\,dcB}$	$33.47\times10^{-3}\pm0.15\times10^{-3\,bA}$	$99.47\times10^{-3}\pm0.55\times10^{-3\,eA}$	$105.33\times10^{-3}\pm2.51\times10^{-3\,dB}$	3.156 ± 0.055^{aA}	3.636 ± 0.038^{bB}	$20.17\times10^{-3}\pm0.21\times10^{-3\,bA}$	$21.06\times10^{-3}\pm1.00\times10^{-3\,aA}$
CHH1	$47.83\times10^{-3}\pm0.47\times10^{-3\,eA}$	$54.53\times10^{-3}\pm1.16\times10^{-3\,eB}$	$66.00\times10^{-3}\pm3.21\times10^{-3\,cA}$	$85.37\times10^{-3}\pm3.57\times10^{-3\,cB}$	3.460 ± 0.025^{bA}	4.490 ± 0.036^{dB}	$20.70\times10^{-3}\pm0.35\times10^{-3\,bA}$	$24.57\times10^{-3}\pm0.75\times10^{-3\,dB}$
AQL33/QL36	$45.37\times10^{-3}\pm0.38\times10^{-3\,dB}$	$35.57\times10^{-3}\pm1.21\times10^{-3\,dbA}$	$89.57\times10^{-3}\pm2.37\times10^{-3\,dB}$	$63.26\times10^{-3}\pm0.53\times10^{-3\,aA}$	4.215 ± 0.013^{cB}	3.957 ± 0.081^{bA}	$19.30\times10^{-3}\pm0.17\times10^{-3\,bB}$	$17.50\times10^{-3}\pm0.46\times10^{-3\,bA}$
Ai4	$37.43\times10^{-3}\pm0.75\times10^{-3\,cA}$	$35.57\times10^{-3}\pm0.31\times10^{-3\,cA}$	$96.47\times10^{-3}\pm1.32\times10^{-3\,eB}$	$87.17\times10^{-3}\pm1.92\times10^{-3\,cA}$	3.480 ± 0.010^{bA}	3.580 ± 0.043^{bB}	$27.03\times10^{-3}\pm0.84\times10^{-3\,eA}$	$26.80\times10^{-3}\pm0.63\times10^{-3\,eA}$
PI563516	$14.07\times10^{-3}\pm0.15\times10^{-3\,aA}$	$19.57\times10^{-3}\pm0.31\times10^{-3\,aB}$	$50.87\times10^{-3}\pm0.77\times10^{-3\,aA}$	$65.74\times10^{-3}\pm0.15\times10^{-3\,bB}$	3.200 ± 0.036^{aB}	2.980 ± 0.075^{aA}	$12.30\times10^{-3}\pm0.15\times10^{-3\,aA}$	$16.10\times10^{-3}\pm0.20\times10^{-3\,aB}$
IS8525	$30.53\times10^{-3}\pm0.25\times10^{-3\,bA}$	$40.37\times10^{-3}\pm0.66\times10^{-3\,dB}$	$58.40\times10^{-3}\pm0.26\times10^{-3\,bA}$	$70.30\times10^{-3}\pm1.18\times10^{-3\,bB}$	3.963 ± 0.051^{bA}	3.593 ± 0.015^{cB}	$24.40\times10^{-3}\pm0.26\times10^{-3\,cA}$	$30.63\times10^{-3}\pm0.06\times10^{-3\,fB}$

a, b, c, d, e, f Values with different letters in the same column are significantly different ($P \leq 0.05$).
A, B Values with different letters in the same genotype and the same mineral are significantly different ($P \leq 0.05$).

doi:10.1371/journal.pone.0148712.t004

The calcium content of the sorghum is also shown in Table 4, and was significantly affected by genotype and temperature ($P \leq 0.05$). The six sorghum genotypes varied significantly ($P \leq 0.05$) in their calcium content, ranging from 50.87×10^{-3} to 99.47×10^{-3} g kg^{-1} flour under OT and 63.26×10^{-3} to 105.33×10^{-3} g kg^{-1} flour under HT. The calcium levels in only two of the sorghum genotypes (AQL33/QL36 and Ai4) were significantly reduced under high growing temperatures, while in the others they were significantly increased ($P \leq 0.05$), also suggesting a genotype influence.

Phosphorus is the major mineral in sorghum and its content in the sorghum grains was significantly influenced by the genotypes and temperature ($P \leq 0.05$). A 1.015 g kg^{-1} flour difference was recorded between AQL/QL36 (the highest level) and PI563516 (the lowest level) under OT. An increase in day temperature from 32 to 38°C resulted in a notable decrease in the phosphorus content in the grains of AQL33/QL36 and PI563516 ($P \leq 0.05$).

Zinc content was significantly affected by genotype and temperature ($P \leq 0.05$), and ranged from $12.33 \times 10^{-3}$to 27.03×10^{-3} g kg^{-1} flour under OT, and from 16.10×10^{-3} to 26.80×10^{-3} g kg^{-1} flour under HT, respectively (Table 4). The zinc content in other sorghum varieties has been reported between 14×10^{-3} and 38×10^{-3} g kg^{-1} [28] Zinc levels were unchanged in Ai4 and CHH2, decreased in AQL33/QL36 ($P \leq 0.05$), but increased in the other 3 genotypes ($P \leq 0.05$) when the growing temperature was increased from 32/21°C to 38/21°C.

As sorghum is one of the staple cereal grains in Africa, sorghum grains may provide the major source of minerals for the local people. In this research, the three sorghum hybrid lines of CHH2 and CHH1with high mineral content are likely suitable genotypes to be grown in a high temperature environment as a good dietary mineral source.

In *vitro iron* availability

The *in vitro* iron dialysability of the whole grain raw flours (as an estimate of availability) is shown in Table 5. Genotype, temperature and their interaction had a significant effect on the *in vitro* iron availability ($P \leq 0.05$) of the uncooked whole grain flours. Under OT the significantly highest *in vitro* iron availability ($P \leq 0.05$) was found in CHH2, while the lowest ($P \leq 0.05$) occurred in IS8525. Values varied from 0.87 to 5.94% for the genotypes grown under OT but ranged from 0.47 to 3.15% under HT, indicating the *in vitro* iron dialysability in most tested sorghum genotypes was decreased, with the exception of genotype Ai4 which was significantly increased, irrespective of tannins and phytate content (Tables 2 and 3). Interestingly, the flour from IS8525 had the lowest levels of *in vitro* iron availability under both growing temperatures. This may be related to the fact that IS8525 had the extremely high levels of tannin, which might exert a significant inhibitory effect on iron absorption [29]. One of the postulated mechanisms of the action of tannins is chelation of pro-oxidant metals like iron [30].

Table 5: Influence of genotype and temperature during plant growth under OT and HT on *in vitro* **iron** dialysability (%) of sorghum whole grain flour and porridge

Genotype	Whole grain flour (OT)	Whole grain flour (HT)	Thick porridge (OT)	Thick porridge (HT)
CHH2	5.94±0.13 ^{fB}	1.79±0.09 ^{cA}	5.76±0.45 ^{fB}	1.67±0.10 ^{cA}
CHH1	4.70±0.07 ^{eB}	3.15±0.17 ^{dA}	4.54±0.15 ^{eB}	3.21±0.11 ^{dA}
AQL33/ QL36	1.54±0.11 ^{bB}	1.10±0.24 ^{bA}	1.47±0.12 ^{bB}	0.89±0.04 ^{bA}
Ai4	2.61±0.06 ^{cA}	3.10±0.08 ^{dB}	2.42±0.23 ^{cA}	3.18±0.23 ^{dB}
PI563516	3.09±0.21 ^{dB}	1.07±0.15 ^{bA}	2.92±0.13 ^{dB}	0.95±0.12 ^{bA}
IS8525	0.87±0.09 ^{aB}	0.47±0.08 ^{aA}	0.83±0.06 ^{aB}	0.44±0.05 ^{aA}

$^{a, b, c, d, e, f}$ Values with different letters in the same temperature column are significantly different ($P \leq 0.05$).
$^{A, B}$ Values with different letters in the same row are significantly different ($P \leq 0.05$).

doi:10.1371/journal.pone.0148712.t005

The *in vitro* iron availability in the porridge varied between 0.83 and 5.76% for the OT samples and between 0.44 and 3.21% for the HT samples ($P \leq 0.05$). The cooking process had no significant effect on the iron dialysability compared to the raw whole grain flours ($P > 0.05$). This could partly be due to the phytate not being affected by the heat treatment (section 3.2), as phytate can chelate irons to form insoluble complexes to decrease iron dialysability. Kruger et al. [7] observed similar results when they prepared thick porridge using four different sorghum genotypes. However, some researchers reported that heat treatment could slightly increase [28] or decrease [31] the *in vitro* iron availability. The *in vitro* iron availability is strongly affected by the levels of polyphenols and phytate, and Towo et al. [2] also proposed that the reduction of the phenolic content without the reduction of phytate would not improve the iron availability. The possible reason was that although cooking can reduce the total measurable tannin content, the amount of tannin bound with iron may have increased during cooking [7]. Some researchers have investigated the effect of reducing phytate in tannin and tannin-free sorghum varieties by fermentation, and reported that only tannin-free sorghum varieties increased their iron availability after phytate reduction [20]. Their results suggested that the *in vitro* iron availability in sorghum grains could be increased by reducing both phytate and tannin content. In this experiment, cooking could significantly decrease tannin content but had no effects on the phytate content, so the iron dialysability was not improved after cooking.

Levels of *in vitro* iron availability in sorghum grains was significantly negatively correlated with tannin (Pearson's correlation: r = 0.463, $P \leq 0.05$) and phytate (r = 0.694, $P \leq 0.05$) content. Other researchers have also confirmed that a significant negative correlation was observed between iron availability and the tannin and phytate content [32, 33].

From the present study, production under HT gave reduced *in vitro* iron dialysability of the raw whole grain and porridge of some sorghum genotypes, so the nutritional values of these foods, from an iron availability view, could decrease under predicted higher temperatures in the future. In terms of sorghum porridge as a staple food for people living in malnourished communities in rural Africa, the sorghum genotype CHH2, which had higher *in vitro* iron availability than the other sorghum genotypes, could be a suitable available iron resource. However, under high temperature production conditions, the sorghum genotypes CHH1 and Ai4 have potential for high *in vitro* iron availability.

CONCLUSIONS

Analysis of the tannin, phytate, mineral (Fe, Ca, P, Zn) contents and *in vitro* iron dialysability of whole grain flours and porridge of six diverse sorghum genotypes of varied grain colours grown under OT and HT provided new evidence of effects of both genotype and temperature of plant growth. Genotype had a significant effect on tannin, phytate, mineral contents and *in vitro*iron dialysability of raw grain flours. The tannin content in all investigated sorghum grains decreased under the higher growth temperature. However, the high growth temperatures had diverse effects on the phytate, mineral contents and *in vitro* iron availability of the sorghum grains, depending on the genotype. This study has the potential to help farmers to select appropriate sorghum genotypes to grow for maximum iron bio-avaiability at elevated production temperatures that may occur in the future. However, more research is required to discover the plant physiological mechanisms behind the changes in tannin, phytate and *in vitro* iron availability under high growth temperatures.

ACKNOWLEDGMENTS

The authors would like to thank Curtin International Postgraduate Research Scholarships (CIPRS).

AUTHOR CONTRIBUTIONS

Conceived and designed the experiments: ZF SKJ. Performed the experiments: GW VS AS. Analyzed the data: GW AS. Contributed reagents/materials/analysis tools: GW AS. Wrote the paper: GW SKJ JFB SJB VS AS ZF.

REFERENCES

1. Afify A, El-Beltagi HS, Abd El-Salam SM, Omran AA. Bioavailability of iron, zinc, phytate and phytase activity during soaking and germination of white sorghum varieties. PloS One. 2011; 6: e25512. doi: 10.1371/journal.pone.0025512. pmid:22003395
2. Towo E, Matuschek E, Svanberg U. Fermentation and enzyme treatment of tannin sorghum gruels: effects on phenolic compounds, phytate and in vitro accessible iron. Food Chem. 2006; 94: 369–376. doi: 10.1016/j.foodchem.2004.11.027.
3. Dykes L, Rooney LW. Sorghum and millet phenols and antioxidants. J Cereal Sci. 2006; 44: 236–251. doi: 10.1016/j.jcs.2006.06.007.
4. Idris WH, AbdelRahaman SM, ElMaki HB, Babiker EE, El Tinay AH. Effect of malt pretreatment on phytate and tannin level of two sorghum (Sorghum bicolor) cultivars. Int J Food Sci Tech. 2006; 41: 1229–1233. doi: 10.1111/j.1365-2621.2006.01190.x.
5. Martınez-Navarrete N, Camacho M, Martınez-Lahuerta J, Martınez-Monzó J, Fito P. Iron deficiency and iron fortified foods—a review. Food Res Int. 2002; 35: 225–231. doi: 10.1016/s0963-9969(01)00189-2
6. Proietti I, Mantovani A, Mouquet-Rivier C, Guyot JP. Modulation of chelating factors, trace minerals and their estimated bioavailability in Italian and African sorghum (Sorghum bicolor (L.) Moench) porridges. Int J Food Sci Tech. 2013; 48: 1526–1532. doi: 10.1111/ijfs.12121.
7. Kruger J, Taylor JR, Oelofse A. Effects of reducing phytate content in sorghum through genetic modification and fermentation on in vitro iron availability in whole grain porridges. Food Chem. 2012; 131: 220–224. doi: 10.1016/j.foodchem.2011.08.063.
8. IPCC, Intergovernmental Panel on Climate Change. IPCC Fourth Assessment Report (AR4). 2007; 11: 17. Available: http://ipcc.ch/publications_and_data/ar4/syr/en/contents.html.
9. Waniska R, Poe J, Bandyopadhyay R. Effects of growth conditions on grain molding and phenols in sorghum caryopsis. J Cereal Sci. 1989; 10: 217–225. doi: 10.1016/S0733-5210(89)80051-7.
10. Li E, Hasjim J, Singh V, Tizzotti M, Godwin ID Gilbert RG. Insights into sorghum starch biosynthesis from structure changes induced by different growth temperatures. Cereal Chem. 2013; 90: 223–230. doi: 10.1094/CCHEM-09-12-0113-R.
11. Nguyen CT, Singh V, van Oosterom EJ, Chapman SC, Jordan DR, Hammer GL. Genetic variability in high temperature effects on seed-set in sorghum. Funct Plant Biol. 2013, 40: 439–448. doi: 10.1071/FP12264.
12. Price ML, Van Scoyoc S, Butler LG. A critical evaluation of the vanillin reaction as an assay for tannin in sorghum grain. J Agr Food Chem. 1978; 26: 1214–1218. doi: 10.1021/jf60219a031.
13. Haug W, Lantzsch HJ. Sensitive method for the rapid determination of phytate in cereals and cereal products. J Sci Food Agr. 1983; 34: 1423–1426. doi: 10.1002/jsfa.2740341217.
14. Luten J, Crews H, Flynn A, Van Dael P, Kastenmayer P, Hurrell R, et al. Interlaboratory trial on the determination of the in vitro iron dialysability from food. J Sci Food Agr. 1996; 72: 415–424. doi: 10.1002/(SICI)1097-0010(199612)72:4<415::AID-JSFA675>3.0.CO;2-X.
15. Singh V, Nguyen CT, van Oosterom EJ, Chapman SC, Jordan DR, Hammer GL. Sorghum genotypes differ in high temperature responses for seed set. Field Crop Res. 2015; 171: 32–40. doi: 10.1016/j.fcr.2014.11.003.
16. Glennie CW, Kaluza WZ, Van Niekerk PJ. High-performance liquid chromatography of procyanidins in developing sorghum grain. J Agr Food Chem. 1981; 29: 965–968. doi: 10.1021/jf00107a020.
17. Cohen SD, Kennedy JA. Plant metabolism and the environment: implications for managing phenolics. Crit Rev Food Sci. 2010; 50: 620–643. doi: 10.1080/10408390802603441.
18. Dykes L, Rooney LW, Waniska RD, Rooney WL. Phenolic compounds and antioxidant activity of sorghum grains of varying genotypes. J Agr Food Chem. 1995; 53: 6813–6818. doi: 10.1021/jf050419e.
19. Dlamini NR, Taylor JR, Rooney LW. The effect of sorghum type and processing on the antioxidant properties of African sorghum-based foods. Food Chem. 2007; 105: 1412–1419. doi: 10.1016/j.foodchem.2007.05.017
20. Wedad HA, El Tinay AH, Mustafa AI, Babiker EE. Effect of fermentation, malt-pretreatment and cooking on antinutritional factors and protein digestibility of sorghum cultivars. Pakistan J Nutr. 2008; 7: 335–341. doi: 10.3923/pjn.2008.335.341.
21. Matuschek E, Towo E, Svanberg U. Oxidation of polyphenols in phytate-reduced high-tannin cereals: effect on different phenolic groups and on in vitro accessible iron. J Agr Food Chem. 2001; 49: 5630–5638. doi: 10.1021/jf0108157.
22. Nyachoti CM. Sorghum tannins: a review. World Poultry Sci J. 1997; 53: 5. doi: 10.1079/wps19970002.
23. Radhakrishnan MR, Sivaprasad J. Tannin content of sorghum varieties and their role in iron bioavailability. J Agr Food Chem. 1980; 28: 55–57. doi: 10.1021/jf60227a038.
24. Dykes L, Rooney LW. Phenolic compounds in cereal grains and their health benefits. Cereal Food World. 2007; 52: 105–111. doi: 10.1094/cfw-52-3-0105.
25. Horvatic M, Balint L. Relationship among the phytic acid and protein content during maize grain maturation. J Agron Crop Sci. 1996; 176: 73–77. doi: 10.1111/j.1439-037x.1996.tb00448.x.
26. Duodu K, Minnaar A, Taylor J. Effect of cooking and irradiation on the labile vitamins and antinutrient content of a traditional African sorghum porridge and spinach relish. Food Chem. 1999; 66: 21–27. doi: 10.1016/

s0308-8146(98)00070-3.

27. Gibson RS, Bailey KB, Gibbs M, Ferguson EL. A review of phytate, iron, zinc, and calcium concentrations in plant-based complementary foods used in low-income countries and implications for bioavailability. Food Nutr Bull. 2010; 31: 134–146. 1177/15648265100312s206 doi: 10.1177/15648265100312s206

28. Kayodé AP, Linnemann AR, Hounhouigan JD, Nout MJ, van Boekel MA. Genetic and environmental impact on iron, zinc, and phytate in food sorghum grown in Benin. J Agr Food Chem. 2006; 54: 256–262. doi: 10.1021/jf0521404.

29. Brune M, Rossander L, Hallberg L. Iron absorption and phenolic compounds: importance of different phenolic structures. Eur J Clin Nutr. 1989; 43: 547–557. pmid:2598894

30. López MA, Martos FC. Iron availability: an updated review. International Journal of Food Sci Nutr. 2004; 55: 597–606. doi: 10.1080/09637480500085820.

31. Hemalatha S, Platel K, Srinivasan K. Influence of heat processing on the bioaccessibility of zinc and iron from cereals and pulses consumed in India. J Trace Elem Med Bio. 2007; 21: 1–7. doi: 10.1016/j.jtemb.2006.10.002.

32. Kayodé APP, Hounhouigan JD, Nout MJR. Impact of brewing process operations on phytate, phenolic compounds and in vitro solubility of iron and zinc in opaque sorghum beer. LWT-Food Scie Technol. 2007; 40: 834–841. doi: 10.1016/j.lwt.2006.04.001.

33. Hassan AB, Osman GA, Rushdi MA, Eltayeb MM, Diab EE. Effect of gamma irradiation on the nutritional quality of maize cultivars (Zea mays) and sorghum (Sorghum bicolor) grains. Pakistan J Nutr. 2009; 8: 167–171. doi: 10.3923/pjn.2009.167.171.

CHAPTER 10

QUANTITATIVE GENETIC ANALYSIS OF AGRONOMIC AND MORPHOLOGICAL TRAITS IN SORGHUM, SORGHUM BICOLOR

Riyazaddin Mohammed, Ashok K. Are, Ramaiah Bhavanasi, Rajendra S. Munghate, Polavarapu B. Kavi Kishor and Hari C. Sharma

Keywords: Sorghum, combining ability, heritability, agronomic traits, morphological traits, GCA, SCA, grain yield

INTRODUCTION

Sorghum [*Sorghum bicolor* (L.) Moench] is an important crop grown primarily in warm and dry climates with a wide range of adaptability to various agro-ecological conditions. It is the fifth most important food crop after wheat, rice, maize, and barley (FAO, 2004), and is widely grown in the semi-arid regions. It is the staple food for 600 million people living in the semi-arid regions. India is the third largest sorghum producer after Nigeria and United States of America, with 6.25 million hectares of area under sorghum cultivation, and with a total production of 5.98 million tonnes (FAOSTAT, 2012).

Information on inheritance of agronomic and morphological traits is useful for improving genotypic performance across environments. In sorghum, both the additive and non-additive type of gene action governs the inheritance of morphological and agronomic traits (Nimbalkar and Bapat, 1992; Umakanth et al., 2002; Mohammed Maarouf, 2009) with considerable amount of G X E interaction (Jayanthi et al., 1996; Dhillon et al., 2006; Aruna et al., 2011a).

Most of the morphological traits in sorghum are associated with one or more economically important traits, and will be helpful in selecting the high yielding sorghum genotypes. Brown midrib increases the fodder quality, while the presence of awns acts as a mechanical barrier to bird damage (Porter et al., 1978;Kullaiswamy and Goud, 1983). Genotypes with tan-colored plants showed resistance to various fungal diseases while the genotypes with closed glumes are resistant to grain mold (Melake-Berhan et al., 1996;Murty, 2000). Although, considerable progress has been made in identifying insect-resistant sorghums (Sharma, 1993;Sharma et al., 2003), but there is little progress in developing insect-resistant high yielding varieties for cultivation by the farmers. This is largely because of the lack of knowledge on inheritance of the agronomic and morphological characteristics associated with insect resistance and grain yield (Sharma et al., 2005; Riyazaddin et al., 2015). The combining ability analysis is useful to understand the nature of gene action, and has been used by the breeders to select the suitable parents for the crossing program. An understanding of the inheritance of morphological and agronomic traits will be helpful in combining the genes for insect resistance and desirable agronomic traits and grain characteristics to increase production and productivity of sorghum. Therefore, we developed a full diallel involving 10 parents to study the inheritance of morphological and agronomic traits. The combining ability studies will be helpful to identify genotypes which can be utilized in the hybridization.

MATERIALS AND METHODS

Experimental Material

Based on *per se* performance of sorghum genotypes in the field against shoot fly, *Atherigona soccata* and molecular diversity, 10 morphologically and genetically diverse sorghum genotypes (Annexure I in Supplementary Material) adapted to the rainy and postrainy seasons were selected and crossed in all possible combinations, which generated 45 direct crosses and 45 reciprocal crosses. These crosses along with the parents were evaluated in a randomized complete block design in three replications during the 2013–14 rainy and postrainy seasons at the International Crops Research Institute for the Semi-Arid Tropics (ICRISAT), Patancheru, Hyderabad, Telangana, India (latitude 17.53°N, longitude 78.27°E, and altitude of 545 m).

Citation: Mohammed R, Are AK, Bhavanasi R, Munghate RS, Kavi Kishor PB and Sharma HC (2015) Quantitative genetic analysis of agronomic and morphological traits in sorghum, Sorghum bicolor. Front. Plant Sci. 6:945. doi: 10.3389/fpls.2015.00945.

Sowing of the test genotypes was carried out using a two cone planter. Each test plot consists of a row length of 2.0 m and a row to row spacing of 75 cm. A distance of 10 cm was maintained in-between the plants within a row. Application of a basal dose of Ammonium phosphate to the field was carried out at 100 kg/ha. Each parent was sown in two rows, and a single row of F_1. Thinning of the test plots was carried out at 7 days after seedling emergence (DAE) and a plant population of 40 plants were retained in a test plot. At 30 DAE earthing up was carried out along with top dressing with urea at 100 kg/ha. During the postrainy season furrow irrigation was given to the experimental material. One set of the replicated test material was grown under protected conditions (application of carbofuran 3G granules in the leaf whorls at 7 days after seedling emergence, and cypermethrin spray after 5 days) to record the agronomic and morphological traits in the undamaged plants during the rainy and postrainy seasons.

Observations

Agronomic Characteristics

Data were recorded on days to 50% flowering, plant height, agronomic score, 100 seed weight, and grain yield. Days to 50% flowering was recorded when half of the panicles and nearly 50% of plants in the plot had attained the anthesis stage. Height of three plants was measured from the base of the plant to the tip of the panicle at physiological maturity in plants selected at random within a test plot. The agronomic desirability of the genotype was recorded at crop maturity on a 1–5 scale (1 = high productive potential, and 5 = poor productive potential). Data on 100 seed weight and grain yield/plot for parents, and grain yield/5 plants for F_1s were recorded after crop harvest.

Morphological Characteristics

Inflorescence exsertion was scored on a 1–4 scale (1 = panicle fully exserted, and 4 = panicle recurved); panicle compactness on a 1–3 scale (1 = loose panicle, and 3 = compact panicle); panicle shape on a 1–4 scale (1 = erect panicle, and 4 = elliptic panicle); glume coverage on a 1–9 scale (1 = 25% grain covered with glumes, and 9 = glumes longer than the grain); awns on 1–2 scale (1 = absence of awns, 2 = presence of awns); grain luster on 1–2 scale (1 = non-lustrous grain, 2 = lustrous grain); and grain color on a 1–5 scale (1 = white colored grain and 5 = buff colored grain) (IBPGR and ICRISAT, 1993).

Statistical Analysis

Analysis of variance (ANOVA) was carried out using GenStat® 13th version (GenStat, 2010). *F*-test was used to test the significance differences between the genotypes, and least significance difference (LSD) for comparing the genotypic means at $P \leq 0.05$. Simple correlation coefficients were calculated to determine the association between the traits studied. Partitioning of the combining abilities (GCA and SCA) was done using the method 1 and model 1 of Griffing (1956), that provides the information on nature of parents, and the hybrid performance, using Windowstat (Indostat Services, 2004) software. The coefficient of variations at phenotypic and genotypic level variation was estimated using the formula adopted by Johnson et al. (1955) and predictability ratio using Baker (1978).

RESULTS

Agronomic Traits

Evaluation of 10 parents and 90 F_1s, including the reciprocals, showed significant differences for all the traits studied across seasons at $P \leq 0.01$. Days to 50% flowering was ranged from 61 to 81 days in the rainy season, 56–78 days in the postrainy season (Table 1). Almost all the crosses flowered early, with few exceptions. Crosses between the parents with early- and late-flowering were early-flowering, indicating dominance of earliness for anthesis in the test material used. The crosses CSV 15 X ICSV 25019, CSV 15 X PS 35805, ICSV 25019 X CSV 15, ICSV 25019 X PS 35805, ICSV 25019 X IS 2123, ICSV 25019 X Swarna, PS 35805 X CSV 15, PS 35805 X ICSV 25019, PS 35805 X IS 2146, PS 35805 X Swarna, Swarna X ICSV 25019, and Swarna X PS 35805 exhibited moderate plant height across seasons. Parents and the crosses with moderate plant height can be exploited in developing the commercial hybrids amenable for mechanical harvesting.

Table 1: Agronomic characteristics of sorghum genotypes (Parents and F_1's) evaluated for resistance to sorghum shoot fly, *A. soccata* **across seasons (ICRISAT,** Patancheru, 2013–14)

Pedigree	Days to 50% flowering		Plant height (cm)		100 seed weight (g)		Grain yield (t/ha)		Agronomic score	
	2013 R	2013 PR	2013 R	2013 PR	2013 R	2013 PR	2013 R	2013 PR	2013 R	2013 PR
PARENTS										
ICSV 700	81	77	309	189	2.30	2.37	0.79	5.02	5.00	3.67
Phule Anuradha	62	63	259	179	2.90	4.23	1.22	6.51	4.67	4.33
M 35-1	75	72	306	187	2.40	3.50	0.66	6.87	5.00	3.67
CSV 15	71	61	254	180	2.50	3.03	1.54	5.23	3.00	3.00
ICSV 25019	65	64	131	109	2.30	2.10	1.90	3.01	2.00	3.00
PS 35805	69	65	121	102	2.20	2.37	1.70	3.12	2.00	3.00
IS 2123	73	72	283	176	2.13	2.50	0.77	5.64	5.00	4.33
IS 2146	68	73	279	181	1.80	2.33	0.88	5.56	5.00	4.67
IS 18551	78	78	313	203	1.70	2.23	0.51	4.23	5.00	3.33
Swarna	67	63	166	138	3.30	3.77	1.89	5.17	2.00	2.00
DIRECT CROSSES										
ICSV 700 X Phule Anuradha	65	75	314	236	2.73	3.77	3.58	15.18	4.33	2.67
ICSV 700 X M 35-1	65	70	321	218	2.63	4.00	3.73	12.51	3.67	3.33
ICSV 700 X CSV 15	75	71	327	209	2.37	3.30	4.73	14.18	2.33	4.00
ICSV 700 X ICSV 25019	67	68	308	199	2.80	3.63	4.80	12.72	3.00	3.00
ICSV 700 X PS 35805	67	67	309	204	2.47	3.07	6.25	12.13	3.33	3.67
ICSV 700 X IS 2123	68	69	303	219	2.50	3.40	4.12	13.73	4.67	4.33
ICSV 700 X IS 2146	69	71	318	208	2.53	3.53	2.57	5.49	4.00	3.67
ICSV 700 X IS 18551	67	70	339	231	2.40	3.30	5.31	11.87	3.67	3.33
ICSV 700 X Swarna	67	68	319	213	3.07	4.03	4.10	10.46	5.00	3.00
Phule Anuradha X M 35-1	64	66	273	202	2.60	4.37	2.93	11.99	4.00	3.33
Phule Anuradha X CSV 15	62	59	302	202	3.00	4.07	1.97	8.94	4.00	3.33
Phule Anuradha X ICSV 25019	61	60	280	183	2.87	3.97	2.58	10.69	3.67	2.67

ICSV 700 X PS 35805	67	67	309	204	2.47	3.07	6.25	12.13	3.33	3.67
ICSV 700 X IS 2123	68	69	303	219	2.50	3.40	4.12	13.73	4.67	4.33
ICSV 700 X IS 2146	69	71	318	208	2.53	3.53	2.57	5.49	4.00	3.67
ICSV 700 X IS 18551	67	70	339	231	2.40	3.30	5.31	11.87	3.67	3.33
ICSV 700 X Swarna	67	68	319	213	3.07	4.03	4.10	10.46	5.00	3.00
Phule Anuradha X M 35-1	64	66	273	202	2.60	4.37	2.93	11.99	4.00	3.33
Phule Anuradha X CSV 15	62	59	302	202	3.00	4.07	1.97	8.94	4.00	3.33
Phule Anuradha X ICSV 25019	61	60	280	183	2.87	3.97	2.58	10.69	3.67	2.67
Phule Anuradha X PS 35805	62	61	280	177	2.97	3.53	5.89	14.08	3.33	3.33
Phule Anuradha X IS 2123	63	64	280	188	2.60	3.83	3.40	12.40	4.33	4.00
Phule Anuradha X IS 2146	64	66	292	187	2.20	4.00	2.89	7.17	4.00	3.67
Phule Anuradha X IS 18551	65	65	310	208	2.27	3.47	4.68	12.70	3.33	4.00
Phule Anuradha X Swarna	63	62	297	199	3.43	4.33	4.76	9.44	3.33	3.33
M 35-1 X CSV 15	64	70	317	211	2.97	3.37	4.47	11.58	3.67	3.00
M 35-1 X ICSV 25019	63	63	304	203	2.77	3.83	3.34	13.94	3.33	3.33
M 35-1 X PS 35805	63	64	296	171	2.73	3.43	4.00	11.51	2.67	3.00
M 35-1 X IS 2123	66	70	296	198	2.57	3.50	3.84	14.34	4.67	4.33
M 35-1 X IS 2146	64	69	293	198	2.20	3.93	1.96	7.17	4.67	4.00
M 35-1 X IS 18551	67	70	322	211	2.37	3.17	4.69	14.47	3.67	3.00
M 35-1 X Swarna	69	67	314	208	2.37	3.13	4.49	11.56	3.33	3.33
CSV 15 X ICSV 25019	63	56	229	149	2.47	3.50	6.94	13.28	1.67	2.33
CSV 15 X PS 35805	65	57	250	158	2.50	3.33	7.16	12.62	2.00	2.67
CSV 15 X IS 2123	66	61	298	184	2.67	3.23	7.95	14.49	4.33	3.67
CSV 15 X IS 2146	64	63	290	197	2.77	3.30	2.88	7.34	3.33	3.67
CSV 15 X IS 18551	65	63	340	214	2.47	2.97	5.72	13.77	3.00	3.00
CSV 15 X Swarna	65	59	308	198	3.10	3.67	4.38	11.11	3.67	2.33
ICSV 25019 X PS 35805	66	60	127	104	1.97	2.43	4.33	7.57	2.00	3.00
ICSV 25019 X IS 2123	64	63	258	167	2.87	3.50	4.95	13.61	4.67	3.33
ICSV 25019 X IS 2146	63	64	273	173	2.80	3.70	2.41	6.99	4.33	3.67
ICSV 25019 X IS 18551	65	65	302	209	2.37	2.90	4.18	12.91	3.33	2.67
ICSV 25019 X Swarna	61	62	187	140	3.27	3.00	7.65	10.16	1.67	2.67
PS 35805 X IS 2123	67	63	276	161	2.73	2.90	3.88	12.48	4.33	3.67
PS 35805 X IS 2146	63	61	261	180	2.53	3.20	2.14	7.20	3.67	3.33
PS 35805 X IS 18551	65	63	318	202	2.43	2.60	4.11	11.42	3.00	3.00
PS 35805 X Swarna	64	61	184	139	2.73	2.80	5.36	10.40	2.00	2.67
IS 2123 X IS 2146	68	71	292	196	1.97	3.10	2.01	6.51	4.67	4.33
IS 2123 X IS 18551	69	73	298	207	2.00	2.50	2.49	10.32	5.00	4.33
IS 2123 X Swarna	67	67	290	198	3.20	3.60	5.07	10.86	4.33	4.33
IS 2146 X IS 18551	67	72	290	211	2.10	3.03	2.23	8.33	4.67	4.67
IS 2146 X Swarna	63	62	297	210	2.63	3.87	3.13	5.40	4.33	3.67
IS 18551 X Swarna	68	65	327	222	2.90	3.50	5.06	12.10	3.67	3.00
RECIPROCAL CROSSES										
Phule Anuradha X ICSV 700	66	64	316	213	2.67	4.17	3.93	10.04	3.67	3.67
M 35-1 X ICSV 700	68	69	317	219	2.67	4.30	6.54	11.72	4.00	3.67
M 35-1 X Phule Anuradha	65	64	283	212	2.57	4.10	3.25	10.90	3.33	4.33
CSV 15 X ICSV 700	69	68	333	206	2.47	3.27	5.82	15.09	2.33	3.00
CSV 15 X Phule Anuradha	64	59	310	218	3.10	4.10	4.68	11.48	3.33	3.67
CSV 15 X M 35-1	63	64	302	212	2.90	4.13	4.14	14.25	3.67	3.67
ICSV 25019 X ICSV 700	68	67	311	198	2.57	3.47	5.02	12.99	3.00	3.00
ICSV 25019 X Phule Anuradha	61	59	282	187	3.30	4.13	5.97	11.13	3.00	2.67
ICSV 25019 X M 35-1	66	69	308	186	2.90	3.50	5.47	15.24	3.67	3.33
ICSV 25019 X CSV 15	64	59	230	158	2.47	3.00	7.46	11.77	2.00	2.67
PS 35805 X ICSV 700	71	66	309	203	2.53	3.07	6.29	12.33	2.67	3.33
PS 35805 X Phule Anuradha	63	68	288	211	2.93	3.27	4.20	17.28	3.33	3.33
PS 35805 X M 35-1	65	64	298	197	2.77	3.53	4.52	14.00	4.67	2.67
PS 35805 X CSV 15	65	57	238	152	2.40	2.90	6.74	10.79	2.83	3.00
PS 35805 X ICSV 25019	67	61	127	104	2.00	2.07	3.10	7.02	2.00	3.00
IS 2123 X ICSV 700	68	70	312	208	2.57	3.47	4.76	12.21	5.00	4.33
IS 2123 X Phule Anuradha	65	67	288	192	2.57	3.63	3.89	11.64	5.00	3.33
PS 35805 X M 35-1	65	64	298	197	2.77	3.53	4.52	14.00	4.67	2.67
PS 35805 X CSV 15	65	57	238	152	2.40	2.90	6.74	10.79	2.83	3.00
PS 35805 X ICSV 25019	67	61	127	104	2.00	2.07	3.10	7.02	2.00	3.00
IS 2123 X ICSV 700	68	70	312	208	2.57	3.47	4.76	12.21	5.00	4.33
IS 2123 X Phule Anuradha	65	67	288	192	2.57	3.63	3.89	11.64	5.00	3.33
IS 2123 X M 35-1	69	68	312	197	2.37	3.50	2.90	13.93	4.33	4.33
IS 2123 X CSV 15	65	64	290	196	2.83	3.20	7.04	13.29	5.00	3.67
IS 2123 X ICSV 25019	67	64	269	161	2.77	3.20	5.09	13.91	4.33	3.67
IS 2123 X PS 35805	67	59	271	171	2.90	3.13	4.60	12.58	4.33	3.67
IS 2146 X ICSV 700	67	71	320	209	2.67	3.17	2.21	5.45	4.33	4.33
IS 2146 X Phule Anuradha	62	64	288	191	2.43	3.73	2.40	6.72	4.67	4.33
IS 2146 X M 35-1	66	68	293	199	2.20	3.70	2.76	7.08	5.00	4.67
IS 2146 X CSV 15	64	57	307	188	2.73	3.37	3.21	6.90	4.33	3.67
IS 2146 X ICSV 25019	64	64	270	166	2.37	3.43	1.96	6.93	4.50	4.67
IS 2146 X PS 35805	64	63	280	169	2.40	3.33	2.10	7.38	4.00	4.33
IS 2146 X IS 2123	68	69	300	199	1.90	3.30	2.32	6.62	4.67	4.67
IS 18551 X ICSV 700	71	71	343	211	2.13	3.20	4.82	10.72	4.00	3.33
IS 18551 X Phule Anuradha	65	65	308	209	2.30	3.43	4.84	14.73	3.67	3.67
IS 18551 X M 35-1	72	70	320	221	2.30	3.23	5.95	13.67	3.33	3.67
IS 18551 X CSV 15	67	64	336	228	2.47	2.97	6.16	12.98	3.33	3.00
IS 18551 X ICSV 25019	66	63	306	198	2.37	2.97	3.12	13.75	3.00	3.00
IS 18551 X PS 35805	69	64	308	207	2.23	2.77	4.28	13.02	3.00	3.00
IS 18551 X IS 2123	73	69	293	208	2.23	2.83	4.00	11.36	5.00	4.33
IS 18551 X IS 2146	68	71	320	197	1.87	2.93	2.35	5.85	5.00	4.33
Swarna X ICSV 700	68	66	318	226	2.90	4.20	5.82	8.62	3.33	3.00
Swarna X Phule Anuradha	61	58	313	204	3.30	4.60	4.73	9.44	3.67	3.33
Swarna X M 35-1	64	62	316	193	3.13	4.47	4.36	10.39	3.33	3.00
Swarna X CSV 15	63	57	299	217	2.77	4.10	4.92	11.58	3.67	2.67
Swarna X ICSV 25019	61	59	187	142	2.83	2.90	7.38	9.46	2.00	2.67
Swarna X PS 35805	65	62	191	149	2.57	3.03	8.70	9.10	1.67	2.33
Swarna X IS 2123	64	63	294	177	2.90	3.73	4.94	10.79	4.33	4.00
Swarna X IS 2146	63	60	287	211	2.60	3.50	2.36	5.92	3.33	3.33
Swarna X IS 18551	65	67	330	230	2.83	3.40	5.27	11.69	3.67	3.00
Mean	66	65	286	191	2.60	3.38	4.10	10.43	3.69	3.46
SE ±	1.13	1.40	6.21	5.83	0.12	0.15	0.71	1.04	0.35	0.40
Vr (99, 198)	10.09**	10.90**	56.80**	22.91**	8.66**	12.87**	6.37**	9.65**	7.48**	2.41**
LSD (P 0.05)	3.14	3.90	17.33	16.26	0.34	0.42	1.97	2.90	0.97	1.12

***F probability significant at P 0.01; R, rainy season; PR, postrainy season.*

Ten crosses exhibited high 100 seed weight and grain yield with good agronomic desirability. Grain yield of the crosses CSV 15 X IS 2123, ICSV 25019 X Swarna, PS 35805 X Swarna, IS 2123 X CSV 15, IS 2123 X ICSV 25019, IS 2123 X Swarna, Swarna X ICSV 25019, and Swarna X IS 2123 was high in the rainy season.

Morphological and Grain Characteristics

All the panicle traits showed significant variability among the genotypes for all the characteristics studied, in both the rainy and postrainy seasons with significant variance ratio ($P \leq 0.01$) (Table 2). The mean scores for inflorescence exsertion were 1.90 and 2.41; for panicle compactness 2.30 and 2.63; and for glume coverage 2.00 and 1.71, respectively, in the rainy and postrainy seasons.

Table 2: Panicle and grain characteristics of sorghum genotypes (Parents and F_1's) evaluated for resistance to sorghum shoot fly, *A. soccata* **across seasons (ICRISAT,** Patancheru, 2013–14)

Pedigree	Inflorescence exsertion		Panicle compactness		Panicle shape	Glume coverage		Awns		Grain luster
	2013 R	2013 PR	2013 R	2013 PR	2013 PR	2013 R	2013 PR	2013 R	2013 PR	2013 PR
PARENTS										
ICSV 700	2.00	2.00	2.00	3.00	4.00	5.00	5.00	2.00	2.00	2.00
Phule Anuradha	2.33	3.00	2.00	3.00	4.00	3.00	1.00	2.00	2.00	2.00
M 35-1	2.00	2.00	2.00	3.00	4.00	3.00	1.00	2.00	2.00	2.00
CSV 15	2.00	1.00	2.00	2.00	2.00	1.00	1.00	1.00	1.00	2.00
ICSV 25019	2.67	2.00	3.00	3.00	4.00	1.00	1.00	1.00	1.00	2.00
PS 35805	3.00	2.00	3.00	3.00	4.00	1.00	1.00	1.00	1.00	2.00
IS 2123	2.33	4.00	3.00	3.00	3.00	3.00	1.00	2.00	2.00	2.00
IS 2146	2.00	4.00	3.00	3.00	3.00	3.00	1.00	2.00	2.00	2.00
IS 18551	2.00	2.00	2.00	3.00	4.00	9.00	9.00	2.00	2.00	1.00
Swarna	1.00	1.00	2.00	2.00	1.00	1.00	1.00	1.00	1.00	2.00
DIRECT CROSSES										
ICSV 700 x Phule Anuradha	2.00	2.67	2.00	3.00	4.00	1.67	1.00	2.00	2.00	2.00
ICSV 700 x M 35-1	2.00	2.33	2.00	3.00	4.00	1.67	1.00	2.00	2.00	2.00
ICSV 700 x CSV 15	1.67	1.33	2.00	2.00	1.33	1.00	1.00	1.00	1.00	2.00
ICSV 700 x ICSV 25019	2.00	1.67	2.00	2.33	2.00	1.00	1.00	1.00	1.00	2.00
ICSV 700 x PS 35805	2.00	1.00	2.00	3.00	4.00	1.67	1.00	1.00	1.00	2.00
ICSV 700 x IS 2123	2.00	4.00	3.00	3.00	3.00	1.67	1.00	2.00	2.00	2.00
ICSV 700 x IS 2146	2.00	3.00	2.67	3.00	3.33	2.33	1.00	2.00	2.00	2.00
ICSV 700 x IS 18551	1.33	2.33	2.00	3.00	4.00	5.67	4.33	2.00	2.00	1.67
ICSV 700 x Swarna	2.00	1.00	2.00	2.00	1.00	1.67	1.67	1.00	1.00	2.00
Phule Anuradha x M 35-1	2.00	3.00	2.00	2.67	3.00	1.00	1.00	2.00	2.00	2.00
Phule Anuradha x CSV 15	1.67	1.67	2.00	2.00	1.33	1.00	1.00	1.00	1.00	2.00
Phule Anuradha x ICSV 25019	2.00	2.33	2.00	2.00	1.00	1.00	1.00	1.00	1.00	2.00
Phule Anuradha x PS 35805	2.00	3.00	2.33	2.33	2.00	1.00	1.00	1.00	1.00	2.00
Phule Anuradha x IS 2123	2.67	4.00	3.00	3.00	3.00	1.67	1.00	2.00	2.00	2.00
Phule Anuradha x IS 2146	2.67	4.00	2.67	3.00	3.00	1.67	1.00	2.00	2.00	2.00
Phule Anuradha x IS 18551	2.00	2.33	2.00	3.00	4.00	3.00	5.00	2.00	2.00	2.00
Phule Anuradha x Swarna	1.67	1.33	2.00	2.00	1.00	1.00	1.00	1.00	1.00	2.00
M 35-1 x CSV 15	1.00	1.00	2.00	2.33	2.67	1.00	1.67	1.00	1.00	2.00
M 35-1 x ICSV 25019	1.00	1.00	2.00	2.00	1.00	1.00	1.00	1.00	1.00	2.00
M 35-1 x PS 35805	1.00	1.67	2.33	3.00	4.00	2.33	1.00	1.00	1.00	2.00
M 35-1 x IS 2123	2.00	4.00	2.67	3.00	3.00	1.67	1.00	2.00	2.00	2.00
M 35-1 x IS 2146	2.33	4.00	2.67	3.00	3.00	1.00	1.00	2.00	2.00	2.00
M 35-1 x IS 18551	1.67	2.33	2.00	3.00	4.00	5.00	4.33	2.00	2.00	2.00
M 35-1 x Swarna	2.00	1.33	2.00	3.00	4.00	3.00	1.00	1.00	1.00	2.00
CSV 15 x ICSV 25019	2.33	3.00	2.00	2.00	1.33	1.00	1.00	1.00	1.00	2.00
CSV 15 x PS 35805	2.33	1.67	2.00	2.00	1.33	1.00	1.00	1.00	1.00	2.00
CSV 15 x IS 2123	2.00	4.00	3.00	3.00	3.33	1.67	1.00	1.00	1.00	2.00
CSV 15 x IS 2146	2.00	2.33	2.67	3.00	3.67	1.00	1.00	1.00	1.00	2.00
CSV 15 x IS 18551	1.33	1.33	2.00	2.00	1.00	4.33	3.67	1.33	1.00	2.00
CSV 15 x Swarna	1.00	1.00	2.00	1.00	2.00	1.67	1.00	1.00	1.00	2.00
ICSV 25019 x PS 35805	3.00	3.00	2.33	3.00	4.00	1.00	1.00	1.00	1.00	2.00
ICSV 25019 x IS 2123	3.00	4.00	3.00	3.00	3.00	1.00	1.00	1.00	1.00	2.00
ICSV 25019 x IS 2146	2.33	4.00	2.67	3.00	3.67	1.00	1.00	1.00	1.00	2.00
ICSV 25019 x IS 18551	1.00	1.00	2.00	2.67	3.00	3.00	3.00	1.00	1.00	2.00
ICSV 25019 x Swarna	1.00	1.33	2.00	2.00	1.00	1.00	1.00	1.00	1.00	2.00
PS 35805 x IS 2123	2.33	4.00	3.00	3.00	3.00	2.33	1.00	1.00	1.00	2.00
PS 35805 x IS 2146	2.33	3.33	3.00	3.00	3.67	1.00	1.00	1.00	1.00	2.00
PS 35805 x IS 18551	1.33	1.67	2.00	2.00	1.00	3.67	3.67	1.00	1.00	1.67
PS 35805 x Swarna	1.00	1.00	2.00	2.00	1.00	1.00	1.00	1.00	1.00	2.00
IS 2123 x IS 2146	3.00	4.00	3.00	3.00	3.00	1.67	1.00	2.00	2.00	2.00
IS 2123 x IS 18551	2.67	4.00	3.00	3.00	3.00	4.33	4.33	2.00	2.00	2.00
IS 2123 x Swarna	2.00	4.00	2.67	3.00	3.67	1.00	1.00	1.00	1.00	2.00
IS 2146 x IS 18551	2.33	3.00	3.00	3.00	3.00	2.33	3.67	2.00	2.00	2.00
IS 2146 x Swarna	1.00	2.00	2.00	3.00	4.00	1.67	1.00	1.00	1.00	2.00
IS 18551 x Swarna	1.00	1.00	2.00	2.00	1.00	4.33	4.33	1.00	1.00	2.00
RECIPROCAL CROSSES										
Phule Anuradha x ICSV 700	2.00	3.00	2.00	3.00	4.00	1.67	1.00	2.00	2.00	2.00
M 35-1 x ICSV 700	1.67	2.67	2.00	3.00	4.00	1.67	1.00	2.00	2.00	2.00
M 35-1 x Phule Anuradha	2.33	2.00	2.00	2.33	2.00	1.00	1.00	2.00	1.67	1.67
CSV 15 x ICSV 700	1.67	1.33	2.00	2.33	2.33	1.67	1.00	1.00	1.00	2.00
CSV 15 x Phule Anuradha	1.33	1.33	2.00	2.00	1.67	1.00	1.00	1.00	1.00	2.00
CSV 15 x M 35-1	1.33	1.33	2.00	2.00	1.33	1.00	1.67	1.00	1.00	2.00
ICSV 25019 x ICSV 700	1.00	1.67	2.00	2.67	3.00	1.67	1.00	1.00	1.00	2.00
ICSV 25019 x Phule Anuradha	1.33	1.33	2.00	2.33	2.00	1.67	1.00	1.00	1.00	2.00
ICSV 25019 x M 35-1	1.33	1.67	2.00	2.67	3.00	1.67	1.00	1.00	1.00	2.00
ICSV 25019 x CSV 15	2.00	2.67	2.00	2.00	1.67	1.00	1.00	1.00	1.00	2.00
PS 35805 x ICSV 700	1.67	1.33	2.00	3.00	4.00	1.00	1.00	1.00	1.00	2.00
PS 35805 x Phule Anuradha	1.67	2.00	2.00	3.00	4.00	1.00	1.00	1.00	1.00	2.00
PS 35805 x M 35-1	1.00	1.33	2.00	3.00	4.00	1.67	1.00	1.00	1.00	2.00
PS 35805 x CSV 15	2.33	2.33	2.00	2.00	1.67	1.00	1.00	1.00	1.00	2.00
PS 35805 x ICSV 25019	3.00	3.00	2.00	3.00	4.00	1.00	1.00	1.00	1.00	2.00

IS 2123 x ICSV 700	2.67	4.00	3.00	3.00	3.00	2.33	1.00	2.00	2.00	2.00
IS 2123 x Phule Anuradha	2.67	4.00	3.00	3.00	3.00	1.67	1.00	2.00	2.00	2.00
IS 2123 x M 35-1	3.00	4.00	3.00	3.00	3.00	1.67	1.00	2.00	2.00	2.00
IS 2123 x CSV 15	2.67	4.00	2.67	3.00	3.00	1.67	1.00	1.00	1.00	2.00
IS 2123 x ICSV 25019	2.00	3.67	3.00	3.00	3.00	1.67	1.00	1.00	1.00	2.00
IS 2123 x PS 35805	2.67	3.67	3.00	3.00	3.00	1.00	1.00	1.00	1.00	2.00
IS 2146 x ICSV 700	2.33	3.67	2.67	3.00	3.33	1.67	1.00	2.00	2.00	2.00
IS 2146 x Phule Anuradha	2.67	4.00	3.00	3.00	3.00	1.67	1.00	2.00	1.67	2.00
IS 2146 x M 35-1	2.67	4.00	3.00	3.00	3.00	1.00	1.00	2.00	2.00	2.00
IS 2146 x CSV 15	1.00	3.33	2.33	3.00	4.00	1.00	1.67	1.00	1.00	2.00
IS 2146 x ICSV 25019	2.33	3.67	3.00	3.00	3.67	1.00	1.00	1.00	1.00	2.00
IS 2146 x PS 35805	2.67	3.67	3.00	3.00	3.33	1.00	1.00	1.00	1.00	2.00
IS 2146 x IS 2123	3.00	4.00	3.00	3.00	3.00	1.67	1.00	2.00	2.00	2.00
IS 18551 x ICSV 700	2.00	2.67	2.00	3.00	4.00	5.67	3.67	2.00	2.00	2.00
IS 18551 x Phule Anuradha	2.33	2.67	2.00	3.00	4.00	4.33	5.00	2.00	2.00	2.00
IS 18551 x M 35-1	2.00	2.33	2.00	3.00	4.00	5.00	5.00	2.00	2.00	2.00
IS 18551 x CSV 15	1.00	1.00	2.00	2.00	1.33	2.33	4.33	1.00	1.00	1.67
IS 18551 x ICSV 25019	1.33	1.00	2.00	2.00	1.00	3.67	3.00	1.00	1.00	2.00
IS 18551 x PS 35805	1.67	1.33	2.00	2.33	2.00	3.67	3.67	1.00	1.00	2.00
IS 18551 x IS 2123	3.00	4.00	3.00	3.00	3.00	4.33	4.33	2.00	2.00	2.00
IS 18551 x IS 2146	3.00	2.33	3.00	3.00	3.00	3.67	3.67	2.00	2.00	2.00
Swarna x ICSV 700	1.33	1.00	2.00	2.00	1.00	1.67	1.00	1.00	1.00	2.00
Swarna x Phule Anuradha	1.00	1.00	2.00	2.00	1.33	1.67	1.00	1.00	1.00	2.00
Swarna x M 35-1	1.33	1.00	2.00	2.33	1.00	1.00	1.00	1.00	1.00	2.00
Swarna x CSV 15	1.00	1.00	2.00	1.00	2.00	1.00	1.00	1.00	1.00	2.00
Swarna x ICSV 25019	1.00	1.00	2.00	2.00	1.00	1.00	1.00	1.00	1.00	2.00
Swarna x PS 35805	1.00	1.00	2.00	2.00	1.00	1.00	1.00	1.00	1.00	2.00
Swarna x IS 2123	1.67	3.00	3.00	3.00	3.33	1.00	1.00	1.00	1.00	2.00
Swarna x IS 2146	1.67	1.00	2.00	3.00	4.00	1.67	1.00	1.00	1.00	2.00
Swarna x IS 18551	1.00	1.00	2.00	2.00	1.00	4.33	5.00	1.00	1.00	2.00
Mean	1.90	2.41	2.30	2.63	2.75	2.00	1.71	1.40	1.35	1.98
SE ±	0.25	0.29	0.12	0.12	0.35	0.43	0.33	0.03	0.05	0.07
Vr	6.49**	15.17**	13.65**	18.62**	9.81**	10.81**	19.91**	208.27**	102.88**	3.18**
LSD (P 0.05)	0.69	0.81	0.33	0.32	0.99	1.20	0.92	0.09	0.13	0.19

**F probability significant at P 0.05; R, rainy season; PR, postrainy season.

Association of the Agronomic and Morphological Traits with Shoot Fly Resistance

Days to 50% flowering, inflorescence exsertion, panicle compactness, glume coverage, and presence of awns were significantly and negatively correlated with shoot fly damage parameters across seasons, with few exceptions (Table 3). Plant height, 100 seed weight, and grain yield showed positive correlation with shoot fly damage across seasons.

Table 3: Association of agronomic and panicle traits with expression of resistance to sorghum shoot fly, *Atherigona soccata* **(ICRISAT,** Patancheru, 2013–14)

Traits	Plants with shoot fly eggs (%)	Number of shoot fly eggs/plant	Shoot fly deadhearts (%)	ORS
Days to 50% flowering	−0.20* (−0.41**)	0.13 (−0.07)	−0.31** (−0.47**)	0.09 (−0.24**)
Plant height	0.13 (−0.01)	0.15 (−0.06)	0.35** (0.07)	−0.07 (−0.02)
100 seed weight	0.21* (0.28**)	−0.03 (−0.04)	0.39** (0.32**)	0.13 (0.15)
Grain yield	0.16* (0.03)	−0.01 (−0.17*)	0.24** (0.05)	0.05 (0.22*)
Agronomic score	−0.03 (−0.43**)	0.09 (−0.17*)	0.02 (−0.41**)	−0.10 (−0.38**)
Inflorescence exsertion	−0.19* (−0.37**)	−0.06 (−0.14)	−0.37** (−0.41**)	−0.18* (−0.31**)
Panicle compactness	−0.13 (−0.55**)	−0.01 (−0.18*)	−0.30** (−0.60**)	−0.28** (−0.45**)
Panicle shape	(−0.48**)	(−0.10)	(−0.49**)	(−0.38**)
Glume coverage	−0.17* (−0.16*)	0.15 (0.08)	−0.18* (−0.44**)	−0.01 (−0.13)
Awns	−0.12 (−0.41**)	0.01 (−0.07)	−0.18* (−0.44**)	0.10 (−0.17*)

*, **Correlation coefficients significant at P 0.05 and P 0.01, respectively; ORS, overall resistance score; The values inside the parentheses are for postrainy season, whereas the values outside the parentheses are for rainy season.

Association between the Agronomic Traits

Agronomic score was positively correlated with days to 50% flowering and plant height, but negatively correlated with grain yield (Table 4). Days to 50% flowering were significantly and positively correlated with plant height, and negatively correlated with 100 seed weight and grain yield. Grain yield was positively correlated with plant height in the postrainy season, and 100 seed weight across seasons. Significant positive correlation was observed between plant height and 100 seed weight in the postrainy season.

Table 4: Association between the agronomic traits in the postrainy season adapted sorghums (ICRISAT, Patancheru, 2013–14)

Traits	Agronomic score	Days to 50% flowering	Plant height	100 seed weight
Days to 50% flowering	0.24** (0.45**)	1.00		
Plant height	0.52** (0.24**)	0.22* (0.41**)	1.00	
100 seed weight	−0.17 (0.01)	−0.51** (−0.23**)	0.04 (0.43**)	1.00
Grain yield	−0.47** (−0.21*)	−0.26** (−0.08)	0.00 (0.36**)	0.36** (0.21*)

*, **Correlation coefficients significant at P 0.05 and P 0.01, respectively; The values inside the parentheses are for postrainy season, whereas the values outside the parentheses are for rainy season.

Association between the Morphological Traits

Significant positive correlation was observed between inflorescence exsertion and panicle compactness, and between awns and the panicle traits across seasons (Table 5). Panicle shape was positively correlated with inflorescence exsertion, and panicle compactness in the postrainy season.

Table 5: Association between the panicle traits in the postrainy season sorghums (ICRISAT, Patancheru, 2013–14)

Traits	Inflorescence exsertion	Panicle compactness	Panicle shape	Glume coverage	Awns
Panicle compactness	0.66** (0.66**)	1.00			
Panicle shape	(0.46**)	(0.86**)	1.00		
Glume coverage	−0.03 (−0.14)	−0.12 (0.05)	(0.07)	1.00	
Awns	0.45** (0.53**)	0.30** (0.52**)	(0.44**)	0.44** (0.29**)	1.00
Grain luster	(0.10)	(0.01)	(−0.03)	(−0.55**)	(−0.13)

*, **Correlation coefficient significant at P 0.05 and P 0.01, respectively; The values inside the parentheses are for postrainy season, whereas the values outside the parentheses are for rainy season.

Combining Ability Analysis

Mean sum of squares for general combining ability of all the traits studied in the rainy season and postrainy seasons were significant at $P = 0.01$ (Table 6). Mean sum of squares due to SCA was significant for all the traits studied, during the rainy and postrainy seasons, except grain luster during rainy season and agronomic score and waxy bloom in the postrainy season indicating the role of both additive and non-additive nature of gene action in controlling most of the morphological and agronomic traits. The mean sum of squares due to reciprocal crosses was significant for days to 50% flowering and 100 seed weight across seasons, inflorescence exsertion during the rainy season; and plant height, panicle compactness, and panicle shape during the postrainy season, suggesting the influence of cytoplasmic factors in the expression of these traits.

Table 6: Analysis of variance (ANOVA) showing mean sum of squares of general, specific and reciprocal combining abilities of F_1(10 X 10) diallel across seasons (ICRISAT, Patancheru, 2013–14)

Source	GCA		SCA		Reciprocal		Error	
	2013 R	2013 PR	2013 R	2013 PR	2013 R	2013 PR	2013 R	2013 PR
Days to 50% flowering	75.23**	168.15**	10.31**	8.40**	2.74**	4.92**	1.27	1.96
Plant height (cm)	14639.30**	5747.19**	1856.50**	494.31**	40.27	69.04**	38.62	33.99
100 seed weight (g)	0.81**	2.07**	0.10**	0.18**	0.02*	0.05**	0.02	0.02
Grain yield (t/ha)	11.80**	44.52**	3.94**	12.95**	0.70	1.08	0.50	1.08
Agronomic score	6.83**	3.06**	0.44**	0.15	0.17	0.11	0.12	0.16
Inflorescence exsertion	2.14**	10.80**	0.34**	0.58**	0.11**	0.11	0.06	0.09
Panicle compactness	1.68**	1.74**	0.08**	0.17**	0.01	0.03**	0.01	0.01
Panicle shape	-	4.88**	-	1.40**	-	0.32**	-	0.13
Glume coverage	17.63**	21.08**	0.70**	0.49**	0.21	0.03	0.19	0.11
Awns	1.91**	1.85**	0.13**	0.13**	0.00	0.00	0.00	0.00
Grain luster	0.02**	0.04**	0.02	0.02**	0.00	0.01	0.00	0.00

*, **F probability significant at P 0.05 and P 0.01, respectively; GCA, general combining ability; SCA, specific combining ability; R, rainy season; PR, postrainy season.

Estimates of General Combining Ability (*gca*), Specific Combining Ability (*sca*) and Reciprocal Effects

gca Effects of Agronomic Traits

gca effects of days to 50% flowering ranged from −2.87 (Phule Anuradha) to 3.36 (ICSV 700) in the rainy season, and from −3.65 (CSV 15) to 4.40 (ICSV 700) in the postrainy season (Table 7). Phule Anuradha (−2.87**), ICSV 25019 (−1.85**), IS 2146 (−0.77**) and Swarna (−1.37**) exhibited significant negative*gca* effects in the rainy season, and Phule Anuradha (−1.61**), CSV 15 (−3.65**), ICSV 25019 (−2.58**), PS 35805 (−2.58**), and Swarna (−2.60**) exhibited significant negative *gca* effects for days to 50% flowering in the postrainy season. ICSV 700 (3.36**, 4.40**, respectively, in the rainy and postrainy season), M 35-1 (0.50*, 2.30**), IS 2123 (1.33**, 1.70**), and IS 18551 (2.46**, 3.17**) across seasons and IS 2146 (1.44**) in the postrainy season showed significant positive *gca* effects for days to 50% flowering.

gca effects for plant height ranged from −44.49 (ICSV 25019) to 32.23 (ICSV 700) in the rainy season, −28.69 (ICSV 25019) to 20.59 (IS 18551) in the postrainy season. ICSV 25019 (−44.49** and −28.69**, respectively, in the rainy and postrainy seasons), PS 35805 (−42.96** and −27.69**) and Swarna (−16.10** and −3.36**) exhibited significant negative *gca* effects for plant height across seasons. ICSV 700, Phule Anuradha, M 35-1, CSV 15, IS 2123, IS 2146, and IS 18551 in the rainy season, and ICSV 700, Phule Anuradha, M 35-1, and IS 18551 in the postrainy season exhibited significant positive *gca* effects for plant height.

gca effects for 100 seed weight ranged from −0.31 (IS 18551) to 0.38 (Swarna) in the rainy and −0.43 (PS 35805) to 0.56 (Phule Anuradha) in the postrainy seasons. PS 35805 (−0.07**), IS 2123 (−0.06**), IS 2146 (−0.26**), and IS 18551 (−0.31**) in the rainy season, and ICSV 25019 (−0.22**), PS 35805 (−0.43**), IS 2123 (−0.16**), and IS 18551 (−0.40**) in the postrainy season exhibited significant negative *gca* effects for 100 seed weight. Whereas, the genotypes Phule Anuradha (0.20**), CSV 15 (0.09**), and Swarna (0.38**) in the rainy season, and ICSV 700 (0.07*), Phule Anuradha (0.56**), M 35-1 (0.33**), and Swarna (0.29**) in the postrainy season showed positive *gca* effects for 100 seed weight.

Table 7: Estimates of general combining ability of agronomic and panicle traits of parents (10 X 10 diallel) across seasons (ICRISAT, Patancheru, 2013–14)

Traits	ICSV 700	Phule Anuradha	M 35-1	CSV 15	ICSV 25019	PS 35805	IS 2123	IS 2146	IS 18551	Swarna
Days to 50% flowering	3.36** (4.40**)	−2.87** (−1.61**)	0.50* (2.30**)	−0.35 (−3.65**)	−1.85** (−2.58**)	−0.44 (−2.58**)	1.33** (1.70**)	−0.77** (1.44**)	2.46** (3.17**)	−1.37** (−2.60**)
Plant height (cm)	32.23** (19.42**)	5.59** (7.92**)	19.43** (10.53**)	5.19** (1.87)	−44.49** (−28.69**)	−42.96** (−27.69**)	3.87** (−2.13)	5.96** (1.53)	31.29** (20.59**)	−16.10** (−3.36**)
100 seed weight (g)	−0.02 (0.07*)	0.20** (0.56**)	0.01 (0.33**)	0.09** (0.01)	0.04 (−0.22**)	−0.07** (−0.43**)	−0.06* (−0.16**)	−0.26** (−0.05)	−0.31** (−0.40**)	0.38** (0.29**)
Grain yield (t/ha)	0.23 (0.44*)	−0.43** (0.52*)	−0.34* (1.27**)	0.90** (0.86**)	0.40** (0.07)	0.48** (0.03)	−0.14 (0.89**)	−1.79** (−3.86**)	−0.06 (0.77**)	0.74** (−0.99**)
Agronomic score	0.13 (0.04)	0.18* (0.11)	0.26** (0.11)	−0.46** (−0.31**)	−0.73** (−0.39**)	−0.75** (−0.31**)	0.96** (0.59**)	0.69** (0.66**)	0.18* (−0.01)	−0.47** (−0.48**)
Inflorescence exsertion	−0.05 (−0.18**)	0.12* (0.17**)	−0.13* (−0.16*)	−0.23** (−0.53**)	0.00 (−0.16*)	0.13* (−0.21**)	0.57** (1.51**)	0.35** (0.96**)	−0.12* (−0.34**)	−0.63** (−1.06**)
Panicle compactness	−0.17** (0.13**)	−0.13** (0.00)	−0.15** (0.13**)	−0.20** (−0.50**)	−0.03 (−0.10**)	0.02 (0.05*)	0.62** (0.37**)	0.43** (0.37**)	−0.13** (0.02)	−0.25** (−0.47**)
Panicle shape	(0.41**)	(0.01)	(0.35**)	(−0.70**)	(−0.24**)	(0.20**)	(0.31**)	(0.58**)	(0.01)	(−0.94**)
Glume coverage	0.35** (0.03)	−0.28** (−0.31**)	−0.05 (−0.27**)	−0.65** (−0.31**)	−0.65** (−0.51**)	−0.55** (−0.44**)	−0.01 (−0.37**)	−0.31** (−0.41**)	2.52** (2.89**)	−0.38** (−0.31**)
Awns	0.24** (0.25**)	0.24** (0.21**)	0.24** (0.23**)	−0.35** (−0.35**)	−0.36** (−0.35**)	−0.36** (−0.35**)	0.24** (0.25**)	0.24** (0.23**)	0.25** (0.25**)	−0.36** (−0.35**)
Grain luster	(0.01)	(0.01)	(0.01)	(0.01)	(0.02)	(0.01)	(0.02)	(0.02)	(−0.13**)	(0.02)

*, **t-test significant at 0.05 and 0.01 probability levels; The values outside the parentheses are for rainy season and inside the parentheses are for postrainy season.

gca effects of sorghum grain yield ranged from −1.79 (IS 2146) to 0.90 (CSV 15) in the rainy season, and −3.86 (IS 2146) to 1.27 (M 35-1) in the postrainy season. The *gca* effects of Phule Anuradha (−0.43**), M 35-1 (−0.34*), and IS 2146 (−1.79**) in the rainy season, and IS 2146 (−3.86**) and Swarna (−0.99**) in the postrainy season exhibited significant negative *gca* effects for grain yield. The genotypes CSV 15 (0.90**), ICSV 25019 (0.40**), PS 35805 (0.48**), and Swarna (0.74**) in the rainy season and ICSV 700 (0.44*), Phule Anuradha (0.52*), M 35-1 (1.27**), CSV 15 (0.86**), IS 2123 (0.89**), and IS 18551 (0.77**) in the postrainy season showed significant positive *gca* effects for grain yield. *gca* effects of agronomic score ranged from −0.75 (PS 35805) to 0.96 (IS 2123) in the rainy season, and −0.48 (Swarna) to 0.66 (IS 2146) in the postrainy season.

gca Effects of Morphological Traits

The *gca* effects of inflorescence exsertion ranged from −0.63 (Swarna) to 0.57 (IS 2123) in the rainy season, and −1.06 (Swarna) to 1.51 (IS 2123) in the postrainy season (Table 7). *gca* effects of panicle compactness ranged from −0.25 (Swarna) to 0.62 (IS 2123) in the rainy season, and −0.50 (CSV 15) to 0.37 (IS 2123 and IS 2146) in the postrainy season. Six genotypes exhibited significant and negative *gca*effects and two genotypes exhibited positive and significant *gca* effects for panicle compactness in the rainy season. Three genotypes exhibited significant negative *gca* effects, while five genotypes exhibited significant positive *gca* effects for panicle compactness in the postrainy season.

gca effects of the panicle shape ranged from −0.94 (Swarna) to 0.58 (IS 2146) in the postrainy season. The general combining ability of glume cover ranged from −0.65 to 2.52 in the rainy season, and −0.51 to 2.89 in the postrainy season. All the genotypes exhibited significant negative *gca* effects except IS 18551 (2.52** and 2.89**) with significant positive *gca* effects for glume coverage across seasons. *gca* effects of awns ranged from −0.36 to 0.25 in the rainy season, and −0.35 to 0.25 in the postrainy season. CSV 15, ICSV 25019, PS 35805, and Swarna exhibited significant negative *gca* effects, while ICSV 700, Phule Anuradha, M 35-1, IS 2123, IS 2146, and IS 18551 exhibited significant positive *gca* effects for presence of awns across seasons.

Specific Combining Ability (*sca*) Effects

sca effects of agronomic traits

The *sca* effects for days to 50% flowering ranged from −3.66 to 3.19 and −3.20 to 3.61 during the rainy and postrainy season, respectively. For plant height the *sca* effects ranged from −71.09 to 45.68 and −30.09 to 22.75, for 100 seed weight from −0.56 to 0.37 and −0.50 to 0.38, for grain yield from −1.24 to 2.65 and −3.24 to 4.70, respectively, in the rainy and postrainy seasons. For agronomic score from −1.02 to 0.91 in the rainy season (Table 8). Ten hybrids in the rainy season and nine hybrids in the postrainy season exhibited significant negative *sca* effects for days to 50% flowering. ICSV 700 X CSV 15 across seasons, and Phule Anuradha X PS 35805, M 35-1 X CSV 15 in the postrainy season showed significant positive *sca* effects for days to 50% flowering.

Table 8: Estimates of specific combining ability effects of agronomic traits of F_1 crosses (10 X 10 diallel) of sorghum across seasons (ICRISAT, Patancheru, 2013–14)

Pedigree	Days to 50% flowering		Plant height (cm)		100 seed weight (g)		Grain yield (t/ha)		Agronomic score
	2013 R	2013 PR	2013 R	2013 PR	2013 R	2013 PR	2013 R	2013 PR	2013 R
ICSV 700 X Phule Anuradha	−0.80	1.30	−8.32*	6.19	−0.06	−0.05	−0.11	1.22	0.00
ICSV 700 X M 35-1	−3.66**	−2.62**	−18.26**	−2.53	0.08	0.37**	1.16*	−0.03	−0.25
ICSV 700 X CSV 15	3.19**	3.16**	7.10	−4.98	−0.23**	−0.18	0.05	2.90**	−1.02**
ICSV 700 X ICSV 25019	−0.15	0.10	36.21**	16.69**	0.09	0.32**	0.21	1.91**	−0.09
ICSV 700 X PS 35805	−0.06	−0.74	34.11**	21.24**	0.01	0.04	1.51**	1.33	−0.07
ICSV 700 X IS 2123	−2.83**	−2.02*	−13.80**	5.13	0.04	0.14	0.29	1.21	0.05
ICSV 700 X IS 2146	−0.90	0.25	−4.79	−3.53	0.25**	−0.06	−0.13	−1.55*	−0.34
ICSV 700 X IS 18551	−2.80**	−1.99*	−7.89	−9.81*	0.02	0.20*	0.83	−0.36	−0.16
ICSV 700 X Swarna	−0.46	−0.05	16.72**	12.47**	0.05	0.38**	−0.07	−0.34	0.82**
Phule Anuradha X M 35-1	0.57	−0.77	−32.19**	−2.14	−0.21**	−0.04	−0.24	−0.77	−0.46*
Phule Anuradha X CSV 15	0.09	−0.99	9.85*	9.30*	0.18*	0.13	−1.21*	−1.60*	0.26
Phule Anuradha X ICSV 25019	−0.41	−1.72	34.53**	14.86**	0.27**	0.32**	0.22	−0.11	0.20
Phule Anuradha X PS 35805	−0.16	3.61**	35.76**	22.75**	0.24**	−0.12	0.92*	4.70**	0.21
Phule Anuradha X IS 2123	−0.60	0.16	−11.09**	−6.70	−0.14	−0.06	0.12	0.19	−0.16
Phule Anuradha X IS 2146	0.50	0.10	−7.02	−11.48**	−0.21*	−0.04	0.79	−0.15	−0.22
Phule Anuradha X IS 18551	−1.06	−1.80	−13.49**	−11.09**	−0.19*	−0.10	1.18*	1.99**	−0.55*
Phule Anuradha X Swarna	−0.06	−1.20	29.75**	6.19	0.21*	0.23*	0.36	−0.52	0.10
M 35-1 X CSV 15	−2.95**	2.76**	−0.64	8.35*	0.25**	0.03	−0.31	0.35	0.18
M 35-1 X ICSV 25019	−0.45	0.70	45.68**	21.69**	0.21*	0.18	0.28	2.82**	0.28
M 35-1 X PS 35805	−2.20**	−0.97	35.01**	10.13**	0.23**	0.20*	0.04	1.03	0.46*
M 35-1 X IS 2123	−0.63	−0.09	−4.89	−2.09	−0.06	−0.05	−0.23	1.55*	−0.41
M 35-1 X IS 2146	−0.70	−0.15	−17.54**	−4.65	−0.13	0.15	0.41	−0.72	0.20
M 35-1 X IS 18551	0.74	−0.39	−15.10**	−5.92	0.05	−0.11	1.64**	1.60*	−0.63**
M 35-1 X Swarna	1.07	−0.29	26.18**	2.47	−0.22**	−0.20*	−0.04	0.27	−0.15
CSV 15 X ICSV 25019	−0.60	−1.19	−16.73**	−10.76**	−0.24**	0.08	1.83**	1.16	−0.66**
CSV 15 X PS 35805	−0.35	−1.69	−3.80	−10.09**	−0.15	0.15	1.50**	0.38	−0.06
CSV 15 X IS 2123	−1.61*	−0.64	−0.65	−0.64	0.14	−0.02	2.65**	1.71*	0.48*
CSV 15 X IS 2146	−1.01	−2.70**	1.70	−2.09	0.34**	−0.01	−0.13	−0.32	−0.08
CSV 15 X IS 18551	−1.91*	−1.27	15.80**	7.75*	0.10	−0.02	1.02*	1.31	−0.24
CSV 15 X Swarna	−0.41	−0.84	28.76**	17.80**	−0.12	0.21*	−1.06*	1.04	0.91**
ICSV 25019 X PS 35805	2.49**	0.75	−71.09**	−30.09**	−0.56**	−0.49**	−1.24**	−3.24**	−0.21
ICSV 25019 X IS 2123	−0.11	−0.70	18.73**	3.80	0.26**	0.34**	0.69	2.37**	0.58*
ICSV 25019 X IS 2146	−0.01	−0.44	24.70**	5.69	0.23**	0.45**	−0.51	0.31	0.77**
ICSV 25019 X IS 18551	−1.08	−1.84*	31.60**	20.52**	0.06	0.17	−0.75	2.05**	0.03
ICSV 25019 X Swarna	−1.91*	0.43	−38.22**	−17.76**	0.06	−0.50**	2.29**	0.29	−0.66**
PS 35805 X IS 2123	−0.36	−3.20**	26.92**	5.02	0.37**	0.22*	−0.17	1.19	0.43
PS 35805 X IS 2146	−1.60*	−1.94*	22.06**	9.69*	0.22**	0.36**	−0.65	0.69	0.20
PS 35805 X IS 18551	−1.16	−2.34*	38.97**	20.63**	0.13	0.13	−0.31	0.99	−0.12
PS 35805 X Swarna	0.00	1.26	−38.93**	−15.98**	−0.24**	−0.33**	1.73**	0.29	−0.64**
IS 2123 X IS 2146	1.30	1.78	0.80	6.91	−0.33**	0.02	0.02	−0.90	−0.67**
IS 2123 X IS 18551	0.74	1.21	−25.08**	−2.15	−0.10	−0.16	−0.65	−1.25	0.17
IS 2123 X Swarna	−0.43	0.48	18.95**	1.80	0.15	0.15	0.32	0.50	0.15
IS 2146 X IS 18551	0.00	1.65	−17.72**	−9.14*	−0.03	0.05	0.07	−0.26	0.28
IS 2146 X Swarna	−1.00	−3.09**	16.32**	21.47**	−0.08	0.06	−0.27	0.07	−0.07
IS 18551 X Swarna	−0.73	0.35	27.64**	17.97**	0.22**	0.18	0.40	1.68*	0.27

*, **t-test significant at P 0.05 and P 0.01 respectively; R, rainy season; PR, postrainy season.

Significant negative *sca* effects for plant height were observed for 14 hybrids in the rainy season, and 10 hybrids in the postrainy season. Fifteen hybrids across seasons, Phule Anuradha X Swarna, ICSV 25019 X IS 2123, ICSV 25019 X IS 2146, PS 35805 X IS 2123, IS 2123 X Swarna in the rainy season, and M 35-1 X CSV 15 in the postrainy season exhibited significant positive *sca* effects for plant height.

Significant negative *sca* effects for grain yield were observed in Phule Anuradha X CSV 15, ICSV 25019 X PS 35805 across seasons, CSV 15 X Swarna in the rainy season, and ICSV 700 X IS 2146 in the postrainy season. Four hybrids across seasons, seven in the rainy season and eight hybrids in the postrainy season, exhibited significant and positive *sca* effects for grain yield. ICSV 700 X CSV 15, Phule Anuradha X M 35-1, Phule Anuradha X IS 18551, M 35-1 X IS 18551, CSV 15 X ICSV 25019, ICSV 25019 X Swarna, PS 35805 X Swarna, IS 2123 X IS 2146 exhibited significant negative *sca* effects while ICSV 700 X Swarna, M 35-1 X PS 35805, CSV 15 X IS 2123, CSV 15 X Swarna, ICSV 25019 X IS 2123, ICSV 25019 X IS 2146 exhibited significant positive *sca* effects for agronomic score in the rainy season.

sca Effects of Morphological Traits

The *sca* effects of the inflorescence exsertion ranged from −0.92 to 0.95 and −0.91 to 1.11, for panicle compactness from −0.52 to 0.37 and −0.67 to 0.50, for glume coverage −1.22 to

0.78 and −1.09 to 0.71, for awns from −0.25 to 0.36 and 0.25 to 0.35, respectively, in the rainy and postrainy seasons. For panicle shape it ranged from −1.46 to 1.60 and grain luster from −0.16 to −0.01 in the postrainy season (Table 9). The genotypes with significant *gca* and/or *sca* for morphological traits can be utilized in developing sorghum cultivars for use by the farmers.

Table 9: Estimates of specific combining ability effects of panicle traits of F_1 **crosses (10 X 10** diallel) of sorghum, across seasons (ICRISAT, Patancheru, 2013–14)

Pedigree	Inflorescence exsertion		Panicle compactness		Panicle shape	Glume coverage		Awns		Grain luster
	2013 R	2013 PR	2013 R	2013 PR	2013 PR	2013 R	2013 PR	2013 R	2013 PR	2013 PR
ICSV 700 X Phule Anuradha	0.02	0.43*	−0.03	0.23**	0.82**	−0.42	−0.43*	0.16**	0.19**	0.01
ICSV 700 X M 35-1	0.10	0.43*	−0.02	0.10	0.49*	−0.65*	−0.46*	0.16**	0.17**	0.01
ICSV 700 X CSV 15	0.03	−0.37	0.03	−0.10	−0.63**	−0.39	−0.43*	−0.25**	−0.25**	0.01
ICSV 700 X ICSV 25019	−0.37*	−0.41*	−0.13	−0.17*	−0.43	−0.39	−0.23	−0.24**	−0.25**	−0.01
ICSV 700 X PS 35805	−0.17	−0.86**	−0.18*	0.18*	0.64**	−0.49	−0.29	−0.24**	−0.25**	0.01
ICSV 700 X IS 2123	−0.10	0.26	0.22**	−0.13	−0.48*	−0.35	−0.36	0.16**	0.15**	−0.01
ICSV 700 X IS 2146	−0.05	0.14	0.07	−0.13	−0.41	−0.05	−0.33	0.16**	0.17**	−0.01
ICSV 700 X IS 18551	−0.08	0.61**	−0.03	0.22**	0.82**	0.78**	−0.63**	0.15**	0.15**	−0.02
ICSV 700 X Swarna	0.43**	−0.17	0.08	−0.30**	−1.23**	−0.32	−0.09	−0.24**	−0.25**	−0.01
Phule Anuradha X M 35-1	0.27	0.08	−0.05	−0.27**	−0.61**	−0.687*	−0.13	0.16**	0.04	−0.16**
Phule Anuradha X CSV 15	−0.30	−0.56**	0.00	−0.13	−0.56*	−0.09	−0.09	−0.25**	−0.21**	0.01
Phule Anuradha X ICSV 25019	−0.37*	−0.59**	−0.17*	−0.37**	−1.03**	0.25	0.11	−0.24**	−0.21**	−0.01
Phule Anuradha X PS 35805	−0.33*	0.13	−0.05	−0.02	0.04	−0.19	0.04	−0.24**	−0.21**	0.01
Phule Anuradha X IS 2123	0.07	−0.09	0.18*	0.00	−0.08	−0.05	−0.03	0.16**	0.19**	−0.01
Phule Anuradha X IS 2146	0.28	0.46*	0.20*	0.00	−0.35	0.25	0.01	0.16**	0.04	−0.01
Phule Anuradha X IS 18551	0.25	0.26	−0.07	0.35**	1.22**	−0.59*	0.71**	0.15**	0.19**	0.14**
Phule Anuradha X Swarna	−0.07	−0.36	0.05	−0.17*	−0.66**	−0.02	−0.09	−0.24**	−0.21**	−0.01
M 35-1 X CSV 15	−0.38*	−0.56**	0.02	−0.10	−0.40	−0.32	0.54*	−0.25**	−0.23**	0.01
M 35-1 X ICSV 25019	−0.62**	−0.76**	−0.15	−0.33**	−0.86**	0.01	0.07	−0.24**	−0.23**	−0.01
M 35-1 X PS 35805	−0.92**	−0.54**	−0.03	0.18*	0.70**	0.58*	0.01	−0.24**	−0.23**	0.01
M 35-1 X IS 2123	0.15	0.24	0.03	−0.13	−0.41	−0.29	−0.06	0.16**	0.17**	−0.01
M 35-1 X IS 2146	0.37*	0.79**	0.22**	−0.13	−0.68**	−0.65*	−0.03	0.16**	0.19**	−0.01
M 35-1 X IS 18551	0.17	0.43*	−0.05	0.22**	0.89**	0.51	0.34	0.15**	0.17**	0.14**
M 35-1 X Swarna	0.52**	−0.02	0.07	0.37**	0.34	0.41	−0.13	−0.24**	−0.23**	−0.01
CSV 15 X ICSV 25019	0.48**	1.11**	−0.10	−0.03	−0.31	0.28	0.11	0.35**	0.35**	−0.01
CSV 15 X PS 35805	0.52**	0.33	−0.15	−0.18*	−0.75**	0.18	0.04	0.35**	0.35**	0.01
CSV 15 X IS 2123	0.08	0.61**	0.08	0.50**	0.80**	0.31	−0.03	−0.25**	−0.25**	−0.01
CSV 15 X IS 2146	−0.53**	−0.01	−0.07	0.50**	1.20**	−0.05	0.34	−0.25**	−0.23**	−0.01
CSV 15 X IS 18551	−0.40*	−0.37	0.00	−0.15*	−0.89**	−0.55	−0.29	−0.10**	−0.25**	−0.02
CSV 15 X Swarna	−0.05	0.18	0.12	−0.67**	0.89**	0.35	−0.09	0.35**	0.35**	−0.01
ICSV 25019 X PS 35805	0.95**	0.96**	−0.15	0.42**	1.29**	0.18	0.24	0.36**	0.35**	−0.01
ICSV 25019 X IS 2123	0.02	0.08	0.08	0.10	0.17	−0.02	0.17	−0.24**	−0.25**	−0.02
ICSV 25019 X IS 2146	0.07	0.63**	0.10	0.10	0.57*	−0.05	0.21	−0.24**	−0.23**	−0.02
ICSV 25019 X IS 18551	−0.63**	−0.91**	−0.17*	−0.22**	−0.53*	−0.55	−1.09**	−0.25**	−0.25**	0.13**
ICSV 25019 X Swarna	−0.28	−0.02	−0.05	−0.07	−0.58*	0.01	0.11	0.36**	0.35**	−0.02
PS 35805 X IS 2123	−0.12	0.13	0.03	−0.05	−0.26	0.21	0.11	−0.24**	−0.25**	−0.01
PS 35805 X IS 2146	0.10	0.34	0.22**	−0.05	−0.03	−0.15	0.14	−0.24**	−0.23**	−0.01
PS 35805 X IS 18551	−0.43**	−0.36	−0.22**	−0.53**	−1.46**	−0.32	−0.49*	−0.25**	−0.25**	−0.02
PS 35805 X Swarna	−0.47*	−0.14	−0.10	−0.22**	−1.01**	−0.09	0.04	0.36**	0.35**	−0.01
IS 2123 X IS 2146	0.17	−0.87**	−0.38**	−0.37**	−0.65**	−0.02	0.07	0.16**	0.17**	−0.02
IS 2123 X IS 18551	0.47**	0.43*	0.18*	−0.02	−0.08	−0.19	0.11	0.15**	0.15**	0.13**
IS 2123 X Swarna	−0.02	0.64**	0.13	0.47**	1.37**	−0.62*	−0.03	−0.24**	−0.25**	−0.02
IS 2146 X IS 18551	0.52**	−0.36	0.37**	−0.02	−0.35	−1.22**	−0.53*	0.15**	0.17**	0.13**
IS 2146 X Swarna	−0.30	−0.81**	−0.52**	0.47**	1.60**	0.35	0.01	−0.24**	−0.23**	−0.02
IS 18551 X Swarna	−0.17	−0.01	0.05	−0.18*	−0.83**	0.18	0.37	−0.25**	−0.25**	0.13**

*, **Significant at P 0.05 and P 0.01 probability levels; R, rainy season; PR, postrainy season.

Reciprocal Combining Ability Effects of Agronomic and Morphological Traits

M 35-1 X ICSV 700, ICSV 25019 X M 35-1, IS 18551 X ICSV 700, IS 18551 X M 35-1, IS 18551 X PS 35805, IS 18551 X IS 2123 in the rainy season, PS 35805 X Phule Anuradha in the postrainy season, and PS 35805 X Phule Anuradha across seasons, exhibited significant negative reciprocal effects for days to 50% flowering (Table 10). CSV 15 X ICSV 700 and Swarna X IS 18551 in the rainy season, six hybrids in the postrainy season, and Swarna X M 35-1 and Swarna X IS 2123 across seasons exhibited significant positive reciprocal effects for days to 50% flowering. The crosses CSV 15 X Phule Anuradha, PS 35805 X Phule Anuradha, PS 35805 X ICSV 700, and Swarna X CSV 15 exhibited significant negative and the crosses Phule Anuradha X ICSV 700, ICSV 25019 X M 35-1, IS 18551 X ICSV 700, and Swarna X IS 2123 showed significant positive reciprocal effects for plant height in the postrainy season.

Table 10: Estimates of reciprocal combining ability effects of reciprocal crosses (10 X 10 diallel) of sorghum across seasons (ICRISAT, Patancheru, 2013–14)

Pedigree	Days to 50% flowering		Plant height (cm)	100 seed weight (g)		Inflorescence exsertion	Panicle compactness	Panicle shape
	2013 R	2013 PR	2013 PR	2013 R	2013 PR	2013 R	2013 PR	2013 PR
Phule Anuradha X ICSV 700	−0.50	5.33**	11.11**	0.03	−0.20*	–	–	–
M 35-1 X ICSV 700	−1.70*	0.33	−0.56	−0.02	−0.15	0.17	–	–
M 35-1 X Phule Anuradha	−0.33	0.83	−5.00	0.02	0.13	−0.17	0.17*	0.50*
CSV 15 X ICSV 700	3.00**	1.50	1.67	−0.05	0.02	–	−0.17*	−0.50*
CSV 15 X Phule Anuradha	−1.00	0.33	−7.78*	−0.05	−0.02	0.17	–	−0.17
CSV 15 X M 35-1	0.67	3.00**	−0.56	0.03	−0.38**	−0.17	0.17*	0.67**
ICSV 25019 X ICSV 700	−0.83	0.50	0.56	0.12	0.08	0.50**	−0.17*	−0.50*
ICSV 25019 X Phule Anuradha	−0.33	0.33	−1.67	−0.20**	−0.08	0.30*	−0.17*	−0.50*
ICSV 25019 X M 35-1	−1.70*	−3.00**	8.89*	−0.07	0.17	−0.17	−0.33**	−1.00**
ICSV 25019 X CSV 15	−0.67	−1.50	−4.44	–	0.25*	0.17	–	−0.17
PS 35805 X ICSV 700	−2.00**	0.67	0.56	−0.03	–	0.17	–	–
PS 35805 X Phule Anuradha	−0.67	−3.67**	−17.22**	0.02	0.13	0.17	−0.33**	−1.00**
PS 35805 X M 35-1	−0.67	−0.33	−12.78**	−0.02	−0.05	–	–	–
PS 35805 X CSV 15	0.33	–	2.78	0.05	0.22*	–	–	−0.17
PS 35805 X ICSV 25019	−0.33	−0.50	–	−0.02	0.18	–	–	–
IS 2123 X ICSV 700	–	−0.67	5.56	−0.03	−0.03	−0.30*	–	–
IS 2123 X Phule Anuradha	−0.67	−1.17	−2.22	0.02	0.10	–	–	–
IS 2123 X M 35-1	−1.33	0.83	0.56	0.10	–	−0.50**	–	–
IS 2123 X CSV 15	0.50	−1.33	−5.56	−0.08	0.02	−0.30*	–	0.17
IS 2123 X ICSV 25019	−1.17	−0.67	2.78	0.05	0.15	0.50**	–	–
IS 2123 X PS 35805	–	1.83*	−5.00	−0.08	−0.12	−0.17	–	–
IS 2146 X ICSV 700	1.17	–	−0.56	−0.02	0.18	−0.17	–	–
IS 2146 X Phule Anuradha	1.00	0.83	−2.22	−0.12	0.13	–	–	–
IS 2146 X M 35-1	−1.17	0.50	−0.56	–	0.12	−0.17	–	–
IS 2146 X CSV 15	–	3.00**	4.45	0.02	−0.03	0.50**	–	−0.17
IS 2146 X ICSV 25019	−0.17	–	3.89	0.20**	0.13	–	–	–
IS 2146 X PS 35805	−0.67	−1.17	5.56	0.07	−0.07	−0.17	–	0.17
IS 2146 X IS 2123	–	0.83	−1.67	0.03	−0.10		–	–
IS 18551 X ICSV 700	−1.80*	−0.50	10.00*	0.13	0.05	−0.30*	–	–
IS 18551 X Phule Anuradha	–	–	−0.56	−0.02	0.02	−0.17	–	–
IS 18551 X M 35-1	−2.50**	–	−5.00	0.03	−0.03	−0.17	–	–
IS 18551 X CSV 15	−1.00	−0.17	−6.67	–	–	0.17	–	−0.17
IS 18551 X ICSV 25019	−0.33	1.00	5.56	–	−0.03	−0.17	0.33**	1.00**
IS 18551 X PS 35805	−1.70*	−0.17	−2.22	0.10	−0.08	−0.17	−0.17*	−0.50*
IS 18551 X IS 2123	−2.00**	2.00*	−0.56	−0.12	−0.17	−0.17	–	–
IS 18551 X IS 2146	−0.50	0.50	7.22	0.12	0.05	−0.30*	–	–
Swarna X ICSV 700	−0.33	0.67	−6.11	0.08	−0.08	0.30*	–	–
Swarna X Phule Anuradha	0.83	1.83*	−2.78	0.07	−0.13	0.30*	–	−0.17
Swarna X M 35-1	2.30**	2.33*	7.22	−0.40**	−0.67**	0.30*	0.33**	1.50**
Swarna X CSV 15	0.67	0.83	−9.45*	0.20*	−0.22*	–	–	–
Swarna X ICSV 25019	−0.33	1.17	−1.11	0.20**	0.05	–	–	–
Swarna X PS 35805	−0.33	−0.33	−5.00	0.08	−0.12	–	–	–
Swarna X IS 2123	1.70*	2.17*	10.56**	0.15	−0.07	0.17	–	0.17
Swarna X IS 2146	0.33	1.00	−0.55	0.02	0.18	−0.30*	–	–
Swarna X IS 18551	1.80*	−1.17	−3.89	0.03	0.05	–	–	–

*, **t-test significant at 0.05 and 0.01 probability levels; ORS, overall resistance score; R, rainy season; PR, postrainy season.

ICSV 25019 X Phule Anuradha in the rainy season and Phule Anuradha X ICSV 700, CSV 15 X M 35-1, Swarna X CSV 15 in the postrainy season, and Swarna X M 35-1 across seasons, exhibited significant negative reciprocal effects for 100 seed weight. IS 2146 X ICSV 25019, Swarna X CSV 15, Swarna X ICSV 25019 in the rainy season and ICSV 25019 X CSV 15, PS 35805 X CSV 15 in the postrainy season exhibited significant positive reciprocal effects for 100 seed weight.

Combining Ability Estimates and Genetic Parameters

Variance due to GCA (σ^2 g) was higher than the variance due to SCA (σ^2 s) for glume cover across seasons (Table 11); agronomic score and waxy bloom in the rainy season and days to 50% flowering and inflorescence exsertion in the postrainy season also showed high *gca* variance, indicating the predominance of additive gene action in controlling the expression of these traits. Plant height and grain yield exhibited higher σ^2 s than the σ^2 g across seasons; days to 50% flowering and inflorescence exsertion in the rainy season, and panicle shape in the postrainy season showed high σ^2 s than the variance due to*gca*, indicating the predominance of non-additive type of gene action in controlling the expression of these traits. The other traits that had similar σ^2 g and σ^2 s exhibited both additive and non-additive type of gene action.

Table 11: Estimates of various genetic parameters for different agronomic and panicle traits of sorghum across seasons (ICRISAT, Patancheru, 2013–14)

Traits	Days to 50% flowering	Plant height (cm)	100 seed weight (g)	Grain yield (t/ha)	Agronomic score	Inflorescence exsertion	Panicle compactness	Panicle shape	Glume coverage	Awns	Grain lustre
$\sigma^2 g$	3.70	730.03	0.04	0.57	0.34	0.10	0.08	(0.24)	0.87	0.10	–
	(8.31)	(285.66)	(0.10)	(2.17)	(0.15)	(0.54)	(0.09)		0.87	0.10	
$\sigma^2 s$	9.05	1817.89	0.08	3.44	0.32	0.27	0.07	(1.27)	0.51	0.12	0.02
	(6.44)	(460.32)	(0.16)	(11.87)		(0.49)	(0.16)		(0.38)	(0.12)	(0.01)
$\sigma^2 r$	0.74 (1.48)	(17.53)	(0.02)	–	–	0.03	(0.01)	(0.1)	–	–	–
$\sigma^2 e$	1.27	38.62	0.02	0.50	0.12	0.06	0.01	(0.13)	0.19	–	–
	(1.96)	(33.99)	(0.02)	(1.08)	(0.16)	(0.09)	(0.01)		(0.11)		
$\sigma^2 a$	7.40	1460.07	0.08	1.13	0.67	0.21	0.17	(0.48)	1.74	0.19	–
	(16.62)	(571.32)	(0.20)	(4.34)	(0.29)	(1.07)	(0.17)		(2.10)	(0.19)	
$\sigma^2 d$	9.05	1817.89	0.08	3.44	0.32	0.27	0.07	(1.27)	0.51	0.12	0.02
	(6.44)	(460.32)	(0.16)	(11.87)		(0.49)	(0.16)		(0.38)	(0.12)	(0.01)
$\sigma^2 p$	18.45	3317.39	0.18	5.17	1.14	0.57	0.25	(1.97)	2.46	0.32	0.02
	(26.50)	(1083.15)	(0.40)	(17.29)	(0.41)	(1.66)	(0.35)		(2.55)	(0.31)	(0.02)
h_{ns}^2	0.40	0.44	0.45	0.22	0.59	0.37	0.68	(0.24)	0.71	0.60	0.10
	(0.63)	(0.53)	(0.51)	(0.25)	(0.71)	(0.65)	(0.50)		(0.82)	(0.60)	(0.17)
h_b^2	0.89	0.99	0.90	0.88	0.87	0.85	0.95	(0.89)	0.92	1.00	1.00
	(0.87)	(0.95)	(0.91)	(0.94)	(0.67)	(0.94)	(0.94)		(0.97)	(0.99)	(0.78)
GCA/SCA ratio	0.41	0.40	0.49	0.16	1.04	0.38	1.26	(0.19)	1.70	0.77	0.06
	(1.29)	(0.62)	(0.64)	(0.18)		(1.09)	(0.55)		(2.73)	(0.75)	(0.14)
Predictability ratio	0.45	0.45	0.5	0.25	0.68	0.43	0.72	(0.27)	0.77	0.61	0.10
	(0.72)	(0.55)	(0.56)	(0.27)		(0.69)	(0.53)		(0.85)	(0.6)	(0.22)

$\sigma^2 g$, gca variance; $\sigma^2 s$, sca variance; $\sigma^2 r$, reciprocal variance; $\sigma^2 e$, environmental/error variance; $\sigma^2 a$, additive variance; $\sigma^2 d$, dominance variance; $\sigma^2 p$, phenotypic variance; h_{ns}^2, narrow-sense heritability; h_b^2, broad-sense heritability; GCA, general combining ability; SCA, specific combining ability. The values in the parentheses are for postrainy season and off the parentheses are for rainy season.

Glume cover showed greater additive (σ^2 a) than the dominance variance (σ^2 d) across seasons. Agronomic score, waxy bloom, and panicle compactness in the rainy season, and days to 50% flowering, plant height, 100 seed weight, and inflorescence exsertion in the postrainy season showed higher additive variance than the dominance variance. Overall resistance score, grain yield, and plant color exhibited higher dominance variance than the additive variance across seasons. Days to 50% flowering, plant height, and inflorescence exsertion in the rainy season and panicle shape in the postrainy season possessed higher dominance variance than the additive variance.

Glume cover and awns exhibited high GCA/SCA ratios across seasons. Agronomic score, waxy bloom, and panicle compactness in the rainy season, and days to 50% flowering, plant height, 100 seed weight, and inflorescence exsertion in the postrainy season exhibited high GCA/SCA ratios, indicating the additive type of gene action controlling the expression of these traits. Panicle compactness, glume cover, and awns showed high predictability ratios across seasons. The predictability ratios for agronomic score, and waxy bloom in the rainy season, and days to 50% flowering, plant height, 100 seed weight, and inflorescence exsertion in the postrainy season were quite high. Heritability estimates of the traits studied ranged from 0.10 to 0.71 (narrow-sense heritability), and 0.85 to 1.00 (broad-sense heritability) in the rainy season, and 0.17 to 0.82 (narrow-sense heritability), and 0.67 to 0.99 (broad-sense heritability) in the postrainy season. Almost all the traits exhibited moderate to high heritability values, except grain yield and grain luster across the seasons. Panicle shape in the postrainy season exhibited low (≤ 0.30) narrow-sense heritability.

DISCUSSION

Significance of *F*-values for all the traits studied indicated the presence of high variability in the parents used for developing the full diallel. Plant height, 100 seed weight and grain yield were associated with susceptibility to shoot fly. Days to 50% flowering, agronomic score, inflorescence exsertion, panicle compactness, panicle shape, glume coverage, and awns were associated with shoot fly resistance.

Association between the shoot fly resistance, morphological, and agronomic traits suggested complex interactions between shoot fly and plant traits. Significance of the GCA and SCA mean sum of squares for all the traits across seasons suggested that both the additive and non-additive nature of gene action for agronomic and panicle characteristics. The significance of reciprocal combining ability effects for days to 50% flowering, plant height, and 100 seed weight, suggesting possible role of cytoplasmic factors in inheritance of these traits. These interactions should be taken into consideration while developing strategies for improving sorghums for shoot fly resistance and high grain yield.

Genotypes with negative GCA effects for days to 50% flowering can be utilized to develop the hybrids with early flowering. To develop hybrids with moderate height, care should be taken to select the parents with moderate plant height. Additive gene action in the rainy season and dominance in the postrainy season for days to 50% flowering and plant height suggested G X E interactions for these traits. This contrasting gene expression in the rainy and postrainy seasons for days to 50% flowering and plant height suggested the season specific breeding of these traits for sorghum improvement. Meng et al. (1998), Rafiq et al. (2002), and Mohammed Maarouf (2009) reported additive gene action for days to 50% flowering, while Erenso (1998) reported additive gene action for plant height. Higher magnitude of SCA variance was reported byManickam and Vijendra Das (1994) and Umakanth et al. (2002) for both the plant traits. High GCA/SCA and predictability ratios for 100 seed weight in the postrainy season indicated the predominance of additive gene action, whereas both additive and non-additive gene action was observed in the rainy season. Grain yield exhibited higher SCA variance suggesting the predominance of dominance (non-additive) type of gene action (Wilson et al., 1978; Singhania, 1980; Amsalu, 1987; Hovny et al., 2000;Umakanth et al., 2002; Girma et al., 2010). However, the importance of both the additive and non-additive gene action was observed for 100 seed weight by Toure et al. (1996).

Knowledge of the genetic architecture of grain yield, and morphological and agronomic traits will be useful for formulating a meaningful breeding strategy for developing improved genotypes. Genetic diversification of sorghum for key traits is important for sustaining the yield gains and mapping QTL underlying agronomically important traits is a key step in understanding their genetic control and for using the tightly linked markers for marker-assisted breeding for crop improvement (Srinivas et al., 2009;Ashok Kumar et al., 2011; Nagaraja Reddy et al., 2013, 2014). Many studies were conducted in identifying the QTL regions of different traits associated with insect resistance as well as the morphological and agronomic traits (Satish et al., 2009; Srinivas et al., 2009; Aruna et al., 2011b; Nagaraja Reddy et al., 2013, 2014). Based on the present inheritance studies of the agronomic and morphological traits and as well as the QTL information available one can effectively plan suitable breeding strategies for sorghum improvement.

Most of the hybrids studied exhibited higher grain yield than the parents even if one of the parent was high yielding, suggesting that one of the parent should possess high grain yield ability for developing high yielding hybrids. This is very critical in sorghum improvement considering the fact that hybrids are preferred over the varieties worldover, barring Africa. Most of the commercial hybrids show 30–40% yield superiority over the best varieties in a given ecology. The panicle trait such as inflorescence exsertion exhibited predominance of additive gene action in the postrainy season, and dominance gene action in the rainy season, while glume cover and presence of awns showed predominance of additive gene action.

The genotypes CSV 15, ICSV 25019, PS 35805, and Swarna exhibited negative *gca* effects for almost all the traits, but positive *gca* effects for grain yield. Hence, these genotypes can be effectively used in breeding the high yielding sorghums. The crosses involving the genotype IS 2146 either as male or female parent showed a decrease in grain yield being a poor combiner coupled with low *per se* mean yield. Phule Anuradha and M 35-1 showed positive *gca* effects for 100 seed weight but negative *gca* effects for grain yield in the rainy season, and positive effects in the postrainy season, suggesting that these genotypes can be effectively utilized for breeding high yielding shoot fly resistant sorghums for the postrainy season. Both M 35-1 and Phule Anuradha are highly adapted to postrainy environments, and are very popular with farmers. ICSV 25019, PS 35805, IS 2123, and IS 18551 exhibited negative *gca* effects for 100 seed weight, but showed positive *gca* effects for grain yield. Hence, these genotypes can be utilized for breeding high yielding sorghums with shoot fly resistance. Though the genotypes CSV 15 and Swarna showed positive *gca* effects for 100 seed weight and grain yield, but these may not be good parents in a shoot fly resistance breeding program. ICSV 700, IS 2123, and IS 18551 showed positive *gca* effects for most of the traits and these can be utilized for improving shoot fly resistance. Higher narrow- and broad-sense heritability estimates suggested high heritability of these traits across environments, and greater role of additive gene action, suggesting that selection is effective for improving these traits. This information can be used for developing high yielding cultivars with insect resistance for sustainable crop production.

CONCLUSIONS

Genotypic response varies across seasons, and hence, it is important to identify genotypes with desirable agronomic traits, insect resistance, and high grain yield for different seasons and locations. The genotypes ICSV 700, Phule Anuradha, M 35-1, ICSV 25019, PS 35805, IS 2123, and IS 18551 exhibiting moderate to high shoot fly resistance and desirable agronomic traits, can be effectively used in sorghum improvement. Both additive and non-additive type of gene action governs the morphological (inflorescence exsertion, panicle compactness, and panicle shape) and agronomic (days to 50% flowering, plant height and 100 seed weight) traits and hence it is important to exploit heterosis breeding for improving agronomic and morphological traits and grain yield in sorghum. The significance of reciprocal effects for some of the traits (days to 50% flowering, plant height, and 100 seed weight) suggested that apart from the direct genetic effects, the cytoplasmic factors also played an important role in inheritance of these traits. An understanding of the association between shoot fly resistance and morphological and agronomic traits will be useful to improve the strategies to develop shoot fly-resistant cultivars with desirable plant types, and season specific adaptation for sustainable crop production.

CONFLICT OF INTEREST STATEMENT

The authors declare that the research was conducted in the absence of any commercial or financial relationships that could be construed as a potential conflict of interest.

ACKNOWLEDGMENTS

We are thankful to the sorghum entomology and breeding staff for their help in carrying out the field experiments, and Bill & Melinda Gates Foundation for financial support through HOPE-Sorghum and Millets project. We also thank, Dr. P. Janila for reviewing and providing constructive suggestions on this manuscript.

REFERENCES

1. Amsalu, A. (1987). *Heterosis and Combining Ability Studies in Grain Sorghum [Sorghum bicolour (L.) Moench]*. M.Sc. thesis, Mahatma Phule Agricultural University, Maharashtra.
2. Aruna, C., Bhagwat, V. R., Madhusudhana, R., Vittal, S., Hussain, T., Ghorade, R. B., et al. (2011b). Identification and validation of genomic regions that affect shoot fly resistance in sorghum [*Sorghum bicolor* (L.) Moench]. *Theor. Appl. Genet.* 122, 1617–1630. doi: 10.1007/s00122-011-1559-y
3. Aruna, C., Bhagwat, V. R., Vittal, S., Hussain, T., Ghorade, R. B., Khandalkar, H. G., et al. (2011a). Genotype X Environment interactions for shoot fly resistance in sorghum [*Sorghum bicolor* (L.) Moench]: response of recombinant inbred lines. *Crop Prot.* 30, 623–630. doi: 10.1016/j.cropro.2011.02.007
4. Ashok Kumar, A., Reddy, B. V. S., Sharma, H. C., Hash, C. T., Srinivasa Rao, P., Ramaiah, B., et al. (2011). Recent advances in sorghum genetic enhancement research at ICRISAT. *Am. J. Plant Sci.* 2, 589–600. doi: 10.4236/ajps.2011.24070
5. Baker, R. J. (1978). Issues in diallel analysis. *Crop Sci.* 18, 533–536. doi: 10.2135/cropsci1978.0011183X001800040001x
6. Melake-Berhan, A, Butler, L. G., Ejeta, G., and Menkir,

A. (1996). Grain mold resistance and polyphenol accumualation in sorghum. *J. Agr. Food Chem.* 44, 2428–2434. doi: 10.1021/jf950580x

7. Dhillon, M. K., Sharma, H. C., Folkertsama, R. T., and Chandra, S. (2006). Genetic divergence and molecular characterisation of sorghum hybrids and their parents for reaction to *Atherigona soccata* (Rondani). *Euphytica* 149, 199–210. doi: 10.1007/s10681-005-9067-2
8. Erenso, D. (1998). *Study of Combining Ability in Lowland Grain Sorghum [Sorghum bicolor (L.) Moench]*. M.Sc. thesis, Alemaya University, Ethiopia.
9. FAO (2004). *Production Yearbook*. Rome: FAO.
10. FAOSTAT (2012). *Crops Primary Equivalent*. Available online at: http://faostat.fao.org. (Retrieved 15th March, 2015).
11. GenStat (2010). *Introduction to GenStat for Windows Genstat, 13th Edn*. Harpenden: Lawes Agricultural Trust, Rothamsted Experimental Station.
12. Girma, M., Amsalu, A., and Ketema, B. (2010). Combining ability for yield and its components in ethiopian sorghum [*Sorghum bicolor*(L.) Moench] landraces. *East Afr. J. Sci.* 4, 34–40. doi: 10.4314/eajsci.v4i1.71522
13. Griffing, B. (1956). Concept of general and specific combining ability in relation to diallel crossing systems. *Aust. J. Biol. Sci.* 9, 463–493.
14. Hovny, M. R. A., El-Naggouly, O. O., and Hassaballa, E. A. (2000). Combining ability and heterosis in grain sorghum [*Sorghum bicolor*(L.) Moench]. *Assiut J. Agr. Sci.* 31, 1–16.
15. IBPGR ICRISAT (1993). *Descriptors for Sorghum [Sorghum bicolor (L.) Moench]. International Board for Plant Genetic Resources*. Rome: International Crops Research Institute for the Semi-Arid Tropics (ICRISAT), Patancheru.
16. Indostat Services (2004). *Windostat*. Hyderabad: Indostat Services.
17. Jayanthi, P. D. K., Reddy, B. V. S., Reddy, D. D. R., Gour, T. B., and Nwanze, K. F. (1996). "Genetics of shoot fly resistance in sorghum hybrids of cytoplasmic male sterile lines," in *Abstracts of Poster Sessions: 2nd International Crop Science Congress: Crop Productivity and Sustainability—Shaping the Future* (New Delhi: National Academy of Agricultural Sciences, Indian Council of Agricultural Research), 152.
18. Johnson, H. W., Robinson, H. F., and Comstock, R. E. (1955). Estimates of genetic and environmental variability in soybean. *Agron. J.*47, 314–318. doi: 10.2134/agronj1955.00021962004700070009x
19. Kullaiswamy, B. Y., and Goud, J. V. (1983). New genes for awning in sorghum [*Sorghum bicolor* (L.) Moench]. *Madras Agr. J.* 70, 355–359.
20. Manickam, S., and Vijendra Das, L. D. (1994). Line X tester analysis in forage sorghum. *Int. Sorghum Millet Newslett.* 35, 79–80.
21. Meng, C. G., An, X. M., Zhang, F. Y., Zheng, J. B., Wang, L. X., and Li, P. L. (1998). Analysis of combining ability of newly developed sorghum male sterile lines. *Acta Agr. Boreali-Sinica* 13, 81–85.
22. Mohammed Maarouf, I. (2009). Line X tester analysis across locations and years in Sudanese x exotic lines of forage sorghum. *J. Plant Breed. Crop Sci.* 1, 311–319. Available online at: http://www.academicjournals.org/article/article1379424511_Mohammed.pdf
23. Murty, D. S. (2000). "Breeding for grain mold resistance in sorghum: opportunities and limitations," in *Technical and Institutional Options for Sorghum Grain Mold Management: Proceedings of an International Consultation, 18–19 May 2000*, eds A. Chandrashekar, R. Bandyopadhyay and A. J. Hall (Patancheru: International Crops Research Institute for the Semi-Arid Tropics (ICRISAT)), 225–227.
24. Nagaraja Reddy, R., Madhusudhana, R., Murali Mohan, S., Chakravarthi, D. V. N., Mehtre, S. P., Seetharama, N., et al. (2013). Mapping QTL for grain yield and other agronomic traits in post-rainy sorghum [*Sorghum bicolor* (L.) Moench]. *Theor. Appl. Genet.* 126, 1921–1939. doi: 10.1007/s00122-013-2107-8
25. Nagaraja Reddy, R., Madhusudhana, R., Murali Mohan, S., Seetharama, N., and Jagannatha Vishnu, P. (2014). Detection and validation of stay-green QTL in post-rainy sorghum involving widely adapted cultivar, M35-1 and a popular stay-green genotype B35. *BMC Genomics* 15:909. doi: 10.1186/1471-2164-15-909
26. Nimbalkar, V. S., and Bapat, D. R. (1992). Inheritance of shoot fly resistance in sorghum. *J. Maharashtra Agric. Univ.* 17, 93–96.
27. Porter, K. S., Axtell, J. D., Lechtenberg, V. L., and Colenbrander, V. F. (1978). Phenotype, fiber composition, and *in vitro* dry matter disappearance of chemically induced brown midrib (bmr) mutants of sorghum. *Crop Sci.* 18, 205–208. doi: 10.2135/cropsci1978.0011183X001800020002x
28. Rafiq, S. M., Thete, R. Y., Madhusudhana, R., and Umakanth, A. U. (2002). Combining ability studies for grain yield and its components in postrainy season sorghum grown in medium-deep and shallow soils. *Int. Sorghum Millets Newslett.* 43, 33–37. Available online at:http://oar.icrisat.org/1106/1/RA_00378.pdf
29. Riyazaddin, M. D., Kavi Kishor, P. B., Ashok Kumar, A., Belum Reddy, V. S., Rajendra, S. M., and Sharma, H. C. (2015). Mechanisms and diversity of resistance to sorghum shoot fly, *Atherigona soccata*. *Plant Breed.* 134, 423–436. doi: 10.1111/pbr.12276
30. Satish, K., Srinivas, G., Madhusudhana, R., Padmaja, P. G., Nagaraja Reddy, R., Murali Mohan, S., et al. (2009). Identification of quantitative trait loci for resistance to shoot fly in sorghum [*Sorghum bicolor* (L.) Moench]. *Theor. Appl. Genet.* 119, 1425–1439. doi: 10.1007/s00122-009-1145-8
31. Sharma, H. C. (1993). Host-Plant Resistance to insects in sorghum and its role in integrated pest management. *Crop Prot.* 12, 11–34. doi: 10.1016/0261-2194(93)90015-B
32. Sharma, H. C., Reddy, B. V. S., Dhillon, M. K., Venkateswaran, K., Singh, B. U., Pampapathy, G., et al. (2005). Host plant resistance to insects in sorghum: present status and need for future research. *Int. Sorghum*

Millets Newslett. 46, 36–43. Available online at:http://oar.icrisat.org/1215/1/ISMN-46_36-43_2005.pdf

33. Sharma, H. C., Taneja, S. L., Kameswara Rao, N., and Prasada Rao, K. E. (2003). *Evaluation of Sorghum Germplasm for Resistance to Insect Pests. Information Bulletin no. 63*. Patancheru: International Crops Research Institute for the Semi-Arid Tropics (ICRISAT).
34. Singhania, D. L. (1980). Heterosis and combining ability studies in grain sorghum. *Indian J. Genet.* 40, 463–471.
35. Srinivas, G., Satish, K., Madhusudhana, R., Nagaraja Reddy, R., Murali Mohan, S., and Seetharama, N. (2009). Identification of quantitative trait loci for agronomically important traits and their association with genic-microsatellite markers in sorghum. *Theor. Appl. Genet.* 118, 1439–1454. doi: 10.1007/s00122-009-0993-6
36. Toure, A., Miller, F. R., and Rosenow, D. T. (1996). Heterosis and combining ability for grain yield and yield components in guinea sorghums. *J. Afr. Crop Sci.* 4, 383–391.
37. Umakanth, A. V., Madhusudhana, K., Madhavi Latha, P., Hema, K., and Kaul, S. (2002). Genetic architecture of yield and its contributing characters in postrainy season sorghum. *Int. Sorghum Millets Newslett.* 43, 37–40. Available online at:http://oar.icrisat.org/1106/1/RA_00378.pdf
38. Wilson, N. D., Weibel, D. E., and McNew, R. W. (1978). Diallel analyses of grain yield, percent protein, and protein yield in grain sorghum. *Crop Sci.* 18, 491–495.

CHAPTER 11

MORPHO-PHYSIOLOGICAL AND PROTEOME LEVEL RESPONSES TO CADMIUM STRESS IN SORGHUM

Swapan Kumar Roy, Seong-Woo Cho, Soo Jeong Kwon, Abu Hena Mostafa Kamal, Sang-Woo Kim, Myeong-Won Oh, Moon-Soon Lee, Keun-Yook Chung, Zhanguo Xin, Sun-Hee Woo

INTRODUCTION

Over many decades, the industrial revolution has resulted in the environment being extensively polluted with different toxic metals and metalloids [1]. Heavy metals are among the most widespread soil contaminants, and they can cause damage to plants by inducing alterations in major physiological and metabolic processes [2,3,4,5,6]. The effects of certain heavy metals, such as cadmium (Cd), copper (Cu), lead (Pb) and zinc (Zn), have become an important concern for plant scientists because of their harmful effects on soil-plants and other cellular systems in the environment [7].

Cd has become a major environmental problem in agricultural systems where Cd has been ranked no. 7 among the top 50 toxic substances [8].Cd is introduced into the environment via many anthropogenic routes including power stations, heating systems, metal-working industries, waste incinerators, urban traffic, cement factories and as a byproduct of phosphate fertilizers [9].

Although Cd is a non-essential element in plants, it is easily taken up by roots and can be loaded into the xylem for transport to the leaves. The accumulated Cd within the plant induces various symptoms of toxicity, such as chlorosis, wilting, reduced growth and cell death [10]. Cd may influence plant physiological and biochemical processes by affecting the concentrations and functions of nutrient minerals that would otherwise have protective roles against the toxic effects of Cd-induced stress [11]. In addition, as a heavy metal, Cd has an adverse effect on the successful growth and development of plant [12].

However, when plants are exposed to high levels of heavy metals, they generate reactive oxygen species (ROS) as a primary response, and the combination of these new ROS with the oxidative stress already present in plants could be an indirect consequence of heavy metal toxicity [13,14]. Moreover, plants have evolved mechanisms to adapt to adverse environmental heavy metal toxicity pollution. To this end, plants produce low molecular weight thiols such as glutathione (GSH), phytochelatin (PCs) and cysteine that have a high affinity for toxic metals [15]. These compounds are produced primarily to be targets for free Cd-ions and GSH in particular increases heavy metal tolerance in plant cells and protects the cells from the oxidative stress that is induced by heavy metals [16] by directly binding metals in the cytosol and vacuole. Furthermore, GSH serves as a key regulator of redox signaling by altering gene expression at both the transcriptional and the translational level [17].

S. bicolor L. is an important crop because it is widely used as a food, feed and energy crop. In addition, sorghum has shown promise as a cereal crop, because it has some non-food uses, particularly in the production of bioethanol [18]. Previous studies demonstrated that sorghum plants were able to accumulate large quantities of heavy metal in the shoots [19,20], as well as highly tolerant to metal pollution [21,22]. Nevertheless, the availability of the sorghum full genome sequence [23] makes it a reasonable C_4 model plant, which could be combined with the C_3 plant models Arabidopsis and rice, to study the gene products involved in adaptation to heavy metal stress. Proteomics techniques present high throughput

Citation: Roy SK, Cho S-W, Kwon SJ, Kamal AHM, Kim S-W, Oh M-W, et al. (2016) Morpho-Physiological and Proteome Level Responses to Cadmium Stress in Sorghum. PLoS ONE 11(2): e0150431. doi:10.1371/journal.pone.0150431.

and large scale methods that can be used for the exploration of proteins in a particular organism, tissue or cell at any given time [24,25,26]. Two-dimensional electrophoresis (2-DE) in combination with mass spectrometry, has paved the way toward identifying the differentially expressed proteins that are produced in responses to various stresses in sorghum [27].

A considerable number of proteomics studies have been performed in plants exposed to abiotic stress conditions, but studies that involve exposure to heavy metals have been limited. Recently, proteomic techniques have been applied in many model and hyperaccumulator plants to identify Cd-regulated proteins at the molecular level [28,29,30,31,32,33,34,35]. Additionally, these studies have revealed that most of the studied plants undergo common changes in several types of functional proteins that are involved in energy and carbohydrate metabolism [28,33,36,37,38,39,40], transcription and translation [33,39,41] and stress-related proteins [33,38,40]. However, no report has been published to describe Cd-stress in C_4 plants using proteomics. During the past decade, only a few studies dealing with the morpho-physiology and interactive effects of a combination of heavy metals on plants have been published [42], and some of these investigations have prompted the development of bioremediation [43] and phytoremediation [44] strategies. In addition, some investigations have been conducted to investigate the accumulated levels of heavy metals in the soil [45] and the growth and physiological characteristics of plants [12].

Taking these previous studies into account, in the present investigation, a high-throughput proteome technique was used to identify the differentially expressed proteins that may potentially be induced by different levels of Cd stress and to the response mechanisms induced by injury resulting from exposure to Cd in *S. bicolor*. Leaves of sorghum seedlings subjected to different levels of Cd were used as the experimental materials, and leaf proteome was performed using a 2-DE approach. The differentially expressed proteins identified by MALDI-TOF/TOF MS were compared to investigate cellular responses and to shed light on the molecular mechanisms underlying Cd-stress.

MATERIALS AND METHODS

Plant Growth Condition and Cadmium Treatment

Sorghum [*Sorghum bicolor* (L.) Moench] inbred line BTx623 (http://www.phytozome.net/sorghum), which is a parent for several mapping populations in sorghum and the genotype for sequencing the sorghum genome, is used to generate the mutant populations [18]. It was developed by Dr. Fred Miller at Texas A & M University and released by Texas A & M University, Texas Agricultural Experiment Station in 1977. The pedigree of BTx623 is (BTx3197*SC170-6-4-4)-7-3-1-3-2-1. The lines are resistant to downy mildew, insecticidal leaf burn and rust zonate leaf spot. The lines have longer and wider stigmatic areas. These lines have the characteristics of tropical adaption, which produce higher yields under the short day length and hot night temperatures. The lines will be useful in forage and sugar sorghum production because of their resistance to downy mildew and other foliage pathogens and because of their sweet juicy stems. Seeds of *S. bicolor* L. (BTX 623) were surface-sterilized and placed in petri dishes containing two layers of filter papers that were moistened with de-ionized water. The seeds placed in petri dishes were then placed in a growth chamber and grown in a controlled environment at 25°C with a light intensity of 8000 Lux and 70% humidity. After 5 days, the seedlings were transplanted to vessel containing Hoagland solution [46]. The solution was aerated daily via air bubbling for 30 min and changed every 2 days.

For the Cd-treatment experiments, three replicates of seven seedlings each were included in both the control and Cd treatment groups. The 10-day-old plants were exposed to Hoagland nutrient solutions, supplemented with 0, 100, 150 μM $CdCl_2$ and grown under the same controlled environmental conditions described above. After 5 days of Cd-treatment, the seedling leaves were harvested and morpho-physiological and proteomic analyses were performed.

Growth Parameters

Shoot lengths (cm), root lengths (cm), shoot fresh weights (g), and root fresh weights (g) were measured from each of the collected samples. The dry weight of the shoot and root (g) was recorded in calibration balance (Ohaus Corporation, Pine Brook, NJ USA) after they were dried in a force oven at 65°C for 72 hours as previously described [47].

Relative Water Content Measurement

To evaluate the water status during the cadmium stress period, the relative water content (RWC) was determined as previously described [47]. Briefly, the plants (three seedlings) were weighed (fresh weight, FW) and, saturated in water for 2 hours and then their turgid weights (TWs) were calculated. The samples were dried in an oven at 65°C for 72 hours and their dry weights (DWs) were then calculated. RWC was determined as follows; $RWC = (FW - DW)/(TW - DW) \times 100$

Determination of cadmium accumulation levels in leaves

The levels of cadmium in the leaves and root tissues were determined as previously described [48]. Briefly, the leaves and root tissues were collected, washed with distilled water, and dried at 105°C for 48 hours. Then, the dried materials were ground into a powder. Approximately 50 mg of this powder was digested in 5 ml of HNO_3 (48%, w/v) at 60°C for 48 hours. After diluting the solution with Milli-Q water (1:20), the cadmium in the solution was measured using inductively coupled plasma-optical emission spectrometry (ICP-OES; Optima 5300 V; Perkin-Elmer, Inc., USA). The concentrations of cadmium in the tissues were calculated as mg per kg dry weight.

Ion Concentration Measurements

The ion concentrations in the leaf tissues were determined as previously described [49]. Briefly, a portion (0.5 g) of the leaves of a sorghum plant was placed in a micro-Kjeldahl flask and 5 ml H_2SO_4 was then added. Filter paper (no.6 or no.7) was used to quantitate the amount of Zn^{2+}, Ca^{2+}, Fe^{2+}, and Cd^{2+}. The concentrations of positive ions, including cadmium, in the leaves and roots of the plants were determined using inductively coupled plasma-optical emission spectrometry (ICP-OES; Optima 5300 V; Perkin-Elmer, Inc., USA).

Physiological Analysis using Confocal Microscopy

For confocal microscopy, sorghum leaves were carefully cut and collected in petri dishes (60 x 15 mm) containing double-distilled water. After rinsing, the leaves were stained with a 100 μM aqueous solution of Dithizone for 30 min. They were then washed twice with double-distilled water for 5 min each. Finally, they were placed on slides, mounted in mounting solution and observed using confocal microscopy (LSM 410; Carl Zeiss, Jena, Germany).

Protein Extraction

Proteins were extracted from the leaves of *S. bicolor* using the trichloroacetic acid (TCA)/acetone method as previously described [50], with minor modifications. Frozen leaf tissues were ground to a fine powder in liquid nitrogen and homogenized in ice-cold 10% trichloroacetic acid and 0.07% 2-mercaptoethanol in acetone. After the suspension was vortexed, it was sonicated for 10 min, incubated for 1 hour at -20°C and centrifuged at 9000 x g for 20 min at 4°C. The supernatant was discarded and the pellet was washed twice with 0.07% 2-mercaptoethanol in acetone. The washed pellet was dried using a speed-Vac concentrator (Hanil Science Medical, Modulspin 31, Seoul, South Korea) for 10 min, resuspended in lysis buffer (8 M urea, 2 M thiourea, 5% CHAPS, and 2 mM tributylphosphine) with vortexing, and then incubated for 1 hour at 25°C. The suspension was then centrifuged at 20000 x g for 20 min at 25°C, and the resulting supernatant containing the protein extract was collected in a 1.5 ml tube. The protein concentrations of the samples were determined using the Bradford method [51], with a spectrophotometer (UV-1700 PharmaSpec; Shimadzu Corporation, Kyoto, Japan) using bovine serum albumin (BSA) as standard.

2-DE Separation

The extracted protein samples were purified using a 2D cleanup kit (GE Healthcare Biosciences Corp, 800 Centennial Ave, Piscataway, USA). Protein extracts (100 μg) were suspended in rehydration buffer containing 8 M urea, 2% CHAPS, 50 mM DTT, 0.2% Bio-Lyte 3/10 ampholyte, and 0.001% Bromophenol Blue (Bio-Rad, Hercules, CA, USA) to attain a final volume of 150 μL. The solutions were directly loaded onto a focusing tray. Iso-electric focusing (IEF) was performed using immobilized pH gradient strips (3–10 NL, 7 cm Bio-Rad, USA) and then actively rehydrated for 12 hours at 50v. IEF was performed using a Protean IEF cell system (Bio-Rad) under the following sequential conditions; 250 V for 15 min with a linear ramp, 4000 V for 1 hour with a linear ramp, and 4000 V at 12000 V/h with a rapid ramp at 20°C. After IEF separation, the strips were incubated in 2 ml of equilibrium I (6 M urea, 2% w/v SDS, 0.375 M Tris- HCl (pH 8.8), 20% v/v glycerol and 2% w/v DTT) for 15 min and then equilibrated again for 15 min in the same buffer described above but with DTT replaced by iodoacetamide (2.5% w/v). For second-dimension electrophoresis, the equilibrated strips were transferred to 12% SDS-polyacrylamide gels with 5% stacking gels sealed with 1% agarose. Electrophoresis (2D) was performed at 25 V for the first 30 min and then 50 V until the bromophenol blue dye reached the bottom of the gel.

Protein Visualization and gel-image Analysis

Gels obtained from 2-D PAGE were stained with Plus One Silver Staining Kit (GE Healthcare Biosciences AB, Uppsala, Sweden). The triplicate gels were scanned under HP Scanjet G 4010. The gel images were computationally analyzed using the Progenesis SameSpot software version 3.0 (Nonlinear Dynamics Ltd.). The intensity of all protein spots was normalized relative to the total abundance of all valid spots. After normalization and background subtraction, a matchset was created for both control gels (three replicates) and Cd-treated gels. To validate the automated spot detection and matching process, the images were edited manually and streaks, speckles, and artifacts were removed. Spot patterns of all gels were matched to each other to quantify each spot after normalization using the local regression model available in Progenesis SameSpots. The average intensities of resolved spots were compared using quantitative, qualitative and statistical functions within the Progenesis SameSpots software. Significant changes between spots were determined using Student's *t*-test for paired observations. Changes with a P value of < 0.05 were considered as being statistically significant.

In gel Digestion

To identify proteins separated by 2-DE, selected protein spots were manually excised from silver stained 2-DE gels. Then the gel slices washed with double distilled water were destained with 100 mM sodium thiosulfate and 30 mM potassium ferricyanide (1:1). The sample was then vortexed for 10 min, washed with distilled water for 3–5 times until completely destained and dehydrated for 10 min with 100% acetonitrile (ACN) and dried by vacuum centrifugation. After destaining, the gel pieces were reduced with 10 mM dithiothreitol in 100 mM NH_4HCO_3 for 1 hour at 56°C and again incubated with 55 mM iodoacetamide in 100 mM NH_4HCO_3 in the dark for 40 min. The gel slices were digested in 100 mM NH_4HCO_3 with 7–8 μL (0.1 μg/μL) trypsin enzyme (Promega Corporation, Madison, WI 53711–5399, USA) and incubated at 37°C for 16 hours. The tryptic peptides were extracted from the gel grains with 5% trifluoroacetic acid (TFA) in 50% acetonitrile 3 times. The solution containing eluted peptide was concentrated up to drying by vacuum centrifugation and the resultant extracts were analyzed by mass spectrometry.

Protein Identification by mass Spectrometry

Differentially expressed protein spots were identified by matrix assisted laser desorption/ionization time-of-flight tandem mass spectrometry (MALDI-TOF/TOF MS). The peptides were eluted with 1μL matrix solution (α-cyano-4-hydroxy-cinnamic acid in 5% TFA, 50% acetonitrile) before application to the target plate. Samples were allowed to be air dry and analyzed by 4700 MALDI-TOF-TOF analyzer (Applied Biosymtems, Foster City, CA, USA). All mass spectra were acquired in the reflection mode with 0–4000 m/z by a 4700 proteomics analyzer (Applied Bio-systems, Framingham, MA, USA). External calibration was performed using a standard peptide mixture of des-Arg bradykinin, angiotensin, Glufibrinopeptide B, adrenocorticotropic hormone (ACTH) clip 1–17, ACTH clip 18–39, and ACTH clip 7–38.

Bioinformatics analysis

The acquired MS and MS/MS spectra were used to identify differentially regulated proteins using Mascot Generic File (MGF) with an in-house licensed MASCOT search engine (Mascot v. 2.4.0, Matrix Science, London, UK) against the viridiplantae within the UniProt database. In the MASCOT search, carbamidomethylation of cysteines was set as a fixed modification and the oxidation of methionines was set as a variable modification. MASCOT was used with the monoisotopic mass selected, a peptide mass tolerance of 100 ppm, and a fragment iron mass tolerance of 2 Da. Trypsin was specified as the proteolytic enzyme with one potential missed cleavage. All proteins identified by high-scoring peptides were considered true matches, and at least two peptide matches. Protein hits were validated if the identification involved at least 10 top-ranking peptides with $p < 0.05$ and also selected false positive rate < 0.05. When those peptides matched multiple members of a protein family, the presented protein was selected based on the highest score with at least two peptide matching.

RESULTS

Morphological Responses of Sorghum Seedlings to Cd Stress

Our investigation was focused on studying the effects of exposing sorghum plants to different concentrations (0 µM, 100 µM and 150 µM) of $CdCl_2$. The metal ions affected growth parameters and caused physiological alterations. These are schematically described in Fig 1. To understand the initial morphological responses induced in the sorghum seedlings by Cd stress, the length, fresh weight and relative water content (RWC) of their roots and shoots were measured at 15 days after planting (Fig 2). The growth and characteristics of the sorghum seedlings were analyzed following treatment with Cd, and morphological characteristics were found to be reduced by Cd stress (Fig 2). The most significant growth inhibition was observed in plants treated with the highest concentrations of Cd^{2+} ions (150 µM).

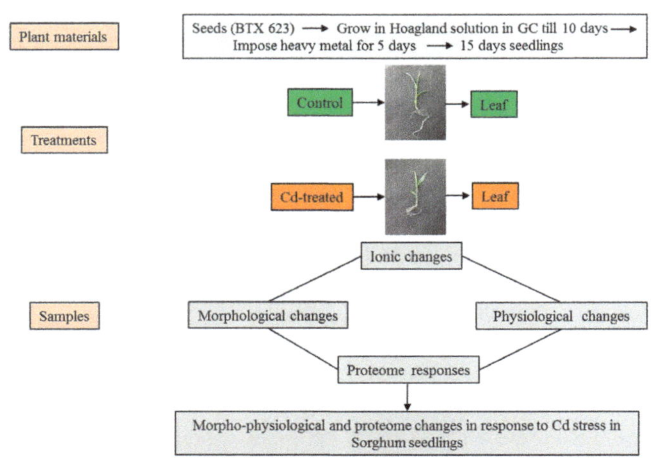

Figure 1: Schematic representation of the experimental setup used to compare Cd-treated sorghum seedlings with untreated plants (Control).

The sorghum seedlings (BTX 623) were grown in growth chambers for 10 days in Hoagland solution. Ten-day-old sorghum seedlings were exposed to different concentrations of cadmium (0 µM, 100 µM, 150 µM $CdCl_2$) for 5 days. The samples were collected from the control and Cd-induced leaves for measurement of morphological and physiological parameters (shoot lengths, fresh weights, relative water content, confocal and ion analysis). For proteome analysis, leaves were excised, pooled, rinsed with de-ionized water, rapidly frozen in liquid nitrogen, and stored at -80°C. Molecular changes were investigated in *S. bicolor* using two-dimensional gel electrophoresis (2-DE) in samples derived from the leaves of both control and Cd-treated seedlings.

http://dx.doi.org/10.1371/journal.pone.0150431.g001

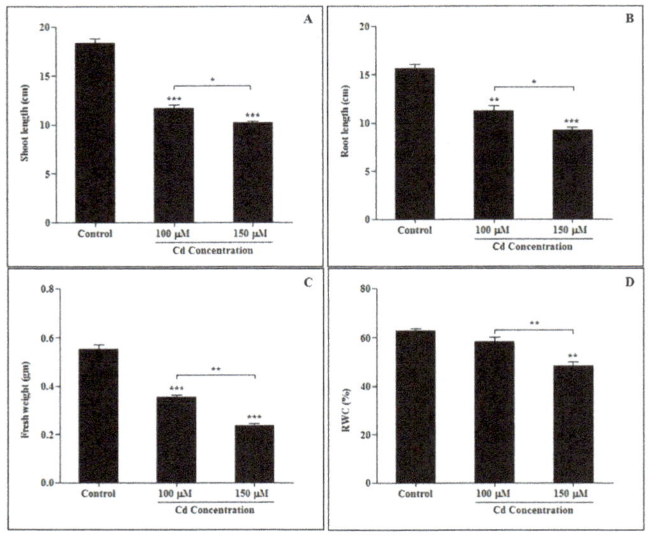

Figure 2: Responses induced by cadmium stress in morphological alterations in sorghum seedlings (A. Shoot length, B. Root length, C. Fresh weight, D. Relative water content) exposed to different concentrations of cadmium.

Ten-day-old sorghum seedlings (BTX 623) were exposed to different concentrations of cadmium (0 µM, 100 µM, 150 µM $CdCl_2$) for 5 days. After 5 days of Cd stress, the leaves and roots were collected from both the control and Cd-treated seedlings. Prior to measuring morphological and physiological parameters, the seedlings were washed with de-ionized water. Three plants were randomly selected for measurements at each time point for each replicate, and the experiment was biologically replicated 3 times. Each bar represents the average ± SE for 3 plants. Significant differences between the control and the cadmium-induced seedlings were determined by performing a one-way analysis of variance (ANOVA) with Tukey's all pairs of column comparison test. Asterisks indicate significant differences at $p < 0.05$.

http://dx.doi.org/10.1371/journal.pone.0150431.g002

Alterations in Accumulated Cadmium levels and ion Concentrations Following Cd Treatment

Cadmium content was determined in the leaves of the experimental plants. High accumulated levels of Cd were observed in leaves of plants that were treated with different concentrations of Cd (Fig 3A–3D). We found that the accumulated level of Cd was significantly increased in plants exposed to the high concentration of Cd (150 μM) compared to the untreated plants (Fig 3A). In the present investigation, Cd^{2+}, Zn^{2+}, Ca^{2+} and Fe^{2+} concentrations were determined in sorghum leaves. The concentrations of Zn^{2+} (Fig 3B) and Ca^{2+} (Fig 3C) ions were decreased, whereas the Fe^{2+} ion concentration was increased when seedling leaves were exposed to Cd^{2+}(Fig 3D). The concentration of cadmium in the plants was significantly increased by exposure to higher concentrations of cadmium.

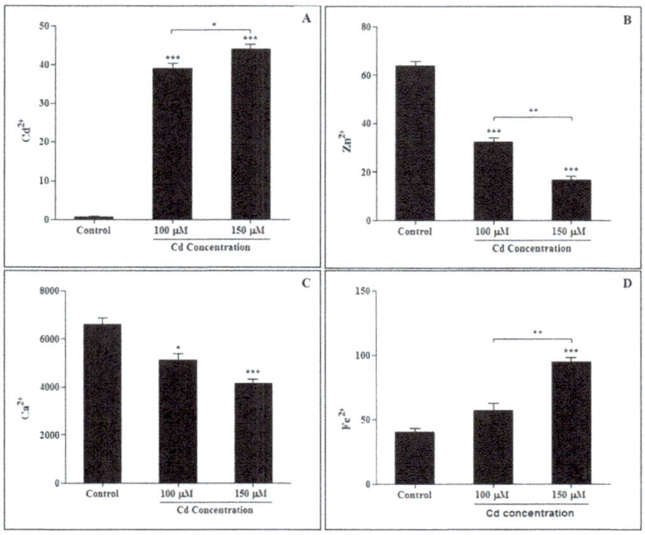

Figure 3: Ion concentrations and cadmium accumulation in sorghum leaves (A. Cd^{2+}, B. Zn^{2+}, C. Ca^{2+}, and D. Fe^{2+}) exposed to cadmium stress.

Ten-day-old sorghum seedlings (BTX 623) were exposed to different concentrations of cadmium (0 μM, 100 μM, 150 μM $CdCl_2$) for 5 days. Values (means ± SD) were determined for 3 independent experiments (n = 3).

http://dx.doi.org/10.1371/journal.pone.0150431.g003

Effects of Absorbed Cd ions as Analyzed using Confocal Microscopy

Confocal microscopy was used to analyze the distribution of Cd absorption in sorghum leaves. Each of the sorghum leaves was cut, stained with dithizone staining solution and then observed using confocal microscopy. Cadmium-dithizone complexes and fluorescence intensity were measured using confocal microscopy. The distribution of cadmium in the sorghum leaves, which was revealed by dithizone staining, is shown in Fig 4A–4D. We found that the degree of absorption of cadmium was increased by its concentration, as was the fluorescence intensity of Cd-dithizone (Fig 4D).

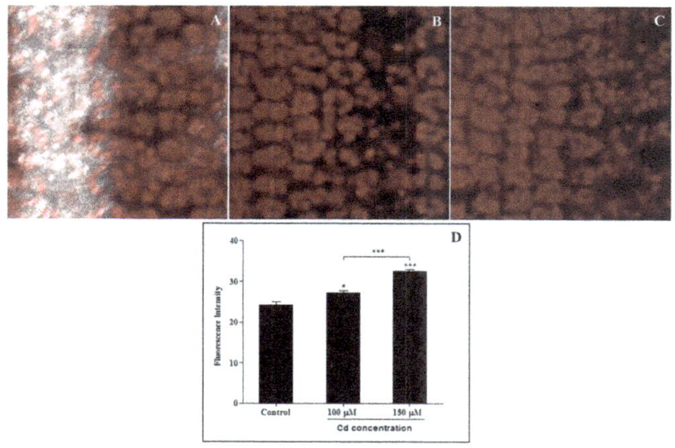

Figure 4: Cadmium distribution and fluorescence intensities in sorghum seedling leaves treated with dithizone staining.

A. Control, B. 100 μM $CdCl_2$, C. 150 μM $CdCl_2$, D. Effect on the fluorescence intensity of cadmium-dithizone complexes. Samples obtained from different cadmium-treated leaves were placed on slides, mounted in mounting solution and observed using confocal microscopy (LSM 410; Carl Zeiss, Jena, Germany).

http://dx.doi.org/10.1371/journal.pone.0150431.g004

Changes in Proteomic Expression Patterns in Sorghum Seedling Leaves in Response to Cd

To explore the underlying mechanisms involved in *S. bicolor* tolerance to Cd stress, the leaf proteome of *S. bicolor* samples was evaluated using two-dimensional electrophoresis. Total soluble proteins were extracted from the leaves of control (0 μmol $CdCl_2$) and Cd-treated (100 μmol and 150 μmol $CdCl_2$) sorghum seedlings. Each experiment was replicated three times, and more than 800 protein spots were reproducibly detected within each sample after silver staining based on our analysis of Progenesis SameSpot software. Quantitative image analysis revealed a total of 33 protein spots that exhibited more than a 1.5-fold changes in intensity (Fig 5). All spots of interests were located, manually excised, and subjected to in-gel tryptic digestion, and the 33 differentially expressed proteins were successfully analyzed using MALDI-TOF/TOF MS analysis. Among these proteins, a total of 15 proteins showed increased expression, and 18 proteins showed decreased expression in the treated samples compared to their levels in untreated seedlings (Fig 6). An enlarged image of a number of the identified gel spots is shown to illustrate the changes observed in the differentially expressed protein spots among the groups (Fig 7). The identified proteins are listed in Table 1.

Ten-day-old sorghum seedlings were exposed to cadmium stress for 5 days. Samples from non-treated control plants and treated plants were collected on the same day. Differences in the intensity of labeling associated with the proteins in both the control and the cadmium-treated samples are shown as clusters. Any statistically significant difference ($p < 0.05$) in labeling intensity was considered to be positive. The protein spot numbers are indicated on the right side of the cluster. The clusters were determined using Genesis software (ver. 1.7.6).

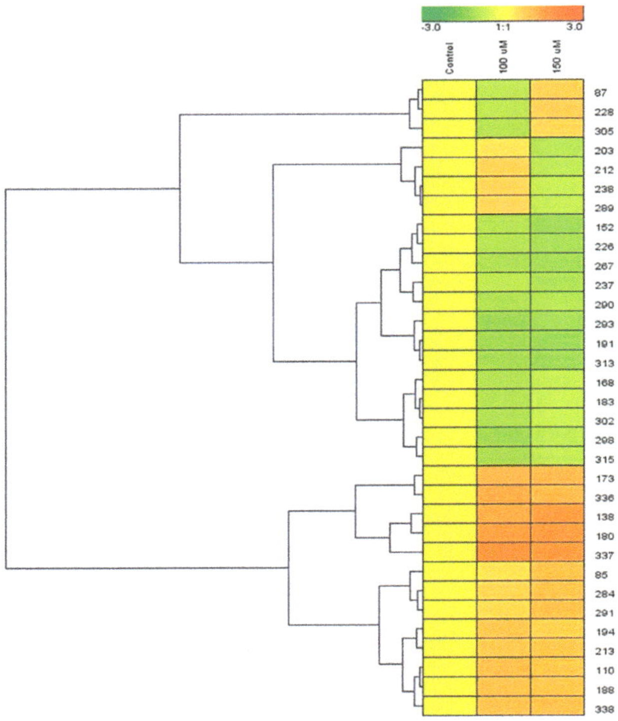

Figure 5: Cluster analysis of cadmium-responsive leaf proteins.

http://dx.doi.org/10.1371/journal.pone.0150431.g005

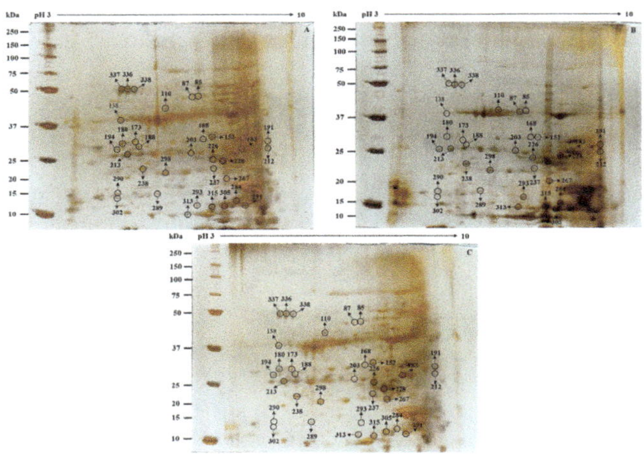

Figure 6. Representative images of gels used in 2-DE analysis of *S. bicolor* **leaves exposed to 0 μM, 100 μM and 150 μM $CdCl_2$.**

Leaf tissues were extracted using TCA-acetone precipitation method as described in the Materials and Methods section. Proteins were extracted from the leaves of 15-day-old seedlings that were treated with cadmium for 5 days. For IEF, 100 μg of proteins was loaded onto pH 3–10 NL IPG strips (7 cm). SDS-PAGE was performed on 12% gels, and the proteins were separated using 2-DE and then stained with silver staining. The differentially expressed protein spots (>1.5-fold difference) are indicated by circles on the 2-D gel map. These spots were found to be statistically significant at a level of 95% per group (Student's *t*-test) using biological and analytical replicates (n = 3). The MW of each protein was determined using standard protein markers.

http://dx.doi.org/10.1371/journal.pone.0150431.g006

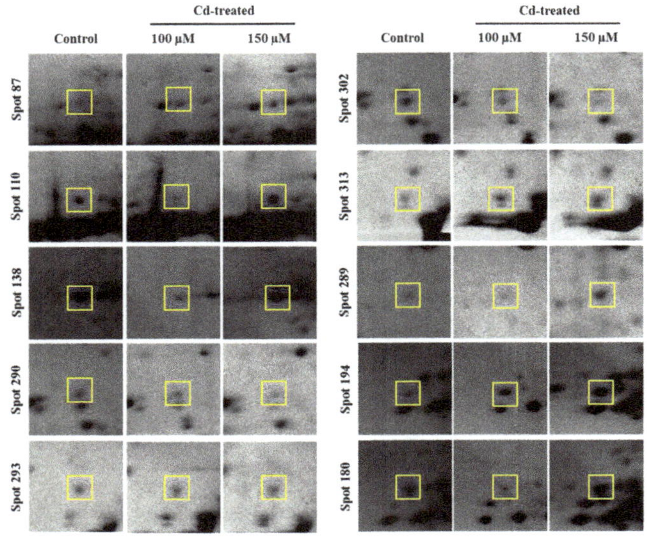

Figure 7: Magnified views of some of the differentially expressed protein spots that correspond to the identified proteins.

The protein spots that were identified using mass spectrometry are indicated by squares and labeled in the figure.

http://dx.doi.org/10.1371/journal.pone.0150431.g007

Table 1: List of differentially identified proteins altered by Cd stress in the leaves of *S. bicolor* **using** MALDI-TOF-TOF mass spectrometry

Spot Number[a]	Accession Number[b]	Protein Description	Protein Score[c]	Molecular Weight[d] (kDa)	Protein Matches[e]	Protein Coverage[f] (%)	Fold Change[g]	
							100 μM/ Control	150 μM/ Control
		Translation						
85	RR2_OSTTA	30S ribosomal protein S2	48	25468	3	17	1.04↑	1.44↓
203	RPOA_PANGI	DNA-directed RNA polymerase subunit alpha	32	40481	2	18	1.07↑	1.47↓
213	RR3_ADICA	30S ribosomal protein S3	35	24952	3	28	1.71↑	1.33↓
238	RK5_MESVI	50S ribosomal protein L5, chloroplastic	32	20808	2	43	1.45↑	1.11↓
		Signal Transduction						
87	CDPK1_ORYSJ	Calcium-dependent protein kinase isoform 1	56	61136	5	34	1.08↓	1.19↑
		Transcription						
152	MATK_DIOEL	Maturase K	32	62406	2	15	1.25↓	1.89↓
183	MATK_NANDO	Maturase K	43	60828	3	25	1.33↓	1.06↓
194	MATK_ARCUU	Maturase K	36	59873	3	23	1.76↓	1.12↓
337	DRB1_ORYSJ	Double-stranded RNA-binding protein 1	37	47483	3	27	3.42↑	2.57↑
		Transport						
110	PPI1_ARATH	Proton pump-interactor 1	46	68951	3	16	1.98↑	1.32↑
168	OEP37_ARATH	Outer envelope pore protein 37	51	38982	3	20	1.32↓	1.16↓
290	HMA2_ARATH	Cadmium/zinc-transporting ATPase	34	103806	5	14	1.35↓	1.47↓
293	PCR7_ARATH	Protein PLANT CADMIUM RESISTANCE 7	38	15620	4	19	1.79↓	1.79↓
		Transcriptional regulation						
188	PP193_ARATH	Pentatricopeptide repeat-containing protein At2g38420	31	52248	2	57	1.89↑	1.36↑
191	C3H14_ARATH	Zinc finger CCCH domain-containing protein 14	37	35656	2	45	1.67↑	1.92↑
		Cell wall reorganization						
237	CADH2_ARATH	Cinnamyl alcohol dehydrogenase 2	68	41339	2	21	1.03↓	1.61↓
		Stress response						
173	C71AO_ARATH	Cytochrome P450	49	56426	4	25	2.29↑	1.83↑
180	C75A2_SOLME	Flavonoid 3',5'-hydroxylase	63	57974	6	39	2.97↑	2.53↑
228	GSTUO_ARATH	Glutathione S-transferase	61	25461	2	32	1.04↓	1.31↑
		Photosynthesis						
305	PSBQ2_ARATH	Oxygen-evolving enhancer protein 3–2	75	24628	3	13	1.27↓	1.07↓
302	RBS1_ORYSI	Ribulose bisphosphate carboxylase small chain	59	19862	3	17	1.43↓	1.03↓
		Metabolism						
Cellular metabolism								
212	CCNB2_MEDSA	G2/mitotic-specific cyclin-2	43	38142	6	47	1.48↑	1.32↑
Energy and carbohydrate metabolism								
226	ATPA_CERDE	ATP synthase subunit alpha	53	55230	2	24	1.30↓	1.75↓
Alcohol metabolic process								
267	ERG14_ARATH	Squalene epoxidase 1	35	58878	3	36	1.47↓	1.82↓
Carbohydrate metabolism								
313	SGGP_ARATH	Haloacid dehalogenase	82	26886	2	29	1.61↓	1.89↓
Photosynthesis and carbohydrate metabolism								
315	CAHC_SPIOL	Carbonic anhydrase, chloroplastic	39	34947	2	19	1.52↓	1.27↓
Glycolysis and carbohydrate metabolism								
338	G3P_ATRNU	Glyceraldehyde-3-phosphate dehydrogenase	49	39415	2	32	1.91↑	1.53↑
Redox homeostasis and defense response								
336	AAT3_ARATH	Aspartate aminotransferase 3, chloroplastic	38	49152	3	14	2.54↑	1.67↑
		Unknown						
138	BRXL1_ARATH	Protein Brevis radix-like 1	67	38152	5	24	2.55↑	2.49↑
284	BMT_GLELI	Bergaptol O-methyltransferase	46	39815	4	31	1.36↑	1.41↑
289	CAPZB_ARATH	Probable F-actin-capping protein subunit beta	42	29200	6	31	1.45↑	1.14↑
291	CX5B2_ARATH	Cytochrome c oxidase subunit 5b-2	37	18856	5	15	1.23↑	1.49↑
298	SLSG0_BRAOA	S-locus-specific glycoprotein	45	51010	2	18	1.72↓	1.30↓

[a] represents the spot number on the 2-DE gels shown in Fig 6

[b] accession number according to the UniProt database

[c] Score of the identified peptides

[d] molecular weight of the identified protein

[e] number of matched peptide, the proteins with >2 matched peptides were considered

[f] protein coverage, the proteins with less than 10% sequence coverage was excluded from the result

[g] fold change of protein quantities in cadmium treated sample against control samples

The symbol ↑ denotes the proteins were up-regulated and ↓ denotes the proteins were down-regulated.

doi:10.1371/journal.pone.0150431.t001

http://dx.doi.org/10.1371/journal.pone.0150431.t001

Functional Classification of Identified Proteins

To increase our understanding of the roles of the proteins involved in Cd stress responses, the identified proteins were categorized into different groups. The 33 differentially expressed protein spots were analyzed using MALDI-TOF/TOF MS and then annotated and classified into functional categories. Gene ontology categories were assigned to all 33 proteins according to their molecular function, cellular component localization and biological processes (Fig 8A–8C). Based on their molecular functions, the proteins were classified into 12 categories. However, among all of the identified proteins, the molecular functional group corresponding to nucleic acid-binding proteins contained the largest number of differentially expressed proteins (Fig 8A), followed by the group containing transferase activity, protein-binding, structural molecule activity and monoxygenase activity-related proteins. Regarding the cellular localization, the identified proteins were classified into 9 categories. Most of the proteins were localized into chloroplasts, followed by membranes, mitochondrion, nucleus, ribosome, chloroplast thylakoid membrane and the cytoplasm (Fig 8B). The proteins were grouped into 11 biological process categories, and most of the proteins were found to be involved in metabolism, followed by translation, transcription and transport (Fig 8C).

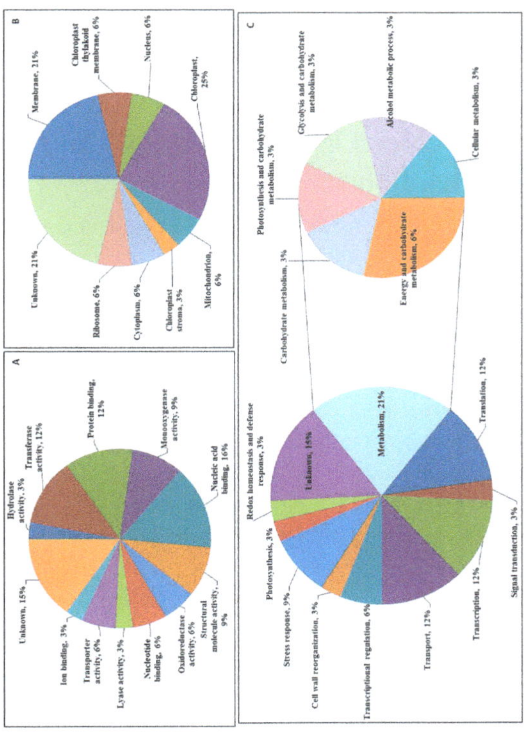

Figure 8: Protein encoding gene functions of 33 differentially expressed proteins identified in the leaves of *S. bicolor*.

The frequency distribution for the identified proteins within functional categories was determined based on their molecular functions (A), cellular localization (B), and their involvement in biological processes (C). Classifications were made using iProClass databases, and the assignment of functions was based on gene ontology.

http://dx.doi.org/10.1371/journal.pone.0150431.g008

Putative Subcellular Localization Model

In the present study, most of the proteins identified in the leaves of sorghum seedlings exposed to Cd stress were expressed in chloroplasts and membranes (Fig 9). Among the 33 differentially expressed proteins proteins, 11 have been observed to be involved in the chloroplast (8 proteins), chloroplast thylakoid membrane (2 proteins) and chloroplast stroma (1 protein).

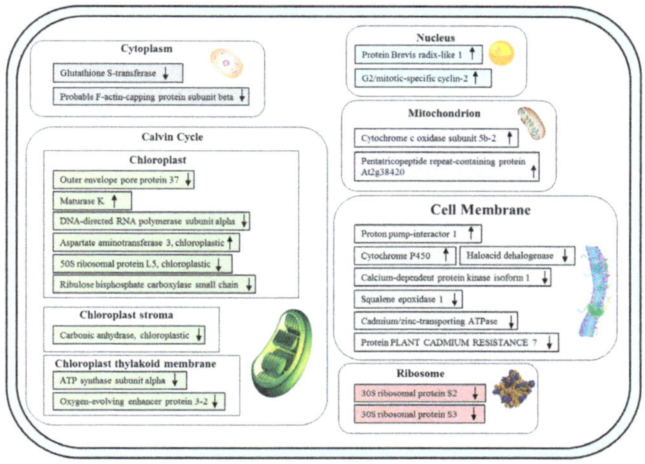

Figure 9: A model of the putative subcellular localization of the leaf proteins identified in *S. bicolor* **seedlings to be affected by exposure to Cd.**

The Cd-responsive proteins are indicated as follows: up-regulated proteins are indicated by "↑", and down-regulated proteins are indicated by "↓". The proteins were categorized with regard to their localization in cellular components using iProClass databases, and the assignment of functions was based on gene ontology.

http://dx.doi.org/10.1371/journal.pone.0150431.g009

Next to the chloroplast, the most common localization was the membrane, with which 7 of the differentially expressed proteins are associated. These data reveal that most of the identified proteins observed to be differentially expressed were decreased in abundance under Cd stress, and the analysis of the subcellular localization of the differentially expressed proteins suggests that the chloroplast and the membrane are very sensitive to Cd stress (Fig 9).

DISCUSSION

Morphological Responses and Changes in Response to Cadmium Accumulation in Sorghum Seedlings Under Cd Stress

Among the many types of environmental stresses, heavy metals severely affect plant growth and development. Many studies have investigated the limiting effect of Cd on fresh and dry mass accumulation, height, root length, leaf area, and other biometric parameters in plants [52,53,54].

In the present study, to investigate the mechanism by which sorghum responds to Cd stress under hydroponic culture conditions, 10-day-old plants were transplanted and cultured in Hoagland solution containing different concentrations of Cd. This morphological investigation was focused on studying the effects of Cd on sorghum plants treated with different concentrations (0 μM, 100 μM and 150 μM $CdCl_2$) of Cd. The leaves of the sorghum seedlings exhibited different accumulated effects when exposed to Cd.

A small number of studies have shown that Cd is a non-essential toxic heavy metal that causes physiological and morphological changes in plants [55]. Cadmium levels were determined in the leaves of the experimental plants in this study. A high level of accumulated Cd was observed in leaves when plants was treated with Cd. We found that the accumulated level of Cd was significantly increased in plants exposed to the highest concentration of Cd (150 μM) compared to untreated plants. In an earlier study in maize, it was shown that the highest level of Cd was obtained in plants exposed to the highest concentration of the combined effect of Cd and Zn ions [56].

Cadmium hyper accumulation has been described in many plant species in previous studies [57,58,59] including sorghum [60]. When *Phytolacca americana* plants were exposed to Cd, the leaves were severely affected by Cd, and metal ions were significantly accumulated at high concentrations of Cd [41].

Ion Concentration Responses to Cd Stress

When plants were exposed to a high concentration of metal, their apoplastic levels altered the aqueous and ionic thermodynamic equilibria; as a result, these plants faced hyperosmotic stress, ionic imbalance, and toxicity. In the present study, the interacting ions (Zn^{2+}, Ca^{2+} and Fe^{2+}) were significantly influenced by treatment with cadmium. However, an angonistic effects was observed between Cd and Zn when tobacco seedlings were subjected to Cd stress [61]. In addition, it was also previously reported that Cd significantly decrease Zn concentrations in plant tissues [62]. To get more insights into the role of Cd-induced Ca changes, the Ca ion was investigated in the present study. Ca ions were decreased significantly in response to Cd exposure. In previous work, it was revealed that Cd induced a significant reduction in the Ca content in the leaves of pea plants [63].

Effects of Absorbed Cd ions Analyzed Using Confocal Microscopy

Confocal microscopy was used to observe the distribution of the Cd that was absorbed into sorghum leaves. Dithizone is a sulfur-containing organic compound. It is a good ligand, and it forms complexes with many metals, including Cd [64]. The distribution of cadmium in sorghum leaves was investigated using dithizone staining in the present study. We found that the degree of absorption of cadmium was dose-dependant. The complex of dithizone with Cd and Pb shows as red, as previously described [65]. The fluorescence intensity of the Cd-dithizone complexes was dramatically increased in our investigation. Confocal analysis in a previous study indicated that the fluorescence intensity of similar complexes was substantially increased in the roots of wheat plants that were exposed to aluminum [49].

Proteome Changes in Plants Exposed to Heavy Metal Stress

Heavy metals such as Cu, Zn, Fe, Co, Mn and Mo act as beneficial mineral element present in small quantities [66], but these metals can be toxic when present at higher concentrations [67]. Some other heavy metals that have no biological function as nutrients and that are also very toxic can restrict plant growth and development. For example, cadmium and other heavy metals such as Pb, Hg induced the production of ROS and the inactivation of enzymes [68].

The results of the present study provide information regarding the stress responses exhibited by actively growing sorghum plants placed in a controlled environment that was coupled with a hydroponic system and exposed to Cd. The present experiments were executed to investigate the molecular mechanism and, more precisely, the alterations in the proteome induced in sorghum plants under Cd stress. These proteomic data, in combination with morphological and physiological analyses provide insights into the mechanisms affected by Cd stress in hydroponically grown sorghum plants.

In the present study, proteomic analysis of the responses induced in sorghum seedling leaves by Cd stress led to the identification of 33 differentially expressed proteins. Some of the identified proteins such as glutathione S-transferase, ribulose bisphosphate carboxylase small chain [69], carbonic anhydrase, glyceraldehyde-3-phosphate dehydrogenase [41], cytochrome P450 [70] have been well characterized in response to Cd stress, while others have not been as well-studied with respect to their roles in stress responses in plants.

Proteins Involved in Photosynthesis and Carbohydrate Metabolism

Cadmium has been shown to induce several alterations and disturbances in physiological processes [71] and to result in the dramatic inhibition of a wide range of metabolic process, including photosynthesis, carbon assimilation and photosynthetic electron transport [72]. In the present study, carbonic anhydrase (Table 1, spot number 315) was reduced by cadmium treatment. This differentially expressed protein is involved in photosynthesis. Carbonic anhydrase (CA) could alter photosystem components and consequently accelerates the deregulation of CO_2 homeostasis, which would, in turn, promote CO_2 depletion and the accumulation of O_2 when leaves are subjected to high doses of Cd [28]. A significant decrease in the abundance of CA was also reported in *Phytolacca americana* [41] and in poplar [69] treated with cadmium. In addition, the down-regulation of CA was also observed in poplar plants exposed to cadmium stress [73], and two isoforms of CA were recently identified to be present at lower levels in Cd-treated plants [74]. Nevertheless, CA is thought to be an important enzyme in many biological functions, especially those that involve carboxylation/decarboxylation [75]. However, several enzymes involved in carbon fixation have also been observed to decrease in abundance, as was previously shown in a study of rice leaves [76].

Proteins Involved in Glycolysis and Carbohydrate Metabolism

Cadmium toxicity may significantly alter the glycolytic pathway and the Cd-induced disorganization of the photosynthetic apparatus, and these effects may have an important impact on the plant's ability to withstand this type of stress. In the present study, spot number 338 was identified as glyceraldehyde-3-phosphate dehydrogenase (GADPH). GADPH is known as an essential enzyme that catalyzes the sixth step of glycolysis, and it assists in breaking down glucose to obtain energy and carbon molecules. The key enzyme in the glycolysis process, GADPH has been observed to be increased in abundance in poplar leaves [73]. The GADPH levels were increased when *A. thaliana* plants were exposed to 10 μM Cd exposure [77], and its level were also increased when *A. thaliana* cells were exposed to different concentrations of Cd [78]. On the contrary, GADPH levels were decreased in the roots of two Cd-tolerant plants, poplar [69] and *B. juncea* [79], after treatment with 20 μM and 250 μM Cd respectively. GADPH was observed to be increased following treatment with both low (10 μM) and high (100 μM) levels of Cd treatment compared to its level in controls in tomato plant roots [80]. However, previous studies of the alterations observed in carbohydrate metabolism-related proteins following exposure to Cd have demonstrated contradictory findings.

Previous research has shown that the metabolic changes induced by Cd are tissue-specific. Consequently, GADPH is induced in leaf tissue but severely decreased its abundance in root tissues in poplar plants [69]. GADPH was also induced in the leaves of rice [31] and poplar plants [74] by treatment with various heavy metals. These results suggest that GADPH protein may play an active role in supplying energy to Cd-treated plants via the glycolytic pathway. Taken together, the previous reports together with the present study, indicate the changes in carbohydrate metabolism upon Cd exposure are dose dependent and plants elevate their energy consumption over energy production when it exposed to Cd stress.

Proteins Involved in Energy and Carbohydrate Metabolism

The results of proteomic analyses have suggested that Cd has a strongly negative impact on proteins related to carbohydrate metabolism [35] and thereby, may limit the efficiency of CO_2fixation and reduce ATP synthesis, resulting in the inhibition of plant growth [81]. In this study, a total of three proteins involved in carbohydrate metabolism were found to be down-regulated by Cd stress, including ATP synthase subunit alpha (Table 1, spot no. 226), ribulose bisphosphate carboxylase (Rubisco) small chain (Table 1, spot no. 302) and haloacid dehalogenase (Table 1, spot no. 313). ATP synthase was observed to be markedly decreased compared to its level in the control plants. Changes in the levels of this protein may cause a decrease in ATP production in response to Cd toxicity [28].

Cellular Metabolism

One protein that was up-regulated by Cd treatment was identified as G2/mitotic-specific cyclin-2 (Table 1, spot no. 212). This protein is involved in cellular metabolism, and it was also previously found to be increased in the leaves of *A. paniculata* by 10 μM Cd [40], suggesting that this protein may play potential role in the regulation of cell growth.

Proteins involved in Photosynthesis

Photosynthesis is an essential metabolic process in plants, and it is also vulnerable to environmental stress. Elevated concentrations of Cd may cause deleterious effects and disturbances in transpiration and photosynthesis [82]. In our investigation, a protein called oxygen-evolving enhancer protein 3–2 (Table 1, spot no. 305) was decreased in abundance in plants exposed to Cd. Changes in the levels of this protein have also been observed in several other studies that showed that the photosystem II complex was greatly suppressed by Cd stress [37].

In addition, Cd harmfully affects protein expression related to primary carbon metabolism in plants and consequently result in the photosynthetic machinery being severely hampered by Cd stress via the degradation of major photosynthesis related proteins, such as Rubisco [37]. The amount of Rubisco (large and small subunits) was dramatically reduced by Cd stress in rice leaves [76] and algae [35]. In addition, there is little evidence for a relationship between stress and haloacid dehydrogenase (HAD). However, HAD has previously been found to be down-regulated in the marine brown algae, *Sargassum fusiforme* [83] by chronic copper stress. Taken together, the results obtained in the present study suggest that Cd may impair ATP production, limit primary carbon metabolism, hamper photosynthetic machineries and inhibit plant growth and development.

Proteins Involved in Stress Response

Three proteins cytochrome P450 (Table 1, spot 173), flavonoid 3', 5'-hydroxylase (Table 1, spot no. 180) and glutathione-S-transferase (GST) protein (Table 1, spot no. 228) were expressed at markedly higher levels in Cd-treated sorghum seedling leaves. Flavonoid 3', 5'-hydroxylase (F3′,5′Hs) belongs to the cytochrome P450 protein family [84]. However, cytochrome P450 (CYPs) is a stress-related protein that belong to the superfamily of proteins containing a heme cofactor. CYPs are also detoxifying enzymes that are mostly found in bacteria, archaea and eukayotes, where they are involved in protection against oxidative stress [85]. Previous evidence has indicated that a putative cytochrome P450 protein was also induced in *Physcia adscendens* [70] in response to short Cd treatments, and a pronounced inhibition of this enzyme was observed when plants were subjected to long exposure to the metal [86,87]. Microarray analysis in rice exposed to cadmium stress demonstrated that several genes encoding cytochrome P450 family proteins and other stress-related proteins were differentially expressed and that stress-related proteins were strongly induced by Cd exposure [88].

In this work, glutathione-S-transferase (GST) (Table 1, spot no. 228) was identified on 2-D gels. The level of GST was found to be increased up to 1.76-fold by Cd treatment (Table 1). GSTs, also known as ligandins, belong to a wide range of protein families [89]. GSTs catalyze the transfer of GSH to a wide variety of hydrophobic, electrophilic and cytotoxic co-substrates [90]. GST proteins are involved in sulfur and GSH metabolism. The level of GST was previously found to be increased in various plant species that were exposed to Cd stress [37,41,78]. However, three isozymes of GST identified in *Schizosaccharomyces pombe*, including GST-I, GST-II and GST-III, were found to be significantly increased by exposure to Cd stress [91]. The levels of GSTs were found to be increased in response to Cd stress, supporting the suggested roles of GST in metal detoxification [92]. Previous results also indicated that these isozymes play an active role in the detoxification of many xenobiotic compounds and they protect cells from oxidative stress [93]. In *Populus tremula*, the overall level of GST activity was higher in roots than in leaves [69]. Moreover, the up-regulation of GSTs levels in the presence of Cd^{2+} was also observed in an earlier investigation [77] in the roots of *Arabidopsis thaliana*.

The increased abundance of GSTs have been demonstrated in other heavy metal stress that involved in antioxidant defense and detoxification systems. Four GSTs (*At*GSTF2, *At*GSTF6,*At*GSTF7 and *At*GSTU19) were also identified to be significantly abundant in copper-treated*Arabidopsis thaliana* seedlings [94]. The abundance of GST enzyme has been increased markedly in both leaves and roots of cu-stressed wheat seedlings [95].

Proteins Involved in Transcriptional Regulation

Heavy metals limit normal plant growth and development and regulate a wide range of genes, resulting in adverse effects on many cellular responses [96]. Regulatory proteins, including pentatricopeptide repeat-containing protein (spot no. 188) and zinc finger CCCH domain-containing protein 14 (Table 1, spot no. 191) were markedly up-regulated by cadmium stress in our investigation. Pentatricopeptide repeat-containing proteins (PPR) are a family of proteins generally found in the plant kingdom, and these predicted proteins are predicted to be targeted to either mitochondria or chloroplasts [97]. Previous report have indicated that PPR protein was significantly induced in the leaves [98] and germinating seedlings of rice [99] exposed to Cd stress. However, the most interesting response to Cd exposure in rice seedling roots was a dramatic accumulation of PPR protein following exposure to low concentrations of Cd and a decrease in its abundance following exposure to a high concentration of Cd [100]. Taken together, the results obtained from these two investigations revealed that these proteins play substantial roles in RNA processing under stress conditions. An increase in the abundance of PPR protein has frequently been observed in strawberries [101] and manchurian ash [102]. These data also suggest that the PPR proteins identified in these organisms are substantially involved in the regulation of transcriptional activity.

Proteins Involved in Transcription

Maturase K (Mat K) proteins (Table 1, spots 152, 183, 194) and double-stranded RNA-binding protein 1 (Table 1, spot no. 337) were identified in our study to show decreased expression during exposure to Cd stress. These proteins are involved in RNA processing, and they exhibited differing expression patterns when seedling leaves were exposed to Cd stress. In previous studies, these proteins have also been observed to be increased in abundance under stress conditions in the leaves of higher plants [103,104]. Maturase K catalyzes intron RNA binding and consequently alters gene expression at the transcriptional level [105]. However, the differential expression of these proteins has been reported in previous studies [32], suggesting that growth and development were also strongly affected by exposure to Cd. Hence, these results suggest that the changes observed in the protein expression of maturase K in plants are species- and stress-type dependent [106].

Proteins Involved in Translation

In the present investigation, proteins were observed that are involved in translation, including 30S ribosomal protein S2 (Table 1, spot no. 85), 30S ribosomal protein S3 (Table 1, spot no. 213), DNA-directed RNA polymerase subunit alpha (Table 1, spot no. 203) and 50S ribosomal protein L5, chloroplastic (Table 1, spot no. 238). Our results show that two proteins (spot no. 85 and spot no. 213) were up-regulated in plants exposed to 100 μM Cd levels but down-regulated when plants were exposed to 150 μM Cd levels. Another protein (Table 1, spot no. 337) was also up-regulated when seedlings were exposed to Cd stress. However, two 30S ribosomal proteins were detected in the leaves of poplar plants [74] that were exposed to Cd stress, and the decreased abundance of these two proteins might indicate that protein synthesis and metabolism are impaired under Cd stress.

In marine cyanobacteria, six subunits of ribosomal proteins were identified, including the 30S ribosomal protein S2. The differential expression of these ribosomal proteins suggests that Cd had a substantial effect on the the character of the proteome [107]. In addition, many ribosomal proteins have been observed to be up-regulated by short-term exposure to Cd, but then later down-regulated in *Schizosaccharomyces pombe* [91]. Moreover, the down-regulation of ribosomal proteins in *Brassica napus* showed that Cd induced toxicity that affected the levels of regulation/protein synthesis-related proteins [108].

Cell wall Reorganization

Cell walls may play an active role in metal tolerance, accumulation and metal binding. Cell wall reorganization has been reported previously in heavy metal toxicity, including Cd [109]. Exposing the leaves of sorghum seedlings to cadmium stress triggered a reorganization of cell walls at both Cd concentrations, as revealed by several changes in protein levels. In the present study, cinnamyl alcohol dehydrogenase 2 (CAD) (Table 1, spot no. 237) was detected, and this protein was markedly decreased by exposure to Cd. Interestingly, several proteins related to cell wall organization were down-regulated by Cd toxicity, but, CAD was found to be particularly increased in abundance in the roots of tomato [80] and rice [110] plants by Cd stress. Hence, the obtained protein (CAD) from the present investigation, related to cell wall organization was down-regulated suggesting that the cell wall is very sensitive to Cd stress.

Proteins Involved in Transport

The down-regulation of cadmium/zinc-transporting ATPase (Table 1, spot no. 290) was identified in our study. The amount of this transporter protein was significantly reduced under stress conditions in cotton [111]. A previous study in *Arabidopsis thaliana* [112] suggested that the mechanism for Cd root-to-shoot translocation relies on the activity of cadmium/zinc-transporting ATPase (HMA 2 and HMA 4). However, plant cadmium resistance 7 (Table 1, spot no. 293) was identified in the present study to be involved in the transport of proteins. However, a previous study observed that this protein mediates cadmium resistance in*Arabidopsis* [113].

Proteins Involved in Redox Homeostasis and Defense Responses

In the present study, aspartate aminotransferase 3, chloroplastic (Table 1, spot no. 336) was identified to be up-regulation by Cd stress. This protein was observed to be increased by exposure to the redox active heavy metal Cr, and has also been shown to be involved in ROX detoxification and defense responses [114].

CONCLUSIONS

The heavy metal Cd is a global problem that seriously limits crops quality and production. Metal uptake, metal accumulation and growth responses that were associated with changes in the proteome in Cd-treated plants have not been characterized in sorghum. By examining the leaves of sorghum seedlings, the present study sheds light on the molecular mechanisms involved in Cd-tolerance in *S. bicolor* and suggests a more active involvement for Cd toxicity in plant morpho-physiological and molecular processes. Cadmium stress causes morpho-physiological alterations and affects the shoot and root metabolic systems. In the ionic study, the concentration of cadmium was significantly increased in the plants that were treated with the highest concentration of Cd. Proteomic analyses revealed Cd-dependant alterations in metabolic processes and in translational and transcriptional regulation-related proteins. The results from these studies were interpreted to demonstrate that Cd stress induce the general inhibition of carbon fixation, impairs ATP production and regulates protein synthesis. In addition, it is tempting to further clarify the role of GST enzymes in C_4 plant cells in responses to Cd toxicity. These findings improve our understanding of the mechanisms that are potentially involved in plant responses to Cd stress at the protein level. Further research is required for a better understanding of cellular and molecular responses to cadmium.

ACKNOWLEDGMENTS

Seeds of *S. bicolor* (BTX 623-Inbred line) were kindly provided by the National Germplasm Resources of USDA-ARS, plant stress and germplasm development unit, USA. The confocal and ICP analysis are supported by the Centre for Research Facilities, Chungbuk National University, Korea.

REFERENCES

1. Anjum NA, Ahmad I, Mohmood I, Pacheco M, Duarte AC, Pereira E, et al. Modulation of glutathione and its related enzymes in plants' responses to toxic metals and metalloids-a review. Environmental and Experimental Botany. 2012;75:307–24. doi: 10. 1016/j. envexpbot. 2011. 07. 002
2. DalCorso G, Farinati S, Maistri S, Furini A. How plants cope with cadmium: staking all on metabolism and gene expression. Journal of Integrative Plant Biology. 2008;50(10):1268–80. doi: 10. 1111/j. 1744-7909. 2008. 00737. x. pmid:19017114
3. Hossain MA, Hasanuzzaman M, Fujita M. Up-regulation of antioxidant and glyoxalase systems by exogenous glycinebetaine and proline in mung bean confer tolerance to cadmium stress. Physiology and Molecular Biology of Plants. 2010;16(3):259–72. doi: 10. 1007/s12298-010-0028-4. pmid:23572976
4. Hossain MA, Hossain MZ, Fujita M. Stress-induced changes of methylglyoxal level and glyoxalase I activity in pumpkin seedlings and cDNA cloning of glyoxalase I gene. Australian Journal of Crop Science. 2009;3(2):53–64.
5. Rascio N, Navari-Izzo F. Heavy metal hyperaccumulating plants: how and why do they do it? And what makes them so interesting? Plant Science. 2011;180(2):169–81. doi: 10. 1016/j. plantsci. 2010. 08. 016. pmid:21421358
6. Villiers F, Ducruix C, Hugouvieux V, Jarno N, Ezan E, Garin J, et al. Investigating the plant response to cadmium exposure by proteomic and metabolomic approaches. Proteomics. 2011;11(9):1650–63. doi: 10. 1002/pmic. 201000645. pmid:21462346
7. Schutzendubel A, Polle A. Plant responses to abiotic stresses: heavy metal-induced oxidative stress and protection by mycorrhization. Journal of Experimental Botany. 2002;53(372):1351–65. pmid:11997381 doi: 10. 1093/jexbot/53. 372. 1351
8. Ostrowski SR, Wilbur S, Chou C-HSJ, Pohl HR, Stevens Y-W, Allred PM, et al. Agency for Toxic Substances and Disease Registry's 1997 priority list of hazardous substances. Latent effects-carcinogenesis, neurotoxicology, and developmental deficits in humans and animals. Toxicology and Industrial Health. 1999;15(7):602–44. pmid:10677885 doi: 10. 1177/074823379901500702
9. Di Toppi LS, Gabbrielli R. Response to cadmium in higher plants. Environmental and Experimental Botany. 1999;41(2):105–30. doi: 10. 1016/s0098-8472(98)00058-6
10. Clemens S. Toxic metal accumulation, responses to exposure and mechanisms of tolerance in plants. Biochimie. 2006;88(11):1707–19. pmid:16914250 doi: 10. 1016/j. biochi. 2006. 07. 003
11. Khan N, Singh S, Nazar R. Activities of antioxidative enzymes, sulphur assimilation, photosynthetic activity and growth of wheat (*Triticum aestivum*) cultivars differing in yield potential under cadmium stress. Journal of Agronomy and Crop Science. 2007;193(6):435–44. doi: 10. 1111/j. 1439-037x. 2007. 00272. x
12. Da-lin L, Kai-qi H, Jing-jing M, Wei-wei Q, Xiu-ping W, Shu-pan Z. Effects of cadmium on the growth and physiological characteristics of sorghum plants. African Journal of Biotechnology. 2013;10(70):15770–6.
13. Mithofer A, Schulze B, Boland W. Biotic and heavy metal stress response in plants: evidence for common signals. FEBS letters. 2004;566(1):1–5. doi: 10. 1016/j. febslet. 2004. 04. 011
14. Wojtaszek P. Oxidative burst: an early plant response to pathogen infection. Biochemical Journal. 1997;322:681–92. pmid:9148737 doi: 10. 1042/bj3220681
15. Yadav S. Heavy metals toxicity in plants: an overview

on the role of glutathione and phytochelatins in heavy metal stress tolerance of plants. South African Journal of Botany. 2010;76(2):167–79. doi: 10. 1016/j. sajb. 2009. 10. 007

16. Semane B, Cuypers A, Smeets K, Van Belleghem F, Horemans N, Schat H, et al. Cadmium responses in *Arabidopsis thaliana*: glutathione metabolism and antioxidative defence system. Physiologia Plantarum. 2007;129(3):519–28. doi: 10. 1111/j. 1399-3054. 2006. 00822. x
17. Zechmann B, Müller M, Zellnig G. Modified levels of cysteine affect glutathione metabolism in plant cells. Sulfur Assimilation and Abiotic Stress in Plants: Springer; 2008. p. 193–206.
18. Xin Z, Wang ML, Barkley NA, Burow G, Franks C, Pederson G, et al. Applying genotyping (TILLING) and phenotyping analyses to elucidate gene function in a chemically induced sorghum mutant population. BMC Plant Biology. 2008;8(1):103. doi: 10. 1186/1471-2229-8-103
19. Epelde L, Mijangos I, Becerril JM, Garbisu C. Soil microbial community as bioindicator of the recovery of soil functioning derived from metal phytoextraction with sorghum. Soil Biology and Biochemistry. 2009;41(9):1788–94. doi: 10. 1016/j. soilbio. 2008. 04. 001
20. Zhuang P, Wensheng S, Zhian L, Bin L, Jintian L, Jingsong S. Removal of metals by sorghum plants from contaminated land. Journal of Environmental Sciences. 2009;21(10):1432–7. doi: 10. 1016/s1001-0742(08)62436-5
21. Angelova V, Ivanova R, Delibaltova V, Ivanov K. Use of sorghum crops for in situ phytoremediation of polluted soils. Journal of Agricultural Science and Technology A. 2011;1(5):693–702.
22. Pinto A, de Varennes A, Goncalves M, Mota A. Sorghum detoxification mechanisms. Journal of Plant Nutrition. 2006;29(7):1229–42. doi: 10. 1080/01904160600767450
23. Paterson AH, Bowers JE, Bruggmann R, Dubchak I, Grimwood J, Gundlach H, et al. The *Sorghum bicolor* genome and the diversification of grasses. Nature. 2009;457(7229):551–6. doi: 10. 1038/nature07723. pmid:19189423
24. Blackstock WP, Weir MP. Proteomics: quantitative and physical mapping of cellular proteins. Trends in Biotechnology. 1999;17(3):121–7. pmid:10189717 doi: 10. 1016/s0167-7799(98)01245-1
25. Pandey A, Mann M. Proteomics to study genes and genomes. Nature. 2000;405(6788):837–46. pmid:10866210 doi: 10. 1038/35015709
26. van Wijk KJ. Challenges and prospects of plant proteomics. Plant Physiology. 2001;126(2):501–8. pmid:11402181 doi: 10. 1104/pp. 126. 2. 501
27. Ngara R, Ndimba BK. Understanding the complex nature of salinity and drought-stress response in cereals using proteomics technologies. Proteomics. 2014;14(4–5):611–21. doi: 10. 1002/pmic. 201300351. pmid:24339029
28. D'Alessandro A, Taamalli M, Gevi F, Timperio AM, Zolla L, Ghnaya T. Cadmium stress responses in *Brassica juncea*: hints from proteomics and metabolomics. Journal of Proteome Research. 2013;12(11):4979–97. doi: 10. 1021/pr400793e. pmid:24074147
29. Hossain Z, Hajika M, Komatsu S. Comparative proteome analysis of high and low cadmium accumulating soybeans under cadmium stress. Amino Acids. 2012;43(6):2393–416. doi: 10. 1007/s00726-012-1319-6. pmid:22588482
30. Junior CAL, de Sousa Barbosa H, Galazzi RM, Koolen HHF, Gozzo FC, Arruda MAZ. Evaluation of proteome alterations induced by cadmium stress in sunflower (*Helianthus annuus* L.) cultures. Ecotoxicology and environmental safety. 2015;119:170–7. doi: 10. 1016/j. ecoenv. 2015. 05. 016. pmid:26004357
31. Lee K, Bae DW, Kim SH, Han HJ, Liu X, Park HC, et al. Comparative proteomic analysis of the short-term responses of rice roots and leaves to cadmium. Journal of Plant Physiology. 2010;167(3):161–8. doi: 10. 1016/j. jplph. 2009. 09. 006. pmid:19853963
32. Li X, Zhou Y, Yang Y, Yang S, Sun X, Yang Y. Physiological and proteomics analyses reveal the mechanism of *Eichhornia crassipes* tolerance to high-concentration cadmium stress compared with *Pistia stratiotes*. PLoS One. 2015; doi: 10. 1371/journal. pone. 0124304.
33. Muneer S, Hakeem KR, Mohamed R, Lee JH. Cadmium toxicity induced alterations in the root proteome of green gram in contrasting response towards iron supplement. International Journal of Molecular Sciences. 2014;15(4):6343–55. doi: 10. 3390/ijms15046343. pmid:24739807
34. Weng Z-X, Wang L-X, Tan F-l, Huang L, Xing J-H, Chen S-P, et al. Proteomic and physiological analyses reveal detoxification and antioxidation induced by Cd stress in*Kandelia candel* roots. Trees. 2013;27(3):583–95. doi: 10. 1007/s00468-012-0811-7
35. Zhang A, Xu T, Zou H, Pang Q. Comparative proteomic analysis provides insight into cadmium stress responses in brown algae *Sargassum fusiforme*. Aquatic Toxicology. 2015;163:1–15. doi: 10. 1016/j. aquatox. 2015. 03. 018. pmid:25827747
36. Hossain Z, Makino T, Komatsu S. Proteomic study of β-aminobutyric acid-mediated cadmium stress alleviation in soybean. Journal of Proteomics. 2012;75(13):4151–64. doi: 10. 1016/j. jprot. 2012. 05. 037. pmid:22652489
37. Kieffer P, Dommes J, Hoffmann L, Hausman JF, Renaut J. Quantitative changes in protein expression of cadmium-exposed poplar plants. Proteomics. 2008;8(12):2514–30. doi: 10. 1002/pmic. 200701110. pmid:18563750
38. Marmiroli M, Imperiale D, Maestri E, Marmiroli N. The response of Populus spp. to cadmium stress: Chemical, morphological and proteomics study. Chemosphere. 2013;93(7):1333–44. doi: 10. 1016/j. chemosphere. 2013. 07. 065. pmid:23981839
39. Semane B, Dupae J, Cuypers A, Noben J-P, Tuomainen M, Tervahauta A, et al. Leaf proteome responses of *Arabidopsis thaliana* exposed to mild cadmium stress. Journal of Plant Physiology. 2010;167(4):247–54. doi:

10. 1016/j. jplph. 2009. 09. 015. pmid:20005002

40. Zeng X-W, Qiu R-L, Ying R-R, Tang Y-T, Tang L, Fang X-H. The differentially-expressed proteome in Zn/Cd hyperaccumulator *Arabis paniculata* Franch. in response to Zn and Cd. Chemosphere. 2011;82(3):321–8. doi: 10. 1016/j. chemosphere. 2010. 10. 030. pmid:21074242

41. Zhao L, Sun Y-L, Cui S-X, Chen M, Yang H-M, Liu H-M, et al. Cd-induced changes in leaf proteome of the hyperaccumulator plant *Phytolacca americana*. Chemosphere. 2011;85(1):56–66. doi: 10. 1016/j. chemosphere. 2011. 06. 029. pmid:21723586

42. Yang X, Long X, Ye H, He Z, Calvert D, Stoffella P. Cadmium tolerance and hyperaccumulation in a new Zn-hyperaccumulating plant species (*Sedum alfredii*Hance). Plant and Soil. 2004;259(1–2):181–9. doi: 10. 1023/b:plso. 0000020956. 24027. f2

43. Arora K, Sharma S. Toxic metal (Cd) removal from soil by AM fungi inoculated sorghum. Asian Journal of Experimental Sciences. 2009;23(1):341–8.

44. Mendoza J, Garrido T, Castillo G, San Martin N. Metal availability and uptake by sorghum plants grown in soils amended with sludge from different treatments. Chemosphere. 2006;65(11):2304–12. pmid:16797672 doi: 10. 1016/j. chemosphere. 2006. 05. 012

45. Faruruwa DM, Yauri UAB, Dangoggo SM. Cadmium, Copper, Lead and Zinc levels in sorghum and millet grown in the city of Kano and its environs. 2013;2(3):82–85

46. Hoagland DR, Arnon DI. The water-culture method for growing plants without soil. Circular California Agricultural Experiment Station. 1950;347 (2nd edit).

47. Barrs H, Weatherley P. A re-examination of the relative turgidity technique for estimating water deficits in leaves. Australian Journal of Biological Sciences. 1962;15(3):413–28.

48. Gong J-M, Lee DA, Schroeder JI. Long-distance root-to-shoot transport of phytochelatins and cadmium in Arabidopsis. Proceedings of the National Academy of Sciences. 2003;100(17):10118–23. doi: 10. 1073/pnas. 1734072100

49. Oh MW, Roy SK, Kamal AHM, Cho K, Cho S-W, Park C-S, et al. Proteome analysis of roots of wheat seedlings under aluminum stress. Molecular Biology Reports. 2014;41(2):671–81. doi: 10. 1007/s11033-013-2905-8. pmid:24357239

50. Huang X, Liu L, Chen J, Zhai Y. Comparative proteomic analysis of the response in resistant and susceptible maize inbred lines to infection by *Curvularia lunata*. Progress in Natural Science. 2009;19(7):845–50. doi: 10. 1016/j. pnsc. 2008. 07. 024

51. Bradford MM. A rapid and sensitive method for the quantitation of microgram quantities of protein utilizing the principle of protein-dye binding. Analytical Biochemistry. 1976;72(1):248–54. doi: 10. 1016/0003-2697(76)90527-3

52. Arduini I, Godbold DL, Onnis A. Cadmium and copper change root growth and morphology of *Pinus pinea* and *Pinus pinaster* seedlings. Physiologia Plantarum. 1994;92(4):675–80. doi: 10. 1034/j. 1399-3054. 1994. 920419. x

53. Foy C, Chaney Rt, White M. The physiology of metal toxicity in plants. Annual Review of Plant Physiology. 1978;29(1):511–66. doi: 10. 1146/annurev. pp. 29. 060178. 002455

54. Malone CP, Miller RJ, Koeppe D. Root growth in corn and soybeans: effects of cadmium and lead on lateral root initiation. Canadian Journal of Botany. 1978;56(3):277–81. doi: 10. 1139/b78-034

55. Chaoui A, Mazhoudi S, Ghorbal MH, El Ferjani E. Cadmium and zinc induction of lipid peroxidation and effects on antioxidant enzyme activities in bean (*Phaseolus vulgaris*L.). Plant Science. 1997;127(2):139–47. doi: 10. 1016/s0168-9452(97)00115-5

56. Kleckerova A, Sobrova P, Krystofova O, Sochor J, Zitka O, Babula P, et al. Cadmium (II) and zinc (II) ions effects on maize plants revealed by spectroscopy and electrochemistry. International Journal of Electrochemical Science. 2011;6(12):6011–31.

57. Baker AJM, Reeves RD, Hajar ASM. Heavy metal accumulation and tolerance in British populations of the metallophyte *Thlaspi caerulescens* J. & C. Presl (Brassicaceae). New Phytologist. 1994;127(1):61–8. doi: 10. 1111/j. 1469-8137. 1994. tb04259. x.

58. Peng K, Li X, Luo C, Shen Z. Vegetation composition and heavy metal uptake by wild plants at three contaminated sites in Xiangxi area, China. Journal of Environmental Science and Health, Part A. 2006;41(1):65–76. doi: 10. 1080/10934520500298838

59. Reeves RD, Schwartz C, Morel JL, Edmondson J. Distribution and metal-accumulating behavior of *Thlaspi caerulescens* and associated metallophytes in France. International journal of Phytoremediation. 2001;3(2):145–72. doi: 10. 1080/15226510108500054.

60. Zancheta ACF, De Abreu CA, Zambrosi FCB, dM Erismann N, Lagôa AMMA. Cadmium accumulation by jack-bean and sorghum in hydroponic culture. International Journal of Phytoremediation. 2015;17(3):298–303. doi: 10. 1080/15226514. 2014. 883492

61. Tkalec M, Stefanic PP, Cvjetko P, Sikic S, Pavlica M, Balen B. The effects of cadmium-zinc interactions on biochemical responses in tobacco seedlings and adult plants. PloS One. 2014;9(1):e87582. doi: 10. 1371/journal. pone. 0087582. pmid:24475312

62. Wu F, Zhang G, Yu J. Interaction of cadmium and four microelements for uptake and translocation in different barley genotypes. Communications in soil science and plant analysis. 2003;34(13–14):2003–20. doi: 10. 1081/css-120023233

63. Sandalio L, Dalurzo H, Gomez M, Romero-Puertas M, Del Rio L. Cadmium-induced changes in the growth and oxidative metabolism of pea plants. Journal of Experimental Botany. 2001;52(364):2115–26. pmid:11604450

64. Seregin I, Kozhevnikova A. Histochemical methods for detection of heavy metals and strontium in the tissues of higher plants. Russian Journal of Plant Physiology.

2011;58(4):721–7. doi: 10. 1134/s1021443711040133

65. Seregin I, Ivanov V. Histochemical investigation of cadmium and lead distribution in plants. Russian Journal of Plant Physiology. 1997;44(6):791–6.

66. Glass D, Raskin I, Ensley B. Phytoremediation of toxic metals: using plants to clean up the environment. Phytoremediation of toxic metals: Using plants to clean up the environment. 2000. doi: 10. 1016/s0168-9452(01)00353-3

67. Monni S, Salemaa M, Millar N. The tolerance of Empetrum nigrum to copper and nickel. Environmental pollution. 2000;109(2):221–9. pmid:15092893 doi: 10. 1016/s0269-7491(99)00264-x

68. Zawoznik MS, Groppa MD, Tomaro ML, Benavides MP. Endogenous salicylic acid potentiates cadmium-induced oxidative stress in *Arabidopsis thaliana*. Plant Science. 2007;173(2):190–7. doi: 10. 1016/j. plantsci. 2007. 05. 004

69. Kieffer P, Schroder P, Dommes J, Hoffmann L, Renaut J, Hausman J-F. Proteomic and enzymatic response of poplar to cadmium stress. Journal of Proteomics. 2009;72(3):379–96. pmid:19367735 doi: 10. 1016/j. jprot. 2009. 01. 014

70. Rustichelli C, Visioli G, Kostecka D, Vurro E, di Toppi LS, Marmiroli N. Proteomic analysis in the lichen *Physcia adscendens* exposed to cadmium stress. Environmental Pollution. 2008;156(3):1121–7. doi: 10. 1016/j. envpol. 2008. 04. 010. pmid:18514371

71. Gallego SM, Pena LB, Barcia RA, Azpilicueta CE, Iannone MF, Rosales EP, et al. Unravelling cadmium toxicity and tolerance in plants: insight into regulatory mechanisms. Environmental and Experimental Botany. 2012;83:33–46. doi: 10. 1016/j. envexpbot. 2012. 04. 006

72. Benavides MP, Gallego SM, Tomaro ML. Cadmium toxicity in plants. Brazilian Journal of Plant Physiology. 2005;17(1):21–34. doi: 10. 1590/s1677-04202005000100003

73. Kieffer P, Planchon S, Oufir M, Ziebel J, Dommes J, Hoffmann L, et al. Combining proteomics and metabolite analyses to unravel cadmium stress-response in poplar leaves. Journal of Proteome Research. 2008;8(1):400–17. doi: 10. 1021/pr800561r

74. Sergeant K, Kieffer P, Dommes J, Hausman J-F, Renaut J. Proteomic changes in leaves of poplar exposed to both cadmium and low-temperature. Environmental and Experimental Botany. 2014;106:112–23. doi: 10. 1016/j. envexpbot. 2014. 01. 007

75. Moroney J, Bartlett S, Samuelsson G. Carbonic anhydrases in plants and algae. Plant, Cell & Environment. 2001;24(2):141–53. doi: 10. 1111/j. 1365-3040. 2001. 00669. x

76. Hajduch M, Rakwal R, Agrawal GK, Yonekura M, Pretova A. High-resolution two-dimensional electrophoresis separation of proteins from metal-stressed rice (*Oryza sativa* L.) leaves: drastic reductions/fragmentation of ribulose-1, 5-bisphosphate carboxylase/oxygenase and induction of stress-related proteins. Electrophoresis. 2001;22(13):2824–31. pmid:11545414 doi: 10. 1002/1522-2683(200108)22:13<2824::aid-elps2824>3. 0. co;2-c

77. Roth U, von Roepenack-Lahaye E, Clemens S. Proteome changes in *Arabidopsis thaliana* roots upon exposure to Cd^{2+}. Journal of Experimental Botany. 2006;57(15):4003–13. pmid:17075075 doi: 10. 1093/jxb/erl170

78. Sarry JE, Kuhn L, Ducruix C, Lafaye A, Junot C, Hugouvieux V, et al. The early responses of *Arabidopsis thaliana* cells to cadmium exposure explored by protein and metabolite profiling analyses. Proteomics. 2006;6(7):2180–98. pmid:16502469 doi: 10. 1002/pmic. 200500543

79. Alvarez S, Berla BM, Sheffield J, Cahoon RE, Jez JM, Hicks LM. Comprehensive analysis of the *Brassica juncea* root proteome in response to cadmium exposure by complementary proteomic approaches. Proteomics. 2009;9(9):2419–31. doi: 10. 1002/pmic. 200800478. pmid:19343712

80. Rodriguez-Celma J, Rellan-Alvarez R, Abadia A, Abadia J, Lopez-Millan A-F. Changes induced by two levels of cadmium toxicity in the 2-DE protein profile of tomato roots. Journal of Proteomics. 2010;73(9):1694–706. doi: 10. 1016/j. jprot. 2010. 05. 001. pmid:20621698

81. Fagioni M, Zolla L. Does the different proteomic profile found in apical and basal leaves of spinach reveal a strategy of this plant toward cadmium pollution response? Journal of Proteome Research. 2009;8(5):2519–29. doi: 10. 1021/pr8011182. pmid:19290619

82. Cvjetko P, Zovko M, Balen B. Proteomics of heavy metal toxicity in plants. Arhiv za Higijenu Rada i Toksikologiju. 2014;65(1):1–17. doi: 10. 2478/10004-1254-65-2014-2443. pmid:24526604

83. Zou H-X, Pang Q-Y, Zhang A-Q, Lin L-D, Li N, Yan X-F. Excess copper induced proteomic changes in the marine brown algae *Sargassum fusiforme*. Ecotoxicology and Environmental Safety. 2015;111:271–80. doi: 10. 1016/j. ecoenv. 2014. 10. 028. pmid:25450944

84. Bogs J, Ebadi A, McDavid D, Robinson SP. Identification of the flavonoid hydroxylases from grapevine and their regulation during fruit development. Plant Physiology. 2006;140(1):279–91. pmid:16377741 doi: 10. 1104/pp. 105. 073262

85. Werck-Reichhart D, Hehn A, Didierjean L. Cytochromes P450 for engineering herbicide tolerance. Trends in Plant Science. 2000;5(3):116–23. pmid:10707077 doi: 10. 1016/s1360-1385(00)01567-3

86. Bozcaarmutlu A, Arinc E. Effect of mercury, cadmium, nickel, chromium and zinc on kinetic properties of NADPH-cytochrome P450 reductase purified from leaping mullet (*Liza saliens*). Toxicology *in vitro*. 2007;21(3):408–16. pmid:17113746 doi: 10. 1016/j. tiv. 2006. 10. 002

87. Williams TD, Diab AM, George SG, Godfrey RE, Sabine V, Conesa A, et al. Development of the GENIPOL European flounder (*Platichthys flesus*) microarray and determination of temporal transcriptional responses to cadmium at low dose. Environmental Science &

Technology. 2006;40(20):6479–88. doi: 10. 1021/es061142h

88. Ogawa I, Nakanishi H, Mori S, Nishizawa NK. Time course analysis of gene regulation under cadmium stress in rice. Plant and Soil. 2009;325(1–2):97–108. doi: 10. 1007/s11104-009-0116-9

89. Dixon DP, Lapthorn A, Edwards R. Plant glutathione transferases. Genome Biol. 2002;3(3):3004. 1–. 10. doi: 10. 1016/j. plantsci. 2004. 11. 009

90. Nepovim A, Podlipna R, Soudek P, Schroder P, Vanek T. Effects of heavy metals and nitroaromatic compounds on horseradish glutathione S-transferase and peroxidase. Chemosphere. 2004;57(8):1007–15. pmid:15488591 doi: 10. 1016/j. chemosphere. 2004. 08. 030

91. Bae W, Chen X. Proteomic study for the cellular responses to Cd^{2+} in*Schizosaccharomyces pombe* through amino acid-coded mass tagging and liquid chromatography tandem mass spectrometry. Molecular & Cellular Proteomics. 2004;3(6):596–607. doi: 10. 1074/mcp. m300122-mcp200

92. Marrs KA. The functions and regulation of glutathione S-transferases in plants. Annual Review of Plant Biology. 1996;47(1):127–58. doi: 10. 1146/annurev. arplant. 47. 1. 127

93. Shin YH, Park E-H, Fuchs JA, Lim C-J. Characterization, expression and regulation of a third gene encoding glutathione S-transferase from the fission yeast. Biochimica et Biophysica Acta (BBA)-Gene Structure and Expression. 2002;1577(1):164–70. doi: 10. 1016/s0167-4781(02)00422-0

94. Smith AP, DeRidder BP, Guo W-J, Seeley EH, Regnier FE, Goldsbrough PB. Proteomic analysis of Arabidopsis glutathione S-transferases from benoxacor-and copper-treated seedlings. Journal of Biological Chemistry. 2004;279(25):26098–104. pmid:15069083 doi: 10. 1074/jbc. m402807200

95. Li G, Peng X, Xuan H, Wei L, Yang Y, Guo T, et al. Proteomic analysis of leaves and roots of common wheat (*Triticum aestivum* L.) under copper-stress conditions. Journal of Proteome Research. 2013;12(11):4846–61. doi: 10. 1021/pr4008283. pmid:24074260

96. Gould GW, Cuenda A, Thomson F, Cohen P. The activation of distinct mitogen-activated protein kinase cascades is required for the stimulation of 2-deoxyglucose uptake by interleukin-1 and insulin-like growth factor-1 in KB cells. Biochemical Journal. 1995;311:735–8. pmid:7487926 doi: 10. 1042/bj3110735

97. Small ID, Peeters N. The PPR motif-a TPR-related motif prevalent in plant organellar proteins. Trends in Biochemical Sciences. 2000;25(2):45–7. doi: 10. 1016/s0968-0004(99)01520-0

98. Yan S-P, Zhang Q-Y, Tang Z-C, Su W-A, Sun W-N. Comparative proteomic analysis provides new insights into chilling stress responses in rice. Molecular & Cellular Proteomics. 2006;5(3):484–96. doi: 10. 1074/mcp. m500251-mcp200

99. Ahsan N, Lee S-H, Lee D-G, Lee H, Lee SW, Bahk JD, et al. Physiological and protein profiles alternation of germinating rice seedlings exposed to acute cadmium toxicity. Comptes Rendus Biologies. 2007;330(10):735–46. pmid:17905393 doi: 10. 1016/j. crvi. 2007. 08. 001

100. Kim YK, Lee MY. Proteomic analysis of differentially expressed proteins of rice in response to cadmium. Journal of the Korean Society for Applied Biological Chemistry. 2009;52(5):428–36. doi: 10. 3839/jksabc. 2009. 075

101. Fang XP, Ma HS, Lu DZ, Yu H, Lai W, Ruan S. Comparative proteomics analysis of proteins expressed in the I-1 and I-2 internodes of strawberry stolons. Proteome Science. 2011;9:26. doi: 10. 1186/1477-5956-9-26. pmid:21569547

102. Liu C-P, Yang L, Shen H-L. Proteomic analysis of immature *Fraxinus mandshurica*cotyledon tissues during somatic embryogenesis: effects of explant browning on somatic embryogenesis. International Journal of Molecular Sciences. 2015;16(6):13692–713. doi: 10. 3390/ijms160613692. pmid:26084048

103. Chen J, Hu W-J, Wang C, Liu T-W, Chalifour A, Shen Z-J, et al. Proteomic analysis reveals differences in tolerance to acid rain in two broad-leaf tree species, *Liquidambar formosana* and *Schima superba*. 2014; doi: 10. 1371/journal. pone. 0102532.

104. Hu W-J, Chen J, Liu T-W, Simon M, Wang W-H, Wu F-H, et al. Comparative proteomic analysis of differential responses of *Pinus massoniana* and *Taxus wallichiana* var. mairei to simulated acid rain. International Journal of Molecular Sciences. 2014;15(3):4333–55. doi: 10. 3390/ijms15034333. pmid:24625662

105. Ji X, Gai Y, Zheng C, Mu Z. Comparative proteomic analysis provides new insights into mulberry dwarf responses in mulberry (*Morus alba* L.). Proteomics. 2009;9(23):5328–39. doi: 10. 1002/pmic. 200900012. pmid:19834890

106. Pandey A, Chakraborty S, Datta A, Chakraborty N. Proteomics approach to identify dehydration responsive nuclear proteins from chickpea (*Cicer arietinum* L.). Molecular & Cellular Proteomics. 2008;7(1):88–107. doi: 10. 1074/mcp. m700314-mcp200

107. Cox AD, Saito MA. Proteomic responses of oceanic *Synechococcus* WH8102 to phosphate and zinc scarcity and cadmium additions. Frontiers in Microbiology. 2013;4:387. doi: 10. 3389/fmicb. 2013. 00387. pmid:24381569

108. Ali B, Gill RA, Yang S, Gill MB, Farooq MA, Liu D, et al. Regulation of cadmium-induced proteomic and metabolic changes by 5-aminolevulinic acid in leaves of*Brassica napus* L. PLoS One. 2015; doi: 10. 1371/journal. pone. 0123328.

109. Kovacik J, Klejdus B. Dynamics of phenolic acids and lignin accumulation in metal-treated *Matricaria chamomilla* roots. Plant Cell Reports. 2008;27(3):605–15. pmid:18066553 doi: 10. 1007/s00299-007-0490-9

110. Aina R, Labra M, Fumagalli P, Vannini C, Marsoni M, Cucchi U, et al. Thiol-peptide level and proteomic changes in response to cadmium toxicity in *Oryza sativa* L. roots. Environmental and Experimental Botany. 2007;59(3):381–92. doi: 10. 1016/j. envexpbot. 2006. 04. 010

111. Deeba F, Pandey AK, Ranjan S, Mishra A, Singh R, Sharma Y, et al. Physiological and proteomic responses of cotton (*Gossypium herbaceum* L.) to drought stress. Plant Physiology and Biochemistry. 2012;53:6–18. doi: 10. 1016/j. plaphy. 2012. 01. 002. pmid:22285410

112. Wong CKE, Cobbett CS. HMA P-type ATPases are the major mechanism for root-to-shoot Cd translocation in *Arabidopsis thaliana*. New Phytologist. 2009;181(1):71–8. doi: 10. 1111/j. 1469-8137. 2008. 02638. x. pmid:19076718

113. Song W-Y, Martinoia E, Lee J, Kim D, Kim D-Y, Vogt E, et al. A novel family of cys-rich membrane proteins mediates cadmium resistance in Arabidopsis. Plant Physiology. 2004;135(2):1027–39. pmid:15181212 doi: 10. 1104/pp. 103. 037739

114. Wang R, Gao F, Guo B-Q, Huang J-C, Wang L, Zhou Y-J. Short-term chromium-stress-induced alterations in the maize leaf proteome. International Journal of Molecular Sciences. 2013;14(6):11125–44. doi: 10. 3390/ijms140611125. pmid:23712354

CHAPTER 12

SILICON MODERATED THE K DEFICIENCY BY IMPROVING THE PLANT-WATER STATUS IN SORGHUM

Daoqian Chen, Beibei Cao, Shiwen Wang, Peng Liu, Xiping Deng, Lina Yin & Suiqi Zhang

INTRODUCTION

Potassium (K) is the most abundant cation in plants. It is involved in various important cellular processes, such as the stabilization of protein synthesis, activation of enzymes, and neutralization of negative charges on proteins1·2. In addition, it also plays a key role in osmotic processes that contribute to cellular turgor, photosynthesis, and transpiration3. K deficiency can disturb these activities directly, and inhibit plant growth and development2·4·5. K is also essential for high-yield crop production, and it can be a limiting factor for such crops under certain environmental conditions6·7. The uptake of silicon (Si), which is one of the most abundant elements in the Earth's crust, is beneficial for increasing plant resistance to various stresses8·9·10. Si application has also been widely reported to alleviate plant nutrient deficiency, including alleviating K deficiency4·11·12·13.

Compared with the benefits of Si under other stresses, the performance and underlying mechanism of Si in alleviating K deficiency has rarely been investigated4. To date, only one study has reported that Si application alleviated K-deficiency-induced growth inhibition in soybean. Several studies also reported that Si can affect tissue K concentration under salt stress conditions14·15. These previous studies suggested that Si can influence the K absorption and/or distribution under certain conditons. K deficiency disturbs plant growth and development in various ways. It is not clear if there are other ways in which Si participates in regulating K deficiency. A better understanding of the interaction between Si application and K deficiency will provide important information about Si and K fertilizer use in crop production.

The Plant-water status involves the balance of the difference in water uptake through the roots and transpiration in leaves; maintaining water status is the basis of metabolic activities in tissues15·16. K^+ is involved in regulating the plant-water status in various ways. First, K^+ plays an important role in stomatal opening and closing, which controls leaf transpiration and plant-water status17·18. Second, at the whole-plant level, K^+ accumulation in the root xylem vessels is also involved in osmotic water absorption19. Third, several recent reports have suggested that xylem K^+ plays a critical role in modifying xylem hydraulic conductance and water relations in plants5·20. In contrast, it has been widely reported that K deficiency affects the plant-water status in different ways. Severe K deficiency generally induces stomatal closure and lowers transpiration rates21. As the main osmotic solute, K deficiency also affects the osmotic potential gradient and decreases water uptake20·22. K deficiency was found to reduce aquaporin activity, thereby suppressing root hydraulic conductance and water supply to the stem and leaf transpiration21. K fertilizer, however, was found to increase plant hydraulic conductance and transpiration20. In sum, K is involved in regulating the plant water status, and severe K deficiency causes tissue dehydration21.

There are no case studies showing that Si application affects the plant-water status under K deficiency stress, but Si was found to be involved in regulating the plant-water status in response to other stresses. Romero-Aranda *et al.*23 demonstrated that Si application enhanced the tomato leaf water content under salt stress. Recently, Liu *et al.*24·25reported that enhanced root hydraulic conductance by aquaporin regulation is responsible for Si-alleviated plant-water status under drought and salt stresses. Under normal growth conditions, Si application was also found to affect the plant-water status15 and up-regulate aquaporin gene *(PIPs)* expression25. In addition, the potential effect of Si application on K^+ absorption and distribution may be involved in regulating the plant water status. Therefore, the

Citation: Daoqian Chen, Beibei Cao, Shiwen Wang, Peng Liu, Xiping Deng, Lina Yin & Suiqi Zhang, Silicon moderated the K deficiency by improving the plant-water status in sorghum, doi:10.1038/srep22882.

close coupling between K deficiency and Si application on the plant-water status suggests that Si may be involved in regulating the plant water-status under K-deficiency conditions.

The aim of this study was to determine whether or not Si improves the plant-water status under K-deficiency conditions. We hypothesized that K deficiency induces a decrease in plant hydraulic conductance, which decreases the water supply for leaf transpiration and plant-water status. Si application increases plant hydraulic conductance, possibly favoring higher gas exchange rates and photosynthesis, thus alleviating the K deficiency. To accomplish the objective of this study, the effect of Si application on leaf relative water content (LRWC), plant transpiration, whole-plant hydraulic conductance (K_{plant}) and root hydraulic conductance (Lp_r), and the effect of aquaporin on transpiration and xylem sap K^+ concentration were measured under K-deficiency and Si application conditions. Moreover, the expression of aquaporin and K^+-related genes were also investigated.

RESULTS

Si application Alleviated the K Deficiency in Sorghum and Improved the plant-Water Status

Extreme K deficiency inhibits plant growth[26]. In the present study, Si application did not affect the sorghum seedling growth under control conditions. However, K deficiency significantly decreased plant growth after 15 days incubation and Si application alleviated the K-deficiency-induced growth inhibition (Fig. 1a, Supplementary Fig. S1). The dry weight of Si-treated plants was 29% ($p < 0.05$) higher than that of untreated plants under K deficiency conditions. At the same time, the root/shoot ratio was not affected by either K deficiency or Si application (Fig. 1b). These results showed that Si application enhanced plants' K deficiency tolerance. Similar to the change in plant dry weight, the stomatal conductance, transpiration rate, and photosynthetic rate were also found to be decreased under K deficiency in various plant species[21,27,28]. In this study, the leaf gas exchange was not influenced by Si under control conditions, but it was significantly inhibited by K deficiency. Si application alleviated the K-deficiency-induced reduction in the photosynthetic rate, stomatal conductance and transpiration rate (Fig. 2). In Si-treated plants, the leaf photosynthetic rate, stomatal conductance and transpiration rate were 139% ($p < 0.05$), 49% ($p < 0.05$) and 94% ($p < 0.05$) higher than that of plants without Si application under K-deficiency conditions (Fig. 2). Water status is reflected by the LRWC[15,16]. K deficiency increases the plant's susceptibility to water stress, and plants with an adequate K supply have better hydrated tissues than those with K deficiency[15,16]. In this study, the LRWC was decreased after 15 days of K deficiency and Si significantly reduced the K-deficiency-induced decrease in LRWC (Fig. 3a). Leaf water potential also decreased after the 15-day K-deficiency treatment, but Si application maintained the leaf water potential at the same level as that of control (Fig. 3b). These results indicated that Si application improved the plant water-status under K-deficiency conditions. Osmotic potential was significantly enhanced after a 15-day K-deficiency treatment, but Si application eliminated the K-deficiency-induced increase in osmotic potential (Fig. 3c). The lower osmotic potential in Si-treated plants suggested that Si could improve the plant-water status by decreasing the osmotic potential.

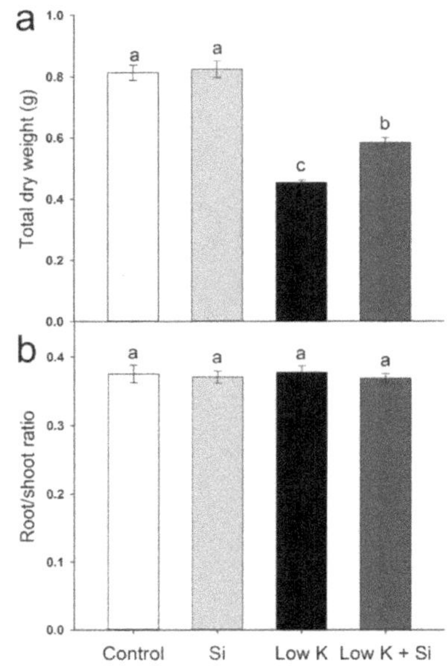

Figure 1: Effects of silicon (Si, 1 mM) on the plant growth of sorghum plants grown under control (3 mM KCl) and low K (0.05 mM KCl) conditions.

(**a**) Total dry weight, (**b**) root/shoot ratio. All parameters were measured after 15 days of treatment. Values are presented as the mean ± SE (n = 24). Different letters in one measure indicate statistically significant differences at $P < 0.05$.

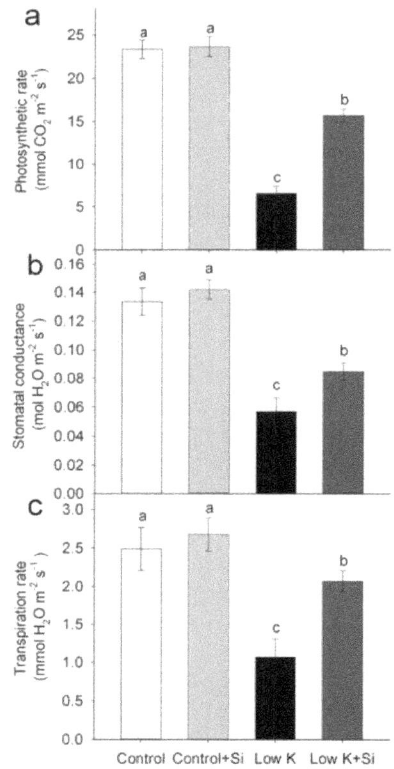

Figure 2: Effects of silicon (Si, 1 mM) on photosynthetic rate (**a**), stomatal conductance (**b**) and transpiration rate (**c**) of sorghum plants grown under control (3 mM KCl) and low K (0.05 mM KCl) conditions. All parameters were measured after 15 days of treatment. Values are presented as the mean ± SE (n = 6). Different letters in one measure indicate statistically significant differences at $P < 0.05$.

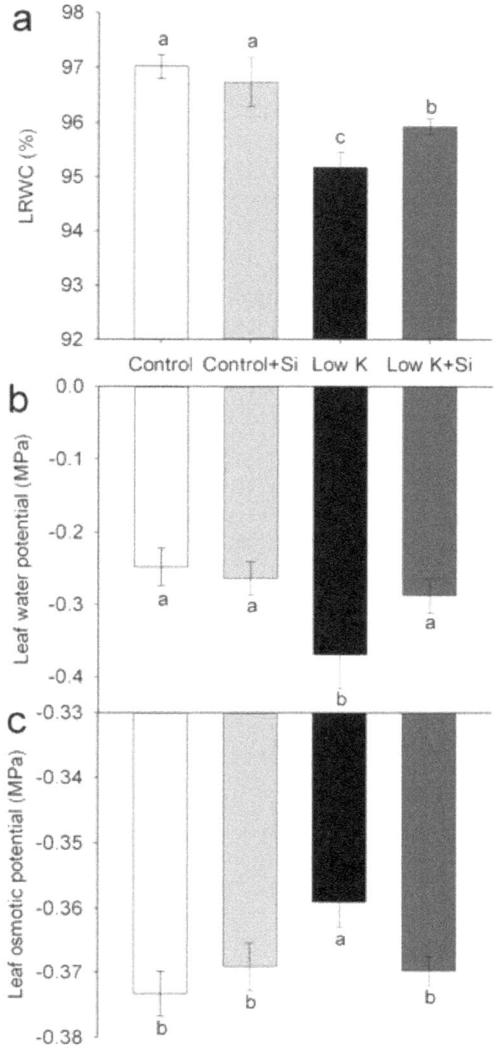

Figure 3:

Effects of silicon (Si, 1 mM) on leaf relative water content (LRWC) (**a**), leaf water potential (**b**) and leaf osmotic potential (**c**) of sorghum plants grown under control (3 mM KCl) and low K (0.05 mM KCl) conditions. All parameters were measured after 15 days of treatment. Values are presented as the mean ± SE (n = 5). Different letters in one measure indicate statistically significant differences at $P<0.05$.

Si Application did not Enhance the K Concentration under the K-deficiency Condition

Activated K^+ uptake in plants by the induction of the expression of high-affinity transporters was considered to be a major mechanism of adaptation to K starvation29·30·31. As shown in Fig. 4, the K^+concentration in both shoots and roots sharply decreased after 15 days of K deficiency. Si application did not influence the K^+ concentration in either shoots or roots under control or K-deficiency conditions. Total K^+ uptake per plant was decreased by K deficiency, and was much lower in plants that did not receive Si treatment (Fig. 4). This result suggested that the increased total K^+ absorption in Si-treated plants under K-deficiency conditions was the consequence of increase in biomass. The same K^+ concentration between the Si-treated and -untreated plants indicates that Si alleviating the K deficiency was not due to a direct increase in K^+ absorption.

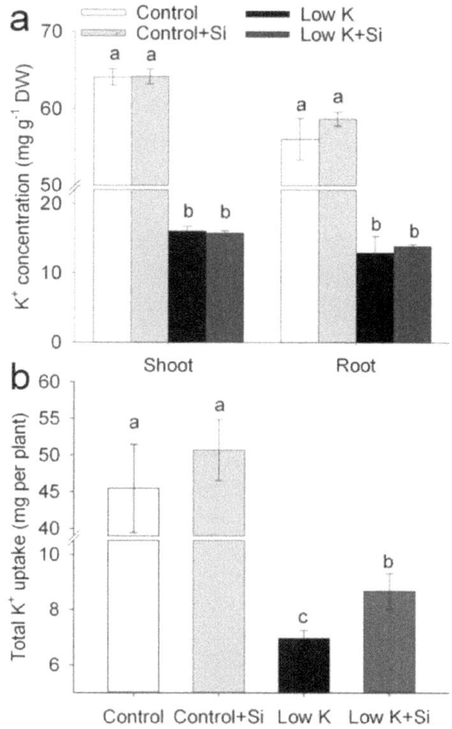

Figure 4: Effects of silicon (Si, 1 mM) on K^+ concentration (**a**) and total K^+ uptake (**b**) of sorghum plants grown under control (3 mM KCl)and low K (0.05 mM KCl) conditions. All parameters were measured after 15 days of treatment. Values are presented as the mean ± SE (n = 3). Different letters in one measure indicate statistically significant differences at $P<0.05$.

Si Improved the plant-water Status by Maintaining the high level of whole-plant Hydraulic Conductance and root Hydraulic Conaductance

The overall plant water uptake is represented by whole-plant hydraulic conductance (K_{plant}), which consists of leaf, stem, and root hydraulic conductance32. In this study, under normal growth conditions, Si application did not change K_{plant}. K deficiency largely decreased $K_{plant}$and Si application significantly reduced this decrease. K_{plant} in plants treated with Si was 31% ($p<0.05$) higher than that of plants without Si under K deficient conditions (Fig. 5). These results showed that Si-treated plants had greater water uptake ability than untreated plants.

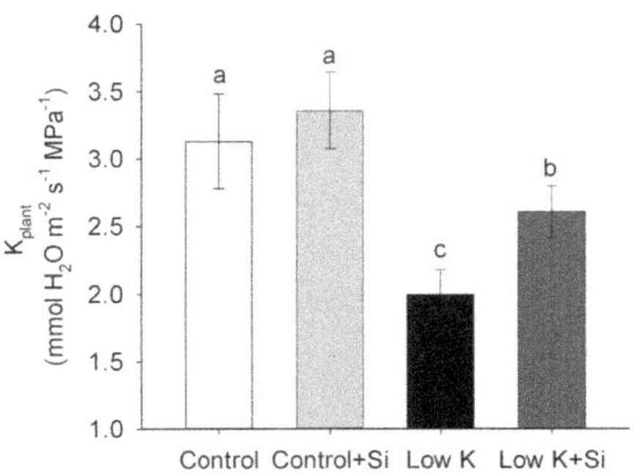

Figure 5: Effects of silicon (Si, 1 mM) on whole-plant hydraulic conductance (K_{plant}) of sorghum plants grown under control (3 mM KCl) and low K (0.05 mM KCl) conditions.

The K_{plant} was calculated by the transpiration rate determined gravimetrically divided by the difference between culture solution and leaf water potential. Values are presented as the mean±SE (n=6). Different letters in one measure indicate statistically significant differences at $P<0.05$.

Root hydraulic conductance (Lp_r) is usually the lowest within the liquid component of the soil–plant–air continuum33, and roots contribute the majority of the whole-plant hydraulic resistance34. In this study, under control conditions, Si application did not change the Lp_r. However, K deficiency significantly decreased Lp_r, and Si application significantly moderated this decrease. Lp_r in Si-treated plants was 35% ($p<0.05$) higher than that in untreated plants under the K deficiency conditions (Fig. 6). These results indicated that the higher Lp_r in Si-treated plants could contribute to the Si-mediated increase in plant water uptake under K-deficiency condition.

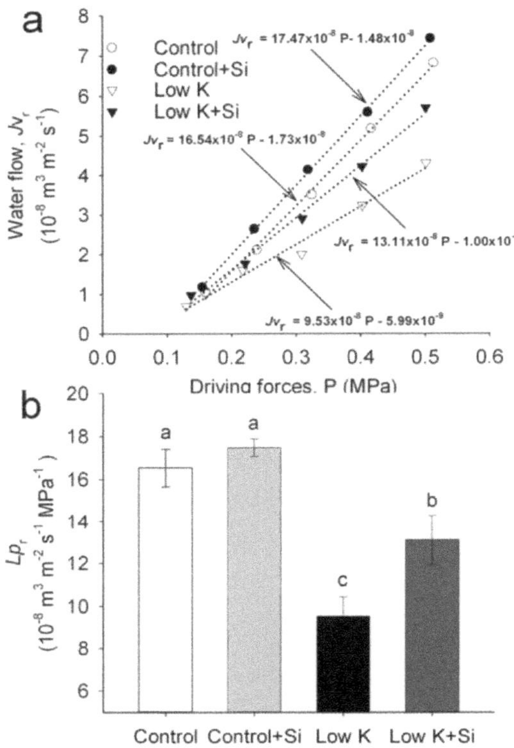

Figure 6: Effects of silicon (Si, 1 mM) on root hydraulic conductance (Lp_r) of sorghum plants grown under control (3 mM KCl) and low K (0.05 mM KCl) conditions.

The Lp_r was measured using a pressure chamber. Exuded sap was collected for 60 s for a given gas pressure and weighed, and the root surface area was then scanned and analyzed. The water flow, Jv_r in $m^3\ m^{-2}\ s^{-1}$, was obtained and Lp_r ($m^3\ m^{-2}\ s^{-1}\ MPa^{-1}$) is calculated from the slope of Jv_r against the driving force P, consisting of P_{gas} and an osmotic gradient (**a**). The mean±SE (n=6) are shown in (**b**). Different letters in one measure indicate statistically significant differences at $P<0.05$.

Si alleviated Lp_r by Enhancing Aquaporin Activity and Xylem sap K^+Concentration

In the process of water uptake by roots, water can move radially toward the xylem along three pathways: the apoplastic, symplastic, and transcellular pathways. The symplastic and transcellular pathways are collectively referred to as the "cell-to-cell pathway"35·36. The cell-to-cell pathway plays an important role in water transport in the roots under certain conditions and is driven by the osmotic gradient between the soil and root xylem sap34. In the present study, the K^+ concentration in xylem sap was largely decreased after 15 days of K deficiency, but Si application significantly reduced the K-deficiency-induced decrease. The xylem sap K^+ concentration in the Si-treated plant was 80% ($p<0.05$) higher than that of untreated plants (Fig. 7a). The xylem sap osmotic potential was significantly enhanced after 15 days of K deficiency. However, under K deficiency conditions, the xylem sap osmotic potential of the plant was 32% ($p<0.05$) lower in Si-treated plants than in untreated plants (Fig. 7b). The low osmotic potential in xylem sap indicates the higher osmotic gradient and driving force between the soil and the root xylem sap in Si-treated plants under the K-deficient condition. The contribution of K^+ to the xylem sap osmotic potential was also investigated and the result showed that K^+ accounts for 38% of the osmotic potential under the control condition with or without Si, 24% of the osmotic potential under K-deficiency conditions with Si treatment, and 17% under K-deficiency conditions without Si. The contribution of K^+ to the osmotic potential in Si-treated plants was 41% ($p<0.05$) higher than that of in Si-untreated plants under the K-deficiency condition (Fig. 7c).

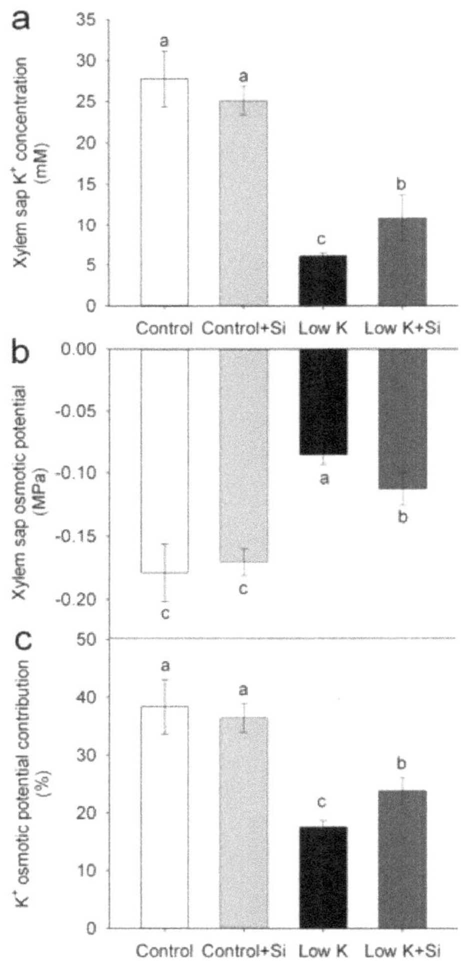

Figure 7: Effects of silicon (Si, 1 mM) on xylem sap K^+ concentration (**a**), osmotic potential (**b**), and K^+osmotic po-

tential contribution (**c**) of sorghum plants grown under control (3 mM KCl) and low K (0.05 mM KCl) conditions. All parameters were measured after 15 days of treatment. Values are presented as the mean ± SE (n = 3). Different letters in one measure indicate statistically significant differences at $P < 0.05$.

In addition to the osmotic gradient, it has been reported that the cell-to-cell pathway can be largely controlled by the activity of aquaporin33. In several studies, aquaporin activity has been assessed using inhibitors such as mercuric chloride ($HgCl_2$), which blocks protein channels and results in a reduction of membrane permeability to water without any adverse effect on the tissue25. In this study, in the absence of $HgCl_2$, the transpiration of Si-treated plants was higher than that of plants without Si treatment under K-deficient condition (Fig. 8). Under control conditions, the presence of $HgCl_2$ decreased the transpiration rate slightly (approximately 6%) in plants that were treated both with and without Si. However, a significant inhibition by $HgCl_2$ in plant transpiration (26% decrease, $p < 0.05$) was found under K-deficient condition. It was also found that the transpirations of plants both with and without Si were decreased by $HgCl_2$ under the K-deficiency condition, but the differences between the Si-treated and -untreated plants were smaller compared with those before $HgCl_2$ addition (Fig. 8). These results showed that aquaporin could be partly responsible for the enhanced Lp_r under the K-deficiency condition.

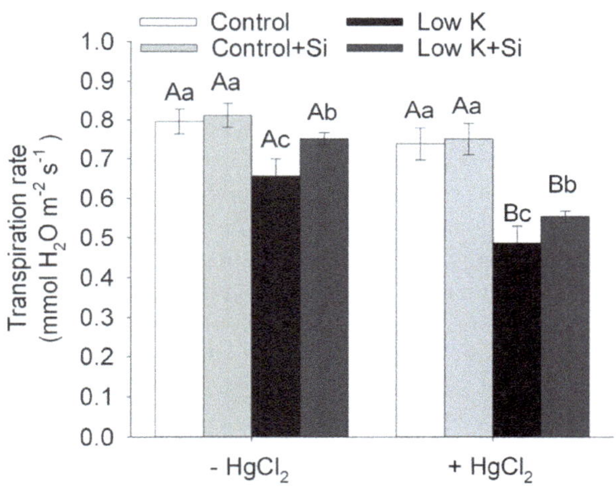

Figure 8: Effect of aquaporin inhibitor ($HgCl_2$) on the transpiration rate of sorghum plants with and without Si application grown under control (3 mM KCl) and low K (0.05 mM KCl) conditions.

The seedlings were divided in two groups, as follows: one measured the normal transpiration rate, and the other was treated with 50 µM $HgCl_2$ for 5 min and then rinsed with distilled water before measuring the transpiration rate in the culture solutions. Values are presented as the mean ± SE (n = 6). Different capital letters and lower cases indicate statistically significant differences at $P < 0.05$ between $HgCl_2$ treatments or K and Si treatments.

Si application Regulated the Expression of Aquaporins genes and K⁺-Related Genes

To investigate the molecular basis for Si alleviation of K-deficiency in plants, we examined the expression levels of aquaporins genes and putative genes homologous to *Arabidopsis* K^+-related genes as reviewed by Wang and Wu37 including those encoding K^+ transporters, K^+channels, regulatory factors, and signaling components in sorghum roots by qRT-PCR (Fig. 9 and Supplementary Table S3). Under control condition, the expression levels of tested genes showed no change by Si application. Under K-deficient condition, mRNA levels of plasma membrane intrinsic protein (*PIP*) *PIP1;3/1;4(2)*, *PIP1;6*, *PIP2;2*, *PIP2;3*decreased up to 2-fold, compared with levels under the control condition, whereas they remained unchanged under K-deficient plus Si treatment. Expression of putative high affinity K^+ transporter (HAK) gene *HAK5.2*, which is regarded as the marker gene in *Arabidopsis*response to K^+ deficiency37 and another K^+-deficiency-induced K^+transporter K uptake permease (KUP) *KUP3.2* increased by 7.0- and 3.3-fold, respectively, at the 12th day under K-deficient condition, whereas they increased less (4.4- and 2.9- fold) under K-deficient plus Si condition. For the putative sorghum *stelar K⁺ outward rectifier* (*SKOR)* gene *SKOR1* and *SKOR2*, those are specifically expressed in the root stele (pericycle and stelar parenchyma) and encode the SKOR channel mediating K^+ secretion from root cortex cells into the xylem38, their expression levels were not affected by K-deficient treatment, whereas they increased significantly by 2.8- and 2.1-fold under K-deficient plus Si condition, respectively.

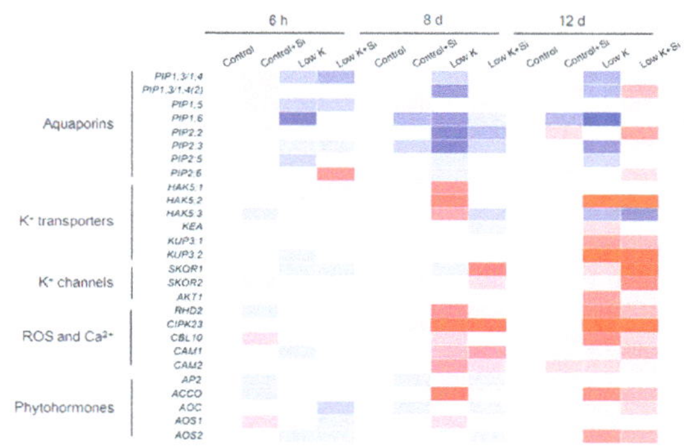

Figure 9: Effects of silicon (Si, 1 mM) on expression of aquaporin genes and putative genes homologous to *Arabidopsis* **K⁺**-related genes in root of sorghum plants with and without Si application grown under control (3 mM KCl) and low K (0.05 mM KCl) conditions.

Each treatment includes three replications and each replication includes three technical replications. Colors indicate change of transcript level in the treated samples with respect to the control samples (red for up-regulation, blue for down-regulation. See color bar at the right of figure).

ROS, Ca^{2+}, ethylene, and jasmonic acid have been considered the major signals in plant responses to K^+-deficiency stress37. In this study, the putative NADPH oxidase gene *root hair defective 2* (*RHD2)*, which is involved in ROS generation39, is significantly up-regulated at the 8th and 12th day under K-deficient condition, whereas the degree of up-regulation was less under K-deficient plus Si condition. Similar changes in mRNA levels were also found in putative sorghum *calcineurin B-like protein* 10 *(CBL10)*, *CBL-interacting serine/threonine-protein kinase 23*(*CIPK23*), and *calmodulin 2 (CaM2)*, those involved in the perception and transduction of K^+-deficiency-induced Ca^{2+} signals, *1-aminocyclopropane-1-carboxylate (ACC) oxidase* (*ACCO)*, encoding the ethylene synthesis enzyme ACC oxidase, and *allene oxide synthase (AOS)*, encoding the jasmonic biosynthetic enzyme allene oxide synthase37.

The up-regulation of the aquaporins genes by Si application under K-deficient condition could be partly responsible for the enhanced aquaporin activity. The up-regulation of the *SKOR* genes and the down-regulation of the *HAK5* gene could contribute to xylem sap K^+concentration under K-deficient condition. Also, Si negatively regulated the Arabidopsis K transporter 1 (AKT1), an inward-rectifying channel, by down regulation of the *CIPK23* gene, and resulted in inhibition of the K^+translocation out of xylem. Moreover, the down-regulation of the K^+-deficiency-responsive genes by Si under K deficient condition suggested that the K deficiency was moderated by Si application, and this moderation was not due to the increase in K uptake by the up-regulation of K^+ transport proteins at the transcriptional level.

DISCUSSION

Severe K deficiency first inhibits plant growth, and leaf necrosis appears in old leaves that had a prolonged exposure time26. In the present study, Si application significantly inhibited the decrease in total dry weight under K-deficiency condition, and leaf necrosis in old leaves was also alleviated in Si-treated plants compared with Si-untreated plants (Fig.1a, Supplementary Fig. S1), indicating that Si application alleviated the K deficiency. The effect of Si on alleviating the K deficiency was also investigated under serial K concentrations (from 6 mM to 0 mM, see Supplementary Fig. S2). The result showed that Si could only alleviate K deficiency when the K concentration was lower than 1.5 mM, and the biggest effect occurred when the K concentration was set between 0.5 mM and 0.01 mM. The results also showed that Si only plays a role when the plant growth was obviously inhibited by K deficiency. Therefore, Si enhanced plant growth specifically under the condition of K deficiency. It is of note that although Si showed a significant effect on alleviating K deficiency, it cannot entirely eliminate the adverse effect of K deficiency under a severe K-deficiency condition.

The major mechanism of adaptation to K starvation in plants is the activation of K^+ uptake through the induction of expression of high affinity transporters29·30·31. A previous study showed that Si application enhanced plant K^+ concentration4. However, in this study, the K^+concentration in both the root and shoot was not affected by Si application. And consistently, the putative *HAK5* and *AKT1*, which mediate almost all K^+ absorption in *Arabidopsis* roots37, were not up-regulated by Si compared with that without Si application under K deficient condition (Fig. 9). Severe K deficiency was also reported to decrease the photosynthetic rate, and maintenance of a high photosynthetic rate is a base of accumulation of more biomass under K deficiency21·27·28. In this study, although the K concentration in plants that were treated with and without Si was low and at the same level under K deficiency, the photosynthetic rate and dry weight in Si-treated plants were much higher than in those without Si, suggesting that the Si-alleviated K-deficiency-induced growth inhibition was not directly due to the increase in K uptake.

Severe K deficiency generally induces stomatal conductance and lowers transpiration rate, which results in a lower photosynthetic rate21·27·28. In this study, plants treated with Si maintained a higher stomatal conductance and transpiration rate, which is coincident with the higher photosynthetic rate (Fig. 2). These results suggest that the Si-improved plant-water status could be the major reason for the increased photosynthetic rate under the K-deficient conditon. Water status, as reflected by LRWC, is the basis of metabolic activity in tissues15·16. It is widely accepted that K deficiency increases a plant's susceptibility to the water stress and that plants with an adequate K supply have better hydrated tissues than those with K deficiency37. In this study, K deficiency decreased the LRWC and water potential, which was moderated by Si application. The results of this study suggest that Si alleviates K deficiency by improving plant-water status.

Plant water-status involves the balance between the differences in water uptake through the roots and the transpiration in leaves15·16. Severe K deficiency has been reported to favor stomatal closure and decrease the transpiration rate22·40. However, the stomatal conductance and transpiration rates were maintained at higher levels in Si-treated plants compared with untreated plants under K-deficiency treatment. This suggests that Si's improvement of the plant-water status is a result of an increase in water uptake, but not because of a reduction in water loss.

The overall plant water uptake is represented by whole-plant hydraulic conductance (K_{plant}), which consists of leaf, stem and root hydraulic conductance32. In this study, K deficiency caused a decrease in K_{plant}that was moderated by Si application (Fig. 5). The seedlings used in the experiment did not form stems. Therefore, K_{plant} was only affected by the root and leaf hydraulic conductance. K_{leaf} was not measured in this study due to technical limitations, but it tends to be similar to leaf water potential41. In the present study, the leaf water potential of sorghum with Si was higher than that without Si under K-deficiency condition, suggesting that plants with Si application maintained a higher K_{leaf} than without Si. Root hydraulic conductance (Lp_r) is usually the lowest within the liquid component of the soil–plant–air continuum33 and roots contribute the majority of the whole plant hydraulic resistance34. In addition, the shoot/root ratio was not affected by Si under the K-deficiency conditon (Fig. 1c). Therefore, the similar extent of Si-enhanced K_{plant} (31%) and Lp_r (35%) (Figs 5 and 7) indicated that Si enhanced the plant water uptake was mainly ascribed to the enhancement of the Lp_r.

Gas exchange rates and photosynthesis are limited by the hydraulic conductance of the liquid water pathway from the soil to the leaves42·43·44·45. In this study, K deficiency caused a significant decrease in the LRWC (Fig. 3). The whole-plant transpiration rate and leaf transpiration was decreased by 18% and 57%, respectively, suggesting that water is a serious limitation for photosynthesis under the K-deficient condition. However, the plant hydraulic conductance was 31% higher in Si-treated plants than Si-untreated plants under the K-deficient condition. Similarly, the whole-plant transpiration rate and leaf transpiration was increased by 14% and 94%, respectively (Figs 2and 8). Meanwhile, the photosynthesis rate of the leaf was enhanced by 139% in Si-treated plants compared to untreated plants. Although there is not a clear quantitative relationship between the plant water conductance and the growth rate or photosynthetic rate under K deficiency, the highly consistent tendency for plant water conductance, transpiration rate, photosynthesis, and dry weight suggests that the alleviation of the plant-water status through the enhancement of the water hydraulic conductance played an important role in Si-alleviated K deficiency.

Lp_r represents the root water uptake capacity and it is determined by root surface, root anatomy and root water permeability46. In the root water uptake, the cell to-cell pathway can be largely controlled by the activity of aquaporins, which respond relatively rapidly and reversibly, causing changes in Lp_r33. In this study, plant transpiration with and without Si is decreased by $HgCl_2$, but the difference between Si-treated and -untreated plants was reduced after $HgCl_2$ treatment, and aquaporin contributed to 29% of the Si-induced increase in transpiration. Liu *et al.*24·25 reported that aquaporin regulation accounts for the mechanism of Si-induced alleviation of the plant-water status under drought and salt stress conditions. However, in the present study, 71% of the Si-induced increase in transpiration remained in addition to the aquaporin-regulated water uptake. Thus, other important approaches that mediate the Si-enhanced water uptake under K-deficiency condition must be considered.

The cell-to-cell pathway plays an important role in water transport in the roots under certain conditions, and is driven by the osmotic gradient between the soil and the root xylem sap34. In this study, the osmotic potential of root xylem sap was 20% lower in Si-treated plants than in Si-untreated plants (Fig. 7). This resulted in an increase in the osmotic gradient between the soil and the root xylem. The osmotic gradient between the cultivated solution and the xylem in plants treated with Si was −0.092 MPa; it was −0.073 MPa in Si-untreated plants under the K-deficiency condition. Thus, the osmotic gradient was 26% higher in Si-treated plants than Si-untreated plants.

In addition to serving as an important component of the xylem osmotic potential, xylem K^+ plays a critical role in modifying the xylem hydraulic conductance and water relations in plants in response to ion-mediated volume changes of pectin in pit membranes47. Nardini *et al.*48 reported that a 30 to 60% increase in stem hydraulic conductance in *Laurus nobilis* is associated with an increase in the xylem sap K^+, from 3 to 12 mM. In another experiment on laurel plants, short-term K fertilization led to an increase in xylem sap K^+ and caused a rapid (within 24 h) 45% increase in the transpiration rate and a 30% increase in plant hydraulic conductance19. Cation-mediated enhancement of xylem hydraulic conductivity has been reported for both herbaceous and woody species49. In this study, the K^+ concentration in the root xylem sap was 80% higher in Si-treated plants than in untreated plants under K-deficiency condition. Although we cannot estimate the contribution of the enhanced K^+ concentration to xylem hydraulic conductance andLp_r, based on the previous study, the huge discrepancy in the xylem sap K concentration between plants treated with and without Si is expected to lead to much higher plant hydraulic conductance and transpiration rates in Si-treated plants.

To date, there is no information available on how Si is involved in regulating important metabolic pathways in plants because of the absence of evidence of organic Si in plant tissue. However, there is some evidence that supports the idea that Si can be involved in regulating the plant stress response. For example, the expression of nearly 4,000 genes was affected in Si-treated *Arabidopsis* that was infected with a fungal pathogen50. Liu *et al.*24 also suggested that reactive oxygen species (ROS) could participate in the Si-regulated aquaporin activity under salt stress. In this study, the aquaporins genes were up-regulated by Si application under K-deficient condition, which could be partly responsible for the enhanced aquaporin activity (Fig. 9and Supplementary Table S3). The *SKOR* genes mediate K^+ secretion from root cortex cells into the xylem. Pilot *et al.* reported that the level of *SKOR* transcripts decreased after K^+ shortage in Arabidopsis51. However, other studies showed that SKOR did not transcriptionally respond to K^+ deprivation52, and also short-term K^+ resupply after long-term K^+ starvation in Arabidopsis53. In addition, *OsSKOR* gene was up-regulated under low-K^+ stress31. These inconsistent results suggest that transcriptional response of *SKOR* to K^+ deficiency may differ among different species and K^+ deficiency treatments. In this study, the expression level of *SKOR* showed no change under low K condition, but the it was up-regulated by Si application under K-deficient condition. In addition, the *HAK5*, *CIPK23* and *RHD2* genes were down-regulated by Si application under K-deficient condition. Those changes contribute to xylem sap K^+ concentration under K-deficient condition (Fig. 9 andSupplementary Table S3). Therefore, it appears that Si can regulate the K-deficiency responses at the transcriptional level directly or indirectly. Recently, the K-deficiency response signals have been clarified in model plants, such as *Arabidopsis*. In particular, it has been shown that ROS, Ca^{2+}, ethylene, and jasmonic acid are involved in plant signaling responses to K^+ deficiency37·54·55·56·57. Although no previous studies have reported how Si is involved in plant signaling regulation under various stresses, the application of Si that influences the ROS generation and elimination has been widely reported24·58. In this study, genes related to ROS metabolism, Ca^{2+} signal perception, and ethylene and jasmonic acid synthesis are down-regulated by Si application under K-deficient condition (Fig. 9 and Supplementary Table S3). Thus, it appears that Si can also influence the plant signaling regulation under K-deficient stress. Therefore, although we do not yet have a clear picture of how Si regulates the K concentration in xylem under K deficiency, based on the present and previous studies, we propose the following model: under the K-deficient condition, Si directly or indirectly influence the plant signaling regulation, which are involved in regulating the K channels and transporter gene expression and post-transcriptional regulation. The regulation of K channels and transporters, especially the outward-rectifying channel SKOR, which mediate K concentration in the xylem under the K-deficient condition, may lead to enhanced K concentration in the xylem.

CONCLUSION

K is widely involved in regulating the plant-water status, and severe K deficiency induces plant dehydration. In this study, Si application alleviated the K deficiency while maintaining higher relative water content, water potential, transpiration and plant hydraulic conductance. Combining the present findings with those of previous studies, a model describing how Si is involved in alleviating K deficiency by moderating plant-water status was proposed (Fig. 10). (1) Under the K-deficient condition, Si activated the expression of*SKOR*, down-regulated the expression of *HAK5* and negatively regulated the AKT1 channel, which resulted in the concentrating of K in xylem. (2) The higher K concentration in the xylem sap contributed to the increase in the osmotic gradient and the xylem hydraulic conductance, leading to enhanced root hydraulic conductance. (3) Si could also participate in regulating the aquaporin activity, resulting in increased root hydraulic conductance. (4) The higher root hydraulic conductance resulted in higher whole-plant hydraulic conductance, which contributed to an increase in water uptake and transport, resulting in an alleviation of K-deficiency-induced plant dehydration, and alleviating the K deficiency. However, the K^+-related genes, whose product are almost all hypothetical proteins, used for expression analysis were gotten from homology searching based on information in *Arabidopsis*, rice and maize. Furthermore, besides transcriptional regulation, posttranscriptional regulation also plays a critical role in plant response to K deficiency, especially for the K^+ channels. Thus, the mechanisms underlying the Si-mediated K^+ concentration in the xylem require further study.

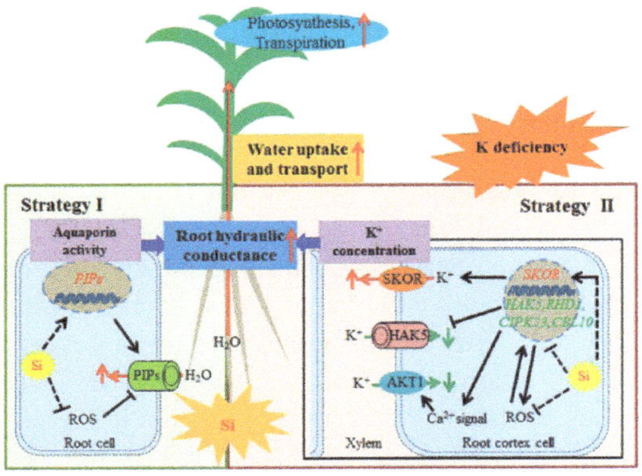

Figure 10: A schematic model for silicon (Si) moderated the K deficiency in sorghum by improving the water uptake and transport.

Two strategies are involved in Si enhance the root hydraulic conductance under K-deficient condition. For strategy I, Si could enhance the aquaporin activity by both up-regulating the expression of aquaporin genes (*PIPs*) and alleviating the decrease of aquaporin activity caused by ROS (Reactive Oxygen Species). For strategy II, Si could activate the K^+translocation to xylem by the activation the expression of *SKOR* gene, which mediate K^+secretion from root cortex cells into the xylem; Si could also down-regulate the expression of *HAK5* gene and may interact with Ca^{2+} and ROS signals to negatively regulate the AKT1 channel to inhibit the K^+ translocation out of xylem (red for up-regulation, green for down-regulation). The enhanced aquaporin activity and K^+ concentration in xylem resulted in increased root hydraulic conductance and higher whole-plant hydraulic conductance, which contributed to an increase in water uptake and transport. The improved water uptake and transport alleviates the K-deficiency-induced plant dehydration and moderates the K deficiency.

MATERIALS AND METHODS

Seedling Cultivation and Si and K Deficiency Treatment

Sorghum cultivar (*Sorghum bicolor* [L.] Moench. cv. Gadambalia) was used in this study. This cultivar can uptake a large amount of Si, and Si application can improve its water status under drought and salt stress conditions24·25. The seedlings were cultivated in a growth chamber that was set to a 14/10 h day/night cycle at a day/night temperature of 28/25 °C with 40 to 50% relative humidity. The amount of photosynthetically active radiation (PAR) to the upper plant was 600 μmol m^{-2} s^{-1}.

Sterilized seeds were germinated in an incubator at 24 °C under dark conditions. After germination for 4 days, uniform seedlings (6 ± 0.5 cm)were selected and transplanted into a plastic container (40 × 28 × 14 cm) with 8 liters of one-quarter strength Hoagland culture solution with no KNO_3, NO_3^- was provided by $Ca(NO_3)_2$ and the K was supplied by KCl. The experiment included four treatments: control, control + Si, low K and low K + Si. The K concentration was 0.05 mM in the low K (K deficiency) treatments and 3 mM in the control (K sufficiency) treatment, supplied by KCl. Under 3 mM K concentration, the growth of seedlings showed no difference compared to that under 6 mM K (normal K concentration in Hoagland solution). The growth of seedlings showed obvious inhibition under the 0.05 mM K condition (Supplemental Fig. S1). The pH of the culture solution was adjusted to 6.053. Si was set to 0 mM in the without Si treatment and 1mM H_2SiO_3 in the Si treatments. H_2SiO_3 was produced by passing sodium silicate solution through a column filled with cation-exchange resins58. The culture solution was renewed every three days and continuously aerated. Fifteen days after transplantation, samples were collected and the parameters listed below were measured.

Dry weight and root/shoot ratio

Fifteen days after transplantation, the plant was harvested. The shoots and roots were separated and dried at 75 °C for 72 h. The total weight and root/shoot ratio were calculated. Each treatment included 10 plants.

Photosynthetic rate, Stomatal Conductance and Transpiration rate

Fifteen days after transplantation, the upper fully expanded leaves were selected to measure the photosynthetic rate, stomatal conductance and transpiration rate using a portable photosynthesis system (Li-6400; LI-COR Inc., Lincoln, NE, USA) between 10:00 AM and 12:00 AM. The leaf was placed in a 6 cm^2 chamber at a photo flux density of 1000 μmol $m^{-2}s^{-1}$. Each treatment included six plants.

Leaf relative Water Content, leaf water Potential and Osmotic Potential

Fifteen days after transplantation, the upper fully expanded leaves were sampled and LRWC was measured and calculated according to Liu *et al.*25. Upper fully expanded leaves were covered with aluminum foil before excision from the plant and the water potential was measured using a pressure chamber (Model 3500, Soil Moisture Corp., Santa Barbara, CA, USA). The leaf osmotic potential of the upper fully expanded leaves was measured using a dew point microvolt meter (Model 5600, Wescor, Logan, UT, USA). Each treatment included five replications.

K^+ Concentration and total K^+ uptake in the tissue

Dried shoot and root samples were ground to powder, weighed and then digested using nitric acid at 320 °C for 5 h. K^+ concentrations were measured using an atomic adsorption spectrometer with a flame photometer (ZL5100, PerkinElmer, Inc., USA)59. The K^+ absorption was calculated based on the dry weight and tissue K^+ concentration. Each treatment included three replications.

Whole-plant Hydraulic Conductance

K_{plant} was calculated according to the following equation32:

$$K_{plant} = \text{Transpiration rate}/(\text{Solution water potential} - \text{Leaf water potential}) \quad (1)$$

The transpiration rate was determined gravimetrically. The leaf water potential was measured as described above. In hydroponic culture, the culture solution water potential was −0.032 MPa in the control, −0.034 MPa in control + Si, −0.018 MPa in low K, and −0.019 MPa in low K + Si treatment groups. Each treatment included six replications.

Transpiration rate in response to aquaporin inhibitor

Aquaporin-mediated water transport was investigated by measuring changes in the transpiration rate in response to an aquaporin inhibitor ($HgCl_2$), according to Knipfer *et al.*60. The seedlings were divided into two groups: one group (6 plants) was used for measuring the transpiration rate, and the other group (6 plants) was treated with 50 µM $HgCl_2$ for 5 min and then rinsed with distilled water (devoid of $HgCl_2$) before measuring the transpiration rate in the culture solutions. The transpiration rate was determined gravimetrically.

Root Hydraulic Conductance

Lp_r based on the root surface area was measured using a pressure chamber (Model 3500, Soil Moisture Corp., Santa Barbara, CA, USA), according to the method of Miyamoto *et al.*61. The roots were enclosed in a steel chamber and pneumatic pressure was applied to the root medium. The pressure in the chamber was increased in steps of 0.1 MPa up to a pressure of 0.5 MPa. Exuded sap was collected for 60 s at a given gas pressure and weighed. The root surface area was then scanned and analyzed using WinRHIZO PRO 2009 software (Regent Inc., Canada). The water flow, Jv_r in $m^3\ m^{-2}\ s^{-1}$, was obtained. The osmotic potential of the collected exuded sap was measured as described above and Lp_r ($m^3 m^{-2}\ s^{-1}\ MPa^{-1}$) was calculated from the slope of Jv_r against driving force P, consisting of P_{gas} and the osmotic gradient, according to the following equation:

$$Jv_r = Lp_r\left(P_{gas} + \sigma_{sr}\Delta\pi\right) \quad (2)$$

In this equation, σ_{sr} is the root reflection coefficient for nutrient salts in the xylem, which was estimated to be $\sigma_{sr} = 0.4$, according to Miyamoto*et al.*56. Each treatment included six plants.

Xylem sap K^+ Concentration and Osmotic Potential

Xylem sap was extracted between 10:00 AM and 12:00 AM according to Liu *et al.*25. Plants were rapidly cut using a razor blade at the base of the root system. The roots were enclosed in a steel chamber and the surface of the cut was washed with distilled water and blotted dry to remove any disrupted and residual cell. The pneumatic pressure applied to the root medium in the chamber was increased to 0.2 MPa until xylem sap flowed out from the excised tip. The first droplet was discarded to avoid contamination, and then 50 µl of sap was collected from each plant to determine the xylem sap K^+ concentration and osmotic potential. The xylem sap osmotic potential was measured using a dew point microvolt meter and the ion concentration was measured by an atomic adsorption spectrometer with a flame photometer as described above. Each treatment included three replications.

Gene expression Analysis

Roots were sampled at the 8th day (visible K^+ starvation symptoms appeared) and the 12th day (significant differences between the plants with or without Si application under K deficiency appeared) after transplantation. In addition, 12 days after transplantation, one group of seedlings (6 plants) under control and control + Si conditions were transplanted to low K and low K+ Si conditions, respectively; roots were sampled also after K deficiency treatment for 6 hours to investigate the short-term starvation responses. The expression of 8 sorghum plasma membrane intrinsic protein (*PIP*) aquaporin genes according to Liu *et al.*25 and 19 putative genes homologous to *Arabidopsis* K^+-related genes based on information in *Arabidopsis*, rice (*Oryza sativa*) and maize (*Zea mays*), were analyzed using quantitative RT-PCR. RNA was isolated and gene expression was analyzed using *Actin1* as constitutive controls25. Each treatment included three replications and each replication included three technical replications. The genes and the sequences of their specific primers are listed in Supplemental Table S1. The protein sequence similarities between putative K^+-related genes and those in*Arabidopsis*, rice and maize are listed in Supplemental Table S2.

Statistical Analysis

Data were analyzed using an analysis of variance (ANOVA) using the Statistical Analysis System (SAS version 8.0) software. The differences between the means were compared using the Tukey-Kramer test ($p < 0.05$). All experiments were repeated at least three times. All plots were created using SigmaPlot12.0 and the different letters indicating statistically significant differences at $P < 0.05$ are indicated in the figures.

REFRENCES

1. Maathuis, F. J. M. Physiological functions of mineral macronutrients. Curr. Opin. Plant Biol. 12, 250–258 (2009).
2. Hafsi, C., Debez, A. & Abdelly, C. Potassium deficiency in plants: effects and signaling cascades. Acta Physiol. Plant. 36, 1055–1070 (2014).
3. Sharma, T., Dreyer, I. & Riedelsberger, J. The role of K+ channels in uptake and redistribution of potassium in the model plant Arabidopsis thaliana. Front. Plant Sci. 4, 224 (2013).
4. Miao, B., Han, X. & Zhang, W. The ameliorative effect of silicon on soybean seedlings grown in potassium-deficient medium. Ann. Bot. 105, 967–973 (2010).
5. Anschutz, U., Becker, D. & Shabala, S. Going beyond

nutrition: Regulation of potassium homoeostasis as a common denominator of plant adaptive responses to environment. J. Plant Physiol. 171, 670–687 (2014).

6. De Boer, A. H. Potassium translocation into the root xylem. Plant Biology 1, 36–45 (1999).
7. Wang, Y. & Wu, W. Genetic approaches for improvement of the crop potassium acquisition and utilization efficiency. Curr. Opin. Plant Biol. 25, 46–52 (2015).
8. Epstein, E. The anomaly of silicon in plant biology. Proc. Natl. Acad. Sci. USA 91, 11–17 (1994).
9. Epstein, E. Silicon. Annu. Rev. Plant Physiol. Plant Mol. Biol. 50, 641–664 (1999).
10. Epstein, E. Silicon: its manifold roles in plants. Ann. Appl. Biol. 155, 155 (2009).
11. Guntzer, F. Benefits of plant silicon for crops: a review. Agron. Sustain. Dev. 32, 201–213 (2012).
12. Pavlovic, J. et al. Silicon alleviates iron deficiency in cucumber by promoting mobilization of iron in the root apoplast. New Phytol. 198, 1096–1107 (2013).
13. Hernandez-Apaolaza, L. Can silicon partially alleviate micronutrient deficiency in plants? a review. Planta 240, 447–458 (2014).
14. Eneji, A. E. et al. Growth and nutrient use in four grasses under drought stress as mediated by silicon fertilizers. J. Plant Nutr. 31, 355–365 (2008).
15. Mali, M. & Aery, N. C. Influence of silicon on growth, relative water contents and uptake of silicon, calcium and potassium in wheat grown in nutrient solution. J. Plant Nutr. 31, 1867–1876 (2008).
16. Flower, D. J. & Ludlow, M. M. Contribution of osmotic adjustment to the dehydration tolerance of water-stressed pigeon pea (Cajanus cajan (L.) millsp.) leaves. Plant Cell Environ. 9, 33–40 (1986).
17. Hsiao, T. C. & Läuchli, A. Role of potassium in plant–water relations in Advances in Plant Nutrition (ed. Tinker, B., Läuchli, A.) Vol. 2, 281–311 (Praeger Scientific, New York, 1986).
18. Shabala, S. Regulation of potassium transport in leaves: from molecular to tissue level. Ann. Bot. 92, 627–634 (2003).
19. Läuchli, A. Mechanism of nutrient fluxes at membranes of the rootsurface and their regulation in the whole plant in Roots, Nutrient and Water Influx, and Plant Growth (ed. Barber, S. A., Bouldin, D. R.) 1–25 (ASA Special Publication, No. 49, Madison, 1984).
20. Oddo, E. et al. Short-term effects of potassium fertilization on the hydraulic conductance of Laurus nobilis L. Tree Physiol. 31, 131–138 (2011).
21. Kanai, S. et al. Potassium deficiency affects water status and photosynthetic rate of the vegetative sink in green house tomato prior to its effects on source activity. Plant Sci. 180, 368–374 (2011).
22. Fournier, J. M., Roldan, A. M., Sanchez, C., Alexandre, G. & Benlloch, M. K+ starvation increases water uptake in whole sunflower plants. Plant Sci. 168, 823–829 (2005).
23. Romero-Aranda, M. R., Jurado, O. & Cuartero, J. Silicon alleviates the deleterious salt effect on tomato plant growth by improving plant water status. J. Plant Physiol. 163, 847–855 (2006).
24. Liu, P. et al. Enhanced root hydraulic conductance by aquaporin regulation accounts for silicon alleviated salt-induced osmotic stress in Sorghum bicolor L. Environ. Exp. Bot. 111, 42–51 (2015).
25. Liu, P. et al. Aquaporin-mediated increase in root hydraulic conductance is involved in silicon-induced improved root water uptake under osmotic stress in Sorghum bicolor L. J. Exp. Bot. 65, 4747–4756 (2014).
26. Mengel, K. & Kirkby, E. A. Principles of Plant Nutrition (International Potash Institute, Bern, 1987).
27. Bednarz, C. W., Oosterhuis, D. M. & Evans, R. D. Leaf photosynthesis and carbon isotope discrimination of cotton in response to potassium deficiency. Environ. Exp. Bot. 39, 131–139 (1998).
28. Tomemori, H., Hamamura, K. & Tabane, K. Interactive effects of sodium and potassium on the growth and photosynthesis of spinach and komatsuna. Plant Prod. Sci. 5, 281–285 (2002).
29. Ashley, M. K., Grant, M. & Grabov, A. Plant responses to potassium deficiencies: a role for potassium transport proteins. J. Exp. Bot. 57, 425–436 (2006).
30. Siddiqi, M. Y. & Glass, A. D. M. Regulation of K+ influx in barley: evidence for a direct control of influx by K+ concentration of root cells. J. Exp. Bot. 38, 935–947 (1987).
31. Ma, T., Wu, W. & Wang, Y. Transcriptome analysis of rice root responses to potassium deficiency. BMC Plant Biol. 12, 161 (2012).
32. Martre, P. et al. Plasma membrane aquaporins play a significant role during recovery from water deficit. Plant Physiol. 130, 2101–2110 (2002).
33. Vandeleur, R. K. et al. The role of plasma membrane intrinsic protein aquaporins in water transport through roots: diurnal and drought stress responses reveal different strategies between isohydric and anisohydric cultivars of grapevine. Plant Physiol. 149, 445–460 (2009).
34. Javot, H. & Maurel, C. The role of aquaporins in root water uptake. Ann. Bot. 90, 301–313 (2002).
35. Steudle, E. Water uptake by roots: effects of water deficit. J. Exp. Bot. 51, 1531–1542 (2000).
36. Steudle, E. Water uptake by plant roots: an integration of views. Plant and Soil 226, 45–56 (2000).
37. Article
38. Wang, Y. & Wu, W. H. Potassium transport and signaling in higher plants. Annu. Rev. Plant Biol. 64, 451–476 (2013).
39. Gaymard, F. et al. Identification and disruption of a plant shaker-like outward channel involved in K+ release into the xylem sap. Cell 94, 647–655 (1998).
40. Shin, R. & Schachtman, D. P. Hydrogen peroxide mediates plant root cell response to nutrient deprivation. Proc. Natl. Acad. Sci. USA 101, 8827–8832 (2004).
41. Mengel, K. & Kirkby, E. A. Principles of Plant Nutrition

(Kluwer Academic Publishers, Dordrecht, 2001).

42. Brodribb, T. J. & Holbrook, N. M. Stomatal closure during leaf dehydration, correlation with other leaf physiological traits. Plant Physiol. 132, 2166–2173 (2003).

43. Brodribb, T. J., Holbrook, N. M., Zwieniecki, M. A. & Palma, B. Leaf hydraulic capacity in ferns, conifers and angiosperms: impacts on photosynthetic maxima. New Phytol. 165, 839–846 (2005).

44. Hubbard, R. M., Ryan, M. G., Stiller, V. & Sperry, J. S. Stomatal conductance and photosynthesis vary linearly with plant hydraulic conductance in ponderosa pine. Plant Cell Environ. 24, 113–121 (2001).

45. Nardini, A. & Salleo, S. Limitation of stomatal conductance by hydraulic traits: sensing or preventing xylem cavitation? Trees 15, 14–24 (2000).

46. Sperry, J. S. Hydraulic constraints on plant gas exchange. Agr. Forest Meteorol. 104, 13–23 (2000).

47. Sutka, M. et al. Natural variation of root hydraulics in Arabidopsis grown in normal and salt-stressed conditions. Plant Physiol. 155, 1264–1276 (2011).

48. Zwieniecki, M. A., Melcher, P. J. & Michele Holbrook, N. M. Hydrogel control of xylem hydraulic resistance in plants. Science 291, 1059–1062 (2001).

49. Nardini, A., Grego, F., Trifilo, P. & Salleo, S. Changes of xylem sap ionic content and stem hydraulics in response to irradiance in Laurus nobilis. Tree Physiol. 30, 628–635 (2010).

50. Jansen, S. et al. Do quantitative vessel and pit characters account for ion-mediated changes in the hydraulic conductance of angiosperm xylem? New Phytol. 189, 218–228 (2011).

51. Fauteux, F., Chain, F., Belzile, F., Menzies, J. G. & Belanger, R. R. The protective role of silicon in the Arabidopsis-powdery mildew pathosystem. Proc. Natl. Acad. Sci. USA 103, 17554–17559 (2006).

52. Pilot, G., Gaymard, F., Mouline, K., Chérel, I. & Sentenac, H. Regulated expression of arabidopsis shaker K+ channel genes involved in K+ uptake and distribution in the plant. Plant Mol. Biol. 51, 773–787 (2003).

53. Gierth, M. & Schroeder, S. I. The potassium transporter AtHAK5 functions in K+ deprivation-induced high-affinity K+ uptake and AKT1 K+ channel contribution to K+ uptake kinetics in arabidopsis roots. Plant Physiol. 137, 1105–1114 (2005).

54. Patrick, A., Rainer, B. & Anna, A. The potassium-dependent transcriptome of arabidopsis reveals a prominent role of jasmonic acid in nutrient signaling. Plant Physiol. 136, 2556–2576 (2004).

55. ISICASPubMedArticle

56. Xu, J. et al. A protein kinase, interacting with two calcineurin B-like proteins, regulates K+ transporter AKT1 in Arabidopsis. Cell 125, 1347–1360 (2006).

57. ISICASPubMedArticle

58. Wang, Y., He, L., Li, H., Xu, J. & Wu, W. Potassium channel alpha-subunit AtKC1 negatively regulates AKT1-mediated K+ uptake in Arabidopsis roots under low-K+ stress. Cell Res. 20, 826–837 (2010).

59. Cherel, I., Lefoulon, C., Boeglin, M. & Sentenac, H. Molecular mechanisms involved in plant adaptation to low K+ availability. J. Exp. Bot. 65, 833–848 (2014).

60. Liu, L., Ren, H., Chen, L., Wang, Y. & Wu, W. A protein kinase, Calcineurin B-Like Protein-Interacting Protein Kinase9, interacts with calcium sensor Calcineurin B-Like Protein3 and regulates potassium homeostasis under low-potassium stress in Arabidopsis. Plant Physiol. 161, 266–277 (2013).

61. Yin, L. N., Wang, S. W., Li, J. Y., Tanaka, K. & Oka, M. Application of silicon improves salt tolerance through ameliorating osmotic and ionic stresses in the seedling of Sorghum bicolor. Acta Physiol. Plant. 35, 3099–3107 (2013).

62. Chen, D. Q., Yin, L. N., Deng, X. P. & Wang, S. W. Silicon increases salt tolerance by influencing the two-phase growth response to salinity in wheat (Triticum aestivum L.). Acta Physiol. Plant. 36, 2531–2535 (2014).

63. Knipfer, T., Besse, M., Verdeil, J. L. & Fricke, W. Aquaporin-facilitated water uptake in barley (Hordeum vulgare L.) roots. J. Exp. Bot. 62, 4115–4126 (2011).

64. Miyamoto, N., Steudle, E., Hirasawa, T. & Lafitte, R. Hydraulic conductivity of rice roots. J. Exp. Bot. 52, 1835–1846 (2001).

CHAPTER 13

GROWTH TEMPERATURE AND GENOTYPE BOTH PLAY IMPORTANT ROLES IN SORGHUM GRAIN PHENOLIC COMPOSITION

GangchengWu, Stuart K. Johnson, Janet F. Bornman, Sarita J. Bennett, MichaelW. Clarke, Vijaya Singh and Zhongxiang Fang

INTRODUCTION

Sorghum (*Sorghum bicolor* (L.) Moench) is an important global cereal crop due to its tolerance to high temperatures and arid conditions. The total global sorghum production was around 62.3 million tonnes in 2013 and more than 50% of that was produced in the African and Asian regions[1]. In these two regions, sorghum grains are used as food, such as flat breads, or porridges[2]. However, in developed countries such as the USA and Australia, sorghum grain is primarily used for animal feed and ethanol production[3].

It has been reported that some sorghum genotypes have high levels of phenolic compounds and strong antioxidant activities[4]. The consumption of foods with high levels of phenolic compounds may assist in prevention of certain age-related diseases[5,6]. Results of epidemiological studies suggested that dietary intake of phenolic abundant foods may reduce the incidence of cardiovascular disease, coronary heart disease and a variety of cancers[7,8]. In addition, the polyphenols, such as 3-deoxyanthocyanidins, specific to sorghum, have reportedly good anti-inflammatory properties[2]. The level of phenolic compounds in sorghum grains vary widely with genotype as previously reported[9,10]. However, most studies only evaluated the polyphenols as a collective group, and information on the differences in specific polyphenols among sorghum genotypes is limited. Though some specific polyphenols in several sorghum genotypes were HPLC analyzed[11], others without authentic standards might not be identified by HPLC without mass spectrometry.

The Intergovernmental Panel on Climate Change (IPCC) Fourth Assessment have predicted that the global temperature will increase by 1.8 to 4.0 °C by 2100[12]. Understanding variability of polyphenol composition under increased temperature is helpful because the concentration and profile of polyphenols can affect the nutritional values of sorghum grain. Published studies investigating how increased temperature influences the phenolic compounds in plants, have mainly focused on fruits and vegetables. Wang and Zheng[13] reported that the total phenolic compounds in strawberry increased under high temperature treatments, whereas higher temperatures significantly decreased total phenolic contents in sweet potato leaves[14]. However, no information is available on the effects of elevated temperatures on phenolic compounds in cereal grains, including sorghum.

Therefore, the purpose of this research was to investigate the effects of two different temperature regimes during plant production on sorghum grain phenolic compounds, analysed using high performance liquid chromatography-diode array detection-electrospray ionization mass spectrometry (HPLC-DAD-ESIMS) in six genotypes. The potential for matching sorghum grains with specific phenolic profiles when grown under high temperature conditions to specific food use are discussed.

RESULTS

Identification of phenolic compounds

The HPLC retention time (Rt), UV absorption maxima and parent and fragment ions of 23 different phenolic compounds positively or tentatively identified in the samples are listed in Table 1. Representative chromatograms of the free polyphenolics in the red-pericarp genotype CHH2 (Fig. 1) are presented. To assist the interpretation of MS fragmentation patterns, the 23 compounds were classified into five groups according to their chemical structures.

Citation: Gangcheng Wu, Stuart K. Johnson, Janet F. Bornman, Sarita J. Bennett, Michael W. Clarke, Vijaya Singh & Zhongxiang Fang, Growth temperature and genotype both play important roles in sorghum grain phenolic composition, doi: 10.1038/srep21835.

Table 1: Identification of free and bound phenolic compounds in sorghum grains by HPLC-DAD-ESIMS and respective standards (std)

Peak No.	Rt[a] (min)	λ_{max} (nm)	m/z $[M-H]^-$	m/z MS^2 (Abundance %)	Tentative identification
1	10.9	280	577	425 (60), 289 (26)	Procyanidin B1
2	13.1	298sh[b], 326sh	253	179 (1), 161 (78), 135 (55)	2-*O*-caffeoylglycerol
3	13.5	280. 310	137	109 (15)	Protocatechuic aldehyde
4	14.4	298sh, 326sh	253	179 (1), 161 (78), 135 (55)	1-*O*-caffeoylglycerol
5	15.0	297, 322	179	135 (100)	Caffeic acid (std)
6	15.8	282	449	287 (8), 151 (100), 135 (41)	Eriodictyol-7-*O*-glucoside
7	16.1	280, 490	269	241 (41), 225 (27), 169 (19), 133 (30)	Luteolinidin (std)
8	17.7	310	237	163 (20), 145 (50), 119 (31)	2-*O*-*p*-coumaroylglycerol
9	18.2	275, 470	253	225 (10), 209 (70), 179 (40), 117 (65)	Apigeninidin (std)
10	19.1	283	433	287 (100), 151 (59)	Eriodictyol deoxyhexoside
11	19.2	310	163	119 (100)	*p*-coumaric acid
12	20.8	295, 325	193	178 (42), 134 (100)	Ferulic acid (std)
13	21.6	288	303	285 (100), 217 (9), 177 (18), 125 (35)	Taxifolin (std)
14	24.3	285	303	285 (10), 177 (18), 125 (35)	Taxifolin isomer
15	26.7	295, 325	193	134 (100)	Ferulic acid isomer
16	28.5	300sh, 326sh	415	253 (100), 179 (100) , 161 (11), 135 (85)	1,2-*O*-dicaffeoylglycerol
17	29.4	300sh, 326sh	415	253 (100), 179 (100) , 161 (11), 135 (85) 135(85),	1,3-*O*-dicaffeoylglycerol
18	30.6	287	287	151(10)	Eriodictyol
19	31.1	252, 347	285	241 (1), 217 (3), 199 (3), 175 (3), 151 (17), 133 (13), 107 (3)	Luteolin (std)
20	33.4	219, 315	399	253 (80), 235 (11), 179 (25), 163 (86), 145 (35), 119 (100),135 (57)	Coumaroyl-caffeoylglycerol
21	34.2	295sh, 325sh	429	253 (70), 235 (11), 193 (100), 175 (32), 161 (53), 135 (41)	Feruloyl-caffeoylglycerol
22	35.9	266, 322	269	225 (10), 201 (30), 183 (10), 149 (32), 117 (100)	Apigenin (std)
23	36.2	295	271	177 (10), 151 (50), 119 (20), 107 (20)	Naringenin (std)

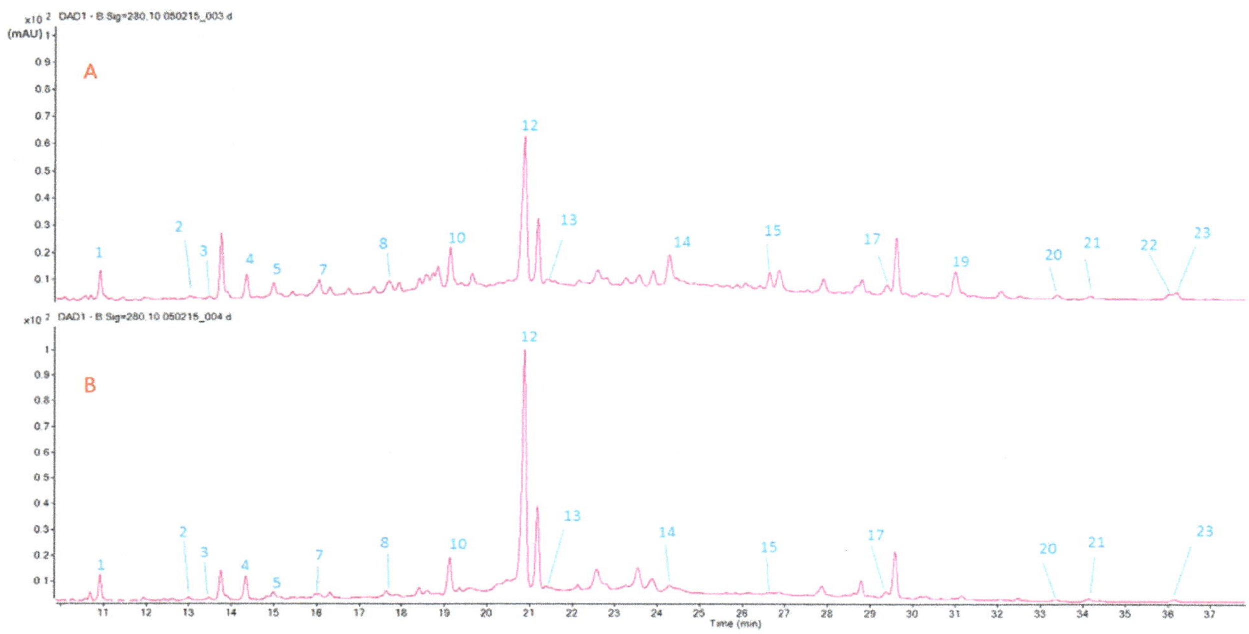

Figure 1: Representative chromatograms of sorghum whole grain phenolic compounds (**A**) Free phenolic compounds from genotype CHH2 grown at OT. (**B**) Free phenolic compounds from CHH2 grown at HT.

Phenolic acids. Phenolic acids consist of hydroxybenzoic acids and hydroxycinnamic acids. Only peak 3 was tentatively identified as a derivative of hydroxybenzoic acids. Its MS spectra showed a pseudomolecular ion $[M-H]^-$ at *m/z* 137 and a prominent fragment ion at *m/z* 109 via loss of a CO group, corresponding to protocatechuic aldehyde. The UV spectra with λ_{max} at 280 and 310 nm were consistent with those of protocatechuic aldehyde reported in the literature15.

Three hydroxycinnamic acids, namely caffeic acid (peak 5), *p*-coumaric acid (peak 11) and ferulic acid (peak 12), were identified, giving the deprotonated molecule $[M-H]^-$ at *m/z* 179, 163, and 193 and distinguished fragment ions at m/z 135, 119 and 135, respectively15. These hydroxycinnamic acids showed the loss of a CO_2 group from the carboxylic acid functional group. Ferulic acid (peak 12) also lost a CH_3group demonstrating a $[M-H-15]^-$ anion at m/z 178 (Table 1). The caffeic and ferulic acids in the samples were further identified by comparison with the HPLC Rt, UV and MS spectra of the authentic standards. Peak 15 had similar UV and MS spectra to ferulic acid, and was tentatively identified as ferulic acid isomer16.

Several derivatives of hydroxycinnamic acids were identified. Peaks 2 and 4 had deprotonated molecule $[M-H]^-$ at *m/z* 253 and fragmentation patterns with ions at m/z 179 [caffeic acid $-H]^-$ and 135 [caffeic acid $-H-CO_2]^-$, suggesting the possible occurrence of a caffeic acid residue. The UV λ_{max} of these two peaks were at 297 and 326 nm, which was the same as that of caffeic acid. The $[M\text{-}H\text{-}74]^-$ ion at *m/z* 179 in peaks 2 and 4 is typical of that produced by the loss of a glycerol residue. Hence, peaks 2 and 4 were tentatively deduced as either 1-*O*-caffeoylglycerol or 2-*O*-caffeoylglycerol17, since it was not possible to ascertain the substituent group positions. Peak 8 had the same UV λ_{max} of 310 nm as that of *p*-coumaric acid. MS spectra showed a deprotonated molecule $[M\text{-}H]^-$ at *m/z* 237 with fragment ions at *m/z* 163, 145 and 119. Fragment ions at *m/z* 163 and 119 appeared to correspond to deprotonated and decarboxylated *p*-coumaric acid, indicating the presence of *p*-coumaric acid. According to the above, this compound which lost the glycerol residue can produce a 163$[M\text{-}H\text{-}74]^-$ anion. Therefore, this compound was tentatively identified as 1-*O*-*p*-coumaroylglycerol or 2-*O*-*p*-coumaroylglycerol17. The UV spectral characteristics of peaks 16 and 17 were similar to that of caffeic acid with λ_{max} at about 326 nm, suggesting derivatives of caffeic acid. They had deprotonated molecule $[M\text{-}H]^-$ at *m/z* 415 and fragmentation patterns with ions at *m/z* 253, 179 [caffeic acid –H]$^-$, 161 [caffeic acid $-H\text{-}H_2O]^-$ and 135 [caffeic acid $-H\text{-}CO_2]^-$, which supported the presence of a caffeic acid residue. The fragment ion of *m/z* at 253 (415-162) could correspond to dehydrated caffeic acid attached to the caffeoylglycerol. From a comparison of literature data of UV and MS spectra, these two peaks were tentatively identified as either 1,2-*O*-dicaffeoyglycerol or 1,3-*O*-dicaffeoyglycerol17. The deprotonated molecule ion $[M\text{-}H]^-$ showed peaks of 20 and 21 at m/z 399 and 429, respectively. Both peaks contained the same ions at*m/z* 253, indicating that these two compounds may also contain a caffeoylglycerol residue. Peak 20 also showed two prominent ions at m/z 163 [*p*-coumaric acid-H]$^-$ and 235 [M-H- *p*-coumaric acid]$^-$, suggesting it could have been derived from the loss of a *p*-coumaric acid residue. For peak 21, some fragment ions at *m/z* 235 and 193 were detected. The ion at *m/z* 235 [M-H-194] indicated the loss of a ferulic acid moiety, while another ion at *m/z* 193, corresponded to deprotonated ferulic acid, suggested the presence of a ferulic acid moiety. After comparisons with the literature17, peaks 20 and 21 were tentatively identified as *p*-coumaroyl-caffeoylglycerol and feruloyl-caffeoylglycerol, respectively.

Flavanols. Peak 1 had a deprotonated molecular ion $[M\text{-}H]^-$ at *m/z* 577 and fragment ions at *m/z* 425 and 289 (Table 1). The ion at *m/z* 425 [M-H-152] can be produced by the removal of a B-ring from the heterocyclic ring of catechin derivatives with the Retro-Diels-Alder rearrangement in the ESI negative ion mode18. The most intense ion at*m/z* 289 $[M\text{-}H\text{-}289]^-$ was considered to be specific for a C—C interflavan linkage in the catechin dimer19. The UV spectra of peak 1 were similar to that of (+)-catechin with λ_{max} at 280 nm. Therefore, this peak was tentatively identified as procyanidin B, which was supported by its MS and UV spectra in the literature20.

Flavones. Most flavones have two characteristic UV absorption bands in the region 250–285 nm and 320–385 nm. Band I(320–385 nm) is produced by the B-ring, while Band II(250–285 nm) by the A-ring21. Peaks 19 and 22 (Fig. 2) were positivity identified as the flavones luteolin and apigenin respectively by comparison with Rt, UV and mass spectral data (Table 1) of the authentic standards.

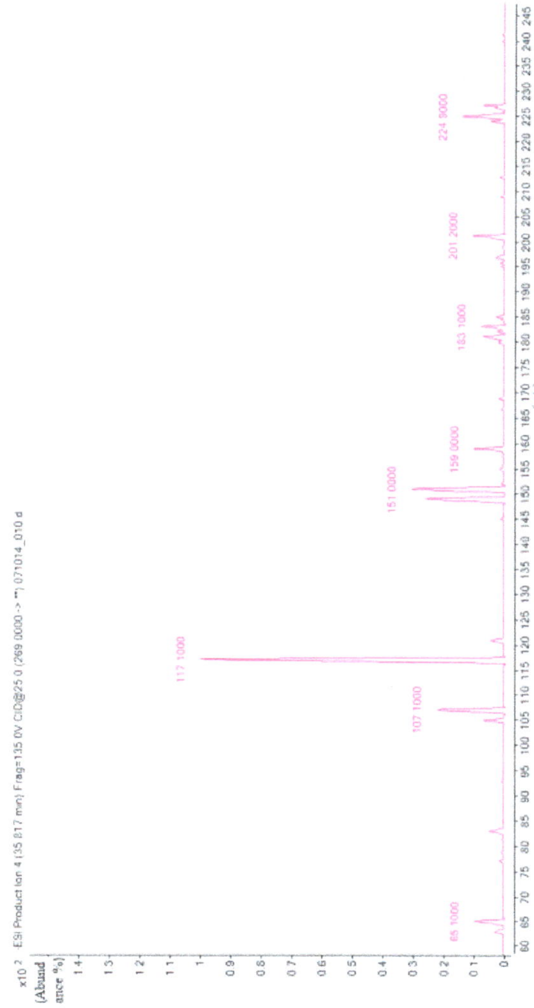

Figure 2: Mass spectra of apigenin of sorghum grains.

Flavanones and dihydroflavonols. Several flavanones were identified in sorghum grains. Peak 23 was positively identified as naringenin by comparing its Rt, UV and mass spectral data with the authentic standard. The deprotonated molecule ion $[M\text{-}H]^-$ of peak 18 was at m/z 287, suggesting this compound might be eriodictyol or dihydrokaempferol with the molecular mass of 288. The peak 18 also showed the characteristic fragmentation ion at *m/z* 151 and UV spectra of eriodictyol, which exhibited a strong absorption in the region of 270–295 nm. Referring to MS and UV spectra of eriodictyol from the literature, peak 18 was tentatively identified as eriodictyol15. Peaks 6 and 10 showed the deprotonated molecule ions $[M\text{-}H]^-$ at *m/z* 449 and 433, respectively. The fragmentation ions at *m/z* 287 and 151, corresponding to an eriodictyol moiety, were also observed. The UV spectra of these two peaks also showed similar absorption in the region 270–295 nm as that for eriodictyol, indicating they might be derivatives of eriodictyol. It is well known that most of the flavonoids can conjugate sugars, and their derivatives are present as the glycosides of *O*- or *C*- forms22. The peak 6 produced a fragmentation ion at *m/z* 287 $[M\text{-}H\text{-}162]^-$ by possible loss of a glucoside, and the same fragmentation ion 287 $[M\text{-}H\text{-}146]^-$ was also produced via the possible loss of a deoxyhexoside from peak 10. Therefore, peaks 6 and 10 were tentatively identified as eriodictyol-7-*O*-glucoside and eriodictyol deoxyhexoside, respectively23.

The compound of peak 13 was positivity identified as the hydroflavonol taxifolin by comparison of Rt, UV and mass spectral data (Table 1) with the authentic standard. Since peak 14 had similar UV and MS spectra to that of peak 13, it was tentatively identified as taxifolin isomer.

3-Deoxyanthocyanidins. Two compounds, corresponding to peaks 7 and 9, exhibited the distinctive UV spectra of 3-deoxyanthocyanidins. Typical 3-deoxyanthocyanidins produce two UV absorption peaks in the 450–560 nm region due to the B ring hydroxy cinnamoyl system and in the 240–280 nm region due to the A ring benzoyl system21. Compared with the Rt, MS and UV spectra of available standards, peaks 7 and 9 were unambiguously identified as luteolinidin and apigeninidin respectively.

Profiles of phenolic compounds

The list of phenolic compounds found in the free and bound forms of the different sorghum genotypes under OT and HT are shown in Table 2. Nearly all phenolic compounds identified across the sample collection (Table 1) were also identified in each individual sorghum sample, except for the white sorghum PI563516, which contained fewer phenolic species. In addition, the range of phenolic compounds in the sorghum grains was larger in free form, and only brown sorghum IS 8525 showed a similar number of phenolic species in both for the free and bound fractions. The phenolic profile in AQL33/QL36, PI563516 and CHH2 genotypes was changed when grown under higher temperatures (Table 2). For example, the free forms of luteolin and apigenin were not detectable in CHH2 genotype under HT though present under OT (Fig. 1A,B).These results indicated that the phenolic profile in sorghum grains was affected strongly by genotype and also by growth temperature.

Table 2: Profiles of phenolic compounds in the six genotypes of sorghum grain grown at OT and HT

Genotype	Peak number[a]			
	OT		HT	
	Free	Bound	Free	Bound
CCH1	1, 2, 3, 4, 5, 6, 7, 8, 10, 12, 13, 15, 16, 17, 19, 20, 21, 23	5, 7, 9, 11, 12, 18, 19, 20, 23	1, 2, 3, 4, 5, 6, 7, 8, 10, 12, 13, 15, 16, 17, 19, 20, 21, 23	5, 7, 9, 11, 12, 18, 19, 20, 23
CCH2	1, 2, 3, 4, 5, 7, 8, 10, 12, 13, 14, 15, 17, 19, 20, 21, 22, 23	5, 7, 9 11, 12, 15, 19, 20, 22, 23	1, 2, 3, 4, 5, 7, 8, 10, 12, 13, 14, 15, 17, 20, 21, 23	5, 7, 9 11, 12, 15, 19, 20, 22, 23
AQL33/QL36	1, 2, 3, 4, 5, 6, 8, 9, 10, 12, 13, 15, 16, 17, 20, 21, 22, 23	5, 7, 9, 11, 12, 17, 18, 19, 20, 22, 23	1, 2, 3, 4, 5, 6, 8, 9, 10, 12, 13, 15, 16, 17, 20, 21, 23	5, 7, 9, 11, 12, 17, 18, 19, 20, 22, 23
Ai4	1, 2, 4, 5, 7, 8, 9, 12, 13, 14, 15, 17, 19, 20, 21, 22, 23	5, 7, 9, 11, 12, 18, 19, 20, 22, 23	1, 2, 4, 5, 7, 8, 9, 12, 13, 14, 15, 17, 19, 20, 21, 22, 23	5, 7, 9, 11, 12, 18, 19, 20, 22, 23
PI563516	1, 2, 4, 5, 7, 8, 9, 12, 16, 17, 20, 21	5, 11, 12, 18	1, 2, 4, 5, 8, 12, 16, 17, 20, 21	5, 11, 12, 18
IS 8525	1, 2, 3, 4, 5, 6, 9, 10, 12, 13, 14, 15, 16, 17, 20, 21, 23	2, 3, 4, 5, 7, 8, 9, 11, 12, 13, 15, 16, 18, 20, 23	1, 2, 3, 4, 5, 6, 9, 10, 12, 13, 14, 15, 16, 17, 20, 21, 23	2, 3, 4, 5, 7, 8, 9, 11, 12, 13, 15, 16, 18, 20, 23

Quantification of free, bound and total individual phenolic compounds

As total levels of phenolic compounds were presented in the previous paper24, individual phenolic compounds were evaluated. In order to more easily evaluate the results of the individual compounds, they were classified into five groups. Each compound was quantified in free and bound fractions and as total (free + bound).

Hydroxycinnamic acids. Three hydroxycinnamic acids were quantified in the sorghum genotypes: caffeic acid, ferulic acid and ferulic acid isomer (Table 3). Free, bound and total fractions of these compounds were significantly influenced when comparing treatments and genotypes ($p \leq 0.05$). HT resulted in significantly lower free caffeic acid concentration than OT across the sorghum genotypes, except for CHH1, which showed higher levels at HT, but show no significant influences on free levels of ferulic acid and ferulic acid isomer of some genotypes (Ai4). HT showed no significant impacts on bound caffeic acid concentration of all genotypes, with the exception of CHH1 with higher levels at HT ($p \leq 0.05$). The total levels of caffeic acid and ferulic acid varied widely ($p \leq 0.05$) among these genotypes under both HT and OT, with PI563516 and IS 8525 having the lowest and highest levels, respectively (Table 3). The total ferulic acid concentration of all genotypes was extremely higher than other two phenolic acids, irrespective of temperature.

Table 3: Free, bound and total hydroxycinnamic acid content (μg/g, dry base (db)) of six genotypes of sorghum grain grown at OT and HT

Hydroxycinnamic acids			Genotypes					
			CHH1	CHH2	AQL33/QL36	Ai4	PI563516	IS 8525
CA	Free	OT	5.98 ± 0.23 ^{aA}	7.19 ± 0.54 ^{bB}	7.86 ± 0.49 ^{bB}	13.12 ± 0.74 ^{cB}	6.54 ± 0.32 ^{aB}	33.72 ± 4.22 ^{dB}
		HT	7.63 ± 0.50 ^{cB}	4.45 ± 0.66 ^{aA}	6.17 ± 0.49 ^{bA}	9.98 ± 0.65 ^{dA}	4.22 ± 0.97 ^{aA}	18.84 ± 1.46 ^{eA}
	Bound	OT	12.14 ± 1.05 ^{cA}	12.92 ± 0.76 ^{cA}	9.37 ± 0.35 ^{bA}	5.56 ± 0.47 ^{aA}	4.80 ± 0.22 ^{aA}	13.89 ± 0.34 ^{cA}
		HT	15.89 ± 0.81 ^{dB}	11.38 ± 1.87 ^{cA}	9.54 ± 0.03 ^{bA}	6.88 ± 0.17 ^{aA}	5.01 ± 1.12 ^{aA}	14.18 ± 0.15 ^{dA}
	Total	OT	18.12 ± 1.27 ^{bA}	20.10 ± 0.21 ^{cB}	17.22 ± 0.83 ^{bB}	18.67 ± 0.27 ^{bB}	11.34 ± 0.54 ^{aB}	47.61 ± 3.89 ^{dB}
		HT	23.51 ± 0.31 ^{cB}	15.83 ± 2.53 ^{bA}	15.71 ± 0.52 ^{bA}	16.68 ± 0.82 ^{bA}	9.23 ± 0.97 ^{aA}	33.02 ± 1.61 ^{dA}
FA	Free	OT	11.06 ± 1.16 ^{cA}	33.16 ± 0.35 ^{dB}	0.63 ± 0.06 ^{aA}	1.06 ± 0.13 ^{bA}	0.53 ± 0.11 ^{aA}	26.82 ± 2.94 ^{dB}
		HT	10.21 ± 0.17 ^{cA}	26.00 ± 3.35 ^{eA}	1.11 ± 0.11 ^{bB}	1.18 ± 0.01 ^{bA}	0.61 ± 0.12 ^{aA}	16.74 ± 5.17 ^{dA}
	Bound	OT	71.07 ± 1.77 ^{bA}	86.09 ± 0.40 ^{cB}	35.43 ± 0.97 ^{aA}	35.85 ± 0.33 ^{aA}	34.88 ± 1.75 ^{aA}	125.50 ± 1.96 ^{dB}
		HT	73.18 ± 0.81 ^{cA}	69.02 ± 0.24 ^{cA}	45.22 ± 0.16 ^{bB}	40.45 ± 1.62 ^{aB}	37.17 ± 0.51 ^{aB}	116.13 ± 5.30 ^{dA}
	Total	OT	82.13 ± 0.61 ^{bA}	119.25 ± 0.05 ^{cB}	36.05 ± 0.91 ^{aA}	36.91 ± 0.45 ^{aA}	35.41 ± 1.84 ^{aA}	152.32 ± 7.06 ^{dB}
		HT	83.39 ± 0.64 ^{cA}	95.02 ± 3.11 ^{dA}	46.32 ± 0.06 ^{bB}	41.62 ± 1.61 ^{aB}	37.78 ± 0.46 ^{aB}	132.88 ± 10.48 ^{eA}
FAI	Free	OT	1.76 ± 0.04 ^{aA}	7.62 ± 0.44 ^{cB}	3.40 ± 0.16 ^{bA}	1.65 ± 0.14 ^{aA}	nd	3.60 ± 0.20 ^{bB}
		HT	2.74 ± 0.13 ^{bB}	4.48 ± 0.97 ^{cA}	4.17 ± 0.48 ^{cA}	1.38 ± 0.07 ^{aA}	nd	1.79 ± 0.38 ^{aA}
	Bound	OT	nd	0.24 ± 0.01 ^{aB}	nd	nd	nd	5.34 ± 0.25 ^{bB}
		HT	nd	0.14 ± 0.02 ^{aA}	nd	nd	nd	4.19 ± 0.35 ^{bA}
	Total	OT	1.76 ± 0.04 ^{aA}	7.86 ± 0.45 ^{cB}	3.40 ± 0.16 ^{bA}	1.65 ± 0.14 ^{aB}	na	9.00 ± 0.63 ^{dB}
		HT	2.74 ± 0.13 ^{bB}	4.62 ± 0.97 ^{cA}	4.17 ± 0.48 ^{cB}	1.38 ± 0.07 ^{aA}	na	5.98 ± 0.73 ^{dA}

3-Deoxyanthocyanidins. Temperature, genotype and their interaction had significant influences on free, bound and total fractions of luteolinidin and apigeninidin accumulation (Table 4) ($p \leq 0.05$). The accumulation trend of free luteolinidin showed a significant decline under HT in three genotypes, with the exception of two genotypes which did not contain the free luteolinidin, and free fraction of luteolinidin cannot be measured in PI 563516 at HT. The free form apigeninidin content changes followed the same pattern of free form luteolinidin under HT. Regardless of the treatment, Ai4 contained the highest free luteolinidin and apigeninidin concentrations under both temperatures. The levels of bound luteolinidin and apigeninidin showed various changes under HT. AQL33/QL36 and CHH2 accumulated the higher levels of bound luteolinidin and apigeninidin under both treatments, respectively.

Table 4: Free, bound and total individual flavonoids content (μg/g, db) of six genotypes of sorghum grain grown at OT and HT

Flavonoids			Genotypes					
			CHH1	CHH2	AQL33/QL36	Ai4	PI563516	IS 8525
3-Deoxyanthocyanidin								
LUT	Free	OT	0.65 ± 0.04 ^{aB}	2.33 ± 0.13 ^{bB}	nd	14.24 ± 0.98 ^{dB}	4.40 ± 0.34 ^{c}	nd
		HT	0.13 ± 0.02 ^{aA}	0.36 ± 0.01 ^{bA}	nd	9.22 ± 0.18 ^{cA}	nd	nd
	Bound	OT	1.92 ± 0.55 ^{bA}	3.43 ± 0.06 ^{cB}	3.88 ± 0.23 ^{cA}	0.69 ± 0.07 ^{aA}	nd	1.62 ± 0.25 ^{bA}
		HT	3.47 ± 0.28 ^{cB}	1.96 ± 0.21 ^{bA}	4.10 ± 0.48 ^{dA}	1.37 ± 0.13 ^{aB}	nd	1.58 ± 0.51 ^{abA}
	Total	OT	2.57 ± 0.51 ^{bA}	5.76 ± 0.08 ^{eB}	3.88 ± 0.23 ^{cA}	14.93 ± 1.05 ^{fB}	4.40 ± 0.34 ^{dB}	1.62 ± 0.25 ^{aA}
		HT	3.60 ± 0.30 ^{cB}	2.31 ± 0.21 ^{bA}	4.10 ± 0.48 ^{cA}	10.59 ± 0.31 ^{dA}	na	1.58 ± 0.51 ^{aA}
API	Free	OT	2.74 ± 0.21 ^{bB}	2.49 ± 0.01 ^{abB}	2.25 ± 0.18 ^{aB}	23.24 ± 1.29 ^{cB}	2.48 ± 0.21 ^{ab}	2.05 ± 0.08 ^{aB}
		HT	0.38 ± 0.02 ^{aA}	0.64 ± 0.08 ^{bA}	0.33 ± 0.01 ^{aA}	14.52 ± 0.71 ^{cA}	nd	0.60 ± 0.02 ^{bA}
	Bound	OT	4.66 ± 0.14 ^{bA}	23.84 ± 1.73 ^{dB}	12.16 ± 1.32 ^{cA}	4.70 ± 0.16 ^{bA}	nd	2.27 ± 0.18 ^{aA}
		HT	7.79 ± 0.08 ^{cB}	16.88 ± 3.79 ^{eA}	11.57 ± 1.10 ^{dA}	6.20 ± 0.31 ^{bB}	nd	2.43 ± 0.65 ^{aA}
	Total	OT	7.40 ± 0.06 ^{cA}	26.33 ± 1.71 ^{eB}	14.41 ± 1.51 ^{dB}	27.93 ± 1.46 ^{eB}	2.48 ± 0.21 ^{a}	4.33 ± 0.26 ^{bB}
		HT	8.16 ± 0.06 ^{bB}	17.52 ± 3.71 ^{dA}	11.89 ± 1.09 ^{cA}	20.72 ± 0.40 ^{eA}	na	3.03 ± 0.63 ^{aA}
Flavones								
LU	Free	OT	0.92 ± 0.17 ^{aA}	10.91 ± 0.57 ^{c}	nd	4.04 ± 0.07 ^{bA}	nd	nd
		HT	0.84 ± 0.10 ^{aA}	nd	nd	5.57 ± 0.35 ^{bB}	nd	nd
	Bound	OT	3.70 ± 0.03 ^{bA}	5.05 ± 0.74 ^{cB}	0.73 ± 0.06 ^{aB}	0.91 ± 0.02 ^{aB}	nd	nd
		HT	3.92 ± 0.12 ^{bA}	3.62 ± 0.13 ^{bA}	0.54 ± 0.03 ^{aA}	0.52 ± 0.03 ^{aA}	nd	nd
	Total	OT	4.62 ± 0.14 ^{bA}	15.96 ± 1.30 ^{cB}	0.73 ± 0.06 ^{aB}	4.95 ± 0.09 ^{bA}	na	na
		HT	4.76 ± 0.22 ^{cA}	3.62 ± 0.13 ^{bA}	0.54 ± 0.03 ^{aA}	6.09 ± 0.37 ^{dB}	na	na
AP	Free	OT	nd	1.42 ± 0.00 ^{b}	0.36 ± 0.08 ^{a}	1.45 ± 0.13 ^{bA}	nd	nd
		HT	nd	nd	nd	2.09 ± 0.37 ^{B}	nd	nd
	Bound	OT	nd	2.61 ± 0.13 ^{cB}	1.80 ± 0.21 ^{bB}	0.86 ± 0.02 ^{aA}	nd	nd
		HT	nd	1.20 ± 0.17 ^{abA}	1.27 ± 0.04 ^{bA}	1.16 ± 0.02 ^{aB}	nd	nd
	Total	OT	na	4.03 ± 0.13 ^{bB}	2.16 ± 0.30 ^{aB}	2.30 ± 0.11 ^{aA}	na	na
		HT	na	1.20 ± 0.17 ^{aA}	1.27 ± 0.04 ^{aA}	3.24 ± 0.40 ^{bB}	na	na
Dihydroflavonol								
TA	Free	OT	1.78 ± 0.15 ^{bA}	7.92 ± 0.82 ^{dB}	2.39 ± 0.05 ^{cB}	1.21 ± 0.01 ^{aA}	nd	12.85 ± 0.78 ^{eA}
		HT	2.35 ± 0.11 ^{cB}	4.15 ± 0.21 ^{dA}	1.85 ± 0.10 ^{bA}	1.55 ± 0.01 ^{aA}	nd	12.36 ± 2.38 ^{eA}
	Bound	OT	nd	nd	nd	nd	nd	41.33 ± 1.10 ^{B}
		HT	nd	nd	nd	nd	nd	31.91 ± 0.03 ^{A}
	Total	OT	1.78 ± 0.15 ^{bA}	7.92 ± 0.82 ^{dB}	2.39 ± 0.05 ^{cB}	1.21 ± 0.01 ^{aA}	nd	54.18 ± 1.88 ^{eB}
		HT	2.35 ± 0.11 ^{cB}	4.15 ± 0.21 ^{dA}	1.85 ± 0.10 ^{bA}	1.55 ± 0.01 ^{aA}	nd	44.27 ± 2.34 ^{eA}
TAI	Free	OT	0.86 ± 0.07 ^{aA}	12.97 ± 0.54 ^{cB}	nd	11.21 ± 0.08 ^{cB}	nd	3.60 ± 0.20 ^{bB}
		HT	1.55 ± 0.04 ^{aB}	5.45 ± 0.48 ^{bA}	nd	9.41 ± 0.28 ^{cA}	nd	1.66 ± 0.50 ^{aA}
	Bound	OT	nd	nd	nd	nd	nd	nd
		HT	nd	nd	nd	nd	nd	nd
	Total	OT	0.86 ± 0.07 ^{aA}	12.97 ± 0.54 ^{cB}	na	11.21 ± 0.08 ^{cB}	na	3.60 ± 0.20 ^{bB}
		HT	1.55 ± 0.04 ^{aB}	5.45 ± 0.48 ^{bA}	na	9.41 ± 0.28 ^{cA}	na	1.66 ± 0.50 ^{aA}
Flavanone								
NAR	Free	OT	1.37 ± 0.53 ^{bA}	2.03 ± 0.16 ^{cB}	0.62 ± 0.08 ^{aA}	1.63 ± 0.04 ^{bcA}	nd	1.97 ± 0.10 ^{cB}
		HT	1.54 ± 0.21 ^{dA}	4.15 ± 0.21 ^{dA}	0.64 ± 0. 01 ^{aA}	1.47 ± 0.06 ^{dA}	nd	1.25 ± 0.03 ^{cA}
	Bound	OT	9.40 ± 1.41 ^{cA}	6.46 ± 0.25 ^{bB}	4.28 ± 0.44 ^{aA}	20.58 ± 0.75 ^{dB}	nd	10.34 ± 0.93 ^{cB}
		HT	9.32 ± 0.91 ^{cdA}	4.87 ± 0.09 ^{aA}	6.45 ± 0.83 ^{bB}	10.54 ± 0.59 d^{A}	nd	8.80 ± 0.38 ^{cA}
	Total	OT	10.76 ± 1.94 ^{cA}	8.48 ± 0.08 ^{bB}	4.90 ± 0.35 ^{aA}	22.21 ± 0.78 ^{eA}	nd	12.31 ± 0.83 ^{dB}
		HT	10.86 ± 1.12 ^{cA}	5.75 ± 0.04 ^{aA}	7.09 ± 0.84 ^{bB}	12.01 ± 0.53 ^{dB}	nd	10.05 ± 0.42 ^{cA}

Flavones. Two flavones: luteolin and apigenin were identified and quantified in the sorghum samples, and are shown in Table 4. Temperature, genotype and the interaction between the two factors had a significant effect on the flavone contents ($p \leq 0.05$). Both free and bound luteolin and apigenin were not detected in the two sorghum genotypes of PI563516 and IS 8525. CHH2 contained detectable levels of free luteolin and apigenin at OT, but these flavones were not detected in this genotype at HT. Bound flavones accumulation also appears to be genotype dependent under HT. Generally, its concentration had no significant changes in CHH1, but significantly decreased in CHH2, AQL33/QL36 and Ai4 under HT. In fact, total luteolin and apigenin contents of CHH2 were higher than other genotypes under OT, but these of Ai4 reached to the highest level under HT. Dihydroflavonol. The accumulation trend of two dihydroflavonols of taxifolin and taxifolin isomer are presented in Table 4 and their concentrations were significantly affected by temperature, genotype and their interaction ($p \leq 0.05$). Compared to other sorghum genotypes, no flavonones in free nor bound form were detectable in PI563516 under both OT and HT. In general, free taxifolin and taxifolin isomer contents showed diverse changes in these genotypes under HT. However, the bound form of taxifolin was only detected in sorghum IS 8525 and was around 3 times higher than its free form in this genotype, but HT significantly reduced its concentration ($p \leq 0.05$). Regardless of treatment, no bound taxifolin isomer was detected across all genotypes. These results suggest that these two compounds were mainly in the free form. The total values of taxifolin were generally higher in IS8525 as compared to others under both treatments.

Flavanones. Several flavanones (Table 1) were identified in the sorghum samples but only naringenin was quantified, due to a lack of available standards. Values of naringenin were significantly affected by genotype, temperature and their interaction ($p \leq 0.05$). Apart from sorghum PI563516, all other sorghum genotypes contained naringenin in free and bound forms (Table 4). The bound naringenin level was significantly higher than free forms in five genotypes, irrespective of treatments. Total naringenin content in CHH2, Ai4 and IS 8525 was significantly lower at HT than OT, but significantly higher in AQL33/QL36 ($p \leq 0.05$).

DISCUSSION

In this study several compounds were identified for the first time in sorghum grain. For example, feruloyl-caffeoylglycerol, which is known to be present in *Ananas comosus* L. leaves[17], was tentatively identified in the sorghum grain. Similarly, two flavonoid compounds of eriodictyol deoxyhexoside and taxifolin isomer were also tentatively identified. It has been reported the presence of catechin in red sorghum PAN 3860[15]. This compound was not however detected in the six sorghum genotypes used in the present study. The two ubiquitous sorghum 3-deoxyanthocyanidins, apigeninidin (yellow) and luteolinidin (orange) were determined in our sorghum genotypes. The two other 3-deoxyanthocyanidins of 5-methoxyluteolinidin and 7-methoxyapigeninidin have not been detected in our samples, but been reported in sorghum genotypes with lemon-yellow pericarp previously[25]. Additionally, certain phenolic acids, such as gallic and cinnamic acids were identified in sorghum grain[26], however these compounds were not detected in the present study. Some phenolic acid esters of glycerol, including dicaffeoylglycerol and caffeoylglycerol, have been found in other sorghum grains[15], while the isomers of these phenolic acid esters of glycerol are reported for the first time in the present study. The results of the study confirm previous reports that the phenolic profile in sorghum grain is strongly affected by the genotype. There were some putative polyphenol peaks that were present in the HPLC-DAD-ESIMS chromatograms (Fig. 1) that could not be identified, likely due to their low UV signals in the DAD or unidentifiable molecular mass fragments.

The free, bound and total content of individual compounds in the sorghum genotypes in the present study was found to be in the range as previously reported[2,25,26,27]. Most of the phenolic compounds in sorghum grains were found in both free and bound forms (Fig. 1, Tables 3 and 4). For several individual compounds, such as caffeic acid, ferulic acid and naringenin, they mainly occurred in the bound form, which was in agreement with other studies[15,28]. It has been suggested that dietary intake of free or soluble conjugated forms rather than bound forms of phenolic compounds may have health benefits of protection against cardiovascular disease and certain types of cancer, since their gastrointestinal absorption is more rapid[24,29]. However, epidemiological studies also indicate that consuming whole grains with bound phenolic compounds could reduce the risk of colon cancer and other digestive cancers. This may be related to the bound phenolic compounds surviving stomach and intestinal digestion and then being degraded by the microflora and absorbed in the colon to prevent colon cancer[30].

The profiles and levels of phenolic compounds cannot be simply attributed to the sorghum grain colour. It has been reported that five different red sorghum genotypes were planted under the same growth condition, but some individual phenolics, such as luteolin, were only detected in red 99LGWO502. In the present study, luteolin was absent in red sorghum AQL33/QL36 in comparison with other red genotypes. White sorghum PI563516 contained the least amount of flavonoids, with some common flavonoids, such as luteolin, apigenin and naringenin, being absent in this genotype (Table 2). In additon, luteolinidin and apigeninidin were present in white sorghum 02CA47969, but the levels were lower than in PI563516 in the present study. It has been reported that the biosynthesis of phenolic compounds in plants is regulated by enzymes such as phenylalanine ammonia-lyase (PAL), chalcone isomerase (CHI) and UDP glucose flavonoid-glucosyltransferase (UFGT)[31]. Enzyme activities, which have potential influences on phenolic biosynthesis, varied among genotypes. Mccallum and Walker[32]reported that red wheat genotypes contained higher levels of both PAL and CHI than white genotypes during early grain development, which can lead to higher synthesis of flavonoids. The similar results were also observed in two different apple genotypes, and PAL, CHI and UFGT activities of red-skinned genotype was several-fold higher than green-skinned genotype at all stages, so a greater accumulation of flavonoids in red-skinned genotype[33]. In the present study, similar results were obtained in sorghum grains as well. The total 3-deoxyanthocyanidins, flavones, dihydroflavonol and flavanone concentrations of colored sorghum were higher than white sorghum, indicating that these PAL, CHI and UFGT enzyme activities might be higher in colored sorghum.

The impact of high temperature on sorghum grain phenolic compounds was diverse across sorghum genotypes. The possible reason was that the phenolic metabolism pathway was also influenced by temperature. As the first step in the biosynthesis of phenolic compounds was catalyzed by PAL, the PAL activity might have correlation with phenolic levels[32]. Rivero *et al.*[34] reported that high temperature could increase the PAL activity in tomato, and the total phenolic compounds were increased under high temperature stress. Based on the present study, it can be proposed that the phenolic changes under high temperature in sorghum grain may be caused by PAL activity change. For individual polyphenols, their changes might be related to some specific enzyme activity changes. For example, anthocyanin concentration of grape berries grown at a constant 30 °C was lower than those grown at 30 °C/15 °C (day/night), and anthocyanin accumulation correlated strongly with UFGT activity35. However, the biosynthetic pathways of 3-deoxyanthocyanidin in sorghum may be different from anthocyanin biosynthesis in other plants, and the enzymes involved in 3-deoxyanthocyanidin synthesis have still remained unexplored36. In the present research, it was noted that the abundance of free luteolinidin and apigeninidin was markedly decreased under high temperature, and these changes might be related to enzymes. Further research is required to explore enzymes in the metabolic pathway of 3-deoxyanthocyanidin of sorghum grains under high temperature conditions.

The antioxidant activity of total phenolic compounds of sorghum grain determined by 2′–azinobis (3-ethylbenzothiazoline-6-sulfonic acid) diammonium salt (ABTS) and 2-2-diphenyl-1-picrylhydrazyl (DPPH) assays were presented in the previous paper24, and their antioxidant activity was contributed by profiles and levels of individual polyphenols. Chandrasekara and Shahidi37 reported that two millet grain genotypes: Proso and Pearl with less complex profiles of phenolics showed the lowest antioxidant activity. Similar changes in sorghum grain were also presented in the present experiment. PI563516 had lower antioxidant activity than others24, and the profile of phenolics was the least complex (Table 2). For individual phenolic acids, total antioxidant activity of ferulic acid is higher than caffeic acid. Additionally, not all individual flavonoids show high antioxidant activity. For example, apigenidin, luteolin and taxifolin have higher antioxidant activity than naringenin and apigenin21. Therefore, the levels of individual phenolic compounds might decide the antioxidant activity of sorghum grain. Durum wheat pasta with different levels of red sorghum flour or white sorghum flour was made, and it was found that the wheat pasta with red sorghum flour which contained high levels of individual phenolic acids showed higher antioxidant capacity than white sorghum flour11. In this research, IS 8525, CHH1 and CHH2 had higher levels of ferulic acid, caffeic acid than others, and higher antioxidant activity was obtained in these genotypes as well24. As the antioxidant activity of sorghum grain had a strong correlation with phenolic content, HT can also change the antioxidant activity of sorghum grain. Individual polyphenols of IS8525 significantly declined under HT, and total antioxidant activity of it decreased24. Compared to other research, the individual phenolics of strawberry increased when growth temperature increased from 18/12 °C to 30/22 °C (day/night), resulting in a significant increase in antioxidant capacity13. The phenolic changes of sorghum grain under temperature treatments are important for humans, due to their free radical scavenging and biological properties. In this research, the sorghum genotype IS 8525 had extremely high levels of ferulic acid, caffeic acid and taxifolin compared to others under both temperatures, suggesting that this genotype may be suitable for use in health-functional food formulation.

In conclusion, with the use of the rapid analytical procedure of HPLC-DAD-ESIMS, the phenolic compounds in 6 sorghum genotypes under two different growth temperatures were determined. The profiles and contents of phenolic compounds in sorghum grains were significantly influenced by the genotype, growth temperature and their interaction. This study has attempted to further our understanding of the variations of polyphenols in sorghum under the projected higher temperature conditions in certain regions, as well as indicating the potential for particular sorghums to be used as a nutritious food source depending on growth temperature and genotype.

METHODS

Sorghum genotypes

Six sorghum genotypes including three red pericarp hybrid lines (CCH1, CCH2 and AQL33/QL36) and three inbred lines (Ai4 red pericarp; PI563516, white pericarp; IS 8525, brown pericarp) were used in this study (Fig. 3).

Figure 3: Six genotypes of mature sorghum grain under OT and HT growth conditions.

Experimental design

The details of sorghum growing conditions have been described previously38. Briefly, two separate growth chambers were used to control all environmental factors and provide two temperature regimes: optimal temperature (OT, 32 °C day/21 °C night) and high temperature (HT, 38 °C day/21 °C night) at the Controlled Environment Facility, Queensland Bioscience Precinct, The University of Queensland, Brisbane, Australia, 2012. The night temperature was gradually increased at a rate of 3 °C/hour to the maximum temperatures commencing from 1 h after the light period began. The maximum temperatures were kept constant for 7 h and 3 h in the OT and HT chambers, respectively. The temperature was then decreased at a rate of 3 °C/hour until the minimum temperature was obtained in both chambers38. Therefore, the main environmental difference between the two treatments was only the temperature in the middle of the light period. The experiment was a completely randomised block design (using 6 genotypes) within two temperature treatments and three replications (individual plants). Grains were harvested at maturity, manually cleaned and air-dried until a moisture content of around 10% was reached. The dried samples from each replicate were individually vacuum-packed in moisture proof packaging and kept at −20 °C in the dark until analysis.

Extraction of free and bound phenolic compounds

The whole grain samples were milled (CEMOTEC 1090, Foss Tecator, Hoganäs, Sweden) to pass 100% through a 500 µm sieve. The free and bound phenolic compounds were extracted15. All extractions were evaporated to dryness using a rotary vacuum evaporator. The resulting solid was re-dissolved in methanol to a final volume of 10 ml and stored under N_2 at −20 °C in the dark until analysis.

HPLC-DAD-ESIMS analyses

An Agilent 1290 UHPLC system equipped with binary pump, autosampler, thermostated column compartment, degasser, and diode array detector (DAD) was coupled to an Agilent 6460 LC-QQQ LC-MS/MS system (Agilent Technologies, Palo Alto, CA, USA). The separation of phenolic compounds was performed on a Kinetex XB-C 18 reversed phase-HPLC column (5 µm, 250 × 4.6 mm, Phenomenex, Torrance, CA, USA). The DAD was set to scan between 190 and 600 nm at steps of 2 nm. Solvent A consisted of 0.1% formic acid in LC-MS grade water (Honeywell Burdick & Jackson, Gillman, SA, Australia), and solvent B was LC-MS grade acetonitrile (Honeywell Burdick & Jackson, Gillman, SA, Australia). The sample injection volume was 5 µl with the following linear gradient elution: 5%–15% B (5 min), 15%–50% B (40 min), 50%–70% B (2 min), 70%–100% B (1 min), 100% B (7 min), 100%-5% B (1 min), 5% B (9 min). The flow rate was 0.5 ml/min.

Mass spectra were performed in the ESI negative mode with a scan time of 2000 MS under the following conditions: gas (N2) 5 L/min at 300 °C, nebulizer 45 psi, sheath gas (N_2) 11 L/min at 250 °C, capillary voltage −3.5 kV and nozzle voltage −500 V. Phenolic compounds were detected by full scan ranging from *m/z* 50 to 1300.

Identification and quantification of polyphenols

The phenolic compounds in sorghum samples were identified by their UV-Vis and ESIMS spectra. The authentic standards were also used to identify phenolic compounds based on the chromatographic comparisons. The external standard method was used to quantify phenolic compounds in samples under the above HPLC-DAD conditions. Standards were dissolved in methanol (1 mg/ml each), and diluted to several concentrations ranging from 0.001 to 0.1 mg/ml. Data acquisition, peak integration and calibrations were performed with the Agilent ChemStation software. Peak areas compared with calibration curves of the respective standards were used to calculate levels of phenolic compounds, and results were expressed as µg/g sample (dry basis, db). Quantification of taxifolin isomer and ferulic acid isomer were calculated using the calibration curve for taxifolin and ferulic acid, respectively.

Statistical analysis

All data were reported as means ± SD of triplicate independent experiments analysed in triplicate. The main effects of genotype and temperature and their interaction were investigated by two-way or one way ANOVA with SPSS Statistics V20 (IBM Corp., Armonk, NY, USA). Significant differences were considered when $p \leq 0.05$.

ACKNOWLEDGEMENTS

The Authors would like to thank Curtin International Postgraduate Research Scholarship (CIPRS). MC wishes to acknowledge funding from the National Collaborative Research Infrastructure Strategy through Metabolomics Australia and Bioplatforms Australia.

REFERENCES

1. Food and Agriculture Organization of the United Nations. (2013) Available at: http://faostat3.fao.org/browse/Q/QC/E. (Accessed on 1st December 2015).
2. Taleon, V., Dykes, L., Rooney, W. & Rooney, L. Environmental effect on flavonoid concentrations and profiles of red and lemon-yellow sorghum grains. J. Food Compos. Anal. 34, 178–185 (2014).
3. Mahasukhonthachat, K., Sopade, P. & Gidley, M. Kinetics of starch digestion in sorghum as affected by particle size. J. Food Eng. 96, 18–28 (2010).
4. Stefoska-Needham, A., Beck, E. J., Johnson, S. K. & Tapsell, L. C. Sorghum: an underutilized cereal whole grain with the potential to assist in the prevention of chronic disease. Food Rev. 31, 401–437 (2015).
5. Visioli, F. et al. Polyphenols and human health: a prospectus. Crit Rev. Food Sci. 51, 524–546 (2011).
6. Proteggente, A. R. et al. The antioxidant activity of regularly consumed fruit and vegetables reflects their phenolic and vitamin C composition. Free Radical Res. 36, 217–233 (2002).
7. Hertog, M. G., Feskens, E. J., Kromhout, D., Hollman, P. & Katan, M. Dietary antioxidant flavonoids and risk of coronary heart disease: the Zutphen Elderly Study. Lancet. 342, 1007–1011 (1993).
8. Hertog, M. G., Feskens, E. J., Hollman, P. C., Katan, M. B. & Kromhout, D. Dietary flavonoids and cancer risk in the Zutphen Elderly Study. Nutr. Cancer. 22, 175–184 (1994).
9. Dykes, L., Rooney, L. W., Waniska, R. D. & Rooney, W. L. Phenolic compounds and antioxidant activity of sorghum grains of varying genotypes. J. Agric. Food Chem. 53, 6813–6818 (2005).

10. Dicko, M. H., Gruppen, H., Traoré, A. S., van Berkel, W. J. & Voragen, A. G. Evaluation of the effect of germination on phenolic compounds and antioxidant activities in sorghum varieties. J. Agric. Food Chem. 53, 2581–2588 (2005).
11. Khan, I., Yousif, A., Johnson, S. K. & Gamlath, S. Effect of sorghum flour addition on resistant starch content, phenolic profile and antioxidant capacity of durum wheat pasta. Food Res Int. 54, 578–586 (2013).
12. Intergovernmental Panel on Climate Change. IPCC Fourth Assessment Report (AR4). (2007) Available at: http://ipcc.ch/publications_and_data/ar4/syr/en/contents.html. (Accessed on 20th December 2015).
13. Wang, S. Y. & Zheng, W. Effect of plant growth temperature on antioxidant capacity in strawberry. J. Agric. Food Chem. 49, 4977–4982 (2001).
14. Islam, M. S., Yoshimoto, M., Ishiguro, K., Okuno, S. & Yamakawa, O. Effect of artificial shading and temperature on radical scavenging activity and polyphenolic composition in sweetpotato (Ipomoea batatas L.) leaves. J. Am Soc. Hortic Sci. 128, 182–187 (2003).
15. Svensson, L., Sekwati-Monang, B., Lutz, D. L., Schieber, A. & Gänzle, M. G. Phenolic acids and flavonoids in nonfermented and fermented red sorghum (Sorghum bicolor (L.) Moench). J. Agric. Food Chem. 58, 9214–9220 (2010).
16. Luthria, D. L. & Liu, K. Localization of phenolic acids and antioxidant activity in sorghum kernels. J. Funct Foods. 5, 1751–1760 (2013).
17. Ma, C., Xiao, S., Li, Z., Wang, W. & Du, L. Characterization of active phenolic components in the ethanolic extract of Ananas comosus L. leaves using high-performance liquid chromatography with diode array detection and tandem mass spectrometry. J.Chromatogr A. 1165, 39–44 (2007).
18. Sun, W. & Miller, J. M. Tandem mass spectrometry of the B-type procyanidins in wine and B-type dehydrodicatechins in an autoxidation mixture of (+)-catechin and (−)-epicatechin. J. Mass Spectrom. 38, 438–446 (2003).
19. Guyot, S., Vercauteren, J. & Cheynier, V. Structural determination of colourless and yellow dimers resulting from (+)-catechin coupling catalysed by grape polyphenoloxidase. Phytochemistry. 42, 1279–1288 (1996).
20. Sun, J., Liang, F., Bin, Y., Li, P. & Duan, C. Screening non-colored phenolics in red wines using liquid chromatography/ultraviolet and mass spectrometry/mass spectrometry libraries. Molecules. 12, 679–693 (2007).
21. Rice-Evans, C. A., Miller, N. J. & Paganga, G. Structure-antioxidant activity relationships of flavonoids and phenolic acids. Free Radic Biol. Med. 20, 933–956 (1996).
22. Stobiecki, M. Application of mass spectrometry for identification and structural studies of flavonoid glycosides. Phytochemistry. 54, 237–256 (2000).
23. Fang, Z., Zhang, M. & Wang, L. HPLC-DAD-ESIMS analysis of phenolic compounds in bayberries (Myrica rubra Sieb. et Zucc.). Food Chem. 100, 845–852 (2007).
24. Wu, G. et al. Effect of genotype and growth temperature on sorghum grain physical characteristics, polyphenol content and antioxidant activity. Cereal Chem. submitted.
25. Dykes, L., Peterson, G. C., Rooney, W. L. & Rooney, L. W. Flavonoid composition of lemon-yellow sorghum genotypes. Food Chem. 128, 173–179 (2001).
26. Hahn, D., Faubion, J. & Rooney, L. Sorghum phenolic acids, their high performance liquid chromatography separation and their relation to fungal resistance. Cereal Chem. 60, 255–259 (1983).
27. Taleon, V., Dykes, L., Rooney, W. L. & Rooney, L. W. Effect of genotype and environment on flavonoid concentration and profile of black sorghum grains. J. Cereal Sci. 56, 470–475 (2012).
28. Dykes, L. & Rooney, L. W. Sorghum and millet phenols and antioxidants. J. Cereal Sci. 44, 236–251 (2006).
29. Bravo, L. Polyphenols: chemistry, dietary sources, metabolism, and nutritional significance. Nutr Rev. 56, 317–333 (1998).
30. Adom, K. K. & Liu, R. H. Antioxidant activity of grains. J. Agric. Food Chem. 50, 6182–6187 (2002).
31. Cohen, S. D. & Kennedy, J. A. Plant metabolism and the environment: implications for managing phenolics. Crit Rev. Food Sci. 50, 620–643 (2010).
32. Mccallum, J. A. & Walker, J. R. Phenolic biosynthesis during grain development in wheat (Triticum aestivum L.) III. Changes in hydroxycinnamic acids during grain development. J. Cereal Sci. 13, 161–172 (1991).
33. Lister, C. E., Lancaster, J. E. & Walker, J. R. Developmental Changes in Enzymes Biosynthesis in the Skins of Red and. J. Sci. Food. Agric. 71, 330 (1996).
34. Rivero, R. M. et al. Resistance to cold and heat stress: accumulation of phenolic compounds in tomato and watermelon plants. Plant Sci. 160, 315–321 (2001).
35. Mori, K., Sugaya, S. & Gemma, H. Decreased anthocyanin biosynthesis in grape berries grown under elevated night temperature condition. Sci Hortic-amsterdam. 105, 319–330 (2005).
36. Shih, C. H., Chu, I. K., Yip, W. K. & Lo, C. Differential expression of two flavonoid 3′-hydroxylase cDNAs involved in biosynthesis of anthocyanin pigments and 3-deoxyanthocyanidin phytoalexins in sorghum. Plant. Cell Physiol. 47, 1412–1419 (2006).
37. Chandrasekara, A. & Shahidi, F. Determination of antioxidant activity in free and hydrolyzed fractions of millet grains and characterization of their phenolic profiles by HPLC-DAD-ESI-MS n. J. Funct Foods. 3, 144–158 (2011).
38. Nguyen, C. T. et al. Genetic variability in high temperature effects on seed-set in sorghum. Funct Plant Biol. 40, 439–448 (2013).

CHAPTER 14

SUSTAINABLE GRAIN YIELD FOR SORGHUM CULTIVATED ON ACID SOILS

Geraldo Carvalho Jr, Robert Eugene Schaffert, Marcos Malosetti, Joao Herbert Moreira Viana, Cicero Bezerra Menezes, Lidianne Assis Silva, Claudia Teixeira Guimaraes, Antonio Marcos Coelho, Leon V. Kochian, Fred A. van Eeuwijk and Jurandir Vieira Magalhaes

Keywords: Al tolerance, QTL mapping, *Sorghum bicolor*, Alt_{SB}, field trials

INTRODUCTION

At soil pH values below 5.0, aluminum (Al) is solubilized as the rhizotoxic cation Al^{3+}, which damages plant roots and inhibits root growth, reducing water and nutrient uptake, and crop yields (Kochian *et al.* 2004). Approximately 50% of the potentially arable lands worldwide are acidic (von Uexküll and Mutert 1995), mostly in the tropics and subtropics. Agriculture in regions such as sub-Saharan Africa is strongly constrained by Al toxicity (Doumbia *et al.*1993, 1998), and Al tolerance can thus help liberate an enormous food production potential exactly where food security is most tenuous.

Aluminum-induced organic acid release protects the sensitive growing zone in the root apex by chelating Al as nontoxic complexes (Ma *et al.* 2001; Ryan *et al.*2001), leading to sustained root growth in nutrient solution containing Al^{3+}(Kochian *et al.* 2004). Major Al tolerance genes belonging to the Aluminum-Activated Malate Transporter (ALMT), and the Multidrug and Toxic Compound Extrusion (MATE) families were cloned in wheat (*TaALMT1*) (Sasaki *et al.* 2004), sorghum (*SbMATE*) (Magalhaes *et al.* 2007), and barley (*HvAACT1*) (Furukawa *et al.* 2007) via hydroponic-based assessment of Al tolerance.

In sorghum, *SbMATE* underlies Alt_{SB}, a major Al tolerance locus mapped to sorghum chromosome 3, and explains a large proportion of the phenotypic variation for Al tolerance assessed in hydroponics (Magalhaes *et al.* 2004,2007). SbMATE is an Al-regulated (for both gene expression and protein function) root citrate transporter that acts to exclude Al from sensitive regions in the root apex, thus conferring Al tolerance (Magalhaes *et al.* 2007).*TaALMT1*, the major wheat Al tolerance gene, is located on the long arm of wheat chromosome 4D (Raman *et al.* 2005), and encodes an Al-activated malate transporter (Sasaki *et al.* 2004) that likely underlies the major wheat Al tolerance locus, Alt_{BH} (Riede and Anderson 1996). Homologs of those two genes have also been identified in other plant species, including rye (Collins *et al.* 2008; Yokosho *et al.* 2010), maize (Maron *et al.* 2010), wheat (Ryan *et al.*2009), Arabidopsis (Hoekenga *et al.* 2006; Liu *et al.* 2009), rape (Ligaba *et al.*2006), rice (Yokosho *et al.* 2011), rice bean (Liu *et al.* 2013), barley (Zhou *et al.*2013), and soybean (Liang *et al.* 2013).

There is an apparent agreement between sustained root growth under Al stress in hydroponics and yield traits assessed on acid soils (Reid *et al.* 1971; Duncan*et al.* 1983; Culvenor *et al.* 1986; Furlani *et al.* 1991; Baier *et al.* 1995). However, because Al tolerance genes were not isolated based primarily on field assessments of Al tolerance, their breeding value in enhancing grain yield on acid soils remains undetermined. Consequently, although the knowledge of the molecular basis of crop Al tolerance offers an unprecedented repertoire of novel strategies for achieving sustainable yields under Al stress, this potential remains highly underutilized for crop breeding.

In the present study, Al tolerance assessed in hydroponics and in field conditions was explained largely by a major QTL colocalized with Alt_{SB} on chromosome 3, which increased grain yield by 0.6 ton ha^{-1} without any detectable penalty in the absence of Al toxicity. Evidence for a second, novel Al QTL on chromosome 9 based on Al-inhibition of root growth was obtained, suggesting the presence of either a second, distinct Al tolerance gene in the sorghum genome or a gene (or genes) acting with *SbMATE*, leading to Al-activated citrate release.

Citation: Geraldo Carvalho Jr, Robert Eugene Schaffert, Marcos Malosetti, Joao Herbert Moreira Viana, Cicero Bezerra Menezes, Lidianne Assis Silva, Claudia Teixeira Guimaraes, Antonio Marcos Coelho, Leon V. Kochian, Fred A. van Eeuwijk and Jurandir Vieira Magalhaes, "Sustainable Grain Yield for Sorghum Cultivated on Acid Soils," doi: 10.1534/g3.115.025791.

An independent assessment of the Alt_{SB} effect using isogenic hybrids for the Al tolerance locus yielded a female effect of 0.5 ton ha^{-1}, arising from one copy of the tolerant allele at Alt_{SB}. Our results show thatAlt_{SB} is a major asset for sustainable sorghum production on acid, Al toxic soils, and should be used to enhance food security under Al toxicity on acid soils prevalent in developing countries in the tropics and subtropics.

MATERIALS AND METHODS

Plant material

Recombinant inbred line population:

A recombinant inbred line (RIL) population consisting of 90 $F_{7:8}$ progeny derived from a cross between BR007, an Al-sensitive breeding line from the Embrapa Maize and Sorghum breeding program, and SC283, a highly Al-tolerant sorghum line belonging to the guinea race (Magalhaes *et al.* 2004), was developed by the single seed descent method (Johnson and Bernard 1962). Only 3-dwarf types were used to minimize the confounding effect of plant height variation on grain yield.

Development of isogenic hybrid stocks:

The effect of Alt_{SB} on grain yield was also assessed in sorghum hybrids by comparing grain yield in homozygous sensitive genotype designated as (tt), heterozygous (Tt), or homozygous tolerant (TT) hybrids for Alt_{SB}. Isogenic hybrids were developed to allow comparison of Alt_{SB} alleles from different donors within a homogeneous genetic background.

NIL development:

ATF13B and ATF14B are sister nonrestorer lines in A1 cytoplasm derived by selfing a single F_5 plant heterozygous for Alt_{SB} derived from BR007 (tt) × SC283 (TT). Homozygous sensitive and tolerant progeny were selected by progeny testing for Al tolerance and self-pollinated for at least seven generations, giving rise to ATF13B (Al sensitive) and ATF14B (Al tolerant) lines. ATF13B and ATF14B were backcrossed to a source of A1 cytoplasmic male sterility (BR007A) for six generations to produce ATF13A and ATF14A. ATF13A (tt) and ATF14A (TT) are thus near isogenic lines (NILs) fixed for alternativeAlt_{SB} alleles derived from the Al-sensitive and Al-tolerant parents of our RIL population, BR007 and SC283, respectively.

BR012R is an Al-sensitive restorer line (tt), whereas SC549, CMS225, and SC566 are Al-tolerant lines (TT) in increasing level of tolerance (Caniato *et al.*2007, 2011). NILs fixed for Alt_{SB} were developed by introgressing from each Al tolerant donor the respective Alt_{SB} alleles into the BR012R background. The SC566-NIL (BC_3F_3) (Melo *et al.* 2013), and the SC549- and CMS225- NILs ($BC_6F_{3:8}$) are fixed for the donor alleles at Alt_{SB}. The NILs with the genetic background belonging to BR012 were designated here as BR012(SC566), BR012(SC549) and BR012(CMS225).

Hybrid production:

Each of the four isogenic male restorer (R) lines were crossed with the two female (A) lines to generate the hybrids H1 to H8 shown in Supporting Information, which depicts the different male and female parents, and each of the derived isogenic hybrids.

Grain yield differences between H1(tt) and H2(Tt) resulted from substituting one Al-sensitive allele from BR007 with one Alt_{SB} allele donated by the Al-tolerant parent, SC283. The comparisons H1(tt) *vs.* H3(Tt) *vs.* H4 (TT) gives the effect in grain yield of one Alt_{SB} allele donated by SC566 (H3) and two Al tolerant alleles, with one donated by SC283 and one donated by SC566 (H4). Similar comparisons for the SC549 and CMS225 alleles were obtained by contrasting H1(tt) *vs.* H5(Tt) *vs.* H6 (TT), and H1(tt) *vs.* H7(Tt) *vs.* H8 (TT), respectively, with the homozygous tolerant hybrid being always formed by oneAlt_{SB} allele from SC283 in combination with another tolerant allele, from SC549 or CMS225.

Hydroponic evaluation of Al tolerance

Aluminum-inhibition of seminal root growth was used to quantify Al tolerance, using the basal nutrient solution described in Magnavaca *et al.* (1987). Al tolerance was assessed in nutrient solution containing 0 or 148 µM Al added as $AlK(SO_4)_2 \cdot 12H_2O$, which correspond to {0} and {27} µM Al^{3+} activities, respectively (brackets indicate free Al^{3+} activity, as estimated with the speciation software GEOCHEM-PC) (Parker *et al.* 1995). The 27 µM Al^{3+} activity has been shown to adequately differentiate SC283 and BR007 for Al tolerance (Magalhaes *et al.* 2004, 2007; Caniato *et al.* 2007, 2011). Seeds were first scarified with sterile sand to break seed dormancy, surface-sterilized with NaOCl 0.1% for 8 min under continuous agitation, and rinsed eight times with 50 ml of 18MΩ H_2O. Seeds were subsequently transferred to germination paper rolls moistened in 18MΩ H_2O, and germination was allowed to proceed for 4 d in a growth chamber with a light intensity of 550 µmol photons m^{-2} sec^{-1}, and a 12-hr photoperiod with 27±3° day and 20±3° night average temperature.

The seminal root from each seedling was then inserted through the mesh bottoms of polyethylene cups, covered with black beads, and placed in plastic tubs with 8 liters of nutrient solution under continuous aeration. A 24-hr acclimation period was subsequently applied in nutrient solution lacking Al. The *i*nitial *l*ength of each seminal root in *c*ontrol solution (il_c) was measured after the acclimation period, and the *f*inal *l*ength in control solution (fl_c) was recorded after 5 d in nutrient solution lacking Al. Similarly, *i*nitial root *l*ength (il_{Al}) and *f*inal root *l*ength (fl_{Al}) under *Al*-stress were recorded immediately after the acclimation period, and after 5 d of Al exposure, respectively. Accordingly, a control root growth rate ($crgr_{5d}$) was obtained as [$crgr_{5d}=(fl_c)-(il_c)$], and the root growth rate under Al exposure ($Alrgr_{5d}$) was then calculated as [$Alrgr_{5d}=(fl_{Al})-(il_{Al})$]. Relative percent values of net root growth (RNRG) inhibition were obtained as [$RNRG=(Alrgr_{5d}/crgr_{5d})\times 100$]. Each experiment was arranged as a randomized complete block design with two replications and seven plants per replication. The Al tolerant and sensitive NILs, ATF10B and ATF8B, which were derived from the same parents of the RIL population (Magalhaes *et al.* 2007), were included as controls.

Field trials

Development of a field site for Al tolerance phenotyping:

The Al tolerance phenotyping sites were developed at the Embrapa Maize and Sorghum research station in Sete Lagoas (Minas Gerais, Brazil) at 19°27'57" latitude south, and 44°14'49" longitude west, and 766.73 m altitude. Two field sites under a typical Haplustox were characterized for aluminum saturation in the 0–20 cm, and the 20–40 cm soil layers. In the control site, dolomitic limestone containing ~33% CaO and ~10% MgO was applied and incorporated in the 0–20 cm soil layer to achieve low exchangeable Al saturation. No limestone was applied to the Al toxicity site. Two months after liming, the area was divided into rectangular 10 × 10 m and 5 × 5 m grids for the control and Al toxicity site, respectively. Topsoil (0–20 cm) and subsoil (20–40 cm) soil compound samples were collected on the square grid. Accordingly, a total of 51 and 169 points were sampled for each depth interval in the control and Al toxicity site, respectively. Three subsamples located in a maximum radius of 1 m from each sampling grid point were collected, and the compound samples were used for soil analysis. The measurements of soil exchangeable aluminum (Al^{+3}) and exchangeable cations (Ca^{+2}, Mg^{+2}) were performed with KCl 1 mol liter^{-1} as extractor (Donagema *et al.* 2011) using inductively-coupled argon plasma (ICP) emission spectrometry. Percent values of Al saturation were calculated as the ratio of exchangeable Al^{3+} divided by the sum of basic cations plus Al^{+3}.

Field phenotyping of Al tolerance:

Field experiments were conducted from April to September 2007 for the RIL population, and from January to April 2013 for the isogenic hybrids in soils with 2% and 56% Al saturation. For the RIL population, each plot consisted of two 5-m rows, with 0.45 m between rows, and eight plants m^{-1}. In the hybrid experiment, each plot consisted of four 5-m rows, with 0.45 m between rows, and eight plants m^{-1} in a row. In this case, only the two center rows were harvested and considered for subsequent analysis. For both trials, fertilization consisted of 300 kg ha^{-1} of 8-28-16 (NPK) applied at planting, and 160 kg ha^{-1} of urea applied 30 d thereafter. The RIL experiments consisted of an incomplete block design with three replicates, and 10 incomplete blocks in each of the two environments, defined as control site (2% Al saturation), and Al toxicity site (56% Al saturation). A total of 100 genotypes in each replication included 90 RILs and the two parents, BR007 and SC283, with the Al tolerant parent, SC283, repeated nine times. The hybrid experiments consisted of a split plot experiment with the Al treatment (Al *vs.* control) applied at the whole plot level, and the hybrids applied at the sub plot level. At the whole plot level, a completely randomized design was assumed to be applicable with the two treatments (Al, control), each repeated four times. Each of the eight hybrid combinations was randomly allocated to a sub plot within each of the eight whole plots. In total, 64 experimental plots were used: four replicates × two whole plot treatments × eight sub plot treatments. Sorghum grain was manually harvested based on the average physiological grain maturity of the population, which occurred approximately 130 d after seed emergence. Total grain yield in ton ha^{-1} was estimated based on average grain weight per plot.

Statistical analysis of phenotypic data

All statistical analyses described below were performed using GenStat 17 (VSN International 2014), except the map construction, which was done by the Onemap (Margarido *et al.* 2007) software.

Hydroponic data:

A randomized complete block design was used for assessing Al tolerance in hydroponics. Data collected was logarithmically transformed to the base 10 before subsequent analysis (henceforth any reference to RNRG will correspond to log-transformed data). A mixed model taking genotype random was fitted as follows; response = block + genotype + error. The variance components for the genotypic and error variance, σ_g^2 and σ_e^2, respectively, were used to calculate the heritability at a line mean basis; $h^2 = \frac{\sigma_g^2}{\sigma_g^2+\sigma_e^2/r}$, where r corresponds to the number of blocks, which was two. Subsequently, genotype was fitted as a fixed term in order to obtain best linear unbiased estimators (BLUE) for later use in QTL mapping.

Field data:

For the RIL experiment, a multi-environment mixed model analysis was fitted as follows: response = treatment + replicate within treatment + incomplete block within replicate + genotype + error. In this model, the incomplete block, the genotype and the error term were assumed to be normally distributed with treatment dependent variances, *i.e.*, the variances were different between Al and control. For the genotype term, we assumed a covariance between the treatments, *i.e.*, between Al and control. Heritability at a line mean basis, h^2, was estimated for Al and control treatment as $h^2 = \frac{\sigma_g^2}{\sigma_g^2+\sigma_e^2/r}$, where σ_g^2 and σ_e^2 are the genotypic and error variances, respectively, and r corresponds to the number of replicates, which was three. Another mixed model analysis was done with genotypes fixed to produce genotypic BLUEs for multi-trait QTL analysis.

Split plot analysis of variance was performed for the hybrids experiment to test the effects of substituting alleles at the Alt_{SB} locus. The effect of treatment, Al *vs.* control, was tested against the whole plot error, while all other effects were tested against the subplot error.

Marker analysis and linkage map construction

The same sorghum linkage map as developed and described by Sabadin *et al.* (2012) was used in this study. Accordingly, a total of 344 markers consisting of 255 Diversity Arrays Technology (DArT), 83 SSRs, five Sequence-Tagged Site (STS), and one RFLP marker, were used for map construction. The RFLP marker, ISU52, was genotyped as described in Magalhaes *et al.* (2004), while genotyping with the STS markers, CTG29 and M181, was as described in Caniato *et al.* (2007). The STS fluorescent markers DG1, EM1, M1672, and M9612 were described in Sabadin *et al.* (2012).

QTL analysis

The genotypic means for yield as observed in the field under Al and control conditions, and the RNGR genotypic means as obtained on the hydroponics medium were analyzed together in a multi-trait QTL analysis. A multi-trait QTL analysis facilitates the assessment of pleiotropic action of QTL and will have higher power for QTL detection. For our QTL analysis we followed a mixed model-based protocol as described for multi-trait and multi-environment QTL analyses by (Boer *et al.* 2007; Malosetti *et al.* 2008, 2013; Alimi *et al.* 2013). The protocol consisted of an initial round of simple interval mapping (Lander and Botstein 1989) that was followed by one or more rounds of composite interval mapping (Jansen and Stam 1994; Zeng 1994). From the last round of composite interval mapping, a candidate set of QTL was obtained, from which possibly redundant QTL were removed by a backward elimination procedure, with a final set of QTL as end result.

In the simple interval mapping round, the following mixed model was fitted along the genome for the three responses jointly; responses = QTL + polygenic effects + errors, with an unstructured variance covariance model imposed on the polygenic effects, *i.e.*, polygenic variances were trait specific, and polygenic correlations were unique for each pair of traits. The QTL allele substitution effects were fixed and trait specific. In the simple interval mapping scan, genomic positions were tested one by one for QTL presence by Wald tests. A multiple test correction was applied according to Li and Ji (2005). In composite interval mapping, potential QTL elsewhere in the genome were included as cofactors; responses = cofactors + QTL + polygenic effects + errors. The final multi-trait multi-QTL model was of the form responses = QTL + polygenic effects + errors. The amount of genetic variance explained by a QTL was estimated from the drop in polygenic variance that occurred when the QTL was added to a model with exclusively polygenic effects; responses = polygenic effects + errors.

RESULTS

Spatial variability for Al saturation in the soil

Phenotyping sites were set up to assess Al tolerance in the field. Soil Al saturation was low, near 2% in the superficial layer (0–20 cm) in the control site, whereas much higher Al saturation was observed in the Al toxicity site, reaching 56% on average (Figure 1). The same large differences in Al saturation were observed in the subsoil (20–40 cm), with mean Al saturation values of approximately 15% and 65% for the control and Al toxicity sites, respectively.

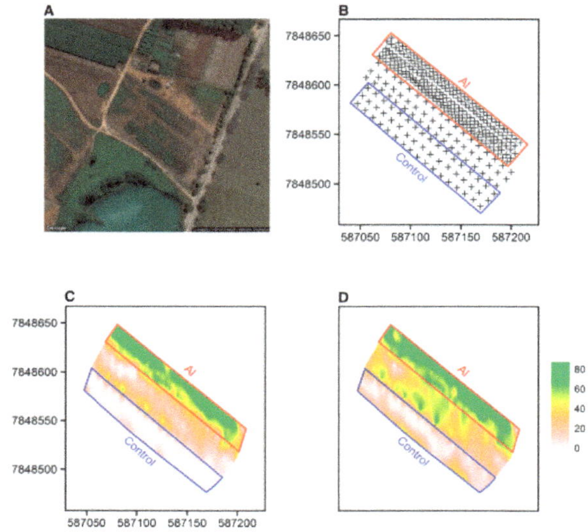

Figure 1: Field phenotyping sites. Satellite image (A) and respective soil sampling points (B). Al saturation (%) spatial variation in the superficial (0–20 cm) (C), and subsuperficial (20–40 cm) (D) soil layers. Adapted fromMenezes *et al.* (2014). *x*- and *y*-axis in panels B, C and D correspond to spatial coordinates in Universal Transverse Mercator (UTM) 23k format. Satellite images were obtained on March 5, 2014.

Assessment of Al tolerance in the parents of the RIL population

Average grain yield in the Al-sensitive parent of our RIL population, BR007, under control conditions was 3.9 ton ha^{-1}, revealing a much higher yield potential compared to the Al-tolerant parent, SC283 (Figure 2). However, upon Al stress, grain yield for BR007 was approximately half of that in the control treatment, while SC283 was able to sustain yield under Al stress. Despite the higher yield potential for BR007 compared to SC283, under Al toxicity, grain yield in SC283 was higher compared to BR007.

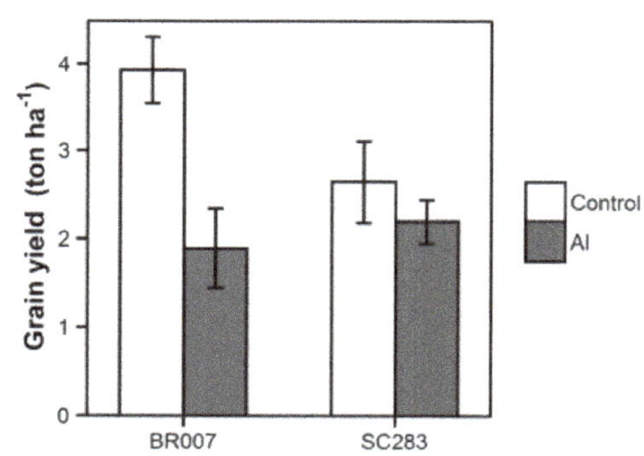

Figure 2: Phenotypic means for grain yield (ton ha^{-1}) under control (2% Al saturation) and Al stress (56% Al saturation) conditions in the field for the parental lines BR007 and SC283. Means are based on 2-year trials. Least significant difference (LSD) = 0.59 (P = 0.05). Error bars are shown.

Genetic complexity of Al tolerance in the RIL population

The primary measure of Al tolerance in hydroponic media, relative net root growth (RNRG), was approximately 10-fold greater for SC283 than for BR007 on average, which was in agreement with previous studies indicating that SC283 is an important Al tolerance standard in sorghum (Duncan *et al.* 1983; Borgonovi *et al.* 1987; Furlani and Clark 1987; Furlani *et al.* 1991). In addition, much greater variation in RNRG was associated with the homozygous tolerant (TT) progeny compared to the homozygous sensitive (tt) progeny for Alt_{SB}, suggesting the presence of minor Al tolerance loci distinct from Alt_{SB} and/or the action of modifiers, with possible involvement in transgressive segregation for Al tolerance.

Compared to Al tolerance in hydroponics, transgressive segregation was much more evident for grain yield performance under high Al saturation, where segregants both below and above the mean of both parents were detected at a reasonable frequency. These results indicate that the genetic architecture underlying grain yield performance under Al toxicity is more complex compared to Al tolerance assessed in hydroponics.

The genetic variance for grain yield was larger in control compared to Al stress conditions in the field, in agreement with previous reports indicating decrease of this variance component under abiotic stress conditions (Johnson and Frey 1967; Ud-Din *et al.* 1992; Kumar *et al.* 2007; Sabadin *et al.* 2012). Broad-sense heritability estimated for the RIL population was high, between 0.7 and 0.9, and it was highest for Al tolerance assessed in hydroponics under strict environmental control. Al toxicity reduced grain yield in the RIL population by approximately 30% and led to slightly lower heritability values compared to control conditions.

Assessment of Al tolerance in isogenic hybrids

In order to assess the Alt_{SB} effect in hybrids, we constructed isogenic hybrid stocks that contrasted for different Alt_{SB} alleles within an otherwise uniform genetic background. Analysis of variance of grain yield detected a significant environmental effect with high Al saturation causing a 23% reduction in grain yield (Table 1). Furthermore, this analysis emphasized a significant average female effect between ATF13A(tt) and ATF14A(TT), which was a striking 0.5 ton ha^{-1} (Table 1). This average effect arises from adding one copy of the Al tolerance allele, T, as tested across the four males to whom the female lines were crossed. Because this average female effect includes comparisons both between 0 (tt), 1 (Tt), and 2 (TT) copies of the Al tolerance allele, it also indicates there is advantage in breeding superior Alt_{SB} alleles into both parents of hybrid stocks due to its additive gene action in field conditions.

Table 1: Means with respective least significant differences ($P=0.05$) for environment and female main effects obtained by analysis of variance of eight isogenic hybrids grown in a split-plot design with four replications and a completely randomized design at the whole plot level

Average Effect		Grain Yield (ton ha^{-1})
Female	ATF13A	3.46
	ATF14A	3.96
	LSD (0.05)	0.45
Environment	Control	4.18
	Al	3.24
	LSD (0.05)	0.90

Genetic architecture of Al tolerance in hydroponics and in the field

Multi-trait analysis revealed a single major QTL for Al tolerance on chromosome 3 assessed by quantifying either RNRG in hydroponics and grain yield in the field under high Al saturation (Figure 3). This QTL was detected at position 184.2 cM, tightly linked to the STS markers, CTG29 and M181 at positions 180.6 and 180 cM, respectively (Figure 4). The Al tolerance QTL assessed in hydroponics, *RNRG3*, explained half of the variance for RNRG, with the SC283 allele positively contributing to Al tolerance. Based on its position,*RNRG3/Gy3* is located in close proximity to Alt_{SB} and *SbMATE* (Figure 4). As observed for *RNRG3*, the positive allele at *Gy3* was also donated by the Al-tolerant parent, SC283, indicating a pleiotropic action of the Alt_{SB} locus on both sustained root elongation in nutrient solution, with an Al^{3+} activity of 27 µM and grain yield on soils with 56% Al saturation. *Gy3* was expressed both in control and Al stress conditions but the effect under Al toxicity was 0.6 ton ha^{-1}, more than threefold greater than the effect estimated in control conditions (Table 2). In addition, the proportion of the phenotypic variation explained under low Al saturation was only 3.5%, contrasting with almost 38% under Al stress conditions.

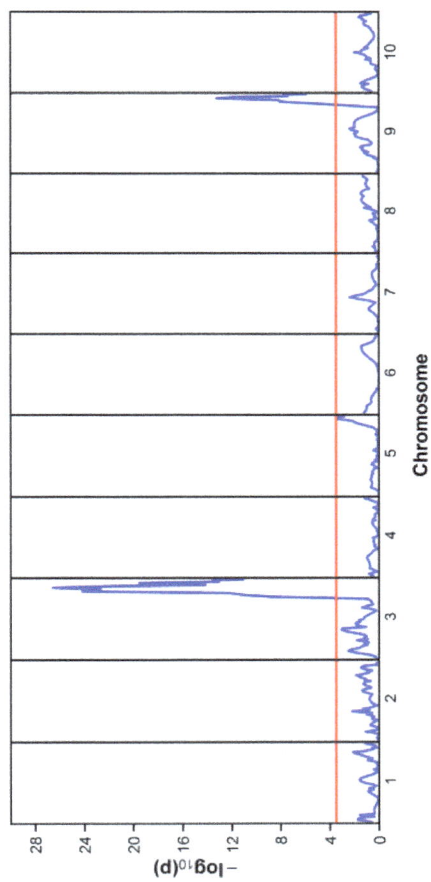

Figure 3: Graphical display of the QTL detected by multi-trait QTL analysis for relative net root growth (RNRG) after 5 d at {27} µM Al^{3+} in nutrient solution, and grain yield (ton ha^{-1}) under 56% Al saturation stress in the field. The associated tail probability of the Wald statistics, P, is expressed as $-\log_{10}(p)$, analogous to the usual LOD score profile. The red horizontal line indicates the significant threshold obtained using a correction for multiple testing following Li and Ji (2005), with a genome wide test level $\alpha=0.05$.

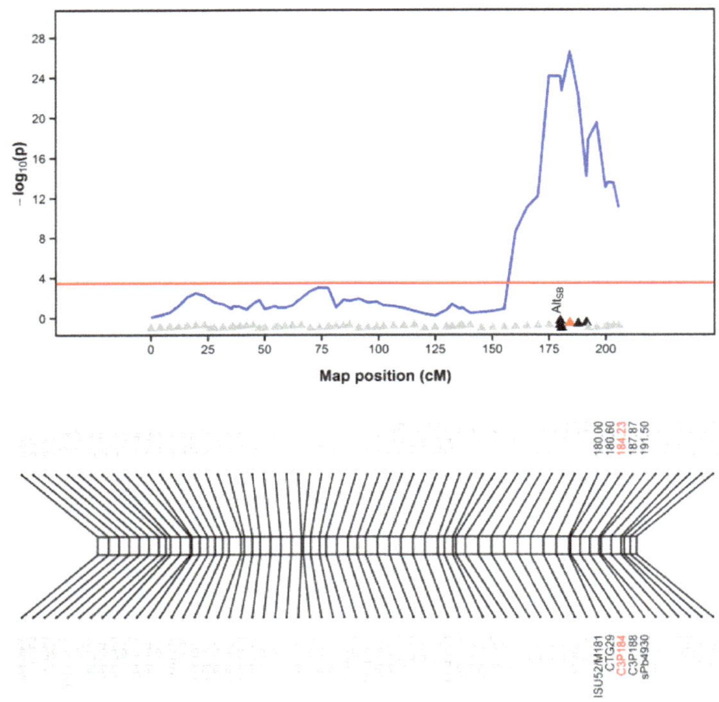

Figure 4: Detailed graphical display of the QTL detected by a multi-trait QTL analysis on chromosome 3 for relative net root growth (RNRG) after 5 d at {27} μM Al^{3+} in nutrient solution and grain yield (ton ha^{-1}) under 56% Al saturation stress in the field. The associated tail probability of the Wald statistics, P, is expressed as $-\log_{10}(p)$, analogous to the usual LOD score profile. The red horizontal line indicates the significant threshold obtained using the Li and Ji (2005) correction with $\alpha = 5\%$.

Table 2: Estimates of QTL effects using multi-trait analysis for relative net root growth ($\log_{10}$ transformed ratio) and grain yield (ton ha^{-1})

Environment[a]	QTL	QTL Position (cM)[b]	Closest Markers	Effects	High Value Allele[c]	%GV	Avse
Hydroponics							
	RNRG3	184.2	ctg29 - sPb4930	0.71	SC283	50.10	(0.08)
	RNRG9	214.0	sPb1826 - sPb4087	0.34	SC283	11.30	(0.07)
				Total		**61.40**	
Field							
Control	*Gy3*	184.2	ctg29 - sPb4930	0.19	SC283	3.50	(0.09)
	Gy9	214.0	sPb1826 - sPb4087	0.66	BR007	44.00	(0.09)
				Total		47.50	
Al	*Gy3*	184.2	ctg29 - sPb4930	0.61	SC283	37.70	(0.10)
	Gy9	214.0	sPb1826 - sPb4087	0.36	BR007	12.70	(0.09)
				Total		**50.40**	

Average standard errors (Avse) for each effect are shown between parentheses. QTL are coded as relative net root growth (RNRG) or grain yield (Gy) numbered according to the chromosome where they were mapped. %GV stands for the percentage of the genetic variance that is explained by a given QTL (the percentage of the genetic variance explained after fitting all significant QTL in the model is shown in bold, following individual %GV estimates).
[a] Control: 2% soil aluminum saturation, Al: 56% soil aluminum saturation.
[b] QTL position corresponds to $-\log_{10}(p)$ peak for a given QTL.
[c] Allele that increases the phenotype.

Our multi-trait analysis also detected a second Al tolerance QTL for both Al tolerance in hydroponics (*RNRG9*) and grain yield (*Gy9*) at position 214.0 cM on chromosome 9 (Figure 3), with *RNRG9* explaining a smaller proportion of the variation for Al tolerance in hydroponics (Table 2). As was the case for *RNRG3*, the positive allele at *RNRG9* was also donated by the Al-tolerant parent, SC283. In turn, the grain yield QTL, *Gy9*, explained a much larger portion of the genetic variance in control compared to Al toxicity conditions, and the allele increasing grain yield was donated by the Al-sensitive line, BR007, both under high and low Al saturation.

DISCUSSION

Our first step in linking hydroponics-based assessments of Al tolerance to a possible yield advantage elicited by the sorghum Al tolerance locus, Alt_{SB}, was to design and set up field phenotyping sites where the effect of Al toxicity was as much as possible isolated from that caused by other abiotic stresses that occur jointly on acid soils (Bahia Filho *et al.* 1997). The Al saturation threshold in the field beyond which sorghum yield is reduced is around 20% Al saturation for sorghum growing in Oxisols/Ultisols (Gourley 1987). Based on that, the control (2% Al saturation) and Al toxicity (56% Al saturation) sites were designed to allow for the specific assessment of the impact of genetic determinants underlying Al tolerance on sorghum yield. Our field results for the parental lines supported previous hydroponics-based classifications of SC283 and BR007 as Al-tolerant and -sensitive lines (Magalhaes *et al.* 2004, 2007; Caniatoet *al.* 2007, 2011, 2014), respectively, and that SC283 expresses Al tolerance in the field (Duncan *et al.* 1983; Gourley 1987; Furlani *et al.* 1991). A strong grain yield reduction caused by Al toxicity was observed for the Al-sensitive parent, BR007, showing that Al toxicity strongly overshadows the grain yield potential of Al-sensitive sorghum elite breeding lines.

Based on relative net root growth assessed in hydroponics, the Al tolerance QTL, *RNRG3*, and the grain yield QTL, *Gy3*, were tightly linked to the STS marker, CTG29, on chromosome 3. This marker is also tightly linked to Alt_{SB} both genetically and physically (~160 kb distant from *SbMATE*) and was used to previously positionally clone *SbMATE*, which underlies Alt_{SB} (Magalhaes *et al.* 2007). The detailed genetic map of the Alt_{SB} region is shown in Figure 4 and depicts CTG29 at position 180.6 cM on chromosome 3, while M181, which was the second flanking marker used to positionally clone *SbMATE* (and located in the same bacterial artificial chromosome as *SbMATE*) (Magalhaes *et al.* 2007), cosegregates with Alt_{SB} at position 180 cM, along with the marker originally used to map Alt_{SB}, ISU52 (Magalhaes *et al.* 2004). This indicates that Alt_{SB} underlies both the *RNRG3* and *Gy3* QTL.

The alleles increasing phenotypic expression at *RNRG3* and *Gy3* were both donated by the Al-tolerant parent, SC283, and *Gy3* had a much higher phenotypic effect and explained a much larger proportion of the phenotypic variance under Al toxicity, compared to the control site with low Al saturation. From these results, a yield advantage of 0.6 ton ha^{-1} was determined, arising from substituting two copies of the BR007 allele with two copies of the Al tolerant (SC283) allele at Alt_{SB}, which translates into a 26% grain yield increase under Al toxicity, based on the RIL grain yield mean.

Our studies with isogenic hybrid stocks indicate that Alt_{SB} has an even stronger effect in hybrid stocks, with an on average 0.5 ton ha^{-1} female effect arising from one Al tolerance Alt_{SB} allele. We also explicitly tested for additive *vs.* dominant gene action by introducing a covariate for the number of Al tolerance alleles (T) in the hybrids. We found the coefficient for this covariate to be the same as the average substitution effect estimated by fitting female and male in place of allele dosage (Table 1). Therefore, it is advantageous to introgress Alt_{SB} into both parents due to its additive gene action on grain yield under Al toxicity. Overall, these results make Al tolerance conferred by Alt_{SB} clearly the default trait when breeding for sorghum adaptation to acidic, Al toxic soils, via the development of homozygous or hybrid cultivars.

Although *Gy3* was expressed under low Al saturation, this QTL explained only 3.5% of the phenotypic variation under low Al saturation compared to 37.7% under high Al saturation, with a much lower effect of 0.19 ton ha^{-1} under low Al saturation. Recently, a sorghum genome-wide association mapping study conducted in West Africa, including gene-specific markers developed for *SbMATE*, revealed that *SbMATE* SNPs were highly associated with grain yield. Associations were found especially under low phosphorus (P) conditions in 29 sites in West Africa, explaining up to 16% of the genotypic variance (Leiser *et al.* 2014). Accordingly, this suggests a pleiotropic role of *SbMATE* in providing tolerance to two of the most serious abiotic stresses for sorghum in West Africa, Al toxicity and P deficiency. As discussed in the Leiser *et al.* (2014) study, overexpression of citrate synthesis and malate transporter genes in different species has resulted in both improved Al-tolerance and enhanced P uptake under low P conditions (Delhaize *et al.* 2009; Wang *et al.* 2012; Liang *et al.* 2013), supporting previous hypotheses that Al tolerance and P uptake can be conferred by similar mechanisms (Liao *et al.* 2006; Magalhaes *et al.* 2007). Therefore, because phosphorus binds tightly to Al and Fe oxides in the clay fraction of acid soils (Novais and Smyth 1999), it is possible that the comparatively mild phenotypic expression of *Gy3* under low Al saturation is due to enhancement of P acquisition, which could have been accompanied by a smaller Al tolerance effect due to somewhat higher Al saturation in the subsoil layer of the control site.

As an important intermediate in the Krebs cycle, citrate is essential for plant growth. Hence, the development and the use of root organic acid efflux to protect against Al stress could carry a significant energy cost for the plant, which may negatively affect plant yields (Liu *et al.* 2012). In support of this, in the absence of Al^{3+}, roots of transgenic Arabidopsis lines overexpressing *AtALMT1* or *AtMATE* had rates of constitutive malate and citrate release that were two to ten times higher, respectively, than those of wild-type plants, and root growth was inhibited 20–30% compared with wild-type plants. This was attributed to the carbon loss from roots due to organic acid exudation, which exerted a measurable carbon cost to the plant (Liu *et al.* 2012). If such a penalty occurs in the case of Alt_{SB}, where *SbMATE* expression is driven by its native promoter, the SC283 allele should decrease grain yield under control conditions. This might be expected because, in the absence of Al toxicity in the control site, the extremely low levels of *SbMATE* expression and citrate release that are inherent to the BR007 allele (Magalhaes *et al.* 2007; Melo *et al.* 2013) would result in minimal root loss of carbon, and thus yield advantage compared to the functional SC283 allele if significant citrate efflux occurred due to the SC283 allele under low Al saturation in the soil. As discussed above, the SC283 allele at *Gy3* actually slightly increased grain yield in the control site, demonstrating that no yield penalty arises from Al-induced citrate release elicited by the Al-activated citrate transporter, *SbMATE*. As such, Alt_{SB}-based Al-tolerant cultivars developed for Al toxic soils should perform equally well in soils without Al toxicity, avoiding the time and the cost required to develop cultivars specific to different soil types.

This outcome is expected based on our current knowledge on physiological and molecular regulation of Al tolerance conferred by *SbMATE*. First, no substantial citrate release is observed in the absence of Al^{3+}, in either Al-tolerant or Al-sensitive sorghum lines (Magalhaes *et al.* 2007; Melo *et al.* 2013), which may be due to another Al-binding protein that interacts with *SbMATE*, and is responsible for Al-activated root citrate efflux (Kochian *et al.* 2015). Second, although *SbMATE* is expressed in the absence of Al^{3+}, its expression is localized to the root apex of Al tolerant near-isogenic lines carrying the SC283 allele compared to other root regions (Magalhaes *et al.* 2007), which is consistent with early reports of spatial localization of Al-induced organic acid release acting to protect Al sensitive zones specifically in the root apex (Ryan *et al.* 1993; Kochian *et al.* 2015). Moreover, using near-isogenic lines derived from BR007 and SC283, it has been recently shown that Al-induced *SbMATE* gene and protein expression is specifically localized to the epidermal and outer cortical cell layers of the root distal transition zone in the Al-resistant NIL (Sivaguru *et al.* 2013). This was shown to be the specific site of Al-induced root damage, and the time course of Al induction of *SbMATE* expression in the distal transition zone was precisely coincident with the onset of the recovery of this root region from Al-induced damage. Overall, the body of evidence underlying *SbMATE* function indicates that Al tolerance encoded by this sorghum MATE protein is under tight control to minimize the loss of carbon due to citrate release, preventing a yield penalty in the absence of Al toxicity.

The research findings presented here also uncovered a novel Al tolerance QTL on sorghum chromosome 9, which is consistent with previous reports of incomplete transfer of Al tolerance from donor lines to NILs based solely on*Alt_{SB}* (Caniato *et al.* 2007, 2014). The Al tolerance and grain yield QTL, *RNRG9*and *Gy9*, were found at position 214 cM (confidence interval 207–221 cM) by multi-trait analysis, suggesting the presence of a single pleiotropic QTL affecting both traits. However, differently than *Gy3* and *RNRG3*, where positive alleles at both loci were donated by SC283, it was the SC283 allele that increased Al tolerance in hydroponics for *RNRG9* whereas the Al sensitive allele from BR007 at *Gy9* increased grain yield both in control and Al toxicity conditions in the soil. Interestingly, a major phenology-related QTL with a strong effect on grain yield has been previously found at position 218 cM on chromosome 9 in the same RIL population, with the positive allele also donated by the high-yielding parent, BR007 (Sabadin *et al.* 2012), as was the case for*Gy9* in the present study. This QTL is likely to correspond to *Sb-HT9.1* (Brown*et al.* 2008) as discussed in the Sabadin *et al.* (2012) study. Unlike *Gy3*, *Gy9*had a higher effect on grain yield and explained a much larger proportion of the genetic variance in control compared to Al stress conditions in the field, suggesting that *Gy9* is unspecific with regards to Al tolerance. Based on this, we speculate that *Gy9* and *RNRG9* could possibly be in fact distinct, with the former affecting primarily plant height/flowering time as described by Sabadin*et al.* (2012) while the allele donated by SC283 in *RNRG9* enhances Al tolerance assessed in hydroponics. Under this hypothesis, the lack of detection of an effect for *RNRG9* in terms of grain yield gain under Al toxicity in the field, with the positive allele coming from the Al-tolerant parent, SC283, may be accounted by an overshadowing effect of the likely phenology-related QTL,*Gy9*, or to an insufficient effect of *RNRG9* in providing grain yield advantage under Al toxicity.

For improving Al tolerance based on *Alt_{SB}*, we have recently developed gene-specific markers associated with Al tolerance that allow not only for marker-assisted introgression but which are also amenable for large-scale allele mining approaches based on *Alt_{SB}*, that can be used to identify accessions harboring*Alt_{SB}* directly in the breeder's gene pool, with obvious adaptive advantages (Caniato *et al.* 2014). Based on the findings reported here, molecular breeding based on *Alt_{SB}* is of fundamental importance to improving sorghum grain yields on acid soils with Al toxicity. This should help to unlock food production potential in agricultural frontiers, with many of them ironically located where they are needed most for mankind, such as in the tropics and subtropics, where food security is most tenuous.

ACKNOWLEDGMENTS

The authors acknowledge funding from the CGIAR Generation Challenge Program (project G3007.04), the Embrapa Macroprogram, and the McKnight Foundation Collaborative Crop Research Program. The authors are also thankful to The National Council for Scientific and Technological Development (CNPq) for an MS fellowship granted to G.C., and for support to J.V.M.

REFERENCES

1. Alimi N. A., Bink M. C., Dieleman J. A., Magan J. J., Wubs A. M., et al., 2013 Multi-trait and multi-environment QTL analyses of yield and a set of physiological traits in pepper. Theor. Appl. Genet. 126: 2597–2625.
2. Bahia Filho, A. F. C., R. Magnavaca, R. E. Schaffert, and V. M. C. Alves, 1997 Identification, utilization, and economic impact of maize germplasm tolerant to low levels of phosphorus and toxic levels of exchangeable aluminum in Brazilian soils, pp. 59–70 in Plant–Soil Interactions at Low pH: Sustainable Agriculture and Forestry Production, edited by A. C. Moniz et al. Brazilian Soil Science Society, Campinas/Viçosa, Brazil.
3. Baier A. C., Somers D. J., Gusiafson J. P., 1995 Aluminium tolerance in wheat: correlating hydroponic evaluations with field and soil performances. Plant Breed. 114: 291–296.
4. Boer M. P., Wright D., Feng L., Podlich D. W., Luo L., et al., 2007 A mixed-model quantitative trait loci (QTL) analysis for multiple-environment trial data using environmental covariables for QTL-by-environment interactions, with an example in maize. Genetics 177: 1801–1813.
5. Borgonovi R. A., Schaffert R. E., Pitta G. V. E., Magnavaca R., Alves V. M. C., 1987 Aluminum tolerance in sorghum, pp. 213–221 in Genetic Aspects of Plant Mineral Nutrition, edited by Gabelman W. H., Loughman B. C.. Springer, Dordrecht, The Netherlands.
6. Brown P. J., Rooney W. L., Franks C., Kresovich S., 2008 Efficient mapping of plant height quantitative trait loci in a sorghum association population with introgressed dwarfing genes. Genetics 180: 629–637.
7. Caniato F. F., Guimaraes C. T., Schaffert R. E., Alves V. M., Kochian L. V., et al., 2007 Genetic diversity for aluminum tolerance in sorghum. Theor. Appl. Genet. 114: 863–876.
8. Caniato F. F., Guimaraes C. T., Hamblin M., Billot C., Rami J. F., et al., 2011 The relationship between population structure and aluminum tolerance in cultivated sorghum. PLoS One 6: e20830.
9. Caniato F. F., Hamblin M. T., Guimaraes C. T., Zhang Z., Schaffert R. E., et al., 2014 Association mapping provides insights into the origin and the fine structure of the sorghum aluminum tolerance locus, AltSB. PLoS One 9: e87438.
10. Collins N. C., Shirley N. J., Saeed M., Pallotta M., Gustafson J. P., 2008 An ALMT1 gene cluster controlling aluminum tolerance at the Alt4 locus of rye (Secale cereale L). Genetics 179: 669–682.
11. Culvenor R., Oram R., Fazekas S. G. C., 1986 Variation in tolerance in Phalaris aquatica L. and a related species to aluminium in nutrient solution and soil. Aust. J. Agric. Res. 37: 383–395.
12. Delhaize E., Taylor P., Hocking P. J., Simpson R. J., Ryan P. R., et al., 2009 Transgenic barley (Hordeum vulgare L.) expressing the wheat aluminium resistance gene (TaALMT1) shows enhanced phosphorus nutrition

and grain production when grown on an acid soil. Plant Biotechnol. J. 7: 391–400.

13. Donagema G. K., Campos D. V. B., Calderano S. B., Teixeira W. G., Viana J. H. M., 2011 Manual de Métodos de Análise de Solo, Embrapa Solos, Rio de Janeiro, Brasil.
14. Doumbia M. D., Hossner L. R., Onken A. B., 1993 Variable sorghum growth in acid soils of subhumid West Africa. Arid Soil Res. Rehabil. 7: 335–346.
15. Doumbia M. D., Hossner L. R., Onken A. B., 1998 Sorghum growth in acid soils of West Africa: variations in soil chemical properties. Arid Soil Res. Rehabil. 12: 179–190.
16. Duncan R. R., Clark R. B., Furlani P. R., 1983 Laboratory and field evaluations of sorghum for response to aluminum and acid soil. Agron. J. 75: 1023–1026.
17. Furlani P. R., Clark R. B., 1987 Plant traits for evaluation of responses of sorghum genotypes to aluminum, pp. 247–254 in Genetic Aspects of Plant Mineral Nutrition, edited by Gabelman W. H., Loughman B. C.. Springer, Dordrecht, The Netherlands.
18. Furlani P. R., Quaggio J. A., Gallo P. B., 1991 Differential responses of sorghum to aluminium in nutrient solution and acid soil, pp. 953–958 in Plant-Soil Interactions at Low pH, edited by Wright R. J., Baligar V. C., Murrmann R. P.. Springer, Dordrecht, The Netherlands.
19. Furukawa J., Yamaji N., Wang H., Mitani N., Murata Y., et al., 2007 An aluminum-activated citrate transporter in barley. Plant Cell Physiol. 48: 1081–1091.
20. Gourley L. M., 1987 Finding and utilizing exotic Al-tolerant sorghum germplasm, pp. 293–309 in Sorghum for Acid Soils, edited by Gourley L., Salinas J.. Proceedings of a workshop on evaluating sorghum for tolerance to Al-toxic tropical soils in Latin America, Cali, Colombia.
21. Hoekenga O. A., Maron L. G., Pineros M. A., Cancado G. M., Shaff J., et al., 2006 AtALMT1, which encodes a malate transporter, is identified as one of several genes critical for aluminum tolerance in Arabidopsis. Proc. Natl. Acad. Sci. USA 103: 9738–9743.
22. Jansen R. C., Stam P., 1994 High resolution of quantitative traits into multiple loci via interval mapping. Genetics 136: 1447–1455.
23. Johnson G. R., Frey K. J., 1967 Heritabilities of quantitative attributes of oats (Avena sp.) at varying levels of environmental stress. Crop Sci. 7: 43–46.
24. Johnson, H. W., and R. L. Bernard, 1962 Soybean genetics and breeding, pp. 199–213 in Advances in Agronomy, edited by Norman, A. G. Academic Press, New York.
25. Kochian L. V., Hoekenga O. A., Pineros M. A., 2004 How do crop plants tolerate acid soils? Mechanisms of aluminum tolerance and phosphorous efficiency. Annu. Rev. Plant Biol. 55: 459–493.
26. Kochian L. V., Pineros M. A., Liu J., Magalhaes J. V., 2015 Plant adaptation to acid soils: the molecular basis for crop aluminum resistance. Annu. Rev. Plant Biol. 66: 571–598.
27. Kumar R., Venuprasad R., Atlin G. N., 2007 Genetic analysis of rainfed lowland rice drought tolerance under naturally-occurring stress in eastern India: heritability and QTL effects. Field Crops Res. 103: 42–52.
28. Lander E. S., Botstein D., 1989 Mapping Mendelian factors underlying quantitative traits using RFLP linkage maps. Genetics 121: 185–199.
29. Leiser W. L., Rattunde H. F., Weltzien E., Cisse N., Abdou M., et al., 2014 Two in one sweep: aluminum tolerance and grain yield in P-limited soils are associated to the same genomic region in West African sorghum. BMC Plant Biol. 14: 1–25.
30. Li J., Ji L., 2005 Adjusting multiple testing in multilocus analyses using the eigenvalues of a correlation matrix. Heredity 95: 221–277.
31. Liang C., Pineros M. A., Tian J., Yao Z., Sun L., et al., 2013 Low pH, aluminum, and phosphorus coordinately regulate malate exudation through GmALMT1 to improve soybean adaptation to acid soils. Plant Physiol. 161: 1347–1361.
32. Liao H., Wan H., Shaff J., Wang X., Yan X., et al., 2006 Phosphorus and aluminum interactions in soybean in relation to aluminum tolerance. Exudation of specific organic acids from different regions of the intact root system. Plant Physiol. 141: 674–684.
33. Ligaba A., Katsuhara M., Ryan P. R., Shibasaka M., Matsumoto H., 2006 The BnALMT1 and BnALMT2 genes from rape encode aluminum-activated malate transporters that enhance the aluminum resistance of plant cells. Plant Physiol. 142: 1294–1303.
34. Liu J., Magalhaes J. V., Shaff J., Kochian L. V., 2009 Aluminum-activated citrate and malate transporters from the MATE and ALMT families function independently to confer Arabidopsis aluminum tolerance. Plant J. 57: 389–399.
35. Liu J., Luo X., Shaff J., Liang C., Jia X., et al., 2012 A promoter-swap strategy between the AtALMT and AtMATE genes increased Arabidopsis aluminum resistance and improved carbon-use efficiency for aluminum resistance. Plant J. 71: 327–337.
36. Liu M. Y., Chen W. W., Xu J. M., Fan W., Yang J. L., et al., 2013 The role of VuMATE1 expression in aluminium-inducible citrate secretion in rice bean (Vigna umbellata) roots. J. Exp. Bot. 64: 1795–1804.
37. Ma J. F., Ryan P. R., Delhaize E., 2001 Aluminium tolerance in plants and the complexing role of organic acids. Trends Plant Sci. 6: 273–278.
38. Magalhaes J. V., Garvin D. F., Wang Y., Sorrells M. E., Klein P. E., et al., 2004 Comparative mapping of a major aluminum tolerance gene in sorghum and other species in the poaceae. Genetics 167: 1905–1914.
39. Magalhaes J. V., Liu J., Guimaraes C. T., Lana U. G., Alves V. M., et al., 2007 A gene in the multidrug and toxic compound extrusion (MATE) family confers aluminum tolerance in sorghum. Nat. Genet. 39: 1156–1161.

40. Magnavaca R., Gardner C. O., Clark R. B., 1987 Inheritance of aluminum tolerance in maize, pp. 201–212 in Genetic Aspects of Plant Mineral Nutrition, edited by Gabelman H. W., Loughman B. C.. Springer, Dordrecht, The Netherlands.

41. Malosetti M., Ribaut J. M., Vargas M., Crossa J., Van Eeuwijk F. A., 2008 A multi-trait multi-environment QTL mixed model with an application to drought and nitrogen stress trials in maize (Zea mays L.). Euphytica 161: 241–257.

42. Malosetti M., Ribaut J. M., van Eeuwijk F. A., 2013 The statistical analysis of multi-environment data: modeling genotype-by-environment interaction and its genetic basis. Front. Physiol. 4: 44.

43. Margarido G. R., Souza A. P., Garcia A. A., 2007 OneMap: software for genetic mapping in outcrossing species. Hereditas 144: 78–79.

44. Maron L. G., Pineros M. A., Guimaraes C. T., Magalhaes J. V., Pleiman J. K., et al., 2010 Two functionally distinct members of the MATE (multi-drug and toxic compound extrusion) family of transporters potentially underlie two major aluminum tolerance QTLs in maize. Plant J. 61: 728–740.

45. Melo J. O., Lana U. G., Pineros M. A., Alves V. M., Guimaraes C. T., et al., 2013 Incomplete transfer of accessory loci influencing SbMATE expression underlies genetic background effects for aluminum tolerance in sorghum. Plant J. 73: 276–288.

46. Menezes C. B., Carvalho Junior G. A., Silva L. A., Bernardino K. C., Vander Souza F., et al., 2014 Combining ability of grain sorghum lines selected for aluminum tolerance. Crop Breeding and Applied Biotechnology 14: 42–48.

47. Novais R. F., Smyth T. J., 1999 Fósforo em solo e planta em condiçoes tropicais, Universidade Federal de Viçosa, Viçosa, MG.

48. Parker D. R., Norvell W. A., Chaney R. L., 1995 GEOCHEM-PC—A chemical speciation program for IBM and compatible personal computers, pp. 253–269 in Chemical Equilibrium and Reaction Models, edited by Loeppert R. H., Schwab A. P., Goldberg S.. Soil Science Society of America and American Society of Agronomy, Madison, WI.

49. Raman H., Zhang K., Cakir M., Appels R., Garvin D. F., et al., 2005 Molecular characterization and mapping of ALMT1, the aluminium-tolerance gene of bread wheat (Triticum aestivum L.). Genome 48: 781–791.

50. Reid D. A., Fleming A. L., Foy C. D., 1971 A method for determining aluminum response of barley in nutrient solution in comparison to response in Al-toxic soil. Agron. J. 63: 600–603.

51. Riede C. R., Anderson J. A., 1996 Linkage of RFLP markers to an aluminum tolerance gene in wheat. Crop Sci. 36: 905–909.

52. Ryan P. R., Ditomaso J. M., Kochian L. V., 1993 Aluminium toxicity in roots: an investigation of spatial sensitivity and the role of the root cap. J. Exp. Bot. 44: 437–446.

53. Ryan P. R., Delhaize E., Jones D., 2001 Function and mechanism of organic anion exudation from plant roots. Annu. Rev. Plant Physiol. Plant Mol. Biol. 52: 527–560.

54. Ryan P. R., Raman H., Gupta S., Horst W. J., Delhaize E., 2009 A second mechanism for aluminum resistance in wheat relies on the constitutive efflux of citrate from roots. Plant Physiol. 149: 340–351.

55. Sabadin P. K., Malosetti M., Boer M. P., Tardin F. D., Santos F. G., et al., 2012 Studying the genetic basis of drought tolerance in sorghum by managed stress trials and adjustments for phenological and plant height differences. Theor. Appl. Genet. 124: 1389–1402.

56. Sasaki T., Yamamoto Y., Ezaki B., Katsuhara M., Ahn S. J., et al., 2004 A wheat gene encoding an aluminum-activated malate transporter. Plant J. 37: 645–653.

57. Sivaguru M., Liu J., Kochian L. V., 2013 Targeted expression of SbMATE in the root distal transition zone is responsible for sorghum aluminum resistance. Plant J. 76: 297–307.

58. Ud-Din N., Carver B. F., Clutter A. C., 1992 Genetic analysis and selection for wheat yield in drought-stressed and irrigated environments. Euphytica 62: 89–96.

59. von Uexküll H. R., Mutert E., 1995 Global extent, development and economic impact of acid soils. Plant Soil 171: 1–15.

60. VSN International, 2014 GenStat for Windows, Ed. 17, VSN International, Hemel Hempstead, UK. Available at: www.GenStat.co.uk

61. Wang Y., Xu H., Kou J., Shi L., Zhang C., et al., 2012 Dual effects of transgenic Brassica napus overexpressing CS gene on tolerances to aluminum toxicity and phosphorus deficiency. Plant Soil 362: 231–246.

62. Yokosho K., Yamaji N., Ma J. F., 2010 Isolation and characterisation of two MATE genes in rye. Funct. Plant Biol. 37: 296–303.

63. Yokosho K., Yamaji N., Ma J. F., 2011 An Al-inducible MATE gene is involved in external detoxification of Al in rice. Plant J. 68: 1061–1069.

64. Zeng Z. B., 1994 Precision mapping of quantitative trait loci. Genetics 136: 1457–1468.

65. Zhou G., Delhaize E., Zhou M., Ryan P. R., 2013 The barley MATE gene, HvAACT1, increases citrate efflux and Al3+ tolerance when expressed in wheat and barley. Ann. Bot. (Lond.) 112: 603–612.

CHAPTER 15

HISTORICAL SYNTHESIS-ANALYSIS OF CHANGES IN GRAIN NITROGEN DYNAMICS IN SORGHUM

Ignacio A. Ciampitti and P. V. Vara Prasad

Keywords: Nitrogen use efficiency, sorghum bicolor, grain yield, nitrogen uptake, grain nitrogen, nutrient partitioning

INTRODUCTION

For the last six decades, US sorghum (*Sorghum bicolor* L. Moench) improvement has been related to targeted modifications in genotype (G component) and management practices (M component), such as (a) fertilization rates, (b) irrigation, and (c) tillage practices (Eghball and Power, 1995; Duvick, 1999; Assefa and Staggenborg, 2010). A long-term study conducted in Texas (1939–1997) documented yield improvements were mainly related to the introduction of new sorghum hybrids, water conditions at planting, better weed (herbicide) control and conservation practices such as zero tillage (Unger and Baumhardt, 1999). One-third (35–40%) of the overall yield improvement could be attributed to the G component and two-thirds to interaction between the M factor by the environment (E) component (Duvick, 1999; Unger and Baumhardt, 1999; Assefa and Staggenborg, 2010). Similar yield gains were also documented by Miller and Kebede (1984). From 1950 to 1999, Mason et al. (2008) documented a slower rate of yield improvement in sorghum relative to maize (*Zea mays* L.). Overall, sorghum yield improvement seemed to be primarily achieved by gains under environmental stress and low yielding environments rather than by modifications or improvements on maximum yield potential (Assefa and Staggenborg, 2010).

Sorghum is a C4 (Kranz leaf anatomy) erect plant constituted from a main stem, leaf canopy, head, and tiller organs. The final tiller number is dependent on the genotype, temperature, and nutrient resources. Final plant size as related to the aboveground portion is dependent on the plant response to photoperiod sensitivity and growing conditions; while the belowground section is composed by extensive fibrous root systems. For the main stem, leaf area increases until full expansion of the flag leaf occurs; a waxy bloom often covered leaf sheath and stem organs. Grains are developed in the head organ (in the uppermost section of the plant) after flowering time. The plant has several uses as for grain, forage, feed, and bioenergy. As related to the nutritional value, sorghum contains approximately 11–13% protein, which is free of gluten presenting an advantage as a food supply from people suffering from celiac problems. Sorghum presents a good nutritional value, with an overall grain composition of 70–80% carbohydrate, 2–5% fat, 1–3% fiber, and 1–2% ash.

Physiologically, sorghum presents a plasticity connected to the capacity of the plant to compensate and adjust its growth based on the resources available at the plant-scale (Heinrich et al., 1983). Superior water use efficiency for grain sorghum as relative to other crops (such as corn and soybean), expressed as higher yield per unit of water, was documented in low-yielding, and water-limited environments (Stone et al., 2006). Historically, sorghum genetic improvement is related to changes in aboveground biomass production (increased leaf to stem ratio and higher leaf mass), longer panicle length, decrease in peduncle length, and superior root mass (Assefa and Staggenborg, 2011). Sorghum yield improvement is tightly connected to changes in number of panicles per unit land area, increased kernel numbers, and increased final total grain weight. Production factors such as non-uniform stands, row spacing, plant population, weed-competition, defoliation, water availability, and N applications directly affect yield components (Stickler and Wearden, 1965; Rajewski et al., 1991; M'Khaitir and Vanderlip, 1992; Norwood, 1992; Larson

Citation: Ciampitti IA and Prasad PVV (2016) Historical Synthesis-Analysis of Changes in Grain Nitrogen Dynamics in Sorghum. Front. Plant Sci. 7:275. doi: 10.3389/fpls.2016.00275.

and Vanderlip, 1994; Limon-Ortega et al., 1998) impacting yield potential of grain sorghum. The E factor (e.g., water and temperature) exerts a large influence; thus, endpoint sorghum productivity may be considered the outcome of a complex G × E × M interaction.

From a plant nutrition perspective, nitrogen (N) is the main nutrient influencing plant growth, aboveground biomass, and yield (Roy and Wright, 1973; Kamoshita et al., 1998; Borrell and Hammer, 2000; Wortmann et al., 2007; van Oosterom et al., 2010; Kaizzi et al., 2012; Mahama et al., 2014). For modern sorghum hybrids, N application improved yields via modification of aboveground biomass, seed number, and grain HI (Mahama et al., 2014). Improvement of NUE (yield to available N ratio) can be understood via dissecting NUE into two components: N recovery efficiency (NRE, plant N uptake to soil N supply) and (NIE, yield to plant N uptake ratio) (Ciampitti and Vyn, 2012). Historical NUE gains in maize were primarily explained by improvements in N uptake, and consequently, NIE (Ciampitti and Vyn, 2012,2013, 2014). Nonetheless, yield improvement has been indirectly accompanied by decreases in grain %N (Duvick and Mickelson, 1997; Ciampitti and Vyn, 2012, 2013). Changes in NIE over time are not yet documented for sorghum. Therefore, detailed review and research on NUE, its components, and grain N pathways is critical and required. For maize, a trade-off was documented for the Remobilized N and Reproductive N (Ciampitti and Vyn, 2013). A scientific knowledge gap exists for better understanding of these processes for sorghum NUE and concomitant yield improvement.

The main objective of this original research paper is to document and understand sorghum NUE and physiological mechanisms related to grain N dynamics. Understanding historical changes of grain N sources (organ size and potential genotypic variation) and physiological strategies of the sorghum for securing this N demand will facilitate parallel improvements in NUE and grain yield.

MATERIALS AND METHODS

Data Standardization

Information gathered from scientific literature over the past several decades was synthesized and presented in Table 1. A data inclusion criteria implemented in a previous synthesis analysis was followed (Ciampitti and Vyn, 2013) for this study. Briefly, information was included in the database if specific criteria were met; primarily focused on completeness of the information (e.g., data on grain yield, plant N content at flowering, plant N fractions = grain and stover N content at maturity). A total of 13 research studies that satisfied all requirements were selected for the analyses, comprising >250 treatment means (Table 1). Unpublished thesis reports, current on-going research (Ciampitti and Prasad, Unpublished) and research studies not yet published in scientific journals (three M.S. and three Ph.D. thesis dissertations) were also included to secure the "unbiasedness" and avoid a misinterpretation (when only significant outcomes are published) of overall population effects (McLeod and Weisz, 2004).

Table 1: Number of study, site/country, author, design, year of experimentation, number of genotypes, and characteristics for each different sorghum experiments

No	Site/ Country	Author	Design	Year	No Genotypes	Main characteristics
1	Kansas, USA	Koch, 1966 (M.S. Thesis)	Split-plot	1964–1965	Six genotypes (three maturity groups)	Effect of maturity and plant density on N uptake and NUE
2	New Delhi, India	Roy and Wright, 1973	Complete factorial in randomized blocks	1967–1968	One genotype 'CSH-I'	Nutrient uptake and partitioning
3	Kansas, USA	Stützel, 1981 (M.S. Thesis)	Alternated rows	1980	Three genotypes (early, medium and late)	Sorghum and millet intercropping: biomass, nutrient uptake, and yield
4	Nebraska, USA	Fernandez, 1987 (Ph.D. Dissertation)	Randomized complete block	1985	Four genotypes	Study of NUE and its physiological mechanisms
5	Texas, USA	Lavelle, 1987 (M.S. Thesis)	Randomized complete block	1983–1984	Four genotypes	NUE among sorghum genotypes (0 and 180 kg ha^{-1})
6	Nebraska, USA	Masi, 1997 (Ph.D. Dissertation)	Randomized complete block	1992–1993	10 genotypes (two hybrids)	Root systems and N uptake in diverse sorghum genotypes
7	Gatton, Australia	Kamoshita et al., 1998	Split-plot	1995–1996	Four genotypes (two early and two late maturing)	Hybrid × Nitrogen levels (0 and 240 kg ha^{-1}) under low N
8	Nebraska, USA	Traore, 1998 (Ph.D. Dissertation)	Randomized complete block	1992–1993 1994–1996	15 genotypes (four hybrids)	Physiological contributions to NUE (genotypes)
9	Warwick, Australia	Borrell and Hammer, 2000	Split-plot	1994–1995	Nine hybrids	N and stay-green characterization
10	Gatton, Australia	van Oosterom et al., 2010	Split-plot	1999	Three genotypes (senescent, stay-green, and RUE*)	Study of reproductive N: three hybrids × N rates (0, 44, and 353 kg ha^{-1})
11	Kansas, USA	Mahama et al., 2014	Split-plot in randomized complete block	2010–2011	12 genotypes (six hybrids)	Hybrid and inbreds, three fertilizer N rates (0, 45, and 90 kg ha^{-1})
12	Udaipur, India	Sumeriya et al., 2014	Split-plot	2004, 2005, 2006	One genotype	Soil moisture conservation practice
13	Kansas, USA	Ciampitti and Prasad, unpublished	Randomized complete block	2014	Four genotypes	Germplasm evaluation: NUE, N uptake and partitioning

**RUE, Indian hybrid with high radiation use efficiency (RUE).*

Geographically, the studies are representing the main sorghum producing areas around the globe such as US Great Plains (Nebraska, Kansas, and Texas), Australia, and India. In relative terms, the database is composed with a frequency of observations (treatment means) of 62% Kansas, 15% Australia, 12% Nebraska, 6% Texas, and 5% India (Table 1). Due to lack of balance geographic number of observations, this factor was not analyzed individually, but included in the historical evaluation (without further separation). The main challenge in evaluating, selecting, and synthesizing information for sorghum NUE was the lack of information for reproductive and non-reproductive N content at varying phenological stages. For exploratory and statistical evaluation (database distribution and normality), the database was arbitrarily divided into three Eras, based on the study year, named as Old Era (1965–1980; n = 87 treatment means); Transient Era (1981–2000; n = 85); and New Era (2001–2014; n = 86). Miller and Kebede (1984) documented a similar genetic yield gain for sorghum from 1960 to 1980. Mason et al. (2008) documented a historical yield improvement from 1980 to 2000. In the last years (2001–2009), a consistent positive yield trend was recorded for main sorghum producing countries around the globe such as Australia, US, India, Brazil, Burkina Faso, Ethiopia, and Mexico (Rakshit et al., 2014), which is related to the interaction between genetic improvement and use of best management practices. For the US Great Plain region, hybrids accounted for only about 15% of the total sorghum seed planted in 1957. But by 1960, sorghum hybrids had rapid acceptance, accounting for 95% of the area (Maunder, 1998, 1999).

Database Description: Parameters Evaluated

Biomass and plant N content was summarized from previously described studies (Table 1). For the purpose of this document, plant biomass refers to the mass accumulation in all aboveground plant fractions (e.g., stem, leaves, and head), excluding the below ground fraction of roots. Similarly, the plant N content considers all aboveground plant fractions and their related N content. Plant N content was calculated by multiplying the biomass by its respective N concentration (dry mass basis), if the plant N content was not explicitly provided in the particular research study. Grain and N harvest indices (HI) were calculated following the equations described by Ciampitti and Vyn (2013). Briefly, if grain HI and/or NHI was not directly reported, it was calculated as follows:

$$\text{Grain HI} = \text{Yield/Plant Biomass (stover : leaf + stem + grain)}$$

$$\text{NHI} = \text{Grain N content/Plant N content}$$

in which the Yield refers to the final grain yield at the end of the season (harvest maturity), adjusted by a constant moisture content (135 g kg^{-1} moisture content) and the plant biomass is expressed in dry weight basis. Both grain and plant N content are expressed in dry weight basis and determined at the end of the season.

Vegetative-stage aboveground biomass (Vegetative Biomass) and N content (Vegetative N) was obtained from all research studies involving biomass and N accumulation from emergence to flowering. Similarly, the reproductive-stage aboveground biomass (Reproductive Biomass) and N content (Reproductive N) involves biomass and N accumulation from flowering until the end of the season. The term named "Remobilized N" was calculated using the balance approach as follows (Ciampitti and Vyn, 2013):

$$\text{Remobilized N} = \text{Vegetative N} - \text{Stover N},$$

in which vegetative N (all plant) at flowering and the stover N at maturity (leaf + stem).

As previously documented by Ciampitti and Vyn (2013), this so-called "balance" approach for estimating N remobilization is less accurate (e.g., sampling error) and more labor intensive (sampling at flowering) than the stable isotopic N method. The balance approach presents constrains related to potential issues related to sampling error due to proper identification of growth stages and temporal-data aggregation (Kichey et al., 2007). Nonetheless, this approach is legitimate and commonly utilized for estimating on-farm N remobilization.

The reproductive N was estimated as follows:

$$\text{Reproductive N} = \text{Plant N} - \text{Vegetative N}$$

in which the plant N (leaf + stem + head, including grain, fractions) at maturity.

Nitrogen internal efficiency was calculated as the yield to the plant N content ratio (Ciampitti and Vyn, 2012) all obtained at harvest maturity. Grain %N was calculated as the grain N content to yield ratio (if the information was not explicitly documented), both obtained at maturity.

$$\text{NIE (grain yield/plant N)} =$$
$$\text{NHI (grain N/plant N)} \times [\text{grain \%N (grain N/grain yield)}]^{-1}.$$

Descriptive Analysis: Database Summary

For all the data, a descriptive analysis was performed involving the total number of observations (and units), and calculation of mean, standard deviation (SD), minimum, 25–75% quartile, median, and maximum (Table 2). Box-plots was calculated for the yield factor across different historical Eras (GraphPad Prism 6; Motulsky and Christopoulos, 2003). For the entire database, histograms were developed using the "hist" function from the R program (R Development Core Team, 2009) to graphically show the grain yield and plant N content distribution. For model analyses (historical comparison between slopes), linear components were tested (F-test, Mead et al., 1993), and selected models were compared with a global fit (GraphPad Prism 6, Motulsky and Christopoulos, 2003). For the entire database, number of observations was not constant since not all parameters were reported or collected in all the studies (essential factors for the synthesis analyses). Grain yield and plant N content (Figure 2A) and NIE versus grain %N (Figure 2B) relationships were documented in per-unit area across historical Eras, with relative NHI or grain yield as data point size. For the associations Remobilized N versus Vegetative N, and Reproductive N versus Remobilized N (Figures 3A,B), the bubble graph technique was implemented (R program, R Development Core Team, 2009) to portray these relationships as a function of a third factor whereby different sizes refer to the grain yield (e.g., larger bubble sizes, high yields). Similar procedure was also implemented for the associations presented in Figure 2.

Table 2: Descriptive summary statistics of the synthesis (1965–2014) Eras relative to sorghum grain yield, Plant N, whole-plant biomass (Biomass) at different phenological stages (flowering and maturity, all on dry basis) and variables related to the partitioning components of yield and N and to the N use efficiency

Parameter	Unit	*n*	Mean	*SD*	Minimum	25% Q	Median	75% Q	Maximum
Grain yield	Mg ha^{-1}	258	5.2	2.4	1.2	3.4	4.9	6.8	20.9
Grain HI	Dimensionless	183	0.42	0.11	0.12	0.34	0.44	0.49	0.68
1000-Grain Weight	g 1000 seeds	154	24.9	3.5	2.6	23.3	25.1	27.2	32.3
Plant density	pl m^{-2}	258	11.0	5.1	3.5	6.8	12.0	13.0	30.0
Vegetative BM	g m^{-2}	220	511	362	34	112	541	791	2037
Plant BM	g m^{-2}	183	1217	505	440	847	1150	1451	4444
Vegetative N	g m^{-2}	258	9.4	5.0	2.4	5.5	8.7	12.1	26.0
Stover N	g m^{-2}	258	5.0	2.9	0.5	2.6	4.3	7.0	13.5
Grain N	g m^{-2}	258	8.8	4.8	1.9	5.0	7.1	13.0	33.6
Plant N	g m^{-2}	258	13.8	6.9	3.6	7.8	12.8	18.7	43.7
NHI	Dimensionless	258	0.64	0.11	0.22	0.57	0.65	0.71	0.96
Remobilized N	g m^{-2}	258	0.6	4.6	−13.7	−1.9	0.5	3.0	12.7
Reproductive N	g m^{-2}	258	4.4	3.3	−5.8	2.1	3.5	6.6	18.3
NIE	g g^{-1}	258	42	14	8	32	37	51	83
Grain %Nc	mg g^{-}	256	16.5	4.9	5.2	12.5	16.9	19.4	37.0

Grain HI, grain harvest index (yield to aboveground biomass –BM- ratio); Vegetative BM, whole-plant aboveground biomass at flowering time; Plant BM, whole-plant aboveground biomass at physiological maturity; Vegetative N, vegetative-stage whole-plant aboveground N content at flowering, Stover N, stover N content (Plant BM minus grain) at physiological maturity; Grain N, grain N content at physiological maturity; Plant N, whole-plant aboveground N content at physiological maturity; NHI (Grain N to Plant N ratio) determined at the end of the season; Remobilized N, reproductive-stage shoot N remobilization; Reproductive N, reproductive-stage whole-plant N content determined after flowering; NIE, Nitrogen Internal Efficiency, calculated as the yield to the Plant N ratio at physiological maturity; Grain %Nc, grain N concentration at physiological maturity.

RESULTS

Descriptive Analysis: Sorghum Database

Historical sorghum yield was 5.2 Mg ha^{-1} with a SD of 2.4 Mg ha^{-1} (25%Q = 3.4 Mg ha^{-1}; 75%Q = 6.8 Mg ha^{-1}) (Table 2). Maximum yield was close to 21 Mg ha^{-1}; while minimum grain yield was of 1.2 Mg ha^{-1}, representing a yield gap of 20 Mg ha^{-1}. Historical yield changes are not necessarily reflecting changes in sorghum germplasm or management practices, but strictly representing the observations collected from this analysis. In the descriptive analysis, low yielding means (<3 Mg ha^{-1}) for the Old Era (1965–1980 years; *n* = 87) represented only 6% of the category with an overall yield of 2.7 Mg ha^{-1}; while for the New Era (2000–2014 years; *n* = 87), those data points represented 35%, averaging 2.4 Mg ha^{-1} (Figure 1A). On the opposite yielding range, for medium- to high-yielding treatment means (>7 Mg ha^{-1}), the Old Era presented 18% of observations with an overall yield of 7.5 Mg ha^{-1}; while the New Era has 14% within this category with a mean yield of 8.1 Mg ha^{-1}.

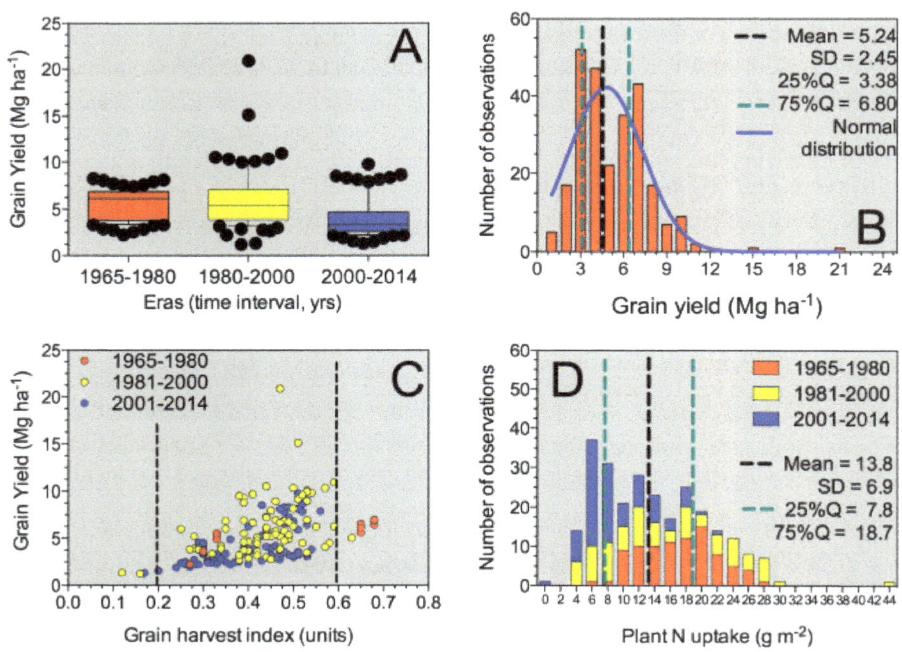

Figure 1: Box-plot for grain yield factor across all Eras (A), histogram for the grain yield parameter (B), yield versus grain NHI relationship (C), and histogram for the plant N uptake at maturity including all Eras evaluated in this synthesis-analysis. For the box-plot (A), the

solid line indicates the mean value for grain yield. The box boundaries indicate the upper and lower quartiles, the whisker caps indicate 90th and 10th percentiles, and the circles indicate observations below and above those percentiles. For grain yield histogram (B), mean, standard deviation (SD), 25–75% quartile and normal distribution (Gaussian fit) was performed. For grain yield versus grain HI (C), information was individualized (identified by color) and separated by Eras (Old Era = 1965–1980; Transient Era = 1981–2000; New Era = 2001–2014). For plant N uptake histogram (D), mean, SD, 25–75% quartile was determined for the entire database of plant N uptake, with Era separation.

Across all Eras, grain yield presented a normal distribution (mean = 4.8 Mg ha^{-1}), with a tendency to high values (positively skewed distribution, skewness = 1.5) and more concentrated, 25% quartile, 25%Q = 3.4 Mg ha^{-1}; 75%Q = 6.8 Mg ha^{-1} (leptokurtic, peaked distribution, kurtosis = 6.1) (Figure 1B). Yield was related (positive response) to the grain HI (grain yield to plant biomass ratio), with more variability in yield as grain HI increases above 0.30 (Figure 1C). Overall grain HI was 0.42 with a broad variation range (0.12 to 0.68 units; Table 2). Grain yields above 10 Mg ha^{-1} presented grain HI values ranging from 0.4 to 0.6 units (Figure 1C). The most recent two Eras (Transient and Modern Eras) did not portray a clear distinction for the grain yield versus grain HI relationship. The seed size, represented by the 1,000-grain weight, varied from 2.6 to 32.3 g with an average of 24.9 g (Table 2), reflecting a "flat" trend with the grain yield trait (data not shown). The head number factor was not tightly connected with the plant density, but instead was influenced by the tillering ability of each hybrid. Overall mean plant density observed was 11 plants m^{-2} broadly ranging (±5 plants m^{-2}), with a distribution greatly concentrated (25–75Q) between 7 and 13 plants m^{-2} (Table 2).

Plant N content and Grain Yield Relationship

Plant N content (aboveground whole-plant N at maturity) varied with the grain yield factor. Mean plant N content was 13.8 g m^{-2} (25 to 75%Q = 7.8 to 18.7 g m^{-2}, respectively) with a distribution governed by the Era classification. For the Old Era (1965–1980 years), plant N content distribution was more flat (platykurtic, less peaked) and biased toward greater plant N content values typically related to greater yield (red columns, Figure 1D). On the opposite side, for the Modern Era (2001–2014 years), plant N content peaked toward lower values, more concentrated around 6 to 8 g m^{-2} (blue columns, Figure 1D). Overall plant biomass (aboveground whole-plant biomass at maturity, plant BM), presented a value of 1217 g m^{-2} (Table 2) with larger plant size connected to higher plant N content.

The grain yield to plant N content relationship (slope = NIE) fitted an exponential growth model (Figure2A). In general, superior grain yield was achieved by increasing plant N content at maturity, regardless of the Eras. High-yielding environments (>8 Mg ha^{-1}) presented a plant N content at or above 25 g m^{-2}(Figure 2A). Overall, NIE was 42 kg kg^{-1}, but broadly ranging, 10-fold variation from min. to max. (8–83 kg kg^{-1}) values (Table 2). Two components are part of the NIE, the NHI and the grain %N both determined at maturity. Superior NIE was negatively and strongly ($R^2 > 0.60$; $n = 258$; $P < 0.001$) related to grain %N; thus, high N efficiency was achieved lowering grain %N at the same time (with yields ranging from 4 to 6 Mg ha^{-1}; Figure 2B). A unique model was fit for the NIE versus grain %N ($Y = \beta2^{*}X^{\beta1}$), with similar allometric coefficients (β1) for the last two Eras (1981–2000, 2001–2014). Historically, a large improvement in the NIE variation accounted by grain %N ($R^2 = 0.35$ for 1965–1980 to $R^2 = 0.55$ for 2001–2014) was documented. Additionally, biomass and N partitioning were weakly associated as reflected by the positive relationship between grain HI and NHI (Figure 2C). On the counterpart of the NIE term, NHI was positively related to grain HI, but larger variation was faced (Figure 2C). Critically, both NHI and grain %N were unrelated (data not shown, $R^2 = 0.01$). In summary, grain %N was the main term accounted for most of the variation (63%, Figure 2B) in NIE as compared to the NHI (10%; Figure2D). Variation in grain %N was ample, eightfold, as compared with the recorded by NHI, fivefold (variation = maximum – minimum; Table 2). In summary, NHI distribution was more concentrated (0.6 to 0.7, 25%–75%Q) as compared to the grain %N (12 to 19 mg g^{-1}, 25–75%Q; Table 2); the broader range on the latter factor contributed to its dominance over the NIE process.

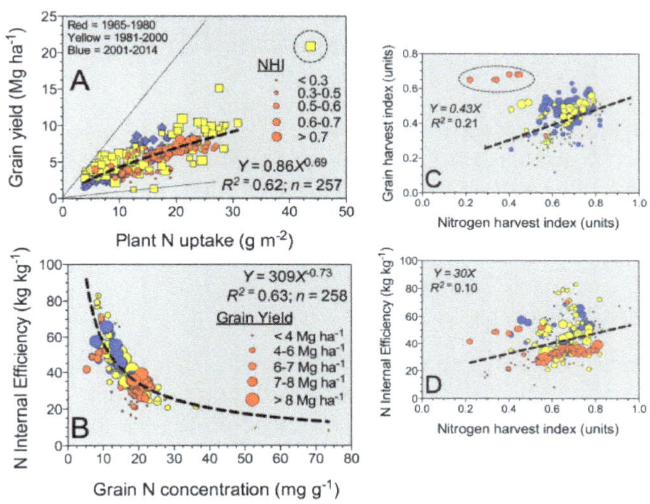

Figure 2: Relationship between yield and plant N uptake (A) and nitrogen (N) internal efficiency (NIE, yield to plant N uptake ratio) versus grain %N (B), grain HI (grain yield to plant biomass, BM) (C), and NIE versus NHI (grain N to plant N content) (D) all determined at maturity, for all Eras (time interval from 1965 to 2014 years). For (A), bubble sizes refer to the different NHI values (larger bubble sizes for greater NHI). For panels (B–D), bubble sizes refer to diverse sorghum grain yield values (larger bubble sizes for high-yielding values).

Nitrogen Use Efficiency (NUE) and Grain N Sources

Grain N sources were dissected to understand the physiological mechanisms underpinning NUE and its component (NIE, plant-related process). The main two mechanisms related to the grain N sources are the (herein "Remobilized N") and the reproductive-stage aboveground plant N uptake (herein "Reproductive N"). For the pooled data, Reproductive N was close to eightfold higher than the Remobilized N (4.4 vs. 0.6 g m^{-2}, respectively) (Table 2). Reproductive N distribution ranged from 2.2 to 6.6 g m^{-2} (25–75%Q), representing a net positive N influx (from flowering till maturity). For the counterpart, Remobilized N ranged from -1.9 to 3.0 g m^{-2} (25–75%Q), negative values were related to accumulation of stover N (N reservoir, stem fraction); while positive values portrayed a net stover N remobilization (Table 2). Overall plant N status attained at flowering is connected to the potential Remobilized N capacity of the plant during the reproductive period. For this study, Remobilized N was related to Vegetative N by a unique linear model, regardless of the Eras evaluated (Figure 3A). Therefore, Remobilized N increased as Vegetative N rose. For the Old (red color) and New (blue color) Eras, slopes of the association between Remobilized N and Vegetative N did not statistically differ ($Y = 0.63X$; F-test; Mead et al., 1993). For the Reproductive N and Remobilized N association, slopes for the Old and New Eras did differ statistically, with a greater trade-off (more negative slope) for the Old ($Y = -0.75X$) relative to the New ($Y = -0.38X$) Eras (F-test; Mead et al., 1993). The N status at flowering (Vegetative N) accounted for a large variation of the Remobilized N ($R^2 = 0.70$); while, the Remobilized N accounted for a small part ($R^2 = 0.13$) of the Reproductive N (Figure 3).

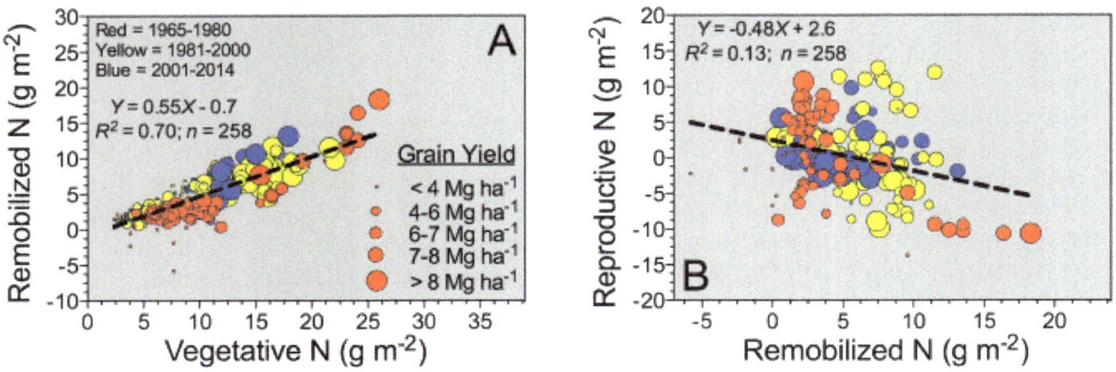

Figure 3: Reproductive-stage shoot N remobilization, from silking till maturity, versus vegetative-stage whole-plant N content (Vegetative N) (A), and Reproductive N versus Remobilized N (B) for the Old (red color: observations from 1965 to 1980; n **= 87), Transient (yellow color: observations from 1981 to 2000;** n **= 84), and the New Eras (blue** color: observations from 2001 and 2014; n **= 87).** For both (A,B), the different sizes of the bubbles correspond to grain yield values ranging from 1.2 to 20.9 Mg ha^{-1}. A unified slope was fitted to represent all Eras (historical analysis).

DISCUSSION

Yield gap from Old to New Era for the high yielding range (Δ = 0.6 Mg ha^{-1}), was similar to the rate of yield increase documented by Mason et al. (2008), 0.67 Mg ha^{-1} from 1950 to 2000 years. The USDA-NASS Commodity Statistics Database (2015) documented a larger historical (1957–2000) yield gain for sorghum under irrigated conditions of 1.1 Mg ha^{-1}. However, Assefa and Staggenborg (2010) documented a lack of yield increase over time (1957–2008) at irrigated sites in Kansas. As connected to yield components, Mason et al. (2008) documented a lesser contribution of seed size to sorghum yield improvement, but primarily governed by head number and grains per head. For the plant density factor, historical changes in plant density were documented in maize (Duvick and Cassman, 1999; Ciampitti and Vyn, 2012), but no changes over time were reported for sorghum crop under irrigated conditions (Assefa and Staggenborg, 2010). Notwithstanding intrarow spacing did diminish over time under dryland conditions (Assefa and Staggenborg, 2010), improvement in plant density factor was not a physiological trait related to sorghum yield gain over time (Assefa and Staggenborg, 2011).

For the plant N content to BM relationship, a proportional gain in N content was documented as plant size (BM) increases. Accordingly, Lemaire et al. (1996) reported similar relationship (and values) between plant N content and plant BM at maturity for sorghum. At a comparable biomass level, sorghum presented higher N content than maize, which seems to be intrinsically related to the species (genotypic difference) (Lemaire et al., 1996). Further research and comparative analysis is needed in order to clarify this point and clearly isolate the effect of plant size and growth/development when comparing maize versus sorghum for plant yields, N content, and NUE.

Grain yield to plant N content relationship, NIE slope, presented a unique model across historical Eras. For sorghum, grain yields ranging from 8 to 9 Mg ha^{-1}, plant N content was close to 25 g m^{-2} even under diverse hybrids (short vs. tall hybrids) (van Oosterom et al., 2010). Similarly to the current synthesized grain yield-to-plant N content data, Vanderlip (1993) reported a plant N content of 20 g m^{-2} for a grain yield of 8.5 Mg ha^{-1}. From the efficiency term, the NIE can be understood as the N physiological efficiency related to the utilization of N by the sorghum plant for yield production. The upper boundary line depicts maximum NIE related to N-limited environments, minimum plant N concentration and/or maximum N partitioning, herein expressed as the NHI (grain N to plant N content). On the opposite boundary, the lower line portrays minimum NIE observations related to yield-limited environments, maximum plant N concentration and/or reduced or minimum NHI (smaller bubble sizes; Figure 2A). The NIE variation seems to be primarily driven by changes in yield (with concomitant changes in plant N content). Similar outcome was synthesized for maize in a historical analysis (Ciampitti and Vyn, 2012, 2013), with also a unique model for all Eras evaluated. Although only a small proportion of the variation in NIE was explained from both grain HI and NHI; a similar behavior previously documented for maize (Ciampitti and Vyn, 2012).

Nitrogen Use Efficiency (NUE) and Grain N Sources

Reproductive N presented negative values in few studies, which can be connected to larger losses via volatilization from leaf organs as compared to the net uptake (Daigger et al., 1976; Stutte et al., 1979;Farquhar et al., 1983; Harper et al., 1987; Francis et al., 1993). Variation on Remobilized N parameter is connected to greater accumulation or net remobilization from the stover organ. van Oosterom et al. (2010) reported that maximum Remobilized N from both stem and rachis (branches of head) were positively associated to their N status (organ N concentration). For maize, Remobilized N was tightly and positively connected with the N status achieved in the plant at flowering (herein "Vegetative N") even under diverse G × E × M across a historical and global evaluation (Ciampitti and Vyn, 2013). A unique linear model fit the relationship between Remobilized N and Vegetative N across historical Eras (Figure3A). The latter confirms that the association did not change in the last 60 years, but the main gain seemed to be related to a greater vegetative-growth ability connected with superior N content at the plant-level.

Increasing N remobilization is a desirable plant trait, but a clear trade-off between Remobilized N and Reproductive N was already recognized in other crops such as maize (Pan et al., 1986; Gallais and Coque, 2005; Ciampitti and Vyn, 2013) and wheat (Kichey et al., 2007; Bogard et al., 2010), and now is also documented for sorghum. Further studies are needed in order to validate that modern sorghum hybrids have a more balanced association between Reproductive N and Remobilized N (herein term "reproductive N trade-off"). Progress in diminishing the strength of the reproductive N trade-off will facilitate NUE improvement from a physiological perspective. In addition, research investigations focused on the non-reproductive organs (primarily stem and leaf) are needed for properly dissecting the role of each organ on the reproductive N trade-off mechanisms. van Oosterom et al. (2010) presented critical information on the role of each organ during the reproductive period, emphasizing the contribution of the stem and leaf organs to the Remobilized N process.

Physiological explanations on the reproductive N trade-off were previously documented and summarized for maize by Ciampitti and Vyn (2013). Briefly, a hypothesis related to the "sink-limitation", as yield is reduced by stress factors, both Remobilized N and Reproductive N are affected via leaf senescence process, affecting the reproductive N trade-off mechanism. The latter scenario depends on the timing of the stress relative to the crop growth. If the stress occurs after flowering, Remobilized N could increase at the expenses of low or negligible Reproductive N (if grain filling duration is shortened). On the counterpart, a vegetative-stress could reduce early-season growth and N content (lower Vegetative N and Remobilized N); thus, Reproductive N could partially compensate low Vegetative N content. For maize,Pan et al. (1984) documented the importance of a balanced Remobilized N (but late-season) and Reproductive N sources. More "balanced" N sources need to be pursued for improving overall sorghum NUE from a physiological viewpoint. Changes in sorghum germplasm could modify NIE and overall NUE.Borrell and Hammer (2000) documented a proportional N allocation to the leaf organs in stay-green versus conventional (senescent) sorghum hybrids before flowering, related to a change in leaf morphology (thickness), consequently increasing leaf N demand. During the reproductive period, higher leaf N demand for stay-green hybrids was expressed as greater Reproductive N and lower Remobilized N, compensating the extra leaf N requirement as relative to senescent hybrids (Borrell and Hammer, 2000). Stay-green trait presented larger Vegetative and Reproductive N; while senescent hybrids have a larger proportion of the grain N content supplied by the Remobilized N process (rapid leaf senescence). Similar outcomes were documented for maize (Ta and Weiland, 1992; Rajcan and Tollenaar, 1999).

Reproductive N balance seems to be related to the "sink demand", primarily driven by the number of heads and grains per head (grain m^{-2}), which is partially explained from the Vegetative N and Remobilized N association (both factors increased as yield boosted, larger bubble sizes) (Figure 3A). van Oosterom et al. (2010) documented a fairly constant N demand right after flowering for three hybrids differing in plant height and stay-green under drought conditions. Early grain N demand (primarily sink strength driven process), even when non-dependent on the plant N supply, could potentially dictate final grain %N under optimal growing conditions. van Oosterom et al. (2010) also documented a similar early-reproductive grain N accumulation rate relative to the final grain %N, 18.2 mg g^{-1}, attained at maturity. Overall, a similar grain %N was synthesized for the current historical study, 16.5 mg g^{-1} on sorghum (Table 2). As postulated by the van Oosterom et al. (2010), sink strength (portrayed as grains m^{-2}) seems to be the primary factor governing N demand during early reproductive. Therefore, if Reproductive N cannot fulfill this grain N demand, then Remobilized N is the main process for satisfying the grain N requirement (Triboï and Triboï-Blondel, 2002). This early grain N deposition is related to structural and metabolic processes (cell division), which seems to be independent of the plant N supply level (e.g., in sorghum, van Oosterom et al., 2010; wheat, *Triticum aestivum* L., Martre et al., 2003). High Remobilized N scenarios from mid-to late-grain filling are the outcome of a grain N demand that cannot be fully sustained with the Reproductive N, which could be indirectly impacted by source-limitations (functional photosynthetic affected via leaf senescence). If the grain N demand is not fulfilled (under N deficiency) with the stem N reservoir, then leaf N remobilization could increase to overcome the stem-to-grain N gap. Similar results were previously documented in sorghum by several authors (Youngquist and Maranville, 1992; Utzurrum et al., 1998; Borrell and Hammer, 2000; van Oosterom et al., 2010). As hypothesized byvan Oosterom et al. (2010), stem N remobilization is connected with its N concentration; thus, N will be translocated until its minimum N concentration is reached (so called "structural N"); from this point afterward, leaf N will become the primary source to meet the grain N demand.

From a genotypic perspective, recent studies were conducted to investigate and map QTLs under varying N levels (Gelli et al., 2016). New insights on differential expression of transcripts related to N metabolism could provide new approaches for further improving NUE. Lastly, from a biochemical perspective, a better understanding of NUE process will assist not only in yield improvement via superior nutrient efficiency but also is connected in the regulation of leaf senescence via the interplay portrayed by the C:N balance (Chen et al., 2015). Therefore, an integrated and multi-disciplinary study on NUE and its processes is needed for fully comprehend the implications on sorghum.

CONCLUSION

In summary, important highlights from this synthesis are: (1) NUE component related to the plant process (NIE), was primarily explained by variations in grain %N; (2) historical changes in NUE were not directly related to physiological modifications of the yield to plant N content ratio, but primarily explained by changes in the grain %N (lower grain %N under superior yields); (3) a strong dependency exists between the Remobilized N and Vegetative N; and (4) an evident trade-off between Remobilized N and Reproductive N. These outcomes are similar to those documented for corn, suggesting similar physiological mechanisms underpinning NUE process across these two crops species with similar photosynthetic pathway (C_4).

The reproductive N trade-off presented in this synthesis-analysis established a platform for the overall N balance sources for sorghum crop. The theory of a "path-dependence" scenario for grain N demand, N utilization, and final NUE can clearly apply when sorghum N balance is dissected in vegetative and reproductive N forces. Greater Vegetative N and more balanced N partition between the non-reproductive organs (primarily leaf vs. stem) could lessen the crop dependency on the leaf N, via utilization of the stem N reservoir, delaying leaf senescence and improving overall functional stay-green. Sustaining Reproductive N during grain filling could assist in minimizing any shortfall between grain N demand and Vegetative N supply. Continuation of N uptake is not only connected to sink-strength, but also to source-limiting factors (also impacting functional stay-green – C supply). Future research studies should focus on elucidating avenues to improve reproductive N trade-off mechanism(s), which can be potentially explored via the use of hybrids (genetics) with high-yield potential and stay-green trait.

AUTHOR CONTRIBUTIONS

IC collected data, designed research, performed research, analyzed, and synthesized data, and wrote the paper. PP analyzed data and wrote the paper.

CONFLICT OF INTEREST STATEMENT

The authors declare that the research was conducted in the absence of any commercial or financial relationships that could be construed as a potential conflict of interest.

ACKNOWLEDGMENTS

Mr. G. Balboa, Mr. D. Romagnoli, Ms. A. Acevedo, Ms. Valentina Crivellini, and Dr. John Sunoj are acknowledged for collecting studies conducted in sorghum in the last decades. Dr. R. Vanderlip is thanked for his helpful comments on an earlier version of the manuscript. This study was supported by the K-State Center for Sorghum Improvement. This is contribution no. 16-120-J from the Kansas Agricultural Experiment Station.

REFERENCES

1. Assefa, Y., and Staggenborg, S. A. (2010). Grain sorghum yield with hybrid advancement and changes in agronomic practices from 1957 through 2008. *Agron. J.* 102, 703–706. doi: 10.2134/agronj2009.0314
2. Assefa, Y., and Staggenborg, S. A. (2011). Phenotypic changes in grain sorghum over the last five decades. *Agron. J. Crop Sci.* 197, 249–257. doi: 10.1111/j.1439-037X.2010.00462.x
3. Bogard, M., Allard, V., Brancourt-Hulmel, M., Heumez, E., Machet, J. M., Jeuffroy, M. H., et al. (2010). Deviation from the grain protein concentration–grain yield negative relationship is highly correlated to postanthesis N uptake in winter wheat. *Exp. J. Bot.* 61, 4303–4312. doi: 10.1093/jxb/erq238
4. Borrell, A. K., and Hammer, G. L. (2000). Nitrogen dynamics and the physiological basis of stay-green in sorghum. *Crop Sci.* 40, 1295–1307. doi: 10.2135/cropsci2000.4051295x
5. Chen, D., Wang, S., Xiong, B., Cao, B., and Deng, X. (2015). Carbon/nitrogen imbalance associated with drought-induced leaf senescence in sorghum bicolor. *PLoS ONE* 10:e0137026. doi: 10.1371/journal.pone.0137026
6. Ciampitti, I. A., and Vyn, T. J. (2012). Physiological perspectives of changes over time in maize yield dependency on nitrogen uptake and associated nitrogen efficiencies: a review. *Field Crops Res.* 133, 48–67. doi: 10.1016/j.fcr.2012.03.008
7. Ciampitti, I. A., and Vyn, T. J. (2013). Grain nitrogen source changes over time in maize: a Review. *Crop Sci.* 53, 366–377. doi: 10.2135/cropsci2012.07.0439
8. Ciampitti, I. A., and Vyn, T. J. (2014). Understanding global and historical nutrient use efficiencies for closing maize yield gaps. *Agron. J.*106, 634–644. doi: 10.2134/agronj14.0025
9. Daigger, L. A., Sander, D. H., and Peterson, G. A. (1976). Nitrogen content of winter wheat during growth and maturation. *Agron. J.* 68, 815–818. doi: 10.2134/agronj1976.00021962006800050033x
10. Duvick, D. N. (1999). "Heterosis: feeding people and protecting natural resources," in *The Genetics and Exploitation of Heterosis in Crops*, eds J. G. Coors and S. Pandey (Madison, WI: Society of Agronomy and Crop Science Society of America, Inc.), 25.
11. Duvick, D. N., and Cassman, K. G. (1999). Post-green revolution trends in yield potential of temperate maize in the Northcentral United States. *Crop Sci.* 39, 1622–1630. doi: 10.2135/cropsci1999.3961622x
12. Duvick, D. N., and Mickelson, H. R. (1997). "What is yield?," in *Proceedings of the Symposium Developing Drought- and Low N-tolerant Maize, 25–29 March 1996*, eds G. O. Edmeades, M. Ban-ziger, and H. R. Pena-Valdivia (El Batan: CIMMYT), 332–335.
13. Eghball, B., and Power, J. F. (1995). Fractal description of temporal yield variability of 10 crops in the United States. *Agron. J.* 87, 152–156. doi: 10.2134/agronj1995.00021962008700020003x

14. Farquhar, G. D., Wetselaar, R., and Weir, B. (1983). "Gaseous nitrogen losses from plants," in *Gaseous Loss of Nitrogen From Plant-Soil Systems*, eds J. R. Freeney and J. R. Simpson (Leiden: Martinus Nijhoff), 159–180.
15. Fernandez, P. G. (1987). *Studies on Physiological Mechanisms of Nitrogen Use Efficiency on Sorghum.* Ph.D. dissertation, University of Nebraska-Lincoln, Lincoln, NE.
16. Francis, D. D., Schepers, J. S., and Vigil, M. F. (1993). Post-anthesis nitrogen loss from corn. *Agron. J.* 85, 659–663. doi: 10.2134/agronj1993.00021962008500030026x
17. Gallais, A., and Coque, M. (2005). Genetic variation and selection for nitrogen use efficiency in maize: a synthesis. *Maydica* 50, 531–547.
18. Gelli, M., Mitchell, S. E., Liu, K., Clemente, T. E., Weeks, D. P., Zhang, C., et al. (2016). Mapping QTLs and association of differentially expressed gene transcripts for multiple agronomic traits under different nitrogen levels in sorghum. *BMC Plant Biol.* 16:16. doi: 10.1186/s12870-015-0696-x
19. Harper, L. A., Sharpe, R. R., Langdale, G. W., and Giddens, J. E. (1987). Nitrogen cycling in a wheat crop: soil, plant and aerial nitrogen transport. *Agron. J.* 79, 965–973. doi: 10.2134/agronj1987.00021962007900060004x
20. Heinrich, G. M., Francis, C. A., and Eastin, J. D. (1983). Stability of grain sorghum yield components across diverse environments. *Crop Sci.* 23, 209–212. doi: 10.2135/cropsci1983.0011183X002300020004x
21. Kaizzi, K. C., Baylebeka, J., Semalulu, O., Alou, I., Zimwanguyizza, W., Nasamba, A., et al. (2012). Sorghum response to fertilizer and nitrogen use efficiency in Uganda. *Agron. J.* 104, 83–90. doi: 10.2134/agronj2011.0182
22. Kamoshita, A., Fukai, S., Muchow, R. C., and Cooper, M. (1998). Sorghum hybrid differences in grain yield and nitrogen concentration under low soil nitrogen availability II. Hybrids with contrasting phenology. *Aust. Agric. J. Res.* 49, 1277–1286. doi: 10.1071/A97130
23. Kichey, T., Hirel, B., Heumez, E., Dubois, F., and Le Gouis, J. (2007). In winter wheat (*Triticum aestivum* L.), postanthesis nitrogen uptake and remobilization to the grain correlates with agronomic traits and nitrogen physiological markers. *Field Crops Res.* 102, 22–32. doi: 10.1016/j.fcr.2007.01.002
24. Koch, D. W. (1966). *Yield, Yield Components, and Nitrogen Accumulation in Grain Sorghum as Affected by Maturity and Plant Density.*MS thesis, Kansas State University, Manhattan, KS.
25. Larson, E. J., and Vanderlip, R. L. (1994). Grain sorghum yield response to nonuniform stand reductions. *Agron. J.* 86, 475–477. doi: 10.2134/agronj1994.00021962008600030003x
26. Lavelle, M. J. (1987). *Variation in Nitrogen Use Efficiency Among Grain Sorghum Genotypes.* MS thesis, Texas Tech University, Lubbock, TX.
27. Lemaire, G., Charrier, X., and Hébert, Y. (1996). Nitrogen uptake capacities of maize and sorghum crops in different nitrogen and water supply conditions. *Agronomie* 16, 231–246. doi: 10.1051/agro:19960403
28. Limon-Ortega, A., Mason, S. C., and Martin, A. R. (1998). Production practices improve grain sorghum and pearl millet competitiveness with weeds. *Agron. J.* 90, 227–232. doi: 10.2134/agronj1998.00021962009000020020x
29. Mahama, G. Y., Prasad, P. V. V., Mengel, D. B., and Tesso, T. T. (2014). Influence of nitrogen fertilizer on growth and yield of grain sorghum hybrids and inbred lines. *Agron. J.* 106, 1623–1630. doi: 10.2134/agronj14.0092
30. Martre, P., Porter, J. R., Jamieson, P. D., and Triboï, E. (2003). Modeling grain nitrogen accumulation and protein composition to understand the sink/source regulations of nitrogen remobilization for wheat. *Plant Phys.* 133, 1959–1967. doi: 10.1104/pp.103.030585
31. Masi, C. E. A. (1997). *Studies of Root Systems and Nitrogen Uptake in Diverse Sorghum Genotypes.* Ph.D. dissertation, University of Nebraska-Lincoln, Lincoln, NE.
32. Mason, S. C., Delon, K., Eskridge, K. M., and Galusha, T. D. (2008). Yield increase has been more rapid for maize than for grain sorghum. *Crop Sci.* 48, 1560–1568. doi: 10.2135/cropsci2007.09.0529
33. Maunder, A. B. (1998). "Why hybrids?," in *Proceedings of the Hybrid Sorghum Workshop*, ed. J. Axtell (Lincoln, NE: INTSORMIL).
34. Maunder, A. B. (1999). "History of cultivar development in the United States: from 'Memoirs of A.B. Maunder-Sorghum breeder' ", in*Sorghum Origin, History, Technology, and Production*, eds C. W. Smith and R. A. Frederiksen (New York, NY: John Wiley & Sons), 191–224.
35. McLeod, B. D., and Weisz, J. R. (2004). Using dissertations to examine the potential bias in child and adolescent clinical trials. *J. Consult. Clin. Psychol.* 72, 235–251. doi: 10.1037/0022-006X.72.2.235
36. Mead, R., Curnow, R. N., and Hasted, A. M. (1993). *Statistical Methods in Agriculture and Experimental Biology.* London: Chapman and Hall.
37. Miller, F. R., and Kebede, Y. (1984). "Genetic contributions to yield gains in sorghum, 1950 to 1980," in *Genetic Contributions to Yield Gains of Five Major Crop Plants*, Vol. 7, ed. W. R. Fehr (Madison, WI: Crop Science Society of American and American Society of Agronomy Special publisher), 1–14.
38. M'Khaitir, Y. O., and Vanderlip, R. L. (1992). Grain sorghum and pearl millet response to date and rate of planting. *Agron. J.* 84, 579–582. doi: 10.2134/agronj1992.00021962008400040007x
39. Motulsky, H. J., and Christopoulos, A. (2003). *Fitting Models to Biological Data Using Linear and Nonlinear Regression: A Practical Guide to Curve Fitting.* San Diego, CA: GraphPad Software. Available at:http://www.graphpad.com/manuals/prism4/RegressionBook.pdf [accessed July 24 2015].

40. Norwood, C. A. (1992). Tillage and cropping system effects on winter wheat and grain sorghum. *Prod. J. Agric.* 5, 120–126. doi: 10.2134/jpa1992.0120

41. Pan, W. L., Kamprath, E. J., Moll, R. H., and Jackson, W. A. (1984). Prolificacy in corn: its effect on nitrate and ammonium uptake and utilization. *Soil Sci. Soc. Am. J.* 48, 1101–1106. doi: 10.2136/sssaj1984.03615995004800050030x

42. Pan, W. L., Kamprath, E. J., Moll, R. H., and Jackson, W. A. (1986). Utilization of previously accumulated and concurrently absorbed nitrogen during reproductive growth in maize. *Plant Physiol.* 82, 247–253. doi: 10.1104/pp.82.1.247

43. R Development Core Team (2009). *R: A Language and Environment for Statistical Computing.* Vienna: R Foundation for Statistical Computing.

44. Rajcan, I., and Tollenaar, M. (1999). Source: sink ratio and leaf senescence in maize: II. Nitrogen metabolism during grain filling. *Field Crops Res.* 60, 255–265. doi: 10.1016/S0378-4290(98)00143-9

45. Rajewski, J. F., Francis, C. A., and Eastin, J. D. (1991). Differential responses to defoliation of grain sorghum yield components and yield related traits. *Crop Sci.* 31, 561–567. doi: 10.2135/cropsci1991.0011183X003100030002x

46. Rakshit, S., Hariprasanna, K., Gomashe, S., Ganapathy, K. N., Das, I. K., Ramana, O. V., et al. (2014). Changes in area, yield gains, and yield stability of sorghum in major sorghum-producing countries, 1970 to 2009. *Crop Sci.* 54, 1571–1584. doi: 10.2135/cropsci2012.12.0697

47. Roy, R. N., and Wright, B. C. (1973). Sorghum growth and nutrient uptake in relation to soil fertility: I. Dry matter accumulation patterns, yield, and N content of grain. *Agron. J.* 65, 709–711. doi: 10.2134/agronj1973.00021962006500050009x

48. Stickler, F. C., and Wearden, S. (1965). Yield and yield components of grain sorghum as influenced by row width and stand density.*Agron. J.* 57, 564–567. doi: 10.2134/agronj1965.00021962005700060015x

49. Stone, L. R., Schlegel, A. J., Khan, A. H., Klocke, N. L., and Aiken, R. M. (2006). Water supply: yield relationships developed for study of water management. *Nat. J. Resour. Life Sci. Educ.* 35, 161–173.

50. Stutte, C. A., Weiland, R. T., and Blem, A. R. (1979). Gaseous nitrogen loss from soybean foliage. *Agron. J.* 71, 95–97. doi: 10.2134/agronj1979.00021962007100010024x

51. Stützel, H. R. E. (1981). *Growth, Nutrient Uptake and Grain Yields of Intercropped Sorghum* (*Sorghum bicolor (L.) Moench*) *and Pearl Millet* (*Pennisetum americanum (L.) Leeke*). MS thesis, Kansas State University, Manhattan, KS.

52. Sumeriya, H. K., Singh, P., and Kaushik, M. K. (2014). Effect of in situ soil moisture conservation practices and its interaction with nutrients in yield, quality and economics of sorghum [*Sorghum bicolor* (L.) Moench]. *Ann. Agri Bio Res.* 19, 234–238.

53. Ta, C. T., and Weiland, R. T. (1992). Nitrogen partitioning in maize during ear development. *Crop Sci.* 32, 443–451. doi: 10.2135/cropsci1992.0011183X003200020032x

54. Traore, A. (1998). *Physiological Contributions to Nitrogen use Efficiency in Grain Sorghum.* Ph.D. dissertation, University of Nebraska, Lincoln, OR.

55. Triboï, E., and Triboï-Blondel, A. M. (2002). Productivity and grain or seed composition: a new approach to an old problem-Invited paper. *Eur. Agron. J.* 16, 136–186.

56. Unger, P. W., and Baumhardt, R. L. (1999). Factors related to dryland grain sorghum yield increases: 1939 through 1997. *Agron. J.* 91, 870–875. doi: 10.2134/agronj1999.915870x

57. USDA-NASS Commodity Statistics Database (2015). Available at: http://www.nass.usda.gov/Statistics_by_Subject/index.php?sector=CROPS [accessed 24 July 2015].

58. Utzurrum, S. B. Jr., Fukai, S., and Foale, M. A. (1998). Effect of late nitrogen application on growth and nitrogen balance of two cultivars of water-stressed grain sorghum. *Aust. Agric. J. Res.* 49, 687–694. doi: 10.1071/A97120

59. van Oosterom, E. J., Chapman, S. C., Borrell, A. K., Broad, I. J., and Hammer, G. L. (2010). Functional dynamics of the nitrogen balance of sorghum. II. Grain filling period. *Field Crops Res.* 115, 29–38. doi: 10.1016/j.fcr.2009.09.019

60. Vanderlip, R. L. (1993). *How a Sorghum Plant Develops.* Manhattan, NY: Kansas State University Agricultural Experiment Station and Cooperative Extension Service.

61. Wortmann, C. S., Mamo, M., and Dobermann, A. (2007). Nitrogen response of grain sorghum in rotation with soybean. *Agron. J.* 99, 808–813. doi: 10.2134/agronj2006.0164

62. Youngquist, J. B., and Maranville, J. W. (1992). Patterns of nitrogen mobilization in grain sorghum hybrids and the relationship to grain and dry matter production. *J. Plant Nutr.* 15, 445–455. doi: 10.1080/01904169209364332

CHAPTER 16

PROSPECTING FOR ENERGY-RICH RENEWABLE RAW MATERIALS: SORGHUM STEM CASE STUDY

Caitlin S. Byrt, Natalie S. Betts, Hwei-Ting Tan, Wai Li Lim, Riksfardini A. Ermawar, Hai Yen Nguyen, Neil J. Shirley, Jelle Lahnstein, Kendall Corbin, Geoffrey B. Fincher, Vic Knauf, Rachel A. Burton

INTRODUCTION

Plants in the *Sorghum* genus are an important source of chemical energy in the form of carbohydrates for animals, humans and biofuels [1, 2]. Cultivated sorghums all belong to *S.bicolor* subsp. *bicolor* and there are five races: bicolor, caudatum, durra, guinea and kafir [3–6]. Two subspecies within *S. bicolor* (*drummondii* and *verticilliflorum* (previously *arundinaceum*) are wild weedy relatives of cultivated sorghums [7, 8]. For example, *S. bicolor* subsp.*verticilliflorum* (Steud., De Wet ex Wiersema & J. Dahlb.; previously classified as *S.arundinaceum*) is a widely distributed common grass found in weedy patches along roadsides in northern Australia [9, 10]. Germplasm diversity marker analysis has revealed that *S. bicolor*subsp. *bicolor* is closely related to wild weedy sorghums and that genetic variation in this subspecies is high [11].

Cultivated varieties of sorghum are commonly grouped according to their end uses, for example, grain sorghum (food and feed), forage sorghum, sweet sorghum (for sugar production) and bioenergy sorghum [12, 13]. There are notable differences in the relative carbon partitioning and morphology between these groups: grain varieties produce large heads of grain rich in starch; sweet sorghums produce a tall, sugar-rich stem; and bioenergy and forage sorghums produce a large amount of vegetative biomass [14].

The composition of the cell wall in the stem tissue varies between genotypes [15, 16] and even within a single stem: the outer rind comprises different tissues than does the inner pith [17–19]. Understanding how sorghum stem cell wall composition affects biomass digestibility is important for improving forage quality and for developing high yielding bioenergy or biofuel crops [1, 20–23]. In general, the amount of sucrose, cellulose and non-cellulosic polysaccharides in mature sorghum biomass is affected by genotype, environmental conditions and photoperiod sensitivity [24, 25].

The most abundant cell wall component in sorghum vegetative tissues is usually cellulose, which is a polymer of (1,4)-β-linked glucosyl residues. Cellulose is synthesised at the plasma membrane by cellulose synthase A (CESA) proteins, which function as subunits of a rosette-shaped complex. Loss of function of CESA proteins tends to result in weak stems and irregular or thin cell walls [20]. Dicotyledonous plants have type I cell walls and the non-cellulosic polysaccharide constituents are pectins and xyloglucans whereas grasses such as sorghum have type II cell walls which contain heteroxylans (arabinoxylan and glucuronoarabinoxylan) and (1,3;1,4)-β-glucan and only a small amount of pectin [20]. In grass heteroxylans, the (1,4)-β-xylan chain is commonly substituted with α-arabinofuranosyl (Ara*f*) residues attached most commonly at the *O*-3 position of the xylosyl residues, but also at the*O*-2 position and occasionally at both *O*-2 and *O*-3 [26]. Ara*f* substituents can be esterified with hydroxycinnamic acids such as ferulic acid and *p*-coumaric acid [20]. (1,3;1,4)-β-Glucan consists of unsubstituted, unbranched chains of (1,3)- and (1,4)-β-glycosyl residues. The random but non-repeating (1,3)-linkages prevent molecules from aggregating, thereby precluding crystallisation and allowing for limited solubility [27]. Cellulose synthase-like genes, such as *CslF* and *CslH*, are involved in synthesising (1,3;1,4)-β-glucan [28, 29].

The amount and type of lignin (a complex polyphenolic cell wall polymer) influences biomass digestibility, and

Citation: Byrt CS, Betts NS, Tan H-T, Lim WL, Ermawar RA, Nguyen HY, et al. (2016) Prospecting for Energy-Rich Renewable Raw Materials: Sorghum Stem Case Study. PLoS ONE 11(5): e0156638. doi:10.1371/journal.pone.0156638.

has been a long-standing barrier to efficient glucose release from woody plant materials [30]. Previous studies in sorghum have investigated reducing biomass recalcitrance by reducing the lignin content [31, 32]. Alternatively, altering the monolignol composition or the ferulate cross linking of lignin to arabinoxylan have been proposed as solutions [33, 34]. Examples also exist where a decrease in cellulose reduces biomass recalcitrance [35]. However, we are yet to understand the relative contribution to recalcitrance that can be attributed specifically to non-cellulosic polysaccharides such as arabinoxylan or (1,3;1,4)-β-glucan.

Genetic variation for cell wall composition in cultivated sorghum has been investigated [15, 17] but a large pool of untapped diversity exists in other species and subspecies [2]. Previously, a population derived from a high-biomass sweet sorghum 'Rio' and a grain sorghum 'BTx623' was used to investigate the genetic basis of leaf and stem biomass composition [15]. Quantitative trait loci (QTL) for structural and non-structural carbohydrate yields were identified that co-localised with QTL for height, flowering time and stand density-tillering. In sorghum, the composition of leaf and stem structural carbohydrates are under separate genetic control; sorghum stem composition is more heritable than leaf composition and contributes more to the total plant biomass than does leaf tissue [15, 36]. Theoretical data from this population predicted that biomass yield would have a greater effect on ethanol yield than biomass composition, but the authors suggest that future work should investigate greater genetic diversity for biomass traits [15].

The highest biomass yields reported for sorghum are from photoperiod-sensitive hybrids [12]. Many wild and forage sorghums are photoperiod-sensitive and remain in a prolonged vegetative state without flowering until the day length drops below 12h 20 min [37]. Breeding strategies have removed photoperiod sensitivity traits to allow sorghum to mature and produce grain in the longer day lengths of temperate areas [37]. In the context of biofuel production from vegetative biomass, the use of photoperiod-sensitive sorghum hybrids with long vegetative growth results in higher biomass yields, however there is limited information about the composition of biomass produced when photoperiod-sensitive lines remain in a vegetative state [12].

In the present study, cell wall composition in the stem of twelve diverse sorghum lines was investigated with a focus on BTx623, Rio, and a partially photoperiod-sensitive wild *S.verticilliflorum* line 'Arun', to explore how differences in amounts and distribution of cell wall components affect stem digestibility.

RESULTS

Variation in Biomass Traits in Diverse *Sorghum* Genotypes

Twelve genetically diverse sorghum lines were grown in controlled greenhouse conditions. At maturity eight photoperiod-insensitive lines were harvested (129 d after planting). Maturity was delayed in photoperiod-sensitive lines. One partially photoperiod-sensitive line was harvested at maturity (159 d after planting), while the remaining photoperiod-sensitive lines remained in a vegetative state and were harvested 248 d after planting. Pith and rind tissue from the third internode above the base of each plant stem was harvested and the amounts of (1,3;1,4)-β-glucan, cellulose, arabinose and xylose were quantified. The amount of mannose, ribose, rhamnose, glucuronic acid and galacturonic acid released from acid-hydrolysed polysaccharides in the pith and rind samples were determined: mannose, ribose and rhamnose accounted for 0–0.1% w/w, glucuronic acid accounted for 0.2–0.5% w/w and galacturonic acid accounted for 0.1–0.9% w/w.

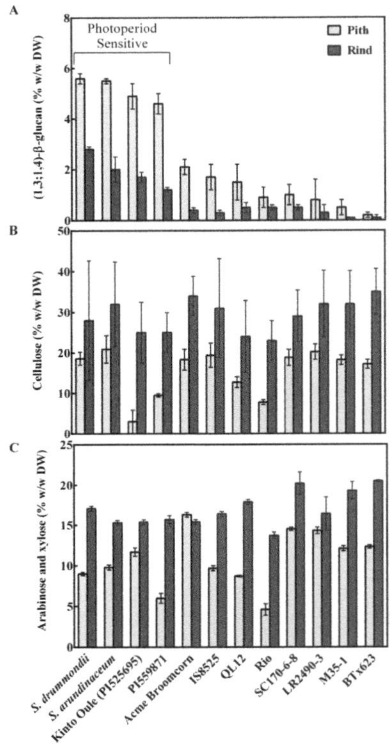

Figure 1: Stem cell wall composition for diverse *Sorghum* **genotypes.**

The amount of (**a**) (1,3;1,4)-β-glucan, (**b**) cellulose and (**c**) arabinose plus xylose in % (w/w) in stem tissues of various genotypes. Mean and standard error for n = 3 biological replicates; three technical replicates per assay. The photoperiod-sensitive lines did not flower with the exception of ssp. *verticilliflorum*, which is a partially photoperiod-sensitive line. The remaining photoperiod-insensitive lines had flowered and reached maturity at harvest.

Large differences were observed between pith and rind composition. The (1,3;1,4)-β-glucan content in the pith was, on average, 2.4% w/w for the 12 genotypes studied compared with 0.9% w/w in the rind (Fig 1a). The rind contained more cellulose, with 29% w/w in rind vs 17% w/w in the pith (Fig 1b), and more arabinose and xylose moieties, with 17% w/w in rind vs 11% w/w, than the pith (Fig 1c). Notably, the stems of photoperiod-sensitive varieties, three of which did not flower, contained approximately five times more (1,3;1,4)-β-glucan at 5.2% w/w in the pith compared with 1.9% w/w in the rind than photoperiod-insensitive varieties (1.1% w/w in pith, 0.3% w/w in rind; Fig 1a).

Carbon Partitioning in Sorghum Plants

To further assess variation in diverse sorghum lines, the morphology of three types of sorghum, a grain (*S. bicolor*

BTx623), a sweet (*S. bicolor* Rio) and a wild (*S. bicolor* ssp. *verticilliflorum* Arun), was investigated. Plant height, tiller diameter and number, wet and dry biomass, water content of whole plants and harvest index for leaf, stem and grain head tissues was recorded at maturity (Fig 2). The three lines displayed many architectural and morphological differences and varied in their biomass production (Fig 2). The sweet and wild lines were similar in height (2.5 m), which was approximately double the height of the grain line. The wild line had thinner tillers but produced many more per plant (13) than the other two lines (1–2 tillers per plant). Water content was similar for all varieties, ranging from 63–68%.

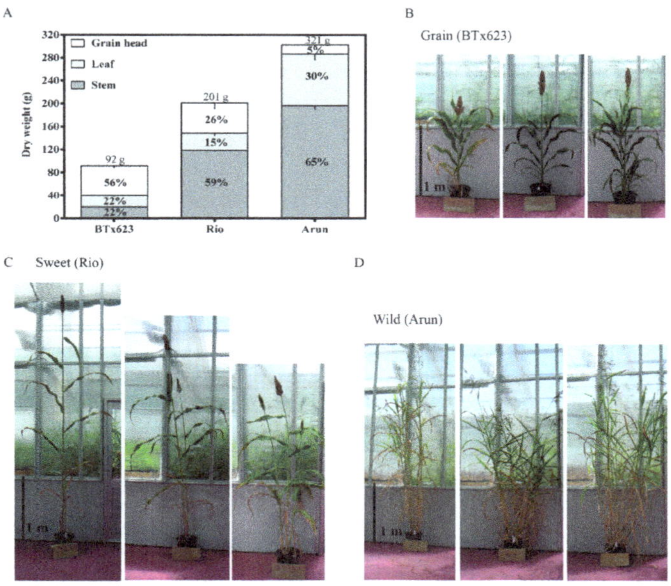

Figure 2: Diversity in morphological and physiological traits in wild, sweet and grain sorghum lines.

Percentage contribution of grain head, leaf and stem to plant dry weight (mean; with SEM indicated by vertical line; n = 3 biological replicates) in mature (**b**) grain (*S. bicolor*'BTx623'), (**c**) sweet (*S. bicolor* 'Rio') and (**d**) wild (*S. bicolor* ssp. *verticilliflorum* 'Arun') sorghum lines. Rio and BTx623 plant were photographed 129 d after planting and Arun 159 d after planting.

Overall biomass production varied considerably. The average dry biomass yield of the wild line (321 g) was 1.6 times greater than the yield of the sweet line and three times greater than that of the grain line. The harvest index, which describes how plants allocate biomass to leaves, stems and grain head, also varied considerably. Stem tissue in the wild and sweet lines accounted for 59–65% of the biomass, whereas in the grain line, the grain head constituted 56% of the dry biomass (Fig 2a).

Images of stem sections that are representative of the pith, rind and epidermal tissues for each line were captured (Fig 3) but reveal no obvious differences in the size of the cells or thickness of the cell walls in the different genotypes. The cell wall composition of each line was therefore analysed to explore biochemical differences.

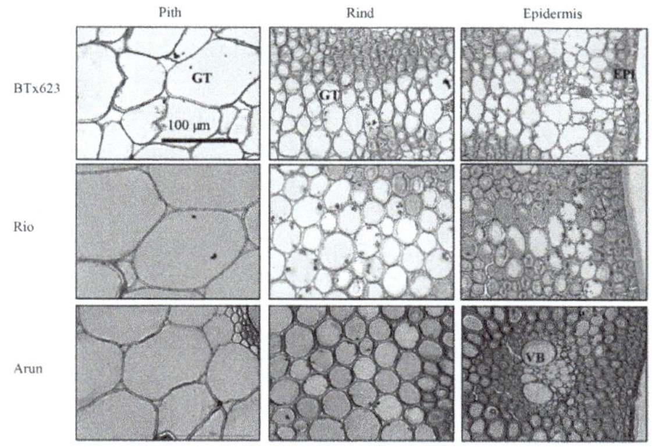

Figure 3: Cell wall characteristics in stems of grain, sweet and wild sorghum lines.

Transmission electron micrographs of transverse sections of pith, rind and epidermal tissues for BTx623, Rio and Arun. Images are all of the same scale, bar indicates 100 µm. Ground tissue (GT), vascular bundle (VB), epidermis (EPI).

Cell Wall Composition in Sorghum Stems

We compared the amount of lignin, acid-insoluble cellulose, arabinose plus xylose and (1,3;1,4)-β-glucan in whole stem or pith and rind tissues of BTx623, Rio and Arun (Fig 4). The amount of arabinose and xylose released from acid-hydrolysed polysaccharides was used as an estimation of the amount of arabinoxylan in the samples, although it is acknowledged that arabinose will also be a constituent of the relatively small amounts of type I and II arabinogalactans in the tissues. The Klason (acid-insoluble) lignin content of whole stem samples was highest in the wild line, with significantly lower amounts in the grain and sweet varieties (Fig 4a). As with the twelve diverse genotypes, pith contained less cellulose and arabinoxylan than rind tissue (Fig 4b–4d). The grain variety and wild line contained the most cellulose and lignin, while the sweet variety contained the least. Again, (1,3;1,4)-β-glucan was much higher in the wild, partially photoperiod-sensitive line (Arun) compared with the two photoperiod-insensitive lines (Fig 4d). Consistent with this finding, when transcript levels of two genes previously reported to be involved in (1,3;1,4)-β-glucan synthesis, namely *SbCslF6* and*SbCslH3* [28, 29, 38], were measured by quantitative PCR, Arun pith had higher transcript levels of *SbCslF6* and *SbCslH3* than did Rio (Fig 5).

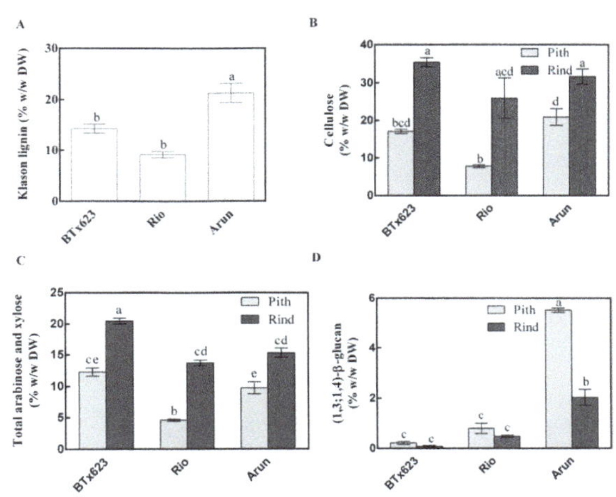

Figure 4: Cell wall components in grain, sweet and wild sorghum lines.

(**a**) Klason lignin, (**b**) Updegraff cellulose, (**c**) total arabinose plus xylose, and (**d**) (1,3;1,4)-β-glucan in sorghum stem sections. Mean and standard error of n = 3 biological replicates, three technical replicates per assay. Different letters indicate statistical differences.

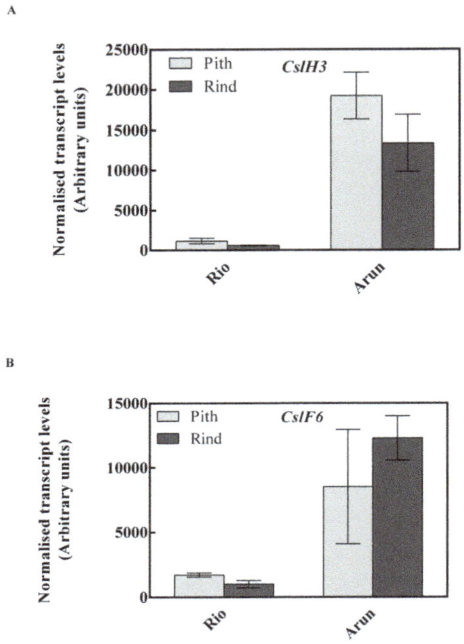

Figure 5: Transcript analysis in sorghum stem sections.

Transcript profile of (a) *SbCslH3* and (**b**) *SbCslF6* in pith, rind and whole stem tissues of Rio and Arun. The transcript levels of these genes in the stem of mature BTx623 were reported previously [38]. Mean and standard errors n = 3 biological replicates. Data are normalised to control genes.

The distribution of all β-(1,4)-linked glucans, such as cellulose and (1,3;1,4)-β-glucan, was observed by fluorescence of transverse tissue sections after staining with Calcofluor-white (Fig 6). BTx623 and Rio displayed a similar staining pattern, with fluorescence visible predominantly in tissues such as the phloem and in patches of parenchyma located in the rind. Arun however, showed bright fluorescence in cell walls throughout the pith and rind (Fig 6).

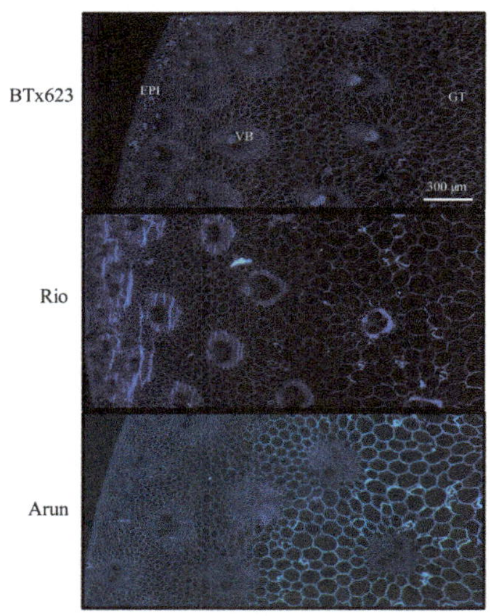

Figure 6: Distribution of glucan polymers in stems of grain, sweet and wild sorghum lines.

Sections of mature sorghum stem were stained with Calcofluor-white and images captured with a dissection microscope and a fluorescence filter. All pictures are of the same scale, bar is 300 µm. Ground tissue (GT), vascular bundle (VB), epidermis (EPI).

The specific distribution of arabinoxylan and (1,3;1,4)-β-glucan was investigated by labelling fixed stem tissues with fluorescent antibodies against these polysaccharides (Fig 7). The relative fluorescence associated with (1,3;1,4)-β-glucan labelling was consistent with the quantitative data (Fig 4d). The variation in fluorescence associated with LM11 between genotypes did not correlate well with the variation in the quantitative data for the sum of arabinose and xylose in each genotype (Fig 4c). However, sections labelled with the LM11 anti-arabinoxylan antibody had brighter fluorescence in rind tissue than pith tissue, and this was consistent with the quantitative data; and pith sections labelled with the BG-1 anti-(1,3;1,4)-β-glucan antibody had brighter fluorescence than rind tissue, consistent with the quantitative data (Fig 4).

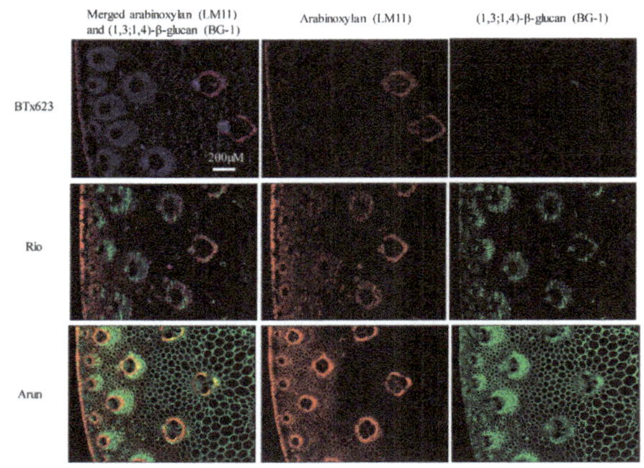

Figure 7: Distribution of (1,3;1,4)-β-glucan and arabinoxylan in sorghum stem sections.

Immunohistological analysis of cell wall epitopes in BTx623, Rio and Arun stems. All panels show transverse sections through mature stems. In merged and single images the (1,4)-β-xylan/arabinoxylan labelled with an LM11 antibody is indicated by red fluorescence and the (1,3;1,4)-β-glucan, labelled with the BG-1 antibody is indicated by green fluorescence. All pictures are of the same scale, bar is 200 µm.

Starch and Glucose Yields from Sorghum Stems

Rio and Arun contained similar amounts of starch in their pith and rind, while BTx623 pith contained much less, with less even still in the rind (Fig 8a). The digestibility of stem samples was investigated by measuring the amount of glucose in the samples after either a commercial cellulase enzyme treatment or a sulfuric acid treatment. The enzyme treatment was more effective than the acid at liberating glucose for all tissues tested, especially the Arun samples (Fig 8b and 8c). With enzymatic digestion, BTx623 pith yielded 15% w/w glucose, while the rind yielded only 9% w/w. Rio biomass released 39% w/w glucose from the pith and 24% w/w from the rind. The highest glucose yield was from Arun pith, which yielded 44% w/w glucose after enzymatic pre-treatment, while rind glucose yields were much lower, at 20% w/w.

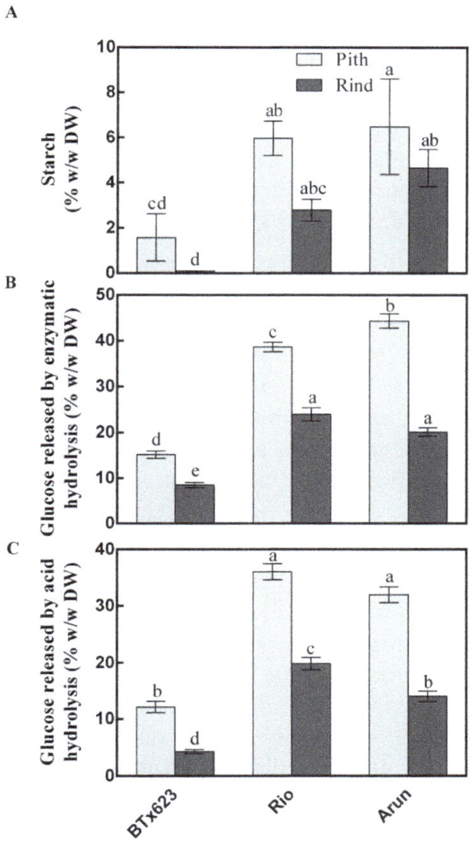

Figure 8: Sorghum stem starch amounts and relative digestibility.

(**a**) Total starch in sorghum stem tissues; (**b**) glucose released after enzymatic digestion; and (**c**) glucose released after sulphuric acid digestion. Mean and standard error of three biological replicates; three technical replicates per assay. Different letters indicate statistical differences.

Theoretical Ethanol Yields from Sorghum Stems

BTx623, Rio and Arun stem biomass and monosaccharide composition data were used to calculate the theoretical ethanol yields following the National Renewable Energy Laboratory Theoretical Ethanol Yield Calculator method (U.S. Department of Agriculture 2013; Table 1). The values represent estimates for stem tissue only; leaf and grain head biomass was not included. If the sugar in harvested stem tissue were efficiently used to produce ethanol, the estimates indicate that up to 10344 L/ha may be produced from Arun stem tissues.

Table 1: Theoretical ethanol yields from stem tissue of wild, sweet and grain sorghum

	Theoretical ethanol yields from stem tissue		
Sorghum	L/Kg	L/plant stem	*L/ha
BTx623	0.20	0.004	728
Rio	0.28	0.033	5969
Arun	0.28	0.057	10344

Calculated using the National Renewable Energy Laboratory Theoretical Ethanol Yield Calculator (http://www1.eere.energy.gov/bioenergy/ethanol_yield_calculator.html) that assumes 1.111 Kg monomeric C6 sugar per 1 Kg polymeric C6 sugar polymer; 1.1363 Kg monomeric C5 sugar per 1 Kg polymeric C5 polymer; 0.51 Kg ethanol produced from 1 Kg sugar; 0.7892 g/L relative density of ethanol. Calculations are for dry weight and composition of mature stem tissue only and do not include ethanol that could be produce from leaf or grain tissues.

*Estimated based on a standard planting density of 180 000 plants/ha and stem dry weight distribution of 1:1 pith:rind. Values for sorghum stem monosaccharide composition used for the calculations are listed in S1 Table.

doi:10.1371/journal.pone.0156638.t001

DISCUSSION

(1,3;1,4)-β-Glucan in Cell Walls of Sorghum Stems

The stems of sorghum varieties investigated in this study varied in the amounts and composition of cell walls in pith and rind tissues (Figs 1 and 4) although no morphological differences could be observed (Fig 3b). The outer rind was generally comprised of higher amounts of cellulose and arabinoxylan compared with the inner pith, which contained higher amounts of (1,3;1,4)-β-glucan (Figs 1, 4, 6 and 7). These data are consistent with the roles played by these stem tissues. The rind has a more structural role, providing rigidity and support for the whole plant and housing the vascular bundles, so it is not surprising that cell walls here consist of molecules such as cellulose and arabinoxylan associated with greater recalcitrance [17–19]. In contrast, the pith has more of a storage role, accumulating excess carbohydrates that buffer source-sink interactions during growth in varying environmental conditions and contributing to plasticity in sink-source dynamics [39]. Storing a carbohydrate buffer in the stem, away from photosynthetic tissues, may also prevent feedback inhibition of photosynthesis, which has been observed when carbohydrates accumulate in the leaves [40,41].

(1,3;1,4)-β-Glucan may accumulate in cell walls so that it can be readily degraded to release glucose for plant growth and reproduction [42, 43]. There is limited information about the importance of (1,3;1,4)-β-glucan as a storage polysaccharide in grasses and how the accumulation of (1,3;1,4)-β-glucan in vegetative tissues influences plant fitness and survival [44]. It is possible that the (1,3;1,4)-β-glucan in sorghum stem tissue is remobilized after the main tillers have reached maturity to supply glucose for the rapid growth of new tillers from the stalk base, the ratoon crop.

Sorghum varieties that are sensitive to photoperiod and do not flower under long day lengths are capable of achieving high levels of biomass production by continuing vegetative growth throughout the season [45]. These sorghum varieties tend to have high yields of lignocellulosic material and some lines produce 80 Mg ha^{-1} (65% moisture; [12]). Four photoperiod-sensitive lines grown as part of the current study had five times more (1,3;1,4)-β-glucan in the stem pith and rind relative to the photoperiod-insensitive lines that flowered (Fig 1a). The photoperiod-sensitive lines may have partitioned excess photoassimilate, not required for producing seed, into the production of (1,3;1,4)-β-glucan in the stem.

Sorghum lines that preferentially store stem glucose as (1,3;1,4)-β-glucan rather than as sucrose, as in sweet sorghum stems, may have improved post-harvest storage characteristics [46] and higher yields of fermentable carbohydrate [47]. Germplasm with high amounts of (1,3;1,4)-β-glucan in the stem, such as the ssp. *verticilliflorum* line identified here, could be incorporated into breeding programs to improve storage and conversion traits for biofuel production. Alternatively, it may be possible to overexpress genes involved in the synthesis of (1,3;1,4)-β-glucan in sorghum [48]. Previously, overexpression of a (1,3;1,4)-β-glucan synthesis gene (*CslF6*) in barley (*Hordeum vulgare*) resulted in sixfold higher amounts of (1,3;1,4)-β-glucan in the leaves but a significant reduction in plant fitness [49]. Recently, senescence-associated promoters were used to overexpress *CslF6* in *Arabidopsis*, resulting in a 42% increase in saccharification without affecting plant health [47]. Another strategy may be to identify sorghum lines with mutations in (1,3;1,4)-β-glucan degradation; for example, elevated levels of (1,3;1,4)-β-glucan were observed in a maize plant that had a defective lichenase [50].

Previous research in barley, wheat and rice has indicated that two types of *Csl* genes, *CslF*and *CslH*, are associated with the synthesis of the non-cellulosic polysaccharide (1,3;1,4)-β-glucan [28, 29, 47, 49, 51]. In sorghum, relatively high expression of *SbCslF6* has been observed in stem compared with other tissues, with an increase in *SbCslH3* expression as stem tissues mature [38]. The current data indicate that an increase in expression of *SbCslF6*and *SbCslH3* in mature stem tissues is associated with the accumulation of (1,3;1,4)-β-glucan (Figs 4d and 5b). Previous studies have suggested that glycan synthases may require the interaction of multiple proteins to translocate the polymer across the membrane [52], raising questions as to whether different CSL isoforms, such as CSLF and CSLH, interact with each other or with CESAs to form an integral catalytic complex. The identification of other factors that influence the regulation of (1,3;1,4)-β-glucan synthesis is also a key target for further research.

Biofuel Production and Genetic Variation in Biomass Yield and Composition

High yields of bioethanol can be produced from sorghum [53]. For example the calculated ethanol yield from the late maturing sweet sorghum 'Lvneng-3', which had particularly high yields of cellulose and non-cellulosic polysaccharides, was up to 13 032 L ha^{-1} [54]. Yields are dependent not only on the amount of fermentable sugar in the plant, but also on the presence, amount and types of inhibitory molecules, such as lignin and ash, and the total amount of biomass produced per plant [55].

A detailed examination of three different types of sorghum varieties—a grain, sweet and wild line—revealed that the plants produce very different amounts of stem, leaf and grain tissues (Fig 2a). The grain line, BTx623, bred to have a shorter stature with high yields of grain, channelled 56% of biomass production to the grain head. Conversely, the sweet line, Rio, bred for long stems full of sucrose, partitioned 59% of its biomass into stem tissue. However, the uncultivated wild Arun line produced the most biomass (321g dry biomass per plant) with 65% biomass in the stem.

The starch content of Arun and Rio stem tissues did not differ, whereas there was more glucose released by enzymatic hydrolysis from Arun than Rio indicating that starch was not the likely source of the difference in glucose (Fig 8). Arun accumulated significantly higher levels of starch in the stem than BTx623, however on a % w/w basis the proportional difference in starch content would be insufficient to account for the proportional differences in glucose released (Fig 8). As starch is readily convertible to glucose during pre-treatment, engineering sorghum stems to increase starch accumulation may be a valid target for the improvement of sorghum as a biofuel crop or industrial raw material. Pre-treating stem tissues from these three varieties either with an enzyme cocktail or with sulfuric acid released up to 44% w/w glucose. The highest yield was from the Arun pith after enzymatic treatment (Fig 8b) with the lower yield from Arun rind (20% w/w) possibly due to the rind having the highest levels of lignin present in the three lines (Fig 4a). However, it is not clear whether lignin content is a factor that influences the release of glucose from (1,3;1,4)-β-glucan during enzymatic treatment.

Data for stem mass and composition were used to calculate theoretical ethanol yields from the stems of the three lines. Previous estimates from whole BTx623 and Rio plants ranged up to 5115 L ha^{-1} and the yield of hybrid plants was up to 8512 L ha^{-1} [15]. The values obtained in this study were very similar for Rio (5977 L ha^{-1}), but much lower for BTx623 (721 L ha^{-1}), presumably because we assessed stem tissue only, which comprises only 22% of the biomass of this grain variety.

The calculated values for the wild Arun stem, 10 344 L ha^{-1}, were greater than the values for most high-yielding recombinant inbred lines from a BTx623 × Rio cross [15]. Recent research in rice indicates that yield from greenhouse-grown plants was a suitable predictor for yield in the field [56], so our results may translate to field-grown crops but future field trials to test this are needed. *S. bicolor* ssp. *verticilliflorum* is a weedy, wild relative of sorghum that grows on roadsides [9], so is an excellent candidate for successful cultivation in remote or marginal areas with little or no production inputs.

The cost of producing biofuel may become increasingly competitive with petroleum fuels if the current trajectory of genetic improvement in bioenergy sorghum continues [57] but it does depend on accurately identifying variation in morphological, structural and physiological traits [36]. Plant size is the main trait influencing bioenergy yield and more effort has so far been invested in increasing overall plant size than in improving the cell wall composition, which is more difficult to screen for in the field [36, 58]. However, the development of new high-throughput methods for screening biomass composition [32] combined with the emerging genomics resources available to the sorghum community [2, 5], make it increasingly feasible to select for genetic variation in structural and physiological traits that enhance biofuel production.

Exploring the diversity in sorghum biomass cell wall composition is important because natural genetic variation will provide the foundation for the improvement of sorghum vegetative tissues as an energy crop [59]. The biomass digestibility and carbohydrate yields from sorghum varieties are linked to the quality and quantity of carbohydrates, for example starch and (1,3;1,4)-β-glucan [45]. The literature currently lacks information about such variation in starch and (1,3;1,4)-β-glucan in diverse sorghum germplasm. Here, we have identified a wild, photoperiod-sensitive sorghum line that accumulates 6.5% starch and 5.5% w/w (1,3;1,4)-β-glucan in the stem pith, relative to the 1% w/w (1,3;1,4)-β-glucan observed in many cultivated varieties. This wild *S. bicolor* ssp. *verticilliflorum* had significantly higher stem digestibility and calculated ethanol yields relative to grain sorghum BTx623 despite having double the lignin content, at theoretical yields of 10344 L ha^{-1}. This heterogeneity in glycan polymers in sorghum stems can be exploited to improve sorghum breeding for forage and biofuel industries.

MATERIALS AND METHODS

Plant Materials

All 12 sorghum genotypes were grown in the Plant Accelerator (Australian Plant Phenomics Facility, Waite Campus, South Australia, Australia) in glass houses under a natural day length 30°C day temperature and 20°C night temperature in October 2011– February 2012 (day length 13:28 h to 12:53 h, respectively). Eight photoperiod-insensitive lines (Rio, Acme Broomcorn, LR2490-3, SC170-6-8, IS8525, BTx623, QL12 and M35-1) were harvested at maturity, 129 d after planting. Ssp. *verticilliflorum* (Arun) is partially photoperiod-sensitive, thus flowering was delayed and this line was harvested 159 d after planting. Three photoperiod-sensitive lines *S. drummondii*, Kinto Oule (PI525695) and PI559871 did not flower and these lines were harvested 248 d after planting. The third internode from the base of the plants was harvested, split manually into pith and rind sections and snap frozen in liquid nitrogen or placed into fixative for histological analysis. To determine plant height, each tiller was measured from the soil surface to the panicle tips. The mean stem diameter of each line was calculated by measuring the diameter at the first internode above the ground and averaging the tillers. For fresh biomass yield, the leaf, head and stem of each line were weighed immediately after harvest and dried at 60°C to constant moisture content. Samples were re-weighed to determine the dry biomass yield.

Biomass Compositional Assays and Digestibility Assays

Tissue for all analyses was harvested, snap frozen in liquid nitrogen and freeze-dried (Labconco Freezone; Adelab Scientific, Adelaide, SA, Australia). Dried material was ground in a 25 mL stainless steel grinding jar with one 7 mm steel ball (Retsch mill MM400, Retsch GmbH; Haan, Germany).

Acid determination of cellulose.

This method was adapted from [60] and [61] as follows; ground dried tissue (20 ± 1 mg) was treated with a mixture (1 mL) of acetic and nitric acid (8:1) in water at 110°C in heating blocks. Samples were washed once with water and twice with ethanol (80%), followed with an acetone (100%) wash before drying to a constant weight in a 45°C oven. The remaining material was recorded as acid-insoluble cellulose. Two technical replicates and three biological replicates were assayed per sample.

(1,3;1,4)-β-glucan analysis.

This method and analysis followed [38]. Two technical replicates and three biological replicates were assayed per sample.

Starch assay.

A small scale version of the Total Starch Assay procedure (AA/AMG method; Megazyme, Ireland) was used on 20 mg flour. The liberated glucose content was quantified using a spectrophotometer (Thermo Fischer, Waltham, MA, USA) at a wavelength of 510 nm. Samples were blanked with water and a maize starch standard (97%) was included in each batch. Two technical replicates and three biological replicates were assayed per sample.

Monosaccharide (mannose, ribose, rhamnose, glucuronic acid, galacturonic acid, glucose, galactose, xylose and arabinose) profiling of sorghum stems.

Acid hydrolysis and reversed-phase high-performance liquid chromatography (RP-HPLC) methodology followed [62, 63]. A 20 mg amount of freeze dried tissue was washed four times with 70% ethanol to remove soluble sugars, acid hydrolysed in 1 M sulfuric acid at 100°C for 3 h and analysed by RP-HPLC as described by [64]. Three biological replicates and three technical replicates were used to assess the variation in monosaccharide content.

Klason lignin.

The method for assaying lignin followed [65] with the following modifications; each 20 mg tissue sample were washed twice with ethanol (80%) and placed at 45°C overnight prior to 12 M sulfuric acid treatment at 35°C in 10 mL glass tubes. Acid-soluble contents were removed from the tissues by washing three times with 10 mL of water. The pellets were diluted in water, filtered through Whatman GF/C55mm glass microfiber filters (Sigma-Aldrich, St Louis, USA) and dried to a constant weight in a 45°C oven. The remaining material was recorded as acid-insoluble lignin. Two technical replicates and three biological replicates were assayed per sample.

Stem digestibility.

The method for assaying digestibility followed [66] with the following modifications; a 25 mg aliquot of ground dried tissue was weighed into 10 mL glass tube and sterilized at 121°C for 20 min with water to prevent bacterial contamination and disrupt cell wall structure. The solution was treated with an enzyme mixture (C2730; Sigma-Aldrich) containing cellulase from*Trichoderma reesei* (≥ 700 units/g) and cellobiase from *Aspergillus niger* (≥ 250 units/g), at 50°C for 72 h to hydrolyze glucans into sugars. The enzyme mix also included a (1,3;1,4)-β-glucanase. The reaction was diluted and incubated with glucose oxidase/peroxidise reagent (GOPOD; Megazyme International, Ireland) at 50°C for 20 min. The absorbance was read at 510 nm using a spectrophotometer (Thermo Fischer) along with a glucose standard and a blank of GOPOD and enzyme mixture. Glucose released from stem tissues by 1 M sulfuric acid was quantified following [53]. Two technical replicates and three biological replicates were assayed per sample.

Theoretical ethanol yields.

These were calculated following [67] using the following standard conversion assumptions; 1.111 kg monomeric C6 sugar per 1 kg polymeric C6 polymer (glucan, fructan); 1.1363 kg monomeric C5 sugar per 1 kg polymeric C5 polymer (xylan, arabinan); 0.51 kg of ethanol produced from 1 kg of sugar; 0.7892 g/L relative density of ethanol.

Histological Analysis and Immunolabelling

Histological analysis followed [62]: Replicate stem segments were harvested at maturity and fixed overnight in formalin–acetic acid–alcohol (FAA) solution containing 50% ethanol, 5% acetic acid and 4% formaldehyde. Samples were serially dehydrated and embedded in LR White resin (London Resin Co., UK) as described by [49]. Sections, 500 nm thick, were cut using a Reichert Ultracut ultramicrotome (Leica, Germany). For immunolabelling, sections were labelled with the (1,4)-β-xylan/arabinoxylan specific rat monoclonal antibody LM11 (McCartney, Marcus & Knox 2005; PlantProbes, Leeds, UK) and the (1,3;1,4)-β-glucan specific mouse primary antibody BG-1 ([68]; Biosupplies, Australia). Secondary antibodies were conjugated to either Alexafluor 488 or 555 (Life Technologies, USA) and samples were examined using a Zeiss M1 AxioImager. Maximum fluorescence intensity measurements were generated using ZEN 2012 software, as described by [69]. When sorghum sections were treated with a lichenase enzyme (Megazyme), (1,3;1,4)-β-glucanase associated fluorescence disappears confirming specificity of (1,3;1,4)-β-glucan labelling [28]. Calcofluor White (Sigma-Aldrich) was used as a general counter-stain to detect β-linked glucose polymers where sections were incubated with 0.1% Calcofluor White and images of labeled sections were taken under fluorescence filter I3 (450–490 nm BP, barrier 515 nm LP) and filter D (335–425 nm BP, barrier 470 nm LP) using a Leica AS LMD Laser Dissection Microscope with a DFC 480 camera.

Quantitative Polymerase Chain Reaction (Q-PCR)

Total RNA was extracted from mature stems using the Direct-zol RNA MiniPrep (Zymo Research, USA). cDNA was synthesized according to [70]. Data is from three biological replicates. Q-PCR analysis and data normalisation using multiple control genes (glyceraldehyde-3-phosphate dehydrogenase (GAPDH), α-tubulin, cyclophilin) followed [38].

Statistical Analysis

Analysis of variance (ANOVA) and Tukey's multiple comparison tests were undertaken using GENSTAT (VSN International) or GraphPad Prism (GraphPad Software Inc., USA).

ACKNOWLEDGMENTS

Thank you to Alison Roberts (University of Rhode Island) for testing glucanase enzymes on labelled sorghum sections to confirm specificity of the (1,3;1,4)-β-glucanase antibody.

AUTHOR CONTRIBUTIONS

Conceived and designed the experiments: CSB GBF VK RAB. Performed the experiments: CSB NSB HTT WLL RAE HYN NJS JL KC. Analyzed the data: CSB NSB RAB. Contributed reagents/materials/analysis tools: GBF VK RAB. Wrote the paper: CSB NSB RAB GBF.

REFERENCES

1. Paterson AH, Bowers JE, Bruggmann R, Dubchak I, Grimwood J, Gundlach H, et al. The Sorghum bicolor genome and the diversification of grasses. Nature. 2009;457(7229):551–6. doi: 10.1038/nature07723. pmid:19189423
2. Mace ES, Tai S, Gilding EK, Li Y, Prentis PJ, Bian L, et al. Whole-genome sequencing reveals untapped genetic potential in Africa's indigenous cereal crop sorghum. Nature Communications. 2013;4. doi: 10.1038/ncomms3320
3. Harlan J, De Wet J. A simplified classification of cultivated Sorghum. Crop Science. 1972;12(2):172–6. doi: 10.2135/cropsci1972.0011183x001200020005x
4. Zheng L-Y, Guo X-S, He B, Sun L-J, Peng Y, Dong S-S, et al. Genome-wide patterns of genetic variation in sweet and grain sorghum (Sorghum bicolor). Genome Biology. 2011;12(11):R114. doi: 10.1186/gb-2011-12-11-r114. pmid:22104744
5. Price HJ, Dillon SL, Hodnett G, Rooney WL, Ross L, Johnston JS. Genome evolution in the genus Sorghum (Poaceae). Annals of Botany. 2005;95(1):219–27. pmid:15596469 doi: 10.1093/aob/mci015
6. Dillon SL, Lawrence PK, Henry RJ. The use of ribosomal ITS to determine phylogenetic relationships within Sorghum. Plant Systematics and Evolution. 2001;230(1–2):97–110. doi: 10.1007/s006060170007
7. Sun Y, Skinner D, Liang G, Hulbert S. Phylogenetic analysis of Sorghum and related taxa using internal transcribed spacers of nuclear ribosomal DNA. Theoretical and Applied Genetics. 1994;89(1):26–32. doi: 10.1007/BF00226978. pmid:24177765
8. Dillon SL, Lawrence PK, Henry RJ. The new use of Sorghum bicolor-derived SSR markers to evaluate genetic diversity in 17 Australian Sorghum species. Plant Genetic Resources. 2005;3(01):19–28. doi: 10.1079/pgr200454
9. Ejeta G, Grenier C. Sorghum and its weedy hybrids. In: Gressel J, editor. Crop Ferality and Volunteerism. Florida, USA: CRC Press, Taylor and Francis; 2005. p. 123–35.
10. Dillon SL, Lawrence PK, Henry RJ, Ross L, Price HJ, Johnston JS. Sorghum laxiflorum and S. macrospermum, the Australian native species most closely related to the cultivated S. bicolor based on ITS1 and ndhF sequence analysis of 25 Sorghum species. Plant Systematics and Evolution. 2004;249(3–4):233–46. doi: 10.1007/s00606-004-0210-7
11. Mace ES, Xia L, Jordan DR, Halloran K, Parh DK, Huttner E, et al. DArT markers: diversity analyses and mapping in Sorghum bicolor. BMC Genomics. 2008;9(26). doi: 10.1186/1471-2164-9-26
12. Rooney WL, Blumenthal J, Bean B, Mullet JE. Designing sorghum as a dedicated bioenergy feedstock. Biofuels, Bioproducts and Biorefining. 2007;1(2):147–57. doi: 10.1002/bbb.15
13. Shakoor N, Nair R, Crasta O, Morris G, Feltus A, Kresovich S. A Sorghum bicolor expression atlas reveals dynamic genotype-specific expression profiles for vegetative tissues of grain, sweet and bioenergy sorghums. BMC Plant Biology. 2014;14(1):35. doi: 10.1186/1471-2229-14-35

14. McBee GG, Miller FR. Carbohydrate and lignin partitioning in sorghum stems and blades. Agronomy Journal. 1990;82(4):687–90. doi: 10.2134/agronj1990.00021962008200040008x

15. Murray SC, Rooney WL, Mitchell SE, Sharma A, Klein PE, Mullet JE, et al. Genetic improvement of sorghum as a biofuel feedstock: II. QTL for stem and leaf structural carbohydrates. Crop Science. 2008;48(6):2180–93. doi: 10.2135/cropsci2008.01.0068

16. Stefaniak TR, Dahlberg JA, Bean BW, Dighe N, Wolfrum EJ, Rooney WL. Variation in biomass composition components among forage, biomass, sorghum-sudangrass, and sweet sorghum types. Crop Science. 2012;52(4):1949–54. doi: 10.2135/cropsci2011.10.0534

17. McBee GG, Creelman RA, Miller FR. Ethanol yield and energy potential of stems from a spectrum of sorghum biomass types. Biomass. 1988;17(3):203–11. doi: 10.1016/0144-4565(88)90114-x

18. Richie W, McBee G. Structural components in sorghum stem biomass. Bioresource technology. 1991;38(1):15–22. doi: 10.1016/0960-8524(91)90216-7

19. Wilson JR, Mertens DR, Hatfield RD. Isolates of cell types from sorghum stems: digestion, cell wall and anatomical characteristics. Journal of the Science of Food and Agriculture. 1993;63(4):407–17. doi: 10.1002/jsfa.2740630406

20. Burton RA, Gidley MJ, Fincher GB. Heterogeneity in the chemistry, structure and function of plant cell walls. Nature Chemical Biology. 2010;6(10):724–32. doi: 10.1038/nchembio.439. pmid:20852610

21. Vancov T, McIntosh S. Mild acid pretreatment and enzyme saccharification of Sorghum bicolor straw. Applied Energy. 2012;92:421–8. doi: 10.1016/j.apenergy.2011.11.053

22. Byrt CS, Betts NS, Farrokhi N, Burton RA, Singh B. Deconstructing plant biomass: cell wall structure and novel manipulation strategies. In: Singh BP, editor. Biofuel Crops: Production, Physiology and Genetics: Centre for Agricultural Bioscience International (CABI); 2013.

23. Vancov T, McIntosh S. Alkali pretreatment of cereal crop residues for second-generation biofuels. Energy & Fuels. 2011;25(7):2754–63. doi: 10.1021/ef200241s

24. Amaducci S, Monti A, Venturi G. Non-structural carbohydrates and fibre components in sweet and fibre sorghum as affected by low and normal input techniques. Industrial Crops and Products. 2004;20(1):111–8. doi: 10.1016/j.indcrop.2003.12.016

25. Betts NS, Fox GP, Kelly AM, Cruickshank AW, Lahnstein J, Henderson M, et al. Non-cellulosic cell wall polysaccharides are subject to genotype ×environment effects in sorghum (Sorghum bicolor) grain. Journal of Cereal Science. 2015;63:64–71. doi: 10.1016/j.jcs.2015.02.007

26. Fincher G, Stone B. Chemistry of nonstarch polysaccharides. In: HCaCEW C. Wrigley, editor. Encyclopedia of Grain Science Oxford, UK: Elsevier; 2004. p. 206–23.

27. Bacic A, Harris PJ, Stone BA. Structure and function of plant cell walls. The biochemistry of plants. 1988;14:297–371. doi: 10.1016/b978-0-08-092615-5.50014-x

28. Burton RA, Wilson SM, Hrmova M, Harvey AJ, Shirley NJ, Medhurst A, et al. Cellulose synthase-like CslF genes mediate the synthesis of cell wall (1, 3; 1, 4)-ß-D-glucans. Science. 2006;311(5769):1940–2. pmid:16574868 doi: 10.1126/science.1122975

29. Doblin MS, Pettolino FA, Wilson SM, Campbell R, Burton RA, Fincher GB, et al. A barley cellulose synthase-like CSLH gene mediates (1, 3; 1, 4)-β-D-glucan synthesis in transgenic Arabidopsis. Proceedings of the National Academy of Sciences. 2009;106(14):5996–6001. doi: 10.1073/pnas.0902019106

30. Weng J-K, Li X, Bonawitz ND, Chapple C. Emerging strategies of lignin engineering and degradation for cellulosic biofuel production. Current opinion in biotechnology. 2008;19(2):166–72. doi: 10.1016/j.copbio.2008.02.014. pmid:18403196

31. Saballos A, Sattler SE, Sanchez E, Foster TP, Xin Z, Kang C, et al. Brown midrib2 (Bmr2) encodes the major 4-coumarate: coenzyme A ligase involved in lignin biosynthesis in sorghum (Sorghum bicolor (L.) Moench). The Plant Journal. 2012;70(5):818–30. doi: 10.1111/j.1365-313X.2012.04933.x. pmid:22313236

32. Martin A, Palmer W, Byrt C, Furbank R, Grof C. A holistic high-throughput screening framework for biofuel feedstock assessment that characterises variations in soluble sugars and cell wall composition in Sorghum bicolor. Biotechnology for Biofuels. 2013;6(1):1–13. doi: 10.1186/1754-6834-6-186

33. Dien BS, Sarath G, Pedersen JF, Sattler SE, Chen H, Funnell-Harris DL, et al. Improved sugar conversion and ethanol yield for forage sorghum (Sorghum bicolor L. Moench) lines with reduced lignin contents. BioEnergy Research. 2009;2(3):153–64. doi: 10.1007/s12155-009-9041-2

34. Sattler SE, Saathoff AJ, Haas EJ, Palmer NA, Funnell-Harris DL, Sarath G, et al. A nonsense mutation in a cinnamyl alcohol dehydrogenase gene is responsible for the sorghum brown midrib6 phenotype. Plant Physiology. 2009;150(2):584–95. doi: 10.1104/pp.109.136408. pmid:19363091

35. Burton RA, Shirley NJ, King BJ, Harvey AJ, Fincher GB. The CesA gene family of barley. Quantitative analysis of transcripts reveals two groups of co-expressed genes. Plant Physiology. 2004;134(1):224–36. pmid:14701917 doi: 10.1104/pp.103.032904

36. Jahn CE, Mckay JK, Mauleon R, Stephens J, McNally KL, Bush DR, et al. Genetic variation in biomass traits among 20 diverse rice varieties. Plant Physiology. 2011;155(1):157–68. doi: 10.1104/pp.110.165654. pmid:21062890

37. Folliard A, Traoré P, Vaksmann M, Kouressy M. Modeling of sorghum response to photoperiod: a threshold–hyperbolic approach. Field crops research. 2004;89(1):59–70. doi: 10.1016/j.fcr.2004.01.006

38. Ermawar RA, Collins HM, Byrt CS, Betts NS, Henderson M, Shirley NJ, et al. Distribution, structure

and biosynthetic gene families of (1, 3; 1, 4)-β-glucan in Sorghum bicolor. Journal of Integrative Plant Biology. 2015;57(4):429–45. doi: 10.1111/jipb.12338. pmid:25661466

39. Grof CP, Byrt CS, Patrick JW. Phloem transport of resources: John Wiley & Sons 2013.

40. Pollock C, Farrar J, Tomos D, Gallagher J, Lu C, Koroleva O. Balancing supply and demand: the spatial regulation of carbon metabolism in grass and cereal leaves. Journal of Experimental Botany. 2003;54(382):489–94. pmid:12508059 doi: 10.1093/jxb/erg037

41. Slewinski TL. Non-structural carbohydrate partitioning in grass stems: a target to increase yield stability, stress tolerance, and biofuel production. Journal of Experimental Botany. 2012;63(13):4647–70. doi: 10.1093/jxb/ers124. pmid:22732107

42. Trafford K, Haleux P, Henderson M, Parker M, Shirley NJ, Tucker MR, et al. Grain development in Brachypodium and other grasses: possible interactions between cell expansion, starch deposition, and cell-wall synthesis. Journal of Experimental Botany. 2013. doi: 10.1093/jxb/ert292

43. Wong S, Shirley N, Little A, Khoo KP, Schwerdt J, Fincher G, et al. Differential expression of the HvCslF6 gene late in grain development may explain quantitative differences in (1,3;1,4)-β-glucan concentration in barley. Molecular Breeding. 2015;35(1):1–12. doi: 10.1007/s11032-015-0208-6

44. Roulin S, Buchala AJ, Fincher GB. Induction of (1→ 3, 1→ 4)-β-D-glucan hydrolases in leaves of dark-incubated barley seedlings. Planta. 2002;215(1):51–9. pmid:12012241 doi: 10.1007/s00425-001-0721-1

45. Snider JL, Raper RL, Schwab EB. The effect of row spacing and seeding rate on biomass production and plant stand characteristics of non-irrigated photoperiod-sensitive sorghum (Sorghum bicolor (L.) Moench). Industrial Crops and Products. 2012;37(1):527–35. doi: 10.1016/j.indcrop.2011.07.032

46. Bennett AS, Anex RP. Production, transportation and milling costs of sweet sorghum as a feedstock for centralized bioethanol production in the upper Midwest. Bioresource Technology. 2009;100(4):1595–607. doi: 10.1016/j.biortech.2008.09.023. pmid:18951018

47. Vega-Sánchez ME, Loqué D, Lao J, Catena M, Verhertbruggen Y, Herter T, et al. Engineering temporal accumulation of a low recalcitrance polysaccharide leads to increased C6 sugar content in plant cell walls. Plant Biotechnology Journal. 2015. doi: 10.1111/pbi.12326

48. Raghuwanshi A, Birch RG. Genetic transformation of sweet sorghum. Plant cell reports. 2010;29(9):997–1005. doi: 10.1007/s00299-010-0885-x. pmid:20535472

49. Burton RA, Collins HM, Kibble NA, Smith JA, Shirley NJ, Jobling SA, et al. Over-expression of specific HvCslF cellulose synthase-like genes in transgenic barley increases the levels of cell wall (1,3; 1,4)-β-D-glucans and alters their fine structure. Plant Biotechnology Journal. 2011;9(2):117–35. doi: 10.1111/j.1467-7652.2010.00532.x. pmid:20497371

50. Pauly M, Hake S, Kraemer FJ, inventorsMaize variety and method of production2011.

51. Nemeth C, Freeman J, Jones HD, Sparks C, Pellny TK, Wilkinson MD, et al. Down-regulation of the CSLF6 gene results in decreased (1, 3; 1, 4)-β-D-glucan in endosperm of wheat. Plant Physiology. 2010;152(3):1209–18. doi: 10.1104/pp.109.151712. pmid:20089768

52. Davis J, Brandizzi F, Liepman AH, Keegstra K. Arabidopsis mannan synthase CSLA9 and glucan synthase CSLC4 have opposite orientations in the Golgi membrane. The Plant Journal. 2010;64(6):1028–37. doi: 10.1111/j.1365-313X.2010.04392.x. pmid:21143682

53. Byrt CS, Grof CP, Furbank RT. C4 plants as biofuel feedstocks: Optimising biomass production and feedstock quality from a lignocellulosic perspective. Journal of Integrative Plant Biology. 2011;53(2):120–35. doi: 10.1111/j.1744-7909.2010.01023.x. pmid:21205189

54. Zhao YL, Dolat A, Steinberger Y, Wang X, Osman A, Xie GH. Biomass yield and changes in chemical composition of sweet sorghum cultivars grown for biofuel. Field Crops Research. 2009;111(1):55–64. doi: 10.1016/j.fcr.2008.10.006

55. Arundale RA, Bauer S, Haffner FB, Mitchell VD, Voigt TB, Long SP. Environment has little effect on biomass biochemical Composition of Miscanthus × giganteus across soil types, nitrogen fertilization, and times of harvest. BioEnergy Research. 2015:1–11. doi: 10.1007/s12155-015-9613-2

56. Tanger P, Vega-Sánchez ME, Fleming M, Tran K, Singh S, Abrahamson JB, et al. Cell wall composition and bioenergy potential of rice straw tissues are influenced by environment, tissue type, and genotype. BioEnergy Research. 2015:1–18. doi: 10.1007/s12155-014-9573-y

57. Mullet J, Morishige D, McCormick R, Truong S, Hilley J, McKinley B, et al. Energy Sorghum—a genetic model for the design of C4 grass bioenergy crops. Journal of Experimental Botany. 2014. doi: 10.1093/jxb/eru229

58. Ragauskas AJ, Williams CK, Davison BH, Britovsek G, Cairney J, Eckert CA, et al. The path forward for biofuels and biomaterials. Science. 2006;311(5760):484–9. pmid:16439654 doi: 10.1126/science.1114736

59. Casa AM, Pressoir G, Brown PJ, Mitchell SE, Rooney WL, Tuinstra MR, et al. Community resources and strategies for association mapping in sorghum. Crop Science. 2008;48(1):30–40. doi: 10.2135/cropsci2007.02.0080

60. Updegraff DM. Semimicro determination of cellulose inbiological materials. Analytical Biochemistry. 1969;32(3):420–4. pmid:5361396 doi: 10.1016/s0003-2697(69)80009-6

61. Burton RA, Gibeaut DM, Bacic A, Findlay K, Roberts K, Hamilton A, et al. Virus-induced silencing of a plant cellulose synthase gene. The Plant Cell Online. 2000;12(5):691–705. doi: 10.2307/3870995

62. Aditya J, Lewis J, Shirley NJ, Tan HT, Henderson M, Fincher GB, et al. The dynamics of cereal cyst nematode infection differ between susceptible and resistant barley cultivars and lead to changes in (1, 3; 1, 4)-β-glucan levels and HvCslF gene transcript abundance. New Phytologist. 2015. doi: 10.1111/nph.13349

63. Pettolino FA, Walsh C, Fincher GB, Bacic A. Determining the polysaccharide composition of plant cell walls. Nature Protocols. 2012;7(9):1590–607. doi: 10.1038/nprot.2012.081. pmid:22864200

64. Corbin KR, Byrt CS, Bauer S, DeBolt S, Chambers D, Holtum JAM, et al. Prospecting for energy-rich renewable raw materials: Agave leaf case study. PLoS ONE. 2015;10(8):e0135382. doi: 10.1371/journal.pone.0135382. pmid:26305101

65. Theander O, Westerlund EA. Studies on dietary fiber. 3. Improved procedures for analysis of dietary fiber. Journal of Agricultural and Food Chemistry. 1986;34(2):330–6. doi: 10.1021/jf00068a045

66. Selig MJ, Tucker MP, Sykes RW, Reichel KL, Brunecky R, Himmel ME, et al. Lignocellulose recalcitrance screening by integrated high-throughput hydrothermal pretreatment and enzymatic saccharification. Industrial Biotechnology. 2010;6(2):104–11. doi: 10.1089/ind.2010.0009

67. Energy USDoEEaR. Theoretical Ethanol Yield Calculator [March 2014]. Available from: http://www.nrel.gov/analysis/models_tools.html.

68. Meikle P, Bonig I, Hoogenraad N, Clarke A, Stone B. The location of (1→ 3)-β-glucans in the walls of pollen tubes of Nicotiana alata using a (1→ 3)-β-glucan-specific monoclonal antibody. Planta. 1991;185(1):1–8. doi: 10.1007/BF00194507. pmid:24186272

69. Chowdhury J, Henderson M, Schweizer P, Burton RA, Fincher GB, Little A. Differential accumulation of callose, arabinoxylan and cellulose in nonpenetrated versus penetrated papillae on leaves of barley infected with Blumeria graminis f. sp. hordei. New Phytologist. 2014;204(3):650–60. doi: 10.1111/nph.12974. pmid:25138067

70. Burton RA, Jobling SA, Harvey AJ, Shirley NJ, Mather DE, Bacic A, et al. The Genetics and Transcriptional Profiles of the Cellulose Synthase-Like HvCslF Gene Family in Barley. Plant Physiology. 2008;146(4):1821–33. doi: 10.1104/pp.107.114694. pmid:18258691

CHAPTER 17

ORGANOSOLV PRETREATMENT OF SORGHUM BAGASSE USING A LOW CONCENTRATION OF HYDROPHOBIC SOLVENTS SUCH AS 1-BUTANOL OR 1-PENTANOL

Hiroshi Teramura, Kengo Sasaki, Tomoko Oshima, Fumio Matsuda, Mami Okamoto, Tomokazu Shirai, Hideo Kawaguchi, Chiaki Ogino, Ko Hirano, Takashi Sazuka, Hidemi Kitano, Jun Kikuchi and Akihiko Kondo

Keywords: Sorghum bagasse, Organosolv pretreatment, Fractionation, 1-butanol, 1-pentanol, Lignin

BACKGROUND

The depletion of fossil fuels and environmental pollution associated with fossil fuel use have increased interest in the utilization of biomass as a feedstock for the production of biofuels and bio-based chemicals [1,2]. Starch-rich biomass is one of the main feedstocks used for these purposes; however, this results in direct competition with global food supplies. Lignocellulosic biomass is a promising alternative feedstock because it is abundant, inexpensive, and renewable. Sorghum (*Sorghum bicolor* L. Moench), which is a highly productive C4 photosynthetic plant, can be used to produce lignocellulosic fractions. Sorghum is also attractive because it uses water efficiently and is drought tolerant [3]. The lignocellulosic fraction produced from sorghum is typically referred to as sorghum bagasse.

The primary components of sorghum bagasse are cellulose, hemicellulose, and lignin [4], each of which can be valorized. For example, cellulose and hemicellulose are used as carbohydrate sources for fermentation, whereas cellulose pulp is used for paper, and lignin is a renewable source of aromatics [5, 6, 7]. The fractionation of sorghum bagasse into its primary components can be carried out in a biorefinery [5, 8]. However, because cellulose is highly crystalline in nature, exhibits complex chemical cross-linking between components, and is recalcitrant to hydrolysis due to the sheathing by hemicellulose and lignin, the fractionation of sorghum bagasse requires appropriate pretreatment [9, 10]. Organosolv pretreatment is a practical methodology for fractionating sorghum bagasse because it facilitates the isolation of high-quality lignin and high-purity cellulose [2, 11, 12]. Organosolv pretreatment involves the use of high-concentration (30–70 %) organic or aqueous organic solvents at temperatures of 100–200 °C, with or without the addition of catalysts [13]. Thus, organosolv pretreatment is expensive and necessitates the recovery of the organic solvents used. Decreasing the concentration of organic solvent necessary for effective pretreatment would thus reduce the cost of fractionation.

As indicated above, organosolv pretreatment has the advantage of enabling the isolation of lignin [2]. The composition of plant cell walls can be characterized by two-dimensional ^{1}H-^{13}C heteronuclear single-quantum coherence nuclear magnetic resonance (2D ^{1}H-^{13}C HSQC NMR) spectroscopy [14, 15]. Thus, 2D-NMR spectroscopy can be used to elucidate the fate of lignin during organosolv pretreatment of sorghum bagasse.

The aim of the present study was to investigate the fractionation effectiveness of five different organic solvents (ethanol, 1-propanol, 2-propanol, 1-butanol, and 1-pentanol) at a relatively low concentration (12.5 %) in organosolv pretreatment of sorghum bagasse. Sulfuric acid was used as the catalyst. The detailed structures of the lignin products obtained in the black liquor and solid fraction of samples pretreated with 1-butanol or 1-pentanol were characterized using 2D ^{1}H-^{13}C HSQC NMR spectroscopy.

Citation: Hiroshi Teramura, Kengo Sasaki, Tomoko Oshima, Fumio Matsuda, Mami Okamoto, Tomokazu Shirai, Hideo Kawaguchi, Chiaki Ogino, Ko Hirano, Takashi Sazuka, Hidemi Kitano, Jun Kikuchi and Akihiko Kondo, "Organosolv pretreatment of sorghum bagasse using a low concentration of hydrophobic solvents such as 1-butanol or 1-pentanol," DOI: 10.1186/s13068-016-0427-z.

RESULTS AND DISCUSSION

The influence of solvent type on organosolv fractionation of sorghum bagasse

Sorghum bagasse was fractionated into three primary components important in biorefinery processes: cellulose, hemicellulose, and lignin. In this study, fractionation was evaluated by comparing pretreatment using five different organic solvents (ethanol, 1-propanol, 2-propanol, 1-butanol, and 1-pentanol) at a low concentration (12.5 %). Pretreatment with no addition of solvent was used as a control.

After treatment at 180 °C for 45 min, samples were centrifuged (Fig. 1). Interestingly, three fractions (solid, liquid, and black liquor) were obtained with 1-butanol or 1-pentanol as the solvent, but only two fractions (solid and liquid) were obtained using ethanol, 1-propanol, or 2-propanol as the solvent and when no solvent was used (control). These results suggest that differences in the physiochemical properties of the solvents affect fractionation. The greater hydrophobicity of 1-butanol and 1-pentanol (partition coefficients: $\log P_{ow} = 0.88$ and 1.51, respectively) compared with ethanol, 1-propanol, and 2-propanol ($\log P_{ow} = -0.31$, 0.25, and 0.05, respectively) resulted in clearer fractionation when using the former solvents. The control condition was equal to dilute acid pretreatment [16]. Thus, hydrophobic 1-butanol and 1-pentanol separated the black liquor fraction from the acid solution even at the low concentration of 12.5 %.

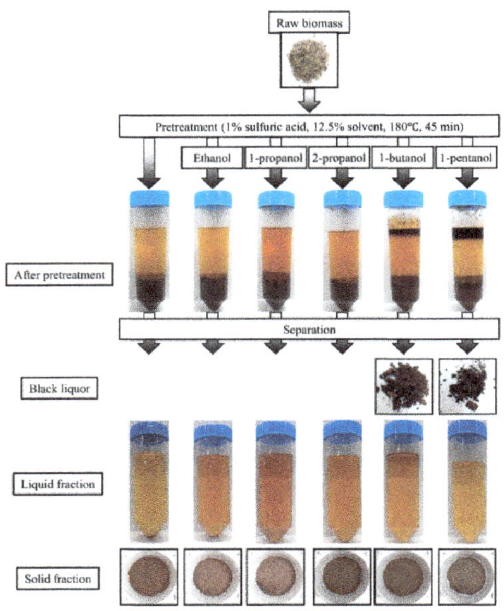

Figure 1: Organosolv pretreatment of sorghum bagasse using 12.5 % solvent concentration. Ethanol, 1-propanol, 2-propanol, 1-butanol, and 1-pentanol were used as the solvent. No addition of solvent was the control. Using ethanol, 1-propanol, and 2-propanol as the solvent and no solvent (control), a solid fraction and a liquid fraction were obtained. Using 1-butanol and 1-pentanol as the solvent, a black liquor fraction was obtained in addition to solid and liquid fractions.

Cellulose-enriched solid fraction

To clarify the effect of solvent type on raw sorghum bagasse, the solid fraction obtained after pretreatment was characterized. Compared with the control, the dry weight of the solid fraction decreased when 1-butanol or 1-pentanol was used as the solvent, but ethanol, 1-propanol, and 2-propanol had no effect on solid fraction dry weight (Fig. 2a). Cellulose is reportedly enriched in the solid fraction of samples pretreated using acid [9], and an increase in the cellulose content in all solid fractions was observed compared with the raw biomass (Fig. 2b). In particular, the cellulose content was higher in the solid fraction when 1-butanol or 1-pentanol was used as the solvent (59.1 and 62.2 %, respectively) compared with ethanol, 1-propanol, 2-propanol, or no solvent (control) (51.6–55.6 %). Cellulose recovery in the solid fraction ranged from 84.4 to 91.1 % when ethanol, 1-propanol, 2-propanol, or 1-butanol was used as the solvent and when no solvent was used (control). Cellulose recovery in the solid fraction was 71.9 % when 1-pentanol was used as the solvent. Similar to the trend with dry weight (Fig. 2a), the acid-insoluble lignin content was lower in the solid fraction when 1-butanol or 1-pentanol was used as the solvent compared with other solvents (Fig. 2b). Accordingly, the greatest increase in glucose yield compared with the control was observed with 1-butanol or 1-pentanol as the solvent, but increased glucose yield was also observed with ethanol, 1-propanol, and 2-propanol (Fig. 2c).

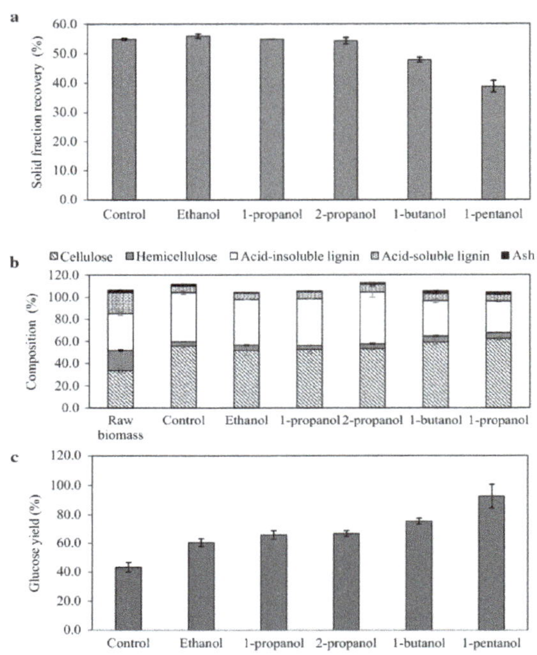

Figure 2: Properties of the solid fraction obtained after organosolv pretreatment. **a** Weight of the solid fraction obtained after pretreatment. **b** Carbohydrate and lignin composition of the solid fraction. **c** Glucose yield from the solid fraction. Organosolv pretreatment of sorghum bagasse was carried out using ethanol, 1-propanol, 2-propanol, 1-butanol, and 1-pentanol as the solvent and no solvent (control).

Lignin removal, as calculated using the equation [(lignin in bagasse)—(lignin in solid fraction)]/(lignin in bagasse), was higher when 1-butanol or 1-pentanol was used as the solvent (64.7–74.3 %), compared with the use of no solvent, ethanol, 1-propanol, or 2-propanol (44.1–49.7 %) (Fig. 2a, b). Lignin, which is degraded by sulfuric acid, is reportedly liberated from the cell wall and forms droplets that attach to cellulose molecules [17]. Hydrophobic alcohols would dissolve these lignin droplets and prevent their attachment to cellulose. The removal of lignin from cellulose reportedly increases the accessibility to hydrolytic enzymes and reduces the degree of irreversible adsorption of the enzyme to lignin [18]. Thus, organosolv pretreatment using 1-butanol or 1-pentanol efficiently removes acid-insoluble lignin, resulting in enhanced enzymatic hydrolysis of cellulose.

Liquid fraction containing hemicellulose-derived sugars

It was expected that hemicellulose-derived sugars, mainly xylose, would be contained in the liquid fraction [2]. The amount of xylose contained in the liquid fraction of samples pretreated with ethanol, 1-propanol, 2-propanol, 1-butanol, 1-pentanol, or with no solvent (control) was similar (Fig. 3a). The xylose recovery in the liquid fraction was higher (90.6–97.4 %) when using ethanol, 1-propanol, 1-pentanol, or no solvent, compared with 2-propanol and 1-butanol (85.5 and 79.2 %, respectively), as calculated using the following equation: (xylose amount in liquid fraction)/[(total xylose)—(xylose amount in solid fraction)]. The reason for the decreased xylose recovery when using 2-propanol or 1-butanol as the solvent is unclear.

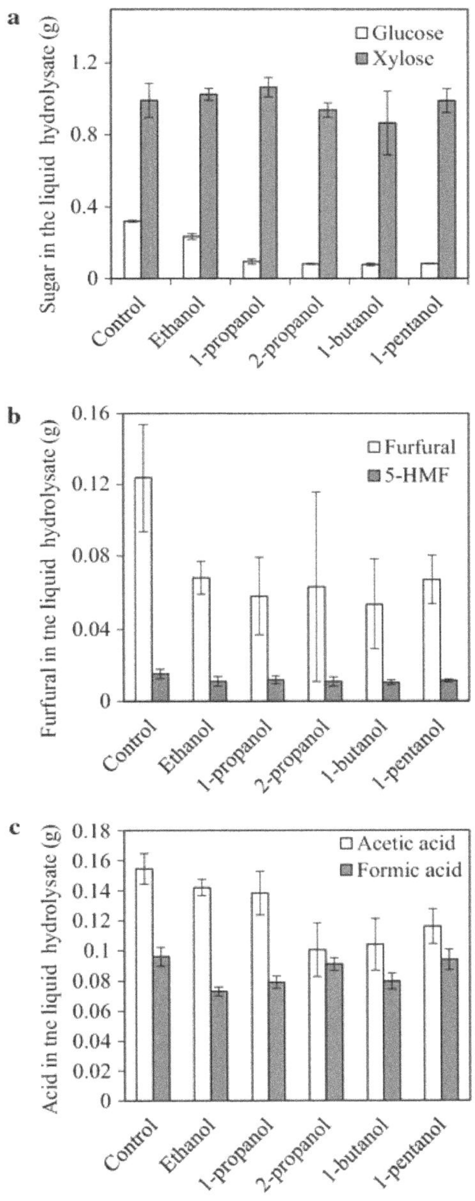

Figure 3: Composition of the liquid fraction. **a** Glucose and xylose. **b** Furfural and 5-HMF. **c** Acetic acid and formic acid. The amount (g) of each component obtained from 6 g of raw biomass is shown.

Compared with the control, the amount of glucose in the liquid fraction was lower in the solvent-pretreated samples (Fig. 3a). Furfural and 5-hydroxymethylfurfural (5-HMF) were contained in the liquid fraction (Fig. 3b). The harsh conditions of pretreatment resulted in the production of similar amounts of acetic and formic acids in the liquid fraction (Fig. 3c). In addition, solvents were blended in the liquid fraction (data not shown). The presence of acetic and formic acids, furfural, 5-HMF, and the pretreatment solvent would inhibit subsequent fermentation. These fermentation inhibitors could be removed, and the xylose and glucose in the liquid fraction could be concentrated, by subsequent application of membrane separation nanofiltration [19, 20].

Lignin contained in the black liquor fraction

A black liquor fraction was obtained when 12.5 % 1-butanol or 1-pentanol was used as the solvent (Fig. 1). Lignin was precipitated from the black liquor fraction by dilution with water [2]. Lignin recovery, as calculated using the equation (lignin in black liquor)/[(lignin in raw biomass)—(lignin in solid fraction)], was 12.5 and 25.6 % using 1-butanol or 1-pentanol as the solvent, respectively, corresponding to 8.1 and 19.1 % lignin content in the raw sorghum bagasse, respectively (Fig. 4). The lignin recovery in the present study was similar to that of a previous study, which reported 5.1–17.2 % lignin recovery from wheat straw pretreated with 50–60 % ethanol at 190–210 °C [5]. However, our lignin recovery was lower than that from rye straw pretreated at 190 °C for 3 h at a 1.3 % acid concentration (35 %) [21]. Therefore, process optimization of lignin recovery with respect to solvent concentration, temperature, time, and acid concentration will be necessary in the future. It is also possible that low lignin recovery was due to the loss of water-soluble products from lignin degradation [21].

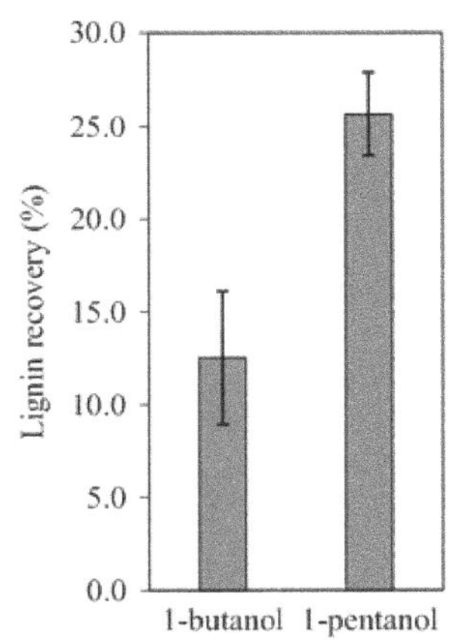

Figure 4: Yield of lignin in black liquor. Lignin recovery was calculated as follows: (lignin in black liquor)/[(lignin in raw biomass)—(lignin in solid fraction)].

Next, the composition and structure of the lignins in the black liquor fraction were analyzed using 2D ^{1}H-^{13}C HSQC NMR [15]. The syringyl, guaiacyl, and *p*-hydroxyphenyl lignin units are important aromatic elements of the plant cell wall [22]. Strong signals related to these lignin components (regions of interest [ROIs] 7–13) were detected in the black liquor, compared with raw sorghum bagasse (Fig. 5). Other strong signals related to aromatic components, such as *p*-coumarate (ROIs 1–6) and ferulate (ROIs 15 and 16), were detected in the black liquor. The only aromatic region signal that was weaker in the black liquor was that related to cinnamyl alcohol (ROI 14). Accordingly, signals related to methoxyl (ROIs 17 and 18) groups were stronger, as methoxyl groups are found in the side chains of syringyl, guaiacyl, and ferulate [23].

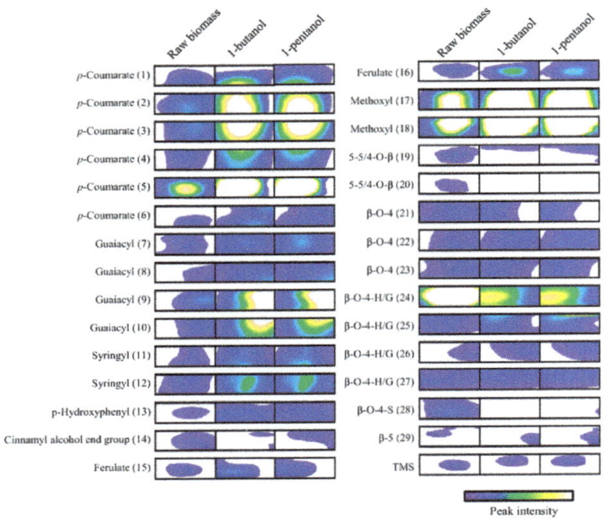

Figure 5: Peaks of raw sorghum bagasse and black liquor. *Left* portion of figure shows NMR peaks of raw biomass. *Middle*and *right* portions of figure show NMR peaks of black liquor obtained after organosolv pretreatment of sorghum bagasse using 1-butanol or 1-pentanol as the solvent, respectively.

In contrast, the signal related to a major interunit structure, β–O–4 (ROI 24), was weaker in the black liquor than in raw sorghum bagasse. This was probably due to partial cleavage of the thermally labile β–O–4 unit in raw sorghum bagasse under heat treatment [24, 25]. Another signal (ROI 25) related to the β–O–4 unit was stronger in the black liquor, but the reason for this difference is unclear. Signals related to other minor interunit structures in raw sorghum bagasse were either stronger or weaker in the black liquor. These results suggest that most of the lignin aromatic components were concentrated in the black liquor in samples pretreated using 1-butanol or 1-pentanol as the solvent; however, the major linkage structure was lost.

Which lignin constituent is most affected by 1-butanol or 1-pentanol was also investigated by comparing 2D NMR signals of solid fractions obtained by pretreatment with no solvent (control), 1-butanol, or 1-pentanol (Fig. 6). Compared with the solid fraction obtained using no solvent (control), signals related to *p*-coumarate (ROIs 2–6) were weaker in solid fractions obtained using 1-butanol or 1-pentanol as the solvent. In addition, the signals related to guaiacyl (ROI 10), syringyl (ROIs 11 and 12), and minor β–O–4 interunits (ROIs 21–23, 26, and 27) in solid fractions obtained using 1-butanol or 1-pentanol as the solvent were weaker compared with the control. Accordingly, because the guaiacyl, syringyl, and β–O–4 units contain a methoxyl side chain, methoxyl-related signals (ROIs 17 and 18) were weaker in solid fractions obtained using 1-butanol or 1-pentanol [23]. The removal of *p*-coumarate, syringyl, and guaiacyl from the solid fraction using 1-butanol or 1-pentanol corresponded with the results of increased aromatic lignin regions in the black liquor fraction and decreased acid-insoluble content in the solid fraction.

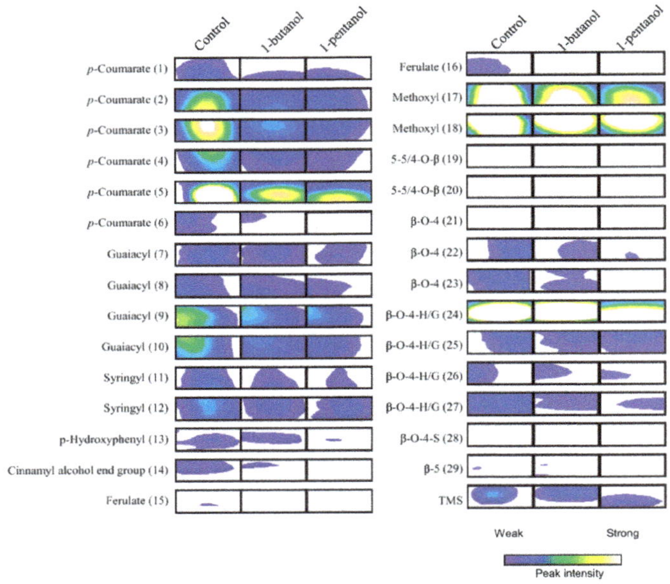

Figure 6: Peak of solid fractions. *Left*, *middle*, and *right* portions of figure show NMR peaks of the solid fractions obtained by organosolv pretreatment of sorghum bagasse using no solvent (control) and 1-butanol and 1-pentanol as the solvent, respectively.

Ethanol fermentation using the solid fraction

The solid fraction obtained after organosolv pretreatment can be utilized as a sugar source for microbial fermentation. The solid fractions obtained after organosolv pretreatment using 1-butanol or 1-pentanol as the solvent were subjected to simultaneous saccharification and fermentation by *S. cerevisiae*. Ethanol production using each of these solid fractions was compared with that obtained with control solid fraction (Fig. 7). As expected, the ethanol production rates obtained with solid fractions of samples pretreated with 1-butanol or 1-pentanol (2.6 and 2.8 g/L/h, respectively) were about 1.6 times higher in the first 9 h than those obtained with the solid fraction of samples pretreated with no solvent (1.6 g/L/h). After 96 h of fermentation, ethanol production from the 1-butanol and 1-pentanol solid fractions was about twice that of the control, reaching 43.1 and 47.2 g/L, respectively. In addition, the theoretical ethanol yield from the 1-butanol and 1-pentanol solid fractions was 59.6 and 62.1 %, respectively, compared with 32.4 % for the control. As suggested previously [26], organosolv delignification and the resultant supply of readily hydrolyzable cellulose substrates results in an increase in the concentration of microbial fermentation products.

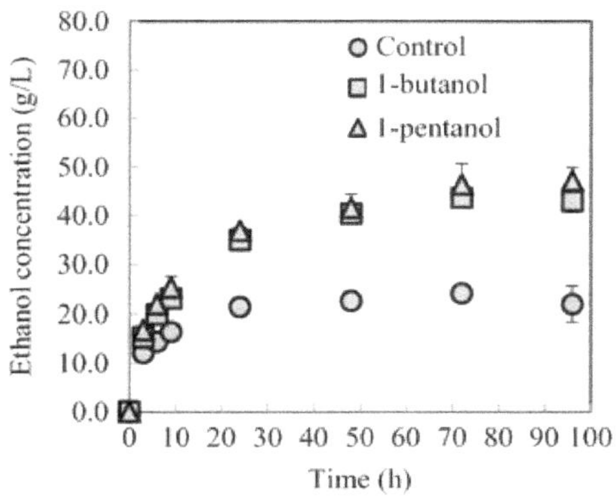

Figure 7: Simultaneous saccharification and fermentation of solid fractions by *Saccharomyces cerevisiae*. Solid fractions were obtained by organosolv pretreatment of sorghum bagasse using 1-butanol (*closed square*) and 1-pentanol (*closed triangle*) as the solvent and no solvent (*closed circle*). Ethanol fermentation was initiated from 200 g/L of solid fraction.

CONCLUSIONS

Sorghum bagasse was fractionated by organosolv pretreatment using various solvents at a low concentration of 12.5 %. The catalyst for the reaction was 1 % sulfuric acid, and the reaction was carried out at 180 °C for 45 min. Pretreatment with the highly hydrophobic solvents 1-butanol and 1-pentanol produced three visible fractions (i.e., a cellulose-enriched solid fraction, a liquid fraction containing hemicellulose-derived xylose, and a black liquor fraction containing lignin). However, only two fractions were produced upon pretreatment with ethanol, 1-propanol, or 2-propanol (i.e., a solid fraction containing cellulose and a liquid fraction containing xylose), and recovery of lignin was difficult. 2D NMR spectroscopy revealed that the black liquor fraction contained lignin aromatic components, such as guaiacyl, syringyl, *p*-hydroxyphenyl, *p*-coumarate, and ferulate. The aliphatic β–O–4 unit of lignin was absent in the black liquor fraction. Pretreatment with 1-butanol or 1-pentanol effectively removed *p*-coumarate, syringyl, and some of the guaiacyl from raw sorghum bagasse. The 1-butanol or 1-pentanol should be recovered after pretreatment and recycled using separation technologies such as pervaporation [27, 28]. We demonstrated here the effective organosolv pretreatment of sorghum bagasse using a low concentration of hydrophobic solvent (1-butanol or 1-pentanol).

METHODS

Plant materials

The hybrid sorghum cultivar Tentaka [29] was grown in 2013 at an experimental field in Okinawa, Japan. Whole plants were harvested at the heading stage and then fully dried in a greenhouse. After removal of the panicles, the culms were ground into a fine powder using a blender (WB-1; TGK, Hachioji, Japan) fitted with a 2-mm screen.

Organosolv pretreatment

Organosolv pretreatment was performed using a laboratory-scale thermostirrer (HHE-19G-U; Koike Precision Instruments, Kanagawa, Japan) having a total volume of 100 mL. Sorghum bagasse powder (6 g) was suspended in 80 mL of aqueous solution. The aqueous solution included 70 mL of 1 % sulfuric acid and 10 mL of solvent (ethanol, 1-propanol, 2-propanol, 1-butanol, or 1-pentanol). An aqueous solution consisting of 80 mL of 1 % sulfuric acid was used as the control. Next, the mixture was treated at 180 °C for 45 min, with agitation at 200 rpm. The optimal temperature and agitation speed for pretreatment were determined according to a previous report [30]. After pretreatment, the mixture was centrifuged at 3500 rpm for 10 min for fractionation. Solid and liquid fractions were obtained under the control condition and with ethanol, 1-propanol, or 2-propanol (Fig. 1). When using 1-butanol or 1-pentanol as the solvent, the supernatant including the liquid and black liquor fractions was recovered and again centrifuged at 3500 rpm for 10 min in order to separate the fractions. The upper phase contained the black liquor fraction, and the lower phase contained the liquid fraction. The solid fraction was washed with deionized water, neutralized to pH 7.0, then dried; the dry weight was measured using an electronic balance (XS105DU; Mettler Toledo, Greifensee, Switzerland). Lignin was separated from the black liquor fraction by precipitation upon dilution with 80 mL of water. The compositions of cellulose, hemicellulose, acid-insoluble lignin, acid-soluble lignin, and ash were determined according to the National Renewable Energy Laboratory method [31].

Sugar analysis

Sugar analysis was performed as described previously [30]. The liquid fraction was neutralized to pH 5.0 by the addition of calcium hydroxide. An aliquot of sample (1.5 μL) was mixed with 1.5 μL of 0.1 % (w/w) Ribitol as internal standard, and then the mixture was dried in a vacuum concentrator (7810010; Labconco, Kansas City, MO, USA). The dried residue was dissolved in 100 μL of 20 mg/mL of methoxyamine hydrochloride in pyridine and incubated at 30 °C for 90 min, after which 50 μL of *N*-methyl-*N*-trimethylsilyltrifluoroacetamide was added and the sample was incubated at 37 °C for 30 min. A 10 μL aliquot of the solution was subjected to gas chromatography-mass spectrometry (GC–MS) (GCMS-2010plus; Shimadzu, Kyoto, Japan) under the following conditions: column, Agilent CP-Sil 8CB-MS (30 m × 0.25 mm); carrier gas, helium; injection temperature, 230 °C; oven temperature, 80 °C at $t = 0$–2 min, then increased to 330 °C at 15 °C/min.

Byproduct analysis

Byproduct analysis was performed as described previously [30]. Acetone (900 μL) was added to 100 μL of liquid fraction and mixed thoroughly. The sample was then centrifuged at 21,880×*g* and room temperature for 10 min. The supernatant (10 μL) was subjected to GC–MS analysis (GC-MS-2010plus; Shimadzu). 5-HMF and furfural were analyzed under the following conditions: column, Agilent CP-Sil 8CB-MS (30 m × 0.25 mm); carrier gas, helium; injection temperature, 250 °C; oven temperature, 50 °C at $t = 0$–5 min, then increased to 280 °C at 20 °C/min. Acetic acid and formic acid were analyzed under the following conditions: column, Agilent DB-FFAP (60 m × 0.25 mm); carrier gas, helium; injection temperature, 250 °C; oven temperature, 100 °C at $t = 0$ to 5 min, then increased to 230 °C at 10 °C/min.

Enzymatic saccharification

Enzymatic saccharification of the solid fraction (10 % dry weight) was performed by adding 0.3 M citrate buffer (pH 4.8) and cellulase (Celic CTec2, Novozyme, Bagsvaerd, Denmark) at a concentration of 6.6 filter paper units (FPU)/g-dry biomass. Tetracycline (40 μg/mL) and cycloheximide (30 μg/mL) were added to prevent microbial growth. The reaction mixture was incubated at 50 °C in a chemi station (PPS-2000, Tokyo Rikakikai, Tokyo, Japan) with agitation at 120 rpm for 72 h. Enzymatic saccharification was stopped by rapid chilling on ice, followed by centrifugation at 21,880×*g* for 10 min at 4 °C. The sugars in the supernatant were analyzed as described above.

Composition analysis

Compositional analysis was performed according to the National Renewable Energy Laboratory (NREL) method [31]. Obtained sugar was analyzed as described above. The amount of acid-soluble lignin was determined by measuring the optical density (OD) at 240 nm.

Solution 2D NMR spectroscopy

Sample preparation for solubilized lignocelluloses was similar to previously reported methods [32, 33]. Briefly, the dried sample was further ground using a Pulverisette 5 ball mill (Fritsch GmbH, Idar-Oberstein, Germany). The ball-milled powder (30 mg) was mixed with 600 μL of dimethyl sulfoxide (DMSO)-d_6:pyridine-d_5 (4:1), heated at 50 °C for 30 min in a Thermomixer Comfort (Eppendorf AG, Hamburg, Germany), then centrifuged at 20,380×*g* for 5 min. The supernatant was transferred to 5-mm □ NMR tubes and subjected to NMR analysis. NMR spectra were recorded on an Avance III HD-600 (Bruker, Billerica, MA, USA) equipped with a 5-mm TXI-cryoprobe operated at 600 MHz for ^{1}H- and 125 MHz for ^{13}C-NMR. The temperature of all NMR samples was maintained at 319 K. The chemical shifts were referenced to the methyl group of DMSO-d_6 at ^{13}C = 40.03 ppm and ^{1}H = 2.582 ppm.

Two-dimensional ^{1}H-^{13}C heteronuclear single-quantum spectra were collected using echo/antiecho gradient selection (the hsqcetgp pulse program in the Bruker library). Forty-eight regions of interest (ROIs) were compared with previously assigned chemical shifts [23, 34].

Liquefaction and ethanol fermentation

Liquefaction of the solid fraction was performed in a 50 mL polypropylene tube (Corning Inc., NY, USA) set in a Thermo Block Rotator SN-06BN heat block (Nissin, Tokyo, Japan), as described previously [35]. The tube was closed with a silicone plug (As One, Osaka, Japan), into which a hole was made using a disposable needle (φ = 0.6 mm) (Termo Corp., Tokyo, Japan). In the tube, the solid fraction (final weight 200 g/L) was mixed with medium containing 50 mM citric acid buffer (pH 5.0) and 6.6 FPU/g-biomass of commercial cellulase (Cellic CTec2; Novozymes Inc., Bagsvaerd, Denmark) and axially rotated at 35 rpm at 50 °C for 2 h.

Saccharomyces cerevisiae strain BY4741 (genotype: *MAT*a *his3Δ1 leu2Δ0 met15Δ0 ura3Δ0*) [36] was purchased from the American Type Culture Collection (ATCC) as ATCC No. 404002. The cells were aerobically propagated for 24 h at 30 °C and 150 rpm in 5 mL of YPD medium [10 g/L yeast extract [Becton, Dickinson and Company, Tokyo, Japan], 20/g L polypeptone (Wako Pure Chemical Industries, Ltd., Osaka, Japan), and 20 g/L glucose], and then cultivated for 24 h in 500 mL of YPD medium. The cells were collected by centrifugation at 3000×*g*for 10 min at 4 °C and washed twice with distilled water. Ethanol fermentation was initiated by the addition of yeast extract (final concentration 10 g/L), peptone (final concentration 20 g/L), and *S. cerevisiae* cells (50 g-wet cells/L, corresponding to 10 g-dry cells/L) after 2 h of liquefaction of the solid fraction. Ethanol fermentation was conducted by axial agitation at 35 rpm and 35 °C for 48 h. Theoretical ethanol yield (%) was calculated as the percent of 0.511 g-ethanol/g-sugar of consumed glucose as follows:

$$\text{Ethanol yield} = \frac{\text{Ethanol produced (g)}}{\text{Glucan in solid fraction (g)}}$$

DECLARATIONS

Authors' contributions

HT, KS, and AK designed this study and drafted the manuscript. HT, TO, and HK-a performed the biomass pretreatment and compositional analysis. MO, TS-h, FM, and JK carried out the NMR analysis and interpreted the NMR results. HK-a and CO analyzed sugars and byproducts. KH, TS-a, and HK-i prepared the sorghum bagasse and helped to revise the manuscript. TO and FM performed enzymatic saccharification. KS and AK performed ethanol fermentation. MO, TS-h, and JK revised the manuscript. All authors read and approved the final manuscript.

Acknowledgements

This work was supported in part by Grants-in-Aid from the NC-CARP project of the Ministry of Education, Culture, Sports, Science and Technology, Japan, and by the Special Coordination Fund for the Promotion of Science and Technology, Creation of Innovation Centers for Advanced Interdisciplinary Research Areas (Innovative Bioproduction Kobe), MEXT, Japan.

REFERENCES

1. Tanaka T, Kondo A. Cell surface engineering of industrial microorganisms for biorefining applications. Biotechnol Adv. 2015;33:1403–11.
2. Zhao X, Cheng K, Liu D. Organosolv pretreatment of lignocellulosic biomass for enzymatic hydrolysis. Appl Microbiol Biotechnol. 2009;82:815–27.
3. Rooney WL, Blumenthal J, Bean B, Mullet JE. Designing sorghum as a dedicated bioenergy feedstock. Biofuels Bioprod Bioref. 2007;1:147–57.
4. Kim M, Day DF. Composition of sugar cane, energy cane, and sweet sorghum suitable for ethanol production at Louisiana sugar mills. J Ind Microbiol Biotechnol. 2011;38:803–7.
5. Wildschut J, Smit AT, Reith JH, Huijgen WJ. Ethanol-based organosolv fractionation of wheat straw for

the production of lignin and enzymatically digestible cellulose. Bioresour Technol. 2013;135:58–66.

6. Zakzeski J, Bruijnincx PCA, Jongerius AL, Weckhuysen BM. The catalytic valorization of lignin for the production of renewable chemicals. Chem Rev. 2010;110:3552–99.

7. Huber GW, Iborra S, Corma A. Synthesis of transportation fuels from biomass: chemistry, catalysts, and engineering. Chem Rev. 2006;106:4044–98.

8. Lan W, Liu CF, Sun RC. Fractionation of bagasse into cellulose, hemicellulose, and lignin with ionic liquid treatment followed by alkaline extraction. J Agric Food Chem. 2011;59:8691–701.

9. Mosier N, Wyman C, Dale B, Elander R, Lee YY, Holtzapple M, Ladisch M. Feature of promising technologies for pretreatment of lignocellulosic biomass. Bioresour Technol. 2005;96:673–86.

10. Sun Y, Cheng J. Hydrolysis of lignocellulosic materials for ethanol production: a re. Bioresour Technol. 2002;83:1–11.

11. Huijgen WJJ, Smit AT, de Wild PJ, den Uil H. Fractionation of wheat straw by prehydrolysis, organosolv delignification and enzymatic hydrolysis for production of sugars and lignin. Bioresour Technol. 2012;114:389–98.

12. Zhang K, Pei Z, Wang D. Organic solvent pretreatment of lignocellulosic biomass for biofuels and biochemicals: a re. Bioresour Technol. 2016;199:21–33.

13. Taherzaden M, Karimi K. Pretreatment of lignocellulosic wastes to improve ethanol and biogas production: a re. Int J Mol Sci. 2008;9:1621–51.

14. Tsuji Y, Vanholme R, Tobimatsu Y, Ishikawa Y, Foster CE, Kamimura N, et al. Introduction of chemically labile substructures into *Arabidopsis* lignin through the use of LigD, the Cα-dehydrogenase from *Sphingobium* sp. strain SYK-6. Plant Biotechnol J. 2015;13:821–32.

15. Mansfield SD, Kim H, Lu F, Ralph J. Whole plant cell wall characterization using solution-state 2D NMR. Nat Protoc. 2012;7:1579–89.

16. Teramura H, Sasaki K, Oshima T, Aikawa S, Matsuda F, Okamoto M, et al. Changes in lignin and polysaccharide components in 13 cultivars of rice straw following dilute acid pretreatment as studied by solution-state 2D ^{1}H-^{13}C NMR. PLoS One. 2015;10:e0128417.

17. Selig MJ, Viamajala S, Decker SR, Tucker MP, Himmel ME, Vinzant TB. Deposition of lignin droplets produced during dilute acid pretreatment of maize stems retards enzymatic hydrolysis of cellulose. Biotechnol Prog. 2007;23:1333–9.

18. Alvira P, Tomás-Pejó E, Ballesteros M, Negro MJ. Pretreatment technologies for an efficient bioethanol production process based on enzymatic hydrolysis: a re. Bioresour Technol. 2010;101:4851–61.

19. Weng YH, Wei HJ, Tsai TY, Lin TH, Wei TY, Guo GL, et al. Separation of furans and carboxylic acids from sugars in dilute acid rice straw hydrolysates by nanofiltration. Bioresour Technol. 2010;101:4889–94.

20. He Y, Bagley DM, Leung KT, Liss SN, Liao BQ. Recent advances in membrane technologies for biorefining and bioenergy production. Biotechnol Adv. 2012;30:817–58.

21. Perez-Cantu L, Schreiber A, Schütt F, Saake B, Kirsch C, Smirnova I. Comparison of pretreatment methods for rye straw in the second generation biorefinery: effect on cellulase, hemicellulase and lignin recovery. Bioresour Technol. 2013;142:428–35.

22. Chylla RA, Van Acker R, Kim H, Azapira A, Mukerjee P, Markley JL, et al. Plant cell wall profiling by fast maximum likelihood reconstruction (FMLR) and region-of-interest (ROI) segmentation of solution-state 2D ^{1}H-^{13}C NMR spectra. Biotechnol Biofuels. 2013;6:45.

23. Kim H, Ralph J. Solution-state 2D NMR of ball-milled plant cell wall gels in DMSO-d6/pyridine-d5. Org Biomol Chem. 2010;8:576–91.

24. Kim JY, Hwang H, Oh S, Kim YS, Kim UJ, Choi JW. Investigation of structural modification and thermal characteristics of lignin after heat treatment. Int J Biol Macromol. 2014;66:57–65.

25. Trajano HL, Engle NL, Foston M, Ragauskas AJ, Tschaplinski TJ, Wyman CE. The fate of lignin during hydrothermal pretreatment. Biotechnol Biofuels. 2013;6:110.

26. Pan X, Gilkes N, Kadla J, Pye K, Saka S, Gregg D, et al. Bioconversion of hybrid poplar to ethanol and co-products using an organosolv fractionation process: optimization of process yield. Biotechnol Bioeng. 2006;94:851–61.

27. Xue C, Du GQ, Chen LJ, Ren JG, Sun JX, Bai FW, et al. A carbon nanotube filled polydimethylsiloxane hybrid membrane for enhanced butanol recovery. Sci Rep. 2014;4:5925.

28. Negishi H, Sakaki K, Ikegami T. Silicalite pervaporation membrane exhibiting a separation factor of over 400 for butanol. Chem Lett. 2010;39:1312–4.

29. Venuto B, Kindiger B. Forage and biomass feedstock production from hybrid forage sorghum and sorghum-sudangrass hybrids. Jpn Soc Grassl Sci. 2008;54:189–96.

30. Matsuda F, Yamasaki M, Hasunuma T, Ogino C, Kondo A. Variation in biomass properties among rice diverse cultivars. Biosci Biotechnol Biochem. 2011;75:1603–5.

31. Sluiter A, Hames B, Ruiz R, Scarlata C, Sluiter J, Templeton D, et al. Determination of structural carbohydrates and lignin in biomass. Golden: Natl Renew Energy Lab; 2011.

32. Meng X, Foston M, Leisen J, DeMartini J, Wyman CE, Ragauskas AJ. Determination of porosity of lignocellulosic biomass before and after pretreatment by using Simons' stain and NMR techniques. Bioresour Technol. 2013;144:467–76.

33. Komatsu T, Kikuchi J. Comprehensive signal assignment of 13C-labeled lignocellulose using multidimensional solution NMR and 13C chemical shift comparison with solid-state NMR. Anal Chem. 2013;85:8857–65.

34. Okushita K, Komatsu T, Chikayama E, Kikuchi J. Statistical approach for solid-state NMR spectra of cellulose derived from a series of variable parameters. Polym J. 2012;44:895–900.

35. Matano Y, Hasunuma T, Kondo A. Display of cellulases

on the cell surface of *Saccharomyces cerevisiae* for high yield ethanol production from high-solid lignocellulosic biomass. Bioresour Technol. 2012;108:128–33.

36. Brachmann CB, Davies A, Cost GJ, Caputo E, Li J, Hieter P, et al. Designer deletion strains derived from *Saccharomyces cerevisiae* S288C: a useful set of strains and plasmids for PCR-mediated gene disruption and other applications. Yeast. 1998;14:115–32.

CHAPTER 18

ETHANOL PRODUCTION FROM SWEET SORGHUM JUICE AT HIGH TEMPERATURES USING A NEWLY ISOLATED THERMOTOLERANT YEAST SACCHAROMYCES CEREVISIAE DBKKU Y-53

Sunan Nuanpeng, Sudarat Thanonkeo, Mamoru Yamada and Pornthap Thanonkeo

Keywords: High temperature fermentation, real-time RT-PCR, Saccharomyces cerevisiae, sweet sorghum, thermotolerant yeast, gene expression

INTRODUCTION

Bioethanol is a clean, renewable, environmental friendly source of fuel energy that can be produced from different feedstocks and conversion technologies. It is one of the most promising substitutes for fossil energy and has high potential to replace petroleum-based fossil fuels [1,2,3]. Bioethanol can be used directly or blended with gasoline (known as "gasohol") to power engines without modification. Development of bioethanol is not only addressing climate change due to the burning of petroleum-based fuels but is also of great significance in protecting national energy security and promoting rural economic growth [3,4].

Approximately 60% of global bioethanol is produced from sugar crops, the remaining 40% is produced from starchy grains [5]. Sweet sorghum (*Sorghum bicolor* L. Moench) is one of the most promising sugar crops for industrial bioethanol production. It is a C4 plant, is similar to sugarcane, and belongs to the grass family. Due to a high photosynthetic efficiency, it is known for high carbon assimilation and the ability to store high levels of extractable sugars in its stalks [6]. Sweet sorghum is one of the most drought-resistant agricultural crops and can be cultivated in nearly all temperate and tropical climate areas in both irrigated and non-irrigated lands [7,8]. Unlike sugarcane, sweet sorghum has a short growing period (3–4 months). Therefore, it can be planted two or three times a year. Its stalks contain both soluble carbohydrates (such as sucrose, glucose, fructose) and insoluble carbohydrates (such as cellulose and hemicellulose) that can be converted into fuel ethanol using a biological fermentation process [9,10]. With respect to the production cost, sweet sorghum has lower production cost than that of sugarcane and sugar beets because it requires less fertilizer. Therefore, considering the potential of sweet sorghum for industry, it is an ideal crop for commercial ethanol production [11]. Although sweet sorghum is considered as an important food resource in some countries, such as India, China, which uses sweet sorghum juice (SSJ) to produce syrup, the Thai government promotes sweet sorghum to be used as an energy crop for large-scale ethanol production together with sugarcane and cassava.

Fermentation at high temperature is a key requirement for effective bioethanol production in tropical countries where average daytime temperatures are usually high throughout the year. The advantages of fermentation at high temperatures are not only an increased the rate of fermentation but also a decreased risk of contamination by mesophilic microorganisms, such as *Williopsis* sp., *Candida* sp., *Zygosaccharomyces* sp., a reduced cost of the cooling system, and the possible use of simultaneous saccharification and fermentation (SSF) when coupled with a continuous stripping system for ethanol recovery. Utilization of a high potential thermotolerant yeast strain is

Citation: Sunan Nuanpeng, Sudarat Thanonkeo, Mamoru Yamada and Pornthap Thanonkeo, Ethanol Production from Sweet Sorghum Juice at High Temperatures Using a Newly Isolated Thermotolerant Yeast Saccharomyces cerevisiae DBKKU Y-53, doi:10.3390/en9040253.

a key to success in ethanol production at high temperatures [12,13]. There are several reports in the literature on the ethanol production at high temperatures using the thermotolerant yeast *Kluyveromyces marxianus* [13,14,15,16,17,18]; however, very few reports have considered the thermotolerant yeast *Saccharomyces cerevisiae* [12,19].

Under stressful conditions, such as heat, ethanol, or osmotic stress, several stress-responsive genes including those encoding for the heat shock proteins (HSPs), enzymes involved in protein degradation, such as ubiquitin ligase, and proteins involved in trehalose and glycogen metabolism in yeast have been reported to be stimulated [20,21]. HSPs play a key role as molecular chaperones by either stabilizing new proteins to ensure correct folding or refolding of proteins to the proper conformation, or degrading misfolded proteins that are damaged by stress conditions. HSPs also help transport proteins across membranes within the cell [22,23]. Trehalose, which is one of the compatible solutes synthesized during adverse environmental conditions, has been reported to protect the cell by replacing water at the surfaces of macromolecules, which holds proteins and membranes in their native conformation [24,25]. Glycogen, which is a reserve carbohydrate in *S. cerevisiae*, has also been reported to be involved with tolerance towards several stresses [26,27]. Although a number of genes responsible for the prevention of protein denaturation in yeast cells have been reported, the molecular mechanism conferring thermotolerance during ethanol fermentation at high temperatures is not fully understood.

In this study, the isolation and screening of highly efficient thermotolerant yeast strains capable of producing high levels of ethanol at high temperatures from SSJ were carried out. The optimum condition for ethanol production for the selected thermotolerant yeast strain was also investigated. Furthermore, to gain a better understanding of the molecular mechanism by which yeast cells adapt to adverse environmental conditions and acquire thermotolerance during high temperature ethanol fermentation, the expression of genes encoding the HSP26, HSP70, HSP90, HSP104, pyruvate kinase, trehalose-6-phosphate synthase, neutral trehalase, glycogen synthase, and ubiquitin ligase was evaluated. This work is the first to demonstrate the physiological changes related to the expression of genes involved in heat-shock response, ATP production, trehalose and glycogen metabolism, and protein degradation in the newly isolated thermotolerant yeast *S. cerevisiae* DBKKU Y-53 during ethanol fermentation at high temperatures.

EXPERIMENTAL SECTION

Isolation of Thermotolerant Yeast Strains

Samples including sugarcane juice, SSJ, rotten fruits, soils from sugarcane and sweet sorghum plantations collected from the Khon Kaen, Udon Thani, Nakhon Ratchasima, Maha Sarakham, Kalasin, Chaiyaphum, and Roi Et provinces of Thailand were used for the isolation of thermotolerant yeasts using the enrichment method described by Limtong *et al.* [13]. YM broth (0.3% yeast extract, 0.3% malt extract, 0.5% peptone, and 1% glucose) supplemented with 4% (*v*/*v*) ethanol was used as a selective medium in this study. After incubation at 35 °C for 3 days on a rotary incubator shaker at 100 rpm, the enriched cultures were then spread on YM agar supplemented with 4% (*v*/*v*) ethanol and subsequently incubated at 35 °C. Pure cultures were collected and maintained on YM agar at 4 °C for short-term storage and in 50% (*v*/*v*) glycerol at −20 °C for long-term storage.

Screening of Thermotolerant Yeast for Ethanol Fermentation at High Temperatures

SSJ containing 100 $g{\cdot}L^{-1}$ total sugars was used as substrate for screening thermotolerant ethanol-producing yeast strains. The ethanol production capability was tested by culturing the isolated thermotolerant yeast strains in 16 × 160 mm test tube containing 10 mL SSJ, which were then incubated at high temperatures (37 °C to 50 °C) on a rotary incubator shaker at 150 rpm. The Durham tube was placed inside the test tube, and strains that produced high levels of CO_2 in the Durham tube were selected for further study.

Identification of The Selected Thermotolerant Yeast Strains

Identification of the yeast strains was carried out using morphological and the D1/D2 domain of the 26S rDNA gene sequencing analysis [28]. Genomic DNA was isolated from the yeast cells using the method described by Harju *et al.* [29]. Amplification of the D1/D2 domain of the 26S rDNA gene was carried out using the specific primers NL-1 (5′-GCA TAT CAA TAA GCG GAG GAA AAG) and NL-4 (5′-GGT CCG TGT TTC AAG ACG G) [30] with the genomic DNA isolated from yeast cells as the template. After the PCR reaction, the amplified product was separated on a 1.0% agarose gel and purified using Invisorb® Fragment CleanUp Kit (Invitek GmbH, Berlin, Germany). All procedures for DNA amplification and purification were carried out according to the manufacturer's instructions. DNA sequencing was performed in the First BASE Laboratories Sdn Bhd (Seri Kembangan, Selangor Darul Ehsan, Malaysia). The sequences of the D1/D2 domain of the 26S rDNA gene were analyzed using GENETYX (Software Development, Tokyo, Japan), whereas homology searching was performed using the FASTA and BLAST programs in the GenBank and DDBJ databases. Phylogenetic analysis was performed using MEGA5 [31], and the tree topologies were analyzed using bootstrap analysis based on the neighbor-joining method [32].

Comparative Study of Ethanol Production by the Selected Thermotolerant Yeast Strains

The ethanol fermentation efficiencies of the selected thermotolerant yeast strains were compared using SSJ with an initial sugar concentration of 220 $g{\cdot}L^{-1}$ as substrate. Batch ethanol fermentation was performed in a 500-mL air-locked flask with a final working volume of 400 mL, and the initial cell concentration was 10^6 $cells{\cdot}mL^{-1}$. Fermentation was carried out at 30 °C, 37 °C, 40 °C, 42 °C and 45 °C under static conditions. During ethanol fermentation, samples were withdrawn at certain time intervals and analyzed. All experiments were performed in duplicated and repeated twice, and the average values ± SD are presented.

Factors Influencing Ethanol Production by the Selected Thermotolerant Yeast

The SSJ was sterilized at 110 °C for 28 min and used as raw material for ethanol production using the selected thermotolerant yeast. In this study, *S. cerevisiae* SC90, which is commonly used in industrial ethanol production in Thailand, was used as a reference strain. The batch ethanol fermentation was carried out in 500-mL Erlenmeyer flasks equipped with an air-lock with a final working volume of 400 mL. The effects of pH (4.0, 4.5, 5.0, 5.5, 6.0), initial cell concentration (1×10^6, 1×10^7, 1×10^8 cells·mL^{-1}), initial sugar concentration (200, 250, 300 g·L^{-1}), and nitrogen source (yeast extract, urea, $(NH_4)_2SO_4$) at various concentrations on ethanol production at high temperatures (37 °C and 40 °C) by the selected thermotolerant yeast were investigated. During ethanol fermentation, samples were withdrawn at certain time intervals and analyzed. All experiments were performed in duplicated and were repeated twice. The average values ± SD are presented.

Ethanol Production in a 2-L Bioreactor

The ethanol production potential of the selected thermotolerant yeast strain was evaluated in a 2-L bioreactor (Biostat®B, B. Braun Biotech, Melsungen, Germany) with a working volume of 1.5 L. The fermentation conditions at the bioreactor scale were selected from the results obtained from the flask-scale trials. The fermentation was carried out at high temperatures (37 °C and 40 °C) with an agitation speed of 100 rpm, and samples were periodically collected and analyzed during ethanol fermentation.

Real-Time RT-PCR Analysis of Gene Expression during Ethanol Fermentation

The expressions of genes encoding HSP26 (*hsp26*), HSP70 (*hsp70*), HSP90 (*hsp90*), HSP104 (*hsp104*), pyruvate kinase (*cdc*), trehalose-6-phosphate synthase (*tps*), neutral trehalase (*nth*), glycogen synthase (*gsy*), ubiquitin ligase (*rsp*) during ethanol fermentation were evaluated using real-time RT-PCR. Cells were grown in SSJ containing 220 g·L^{-1} total sugars at 40 °C until growth reached the exponential phase, and then the cells were harvested by centrifugation at 4 °C and 5000 rpm for 5 min. Total RNA was extracted from the yeast cells using Trizol reagent. The concentration of RNA was measured and adjusted by Nanodrop (Nanodrop Technologies, Wilmington, DE, USA). The real-time RT-PCR amplifications were performed using the Biorad-I-Cycler with the qPCRBIO SyGreen One-Step Lo-ROX (PCRBIOSYSTEMS, London, UK). The reactions (final volume 20 μL) were composed of 1 μL RNA template (100 ng RNA), 0.8 μL of each forward and reverse primer, 1 μL 20X RTase, 10 μL 2X qPCRBIO Sygreen One-Step mix, and 6.4 μL RNase-free water. The thermal cycling conditions for gene amplification were initial denaturation at 95 °C for 2 min, followed by 40 cycles each of denaturation at 95 °C for 5 s and annealing at 55 °C for 30 s. The primers used in this study are listed in Table 1. The actin gene (*act*) was used as an internal control. As a negative control, RNase-free water was used instead of the RNA template. All experiments were independently carried out in triplicate, and data from real-time RT-PCR analysis were determined using the CFX Manager Software (Bio-Rad, Hercules, CA, USA). Calculation of relative gene expression was performed using the comparative critical threshold ($\Delta\Delta C_T$) method in which the amount of the target genes was adjusted to the reference gene (*act* gene) [33].

Table 1: Primer used in this study

Gene Name	Primer Name	Sequence (5′→3′)
hsp26	HSP26-F	AAGGCGGCTTAAGAGGCTAC
	HSP26-R	TGTTGTCTGCATCCACACCT
hsp70	SSA4-F	CGGTTCCAGCCTATTTCAAC
	SSA4-R	TGTCTGAGCAGACGAAGACAG
hsp90	HSP82-F	TTCAAACGACTGGGAAGACC
	HSP82-R	AGCAGCCCTGTTTTGGGTAT
hsp104	HSP104-F	CATATGGAACGTGACTTATCATCTGA
	HSP104-R	ACGGCATTGGAAACAGCTTT
cdc	CDC19-F	TGCTTTGAGAAAGGCTGGTT
	CDC19-R	AAAGCTGGCAAATCGACATC
tps	TPS1-F	TGTCTTCCGTGCAAAGAGTG
	TPS1-R	CTTGTGCATGAAATGGATGG
nth	NTH1-F	CCGTACGAGGACTATGAGTGTTT
	NTH1-R	GCAATTTCGCCTACGTTGTT
gsy	GSY1-F	ACGACTGTGTCGCAAATCACT
	GSY1-R	TGCGGTGACCTCATTAACAG
rsp	RSP5-F	CCTTCTGGCCATACTGCATC
	RSP5-R	CCACCTCCCACTTGAACTGT

Analytical Methods

The yeast cell numbers and total soluble solids of the fermentation broth were determined using a direct counting method using hemacytometer and a hand-held refractometer, respectively [34]. The fermentation broth was centrifuged at 13,000 rpm for 10 min. The supernatant was then determined for total residual sugars using the phenol sulfuric acid method [35]. The ethanol concentration (P, g·L^{-1}) was analyzed using gas chromatography (Shimadzu GC-14B, Kyoto, Japan) using a polyethylene glycol (PEG-20M) packed column with a flame ionization detector. N_2 was used as a carrier gas, and 2-propanol was used as an internal standard. The ethanol yield (Yp/s) was calculated as the actual ethanol produced and was expressed as g ethanol per g glucose utilized (g·g^{-1}). The volumetric ethanol productivity (Qp, g·L^{-1}·h^{-1}) was calculated using the following equations: $Qp = P/t$, where P is the ethanol concentration (g·L^{-1}), and t is the fermentation time (h) giving the greatest ethanol concentration.

RESULTS AND DISCUSSION

Isolation and Selection of Thermotolerant Yeast Strains

Natural habitats are a major source of useful microorganisms for biorefinery production [36]. In this study, soil and plant samples from different locations in Northeastern Thailand were collected, and the isolation of thermotolerant yeasts was performed using the enrichment culture technique as described by Limtong *et al.* [13]. The utilization of this technique under selection pressures has been widely used to isolate several thermotolerant yeast strains [13,17]. It has been reported that the effects of high temperatures on microbial growth have often been aggravated by ethanol concentrations greater than 3% [37]. Therefore, ethanol at 4% (*v*/*v*) and an incubation temperature of 35 °C served as selection pressures for the isolation of thermotolerant yeast strains in this work. As a result, sixty-two yeast isolates were obtained, and most of these yeasts were from soil samples. Their ability to grow and produce ethanol at high temperatures were tested by culturing the yeast strains in SSJ incubated at temperatures between 30 and 50 °C. The results showed that nine isolates of yeast, which were designated as DBKKU Y-53, DBKKU Y-55, DBKKU Y-58, DBKKU Y-102, DBKKU Y-103, DBKKU Y-104, DBKKU Y-105, DBKKU Y-106, and DBKKU Y-107, were able to grow and produce a relatively high level of ethanol at 45 °C. Of these, six isolates (*i.e.*, DBKKUY-102, DBKKUY-103, DBKKUY-104, DBKKUY-105, DBKKUY-106, DBKKUY-107) were capable of growth and ethanol production at 47 °C. Their ability to grow and produce ethanol at high temperatures were comparable to that of *K. marxianus* ATCC 8554, which is known as a thermotolerant yeast. Based on growth performance of the isolated yeast strains at high temperatures (above 40 °C), we speculated that these nine isolates of yeast were thermotolerant yeasts [38]. Therefore, all nine yeast isolates were selected for further study.

Successful isolation and selection of thermotolerant yeasts for ethanol production at high temperatures has been reported by several investigators. According to our literature review, SSJ has never been used directly as the substrate for yeast selection. Almost all previous research studies used glucose as a substrate, e.g., Ballesteros *et al.* [39] reported the screening and selection of thermotolerant yeast strains using glucose as the sole carbon source and found that *K. marxianus* L.G. was the most effective strain for ethanol production at high temperatures. This strain produced an ethanol concentration of approximately 3.76% (*w/v*) when the fermentation was carried out at 42 °C. Banat *et al.* [40] isolated and selected the thermotolerant yeast *K. marxianus* using an enrichment method with glucose as the carbon source and demonstrated that the selected *K. marxianus* strain produced an ethanol concentration of approximately 5.6%–6.0% (*w/v*) at 45–50 °C when using glucose as the substrate and ethanol concentration of approximately 7.5%–8.0% (*w/v*) and 6.5%–7.0% (*w/v*) at 37 °C and 40 °C, respectively, when using molasses as a substrate. Kiran Sree *et al.* [12] screened thermotolerant yeast *S. cerevisiae* VS3 using glucose as the sole carbon source, and they reported that the selected yeast strain produced an ethanol concentration of approximately 75 $g{\cdot}L^{-1}$ from a glucose concentration of 150 $g{\cdot}L^{-1}$ at 40 °C. Edgardo *et al.* [19] used glucose for the selection of thermotolerant yeast strains and obtained a good potential thermotolerant yeast *S. cerevisiae* that was able to produce approximately 75% of the theoretical ethanol yield at 40 °C. In addition to using glucose as a substrate, other carbon sources, such as sugarcane juice [13], xylose [41], sugarcane blackstrap molasses [42], and inulin [43] have also been used for screening and selecting thermotolerant yeasts.

Identification of the Selected Thermotolerant Yeasts

Based on the morphological and physiological characteristics, three isolated yeasts (*i.e.*, DBKKU Y-53, DBKKU Y-55, DBKKU Y-58) were found to be *Saccharomyces*, whereas six isolated yeasts (*i.e.*, DBKKU Y-102, DBKKU Y-103, DBKKU Y-104, DBKKU Y-105, DBKKU Y-106, DBKKU Y-107) were revealed to be *Kluyveromyces*. To confirm this finding, molecular taxonomic analyses based on the nucleotide sequences of the D1/D2 domain of the 26S rDNA gene were carried out. As a result, the nucleotide sequences from the yeast strains DBKKU Y-53, DBKKU Y-55, DBKKU Y-58 and from *S. cerevisiae*(NL45 and NL51) were 99% identical, whereas strains DBKKU Y-102, DBKKU Y-103, DBKKU Y-104, DBKKU Y-105, DBKKU Y-106, DBKKU Y-107, and *K. marxianus* (FJ627963) were 99% identical. Phylogenetic analysis confirmed that strains DBKKU Y-53, DBKKU Y-55, and DBKKU Y-58 were clustered in the same group as *S. cerevisiae* whereas strains DBKKU Y-102, DBKKU Y-103, DBKKU Y-104, DBKKU Y-105, DBKKU Y-106, and DBKKU Y-107 were closely related to *K. marxianus* (Figure 1). Therefore, strains DBKKU Y-53, DBKKU Y-55, and DBKKU Y-58 were identified as *S. cerevisiae*, whereas the other strains were *K. marxianus*.

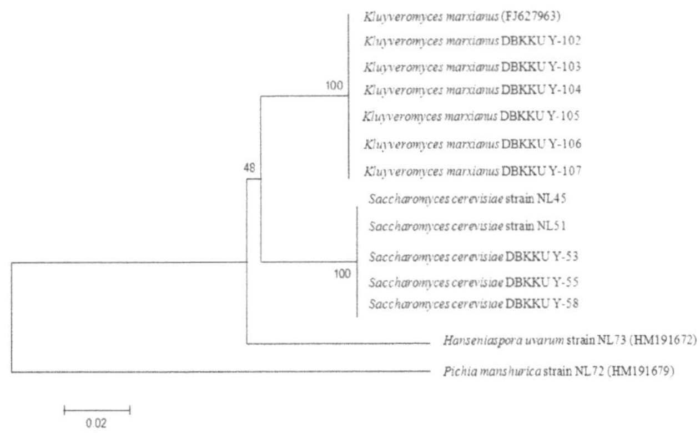

Figure 1: An unrooted phylogenetic tree showing the relationship between the isolated yeasts and the type strain *K. marxianus* and *S. cerevisiae* based on the D1/D2 domain of the 26s rDNA sequence analysis. The tree was built using the neighbor-joining method. Numbers given at the nodes indicate the percentage bootstrap values based on 1000 replications.

Comparative Study on the Ethanol Production by the Selected Thermotolerant Yeasts

A preliminary study of the ethanol production using SSJ by the selected thermotolerant yeasts revealed that only four strains, *i.e.*, DBKKU Y-53, DBKKU Y-102, DBKKU Y-103, and DBKKU Y-104, produced relatively high levels of ethanol at high temperatures after 24 h of fermentation. Therefore, these four strains were selected for further study of the ethanol production from SSJ at various incubation temperatures in a 500-mL flask scale. As shown in Table 2, strain DBKKU Y-53 produced the highest ethanol concentrations and volumetric ethanol productivities at 30 °C, 37 °C, 40 °C and 42 °C compared with the type strain, *K. marxianus* ATCC8554, and the other selected strains tested. At 45 °C, however, the ethanol concentration and volumetric ethanol productivity produced by this strain were lower than of the other strains tested. The ethanol concentrations and volumetric ethanol productivities produced by strains DBKKU Y-102, DBKKU Y-103, and DBKKU Y-104 were not significantly different compared with *K. marxianus* ATCC8554 at 30 °C, 37 °C, and 40 °C. At higher temperatures (42 °C and 45 °C), the ethanol concentrations and volumetric ethanol productivities from these strains were relatively greater than *K. marxianus* ATCC8554. It is evident from these results that increasing the fermentation temperature from 40 °C to 45 °C resulted in a reduction in the ethanol concentrations and productivities. Based on the maximum ethanol concentrations at 37 °C to 42 °C, which is a temperature commonly attained during fermentation in tropical regions, such as Thailand, *S. cerevisiae* DBKKU Y-53 was selected for further analysis.

Table 2: Ethanol production from sweet sorghum juice by isolated yeasts and the type strain, *K. marxianus* ATCC8554, at various temperatures

Strain	30 °C		37 °C		40 °C		42 °C		45 °C	
	P	Q_P	P	Q_P	P	Q_P	P	Q_P	P	Q_P
ATCC 8554	19.55 ± 0.49 ᵃ	0.27 ± 0.01 ᵃ	25.06 ± 1.65 ᵃ	0.35 ± 0.02 ᵃ	25.83 ± 0.17 ᵃ	0.36 ± 0.00 ᵃ	12.45 ± 0.00 ᵃ	0.17 ± 0.00 ᵃ	12.45 ± 0.00 ᵇ	0.17 ± 0.00 ᵃ
DBKKU Y-53	63.67 ± 4.12 ᵇ	1.77 ± 0.11 ᵇ	61.99 ± 4.43 ᵇ	1.72 ± 0.12 ᵇ	58.20 ± 0.54 ᵈ	1.62 ± 0.02 ᵈ	38.77 ± 1.92 ᶜ	0.81 ± 0.04 ᶜ	7.67b ± 1.77 ᵃ	0.16 ± 0.04 ᵃ
DBKKU Y-102	19.39 ± 0.63 ᵃ	0.27 ± 0.01 ᵃ	34.30 ± 8.99 ᵃ	0.48 ± 0.12 ᵃ	29.51 ± 0.34 ᵇ	0.41 ± 0.00 ᵇ	29.46 ± 1.10 ᵇ	0.41 ± 0.02 ᵇ	21.78 ± 1.27 ᶜ	0.30 ± 0.02 ᵇ
DBKKU Y-103	21.89 ± 0.80 ᵃ	0.30 ± 0.01 ᵃ	30.78 ± 0.07 ᵃ	0.43 ± 0.00 ᵃ	31.63 ± 0.05 ᶜ	0.44 ± 0.00 ᶜ	30.70 ± 1.09 ᵇ	0.43 ± 0.02 ᵇ	25.32 ± 0.82 ᶜᵈ	0.35 ± 0.00 ᵇᶜ
DBKKU Y-104	19.24 ± 1.00 ᵃ	0.27 ± 0.01 ᵃ	31.14 ± 1.94 ᵃ	0.43 ± 0.03 ᵃ	27.38 ± 1.23 ᵃ	0.38 ± 0.02 ᵃ	29.98 ± 1.34 ᵇ	0.42 ± 0.02 ᵇ	27.65 ± 2.87 ᵈ	0.38 ± 0.04 ᶜ

P: ethanol concentration produced (g·L^{-1}); Q_P: volumetric ethanol productivity (g·L^{-1}·h^{-1}); Mean values ± SD with different letters in the same column are significant different at $p < 0.05$ based on DMRT analysis.

Factors Influencing Ethanol Production When Using Thermotolerant Yeast S. cerevisiae DBKKU Y-53

There are several factors influencing the growth and ethanol production of yeast, such as the incubation temperature, the pH of the fermentation medium, the cell concentration, the sugar concentration, the nitrogen sources, and the aeration [13,44]. Therefore, the effects of some of the major factors on ethanol production using SSJ by thermotolerant yeast *S. cerevisiae* DBKKU Y-53 were investigated. In this work, the ethanol production efficiency of *S. cerevisiae* DBKKU Y-53 was compared with that of *S. cerevisiae* SC90, which is one of the industrial yeast strains widely used to produce ethanol on a commercial scale in Thailand.

The effect of temperature on ethanol production from SSJ using *S. cerevisiae* DBKKU Y-53 and *S. cerevisiae* SC90 was analyzed, and the results are summarized in Table 3. There was no significantly difference in ethanol concentration produced by both strains at 30 °C. However, at 37 °C and 40 °C, *S. cerevisiae* DBKKU Y-53 produced greater ethanol concentrations than *S. cerevisiae* SC09. The maximum ethanol concentrations produced by *S. cerevisiae* DBKKU Y-53 at 37 °C and 40 °C were 71.73 ± 2.62 g·L^{-1} and 58.14 ± 7.71 g·L^{-1}, respectively. When the incubation temperature was shifted from 40 °C to 42 °C and 45 °C, ethanol concentrations and volumetric ethanol productivities produced by both strains were remarkably decreased. This might be due to the negative effect of high temperature on growth and metabolic processes in yeast cells. It has been reported that high temperature causes a modification of plasma membrane fluidity and a reduction in the effectiveness of the plasma membrane as a semipermeable barrier allowing leakage of essential cofactors and coenzymes required for the activity of enzymes involved in glucose metabolism and ethanol production [44]. Roukas [45] reported that high temperatures caused denaturation of cellular proteins, which resulted in the reduction of cell growth and fermentation activity. Moreover, the reduction in yeast growth at high temperatures was also due to the accumulation of intracellular ethanol, which modifies the cell membrane structure of yeast cell [46]. Our results were in good agreement with Banat *et al.*[40], who observed a reduction in ethanol concentration produced by the thermotolerant yeast *K. marxianus* at 40 °C. Kiran Sree *et al.* [12] reported the ethanol production from glucose (150 g·L^{-1}) using thermotolerant yeast *S. cerevisiae* VS3 and observed a reduction in ethanol concentration when the incubation temperature shifted from 30 °C to 40 °C. Limtong *et al.*[13] demonstrated that the ethanol concentration produced by the newly isolated thermotolerant yeast *K. marxianus*DMKU3-1042 decreased significantly when the incubation temperature was increased from 37 °C to 40 °C. Tofighi *et al.* [47] reported the reduction in cell mass productivity and ethanol fermentation ability of the thermotolerant yeast *S. cerevisiae*when the incubation temperature increased to greater than 40 °C. Recently, Charoensopharat *et al.* [43] found that the ethanol concentration produced by the newly isolated thermotolerant yeast *K. marxianus* decreased dramatically when ethanol fermentation was carried out at temperatures greater than 40 °C.

Table 3: Kinetic parameters of ethanol production from sweet sorghum juice at various temperatures by *S. cerevisiae* DBKKUY-53 and *S. cerevisiae* SC90

Temperature (°C)	*S. cerevisiae* DBKKU Y-53			*S. cerevisiae* SC90		
	P	Q_P	E_y	P	Q_P	E_y
30	82.77 ± 1.99 ᵃ	2.30 ± 0.06 ᵃ	88.40 ± 0.59	83.30 ± 2.26 ᵃ	1.74 ± 0.05 ᵃ	96.47 ± 2.30
37	71.73 ± 2.62 ᵇ	1.49 ± 0.05 ᵇ	87.76 ± 3.22	63.22 ± 2.42 ᵇ	1.32 ± 0.05 ᵇ	98.30 ± 0.46
40	58.14 ± 7.71 ᶜ	1.61 ± 0.20 ᵇ	88.25 ± 2.73	53.68 ± 1.17 ᶜ	1.12 ± 0.02 ᶜ	94.04 ± 0.70
42	32.30 ± 0.49 ᵈ	0.67 ± 0.01 ᶜ	88.20 ± 6.37	32.54 ± 1.56 ᵈ	0.68 ± 0.03 ᵈ	96.59 ± 0.97
45	9.08 ± 4.02 ᵉ	0.19 ± 0.08 ᵈ	62.64 ± 5.60	17.69 ± 2.54 ᵉ	0.37 ± 0.05 ᵉ	98.07 ± 0.58

P: ethanol concentration (g·L^{-1}); Q_P: volumetric ethanol productivity (g·L^{-1}·h^{-1}); E_Y: percentage of ethanol production efficiency (%); Mean values ± SD with different letters in the same column are significant different at $p < 0.05$ based on DMRT analysis.

The pH of the medium is an important factor influencing ethanol yield. Generally, the optimum pH for yeast growth and ethanol production is in the range of 4.0 to 6.0 depending on growth conditions, such as the temperature, the presence of oxygen, the yeast species, and the type of raw material. For instance, the optimum pH for ethanol production from sugarcane juice at high temperatures using the thermotolerant yeast *K. marxianus* DMKU 3-1042 was 5.0 [13]. During ethanol fermentation, the pH of the fermentation medium is almost in the range of 4.0 to 5.5. This pH level typically prevents bacterial contamination during the fermentation process [48]. In this study, the effect of pH on ethanol production by the thermotolerant yeast *S. cerevisiae* DBKKU Y-53 and a reference strain was investigated, and the results are summarized inTable 4 and Table 5. The optimum pH for ethanol production from SSJ by both strains at 37 °C and 40 °C was 5.5, which is in agreement with reports by Ercan *et al.* [49] and Sign and Bishnoi [50]. At a pH less than or greater than 5.5, ethanol concentrations and volumetric ethanol productivities tended to decrease. This might be related to the activity of enzymes involved in the ethanol production pathway. It has been reported that enzymes may be inactivated at a pH level that is less than or greater than the optimum value causing a reduction in cell growth and ethanol fermentation ability [48]. The initial pH of the SSJ was in the range of 5.2–5.5. Therefore, a pH of 5.5, which gave the highest ethanol concentration, was selected for further study

Table 4: Kinetic parameters of ethanol production from sweet sorghum juice under various pHs, cell concentrations, and sugar concentrations by *S. cerevisiae* DBKKUY-53 at 37 °C and 40 °C

Conditions	37 °C			40 °C		
	P	Q_p	E_y	P	Q_p	E_y
pH						
4.0	68.21 ± 3.74 [a]	1.42 ± 0.08 [a]	87.47 ± 3.61	48.79 ± 2.09 [a]	1.02 ± 0.04 [a]	86.48 ± 0.94
4.5	67.80 ± 1.46 [a]	1.41 ± 0.03 [a]	91.27 ± 3.80	49.40 ± 0.13 [a]	1.03 ± 0.00 [a]	89.89 ± 4.43
5.0	71.73 ± 2.62 [a]	1.49 ± 0.05 [a]	87.76 ± 3.22	59.17 ± 5.33 [b]	1.61 ± 0.20 [c]	88.25 ± 2.73
5.5	74.23 ± 1.76 [a]	1.55 ± 0.08 [a]	92.06 ± 5.70	62.78 ± 0.67 [b]	1.31 ± 0.01 [b]	88.43 ± 0.53
6.0	72.14 ± 1.23 [ab]	1.50 ± 0.03 [a]	91.81 ± 0.66	61.98 ± 0.43 [b]	1.72 ± 0.01 [c]	92.35 ± 0.66
Cell concentration (cells· mL^{-1})						
1 × 10^6	74.23 ± 3.62 [a]	1.55 ± 0.08 [a]	92.06 ± 5.70	62.78± 0.67 [a]	1.31 ± 0.01 [a]	88.43 ± 0.53
1 × 10^7	84.06 ± 1.11 [b]	1.75 ± 0.02 [b]	87.72 ± 0.53	77.14 ± 1.58 [b]	2.14 ± 0.04 [b]	87.78 ± 1.37
1 × 10^8	88.17 ± 0.17 [b]	2.94 ± 0.01 [c]	84.73± 0.28	83.35 ± 2.37 [c]	3.47 ± 0.10 [c]	89.07 ± 0.04
Sugar concentration (g· L^{-1})						
200	88.17 ± 0.17 [a]	2.94 ± 0.01 [a]	90.42 ± 0.28	83.35 ± 2.37 [a]	3.47 ± 0.10 [a]	89.07 ± 0.04
250	99.56 ± 0.88 [c]	4.15 ± 0.04 [c]	98.44 ± 0.53	88.97 ± 1.23 [b]	3.71 ± 0.05 [b]	96.53 ± 2.55
300	92.69 ± 0.25 [b]	3.86 ± 0.01 [b]	98.89 ± 0.08	83.25 ± 0.43 [a]	3.47 ± 0.02 [a]	97.40 ± 1.11

P: ethanol concentration (g· L^{-1}); Q_p: volumetric ethanol productivity (g· L^{-1}· h^{-1}); E_y: percentage of ethanol production efficiency (%); Mean values ± SD with different letters in the same column are significant different at $p < 0.05$ based on DMRT analysis.

Table 5: Kinetic parameters of ethanol production from sweet sorghum juice under various pHs, cell concentrations, and sugar concentrations by *S. cerevisiae* SC90 at 37 °C and 40 °C

Conditions	37 °C			40 °C		
	P	Q_p	E_y	P	Q_p	E_y
pH						
4.0	72.90 ± 2.31 [a]	1.22 ± 0.04 [bc]	96.21 ± 2.71	55.08 ± 2.67 [ab]	0.92 ± 0.04 [ab]	97.58 ± 0.28
4.5	70.70 ± 2.55 [a]	1.18 ± 0.04 [c]	97.15 ± 0.52	53.01 ± 1.35 [a]	0.88 ± 0.02 [a]	97.55 ± 1.77
5.0	63.22 ± 2.42 [b]	1.32 ± 0.05 [b]	98.30 ± 0.46	53.68 ± 1.17 [ab]	1.12 ± 0.02 [c]	94.04 ± 0.70
5.5	75.19 ± 2.40 [a]	1.25 ± 0.04 [bc]	95.08 ± 2.92	57.62 ± 0.64 [b]	0.96 ± 0.01 [b]	98.26 ± 1.90
6.0	72.41 ± 1.75 [a]	1.51 ± 0.04 [a]	75.99 ± 0.05	57.09 ± 0.42 [b]	0.96 ± 0.01 [b]	80.29 ± 1.89
Cell concentration (cells· mL^{-1})						
1 × 10^6	75.19 ± 2.40 [a]	1.25 ± 0.04 [a]	95.08 ± 2.92	62.78± 0.67 [a]	1.31 ± 0.01 [a]	88.43 ± 0.53
1 × 10^7	75.16 ± 0.06 [a]	1.57 ± 0.00 [b]	90.50 ± 1.06	77.14 ± 1.58 [b]	2.14 ± 0.04 [b]	87.78 ± 1.37
1 × 10^8	76.41 ± 1.99 [a]	2.55 ± 0.07 [c]	79.98 ± 4.31	83.35 ± 2.37 [c]	3.47 ± 0.10 [c]	89.07 ± 0.04
Sugar concentration (g· L^{-1})						
200	76.41 ± 1.99 [a]	2.55 ± 0.07 [c]	79.98 ± 4.31	70.91 ± 0.96 [a]	2.96 ± 0.04 [a]	88.45 ± 1.81
250	83.05 ± 0.02 [b]	1.73 ± 0.00 [a]	93.79 ± 2.20	73.00 ± 0.28 [a]	2.03 ± 0.01 [b]	94.34 ± 0.21
300	78.98 ± 0.78 [a]	2.19 ± 0.02 [b]	94.79 ± 0.65	71.66 ± 0.96 [a]	1.99 ± 0.03 [b]	90.93 ± 1.15

P: ethanol concentration (g· L^{-1}); Q_p: volumetric ethanol productivity (g· L^{-1}· h^{-1}); E_y: percentage of ethanol production efficiency (%); Mean values ± SD with different letters in the same column are significant different at $p < 0.05$ based on DMRT analysis.

The initial cell concentration affects not only the ethanol yield but also the substrate consumption rate and ethanol fermentation rate. Generally, high initial cell concentrations reduce the lag phase of growth and increase the sugar consumption and ethanol fermentation rate, leading to a high ethanol yield and productivity. In this study, the effect of initial cell concentrations (1 × 10^6, 1 × 10^7, 1 × 10^8 cells·mL^{-1}) on ethanol fermentation from SSJ containing 220 g·L^{-1} of total sugars was investigated, and the results are summarized in Table 4 and Table 5. As a result, increasing the cell concentration resulted in an increase in the ethanol concentration and the volumetric ethanol productivity. The maximum ethanol concentrations and volumetric ethanol productivities produced by *S. cerevisiae* DBKKU Y-53 and SC09 at 37 °C and 40 °C were achieved at an initial cell concentration of 1 × 10^8 cells·mL^{-1}, which is in good agreement with Charoensopharat *et al.*[43] and Laopaiboon *et al.* [51]. High sugar concentrations (more than 20% *w/v*) are not often used in industrial ethanol production because they may reduce the yeast cell viability, the substrate conversion rate, and the ethanol yield [52,53]. However, ethanol production with high sugar concentrations have also been reported, and the fermentation efficiencies vary depending on the yeast species and fermentation conditions. For example, Laopaiboon *et al.* [54] reported a maximum ethanol concentration of 120.68 $g{\cdot}L^{-1}$and volumetric ethanol productivity of 2.01 $g{\cdot}L^{-1}{\cdot}h^{-1}$ when *S. cerevisiae* NP01 was used to produce ethanol from SSJ under a very high gravity fermentation. Charoensopharat *et al.* [43] reported a maximum ethanol concentration of 104.83 $g{\cdot}L^{-1}$ and a volumetric ethanol productivity of 4.37 $g{\cdot}L^{-1}{\cdot}h^{-1}$ when *K. marxianus* DBKKU Y-102 was used to produce ethanol from Jerusalem artichoke tubers at 37 °C during consolidated bioprocessing. To verify the effect of sugar concentration on ethanol production efficiency of the thermotolerant yeast *S. cerevisiae* DBKKU Y-53, SSJ containing various sugar concentrations (200, 250, 300 $g{\cdot}L^{-1}$) was tested. As shown in Table 4 and Table 5, increasing in the sugar concentration from 200 $g{\cdot}L^{-1}$ to 250 $g{\cdot}L^{-1}$resulted in an increase in the ethanol concentration. However, at a sugar concentration of 300 $g{\cdot}L^{-1}$ the ethanol concentration was remarkably decreased, and a large amount of sugar remained in the fermentation broth at both 37 °C and 40 °C. High sugar concentrations have been reported to cause negative effects on cell viability and morphology due to an increase in the osmotic pressure, which leads to a reduction in the cell biomass and ethanol yield [53,55]. The maximum ethanol concentrations produced by *S. cerevisiae* DBKKU Y-53 and SC90 were achieved at a sugar concentration of 250 $g{\cdot}L^{-1}$. Therefore, this sugar concentration was used for subsequent experiments.

Approximately 10% of yeast's dry weight is nitrogen [44]. Therefore, nitrogen is one of the important constituents for cell growth and synthesis of structural and functional proteins involved in metabolic processes. Various types of nitrogen sources have been used to supplement ethanol fermentation medium both in organic (yeast extract, corn steep liquor) and inorganic forms (ammonium sulfate, ammonium nitrate, urea, diammonium phosphate). In this study, the effects on ethanol production of using yeast extract, urea, and ammonium sulfate at different concentrations from SSJ by *S. cerevisiae* DBKKU Y-53 and SC90 were determined, and the results are summarized in Table 6 and Table 7. Supplementation of urea in the SSJ was shown to significantly enhance the ethanol production by both strains of yeast at 37 °C and 40 °C, which is in good agreement with reports from Yue *et al.* [56]. There was no significant difference in ethanol production at 37 °C when comparing yeast extract supplementation and supplementation-free fermentations. However, at 40 °C, supplementation of yeast extract tended to result in a lower ethanol concentration compared with the control condition. Yeast extract has been reported to be a good organic nitrogen source for ethanol production using *S. cerevisiae*. However, its availability to be utilized by yeast cells is depended on the fermentation conditions. For example, Laopaiboon *et al.* [54] reported that supplementation of yeast extract in SSJ improved the ethanol production by *S. cerevisiae* NP01 for very high gravity fermentation. With respect to ammonium sulfate, supplementation of this nitrogen compound resulted in a reduction of the ethanol concentration in all conditions tested compared with the control without nitrogen supplementation. These results clearly indicate that ammonium sulfate was not a good nitrogen source for ethanol production at high temperatures using *S. cerevisiae* DBKKU Y-53 and SC90 when SSJ was used as a substrate. One possibility is that this nitrogen compound was not taken up by the yeast cells during high temperature fermentation. Ter Schure *et al.* [57] and Magasanik and Kaiser [58] reported that the uptake of nitrogen by yeast cells is regulated by the mechanism known as nitrogen catabolite repression (NCR). NCR enables yeast cells to select the best nitrogen sources by repressing the transcription of genes involved in the utilization of the poorer nitrogen [59]. However, the effectiveness of NCR mechanism is influenced by many factors including fermentation temperature and the presence of ethanol. Normally, high temperatures and high ethanol concentrations cause the modification of the plasma membrane. Therefore, the nitrogen sources sensing system, which is mainly located in the plasma membrane of yeast cells, may be affected by these stress conditions, leading to the impairment of ammonium sulfate uptake [60,61]. In this study, urea proved to be the best nitrogen source for ethanol production from SSJ during high temperature fermentation using *S. cerevisiae* DBKKU Y-53 and SC90; therefore, it was chosen for further analysis.

Table 6: Kinetic parameters of ethanol production from sweet sorghum juice with various nitrogen sources by *S. cerevisiae* DBKKUY-53 at 37 °C and 40 °C

Conditions	37 °C			40 °C		
	P	Q_p	E_y	P	Q_p	E_y
Control	99.56 ± 0.88 ^{bcd}	4.15 ± 0.04 ^{a}	98.44 ± 0.53	88.97 ± 1.23 ^{a}	3.71 ± 0.05 ^{a}	95.56 ± 1.83
Yeast extract						
6	101.30 ± 0.84 ^{bc}	3.38 ± 0.03 ^{b}	93.43 ± 1.70	79.19 ± 5.53 ^{bcd}	3.30 ± 0.23 ^{bc}	91.78 ± 3.06
9	100.35 ± 0.43 ^{bcd}	3.35 ± 0.01 ^{b}	92.39 ± 2.90	76.81 ± 8.84 ^{cde}	3.20 ± 0.37 ^{bcd}	94.98 ± 1.66
12	97.05 ± 0.98 ^{d}	3.23 ± 0.03 ^{c}	88.20 ± 2.84	83.76 ± 0.26 ^{abc}	3.49 ± 0.01 ^{ab}	96.01 ± 1.27
Urea						
0.25	97.97 ± 3.29 ^{cd}	2.72 ± 0.09 ^{d}	86.47 ± 1.70	91.18 ± 2.70 ^{a}	3.04 ± 0.09 ^{cde}	95.42 ± 3.93
0.50	102.72 ± 1.94 ^{ab}	2.14 ± 0.04 ^{e}	96.01 ± 1.16	88.27 ± 0.36 ^{a}	2.94 ± 0.01 ^{de}	95.30 ± 3.66
0.75	105.17 ± 1.37 ^{a}	2.19 ± 0.03 ^{e}	94.31 ± 4.16	88.66 ± 0.59 ^{a}	2.96 ± 0.02 ^{de}	87.82 ± 0.26
1.00	99.90 ± 1.41 ^{bcd}	2.78 ± 0.04 ^{d}	85.65 ± 7.36	86.41 ± 2.43 ^{ab}	2.88 ± 0.08 ^{e}	83.82 ± 2.15
$(NH_4)_2SO_4$						
0.25	97.38 ± 0.23 ^{d}	3.25 ± 0.01 ^{c}	94.37 ± 2.64	67.25 ± 1.60 ^{f}	1.40 ± 0.02 ^{g}	94.51 ± 2.69
0.50	96.77 ± 2.41 ^{d}	3.23 ± 0.08 ^{c}	90.37 ± 5.24	69.90 ± 1.02 ^{ef}	1.94 ± 0.01 ^{f}	94.50 ± 1.04
0.75	97.70 ± 0.06 ^{cd}	2.71 ± 0.00 ^{d}	86.76 ± 6.82	70.29 ± 0.59 ^{ef}	1.95 ± 0.01 ^{f}	96.18 ± 1.51
1.00	99.81 ± 1.44 ^{bcd}	2.77 ± 0.04 ^{d}	87.76 ± 6.82	74.27 ± 1.14 ^{def}	2.06 ± 0.02 ^{f}	96.25 ± 0.48

P: ethanol concentration ($g{\cdot}L^{-1}$); Q_p: volumetric ethanol productivity ($g{\cdot}L^{-1}{\cdot}h^{-1}$); E_y: percentage of ethanol production efficiency (%); Mean values ± SD with different letters in the same column are significant different at $p < 0.05$ based on DMRT analysis.

Table 7: Kinetic parameters of ethanol production from sweet sorghum juice with various nitrogen sources by *S. cerevisiae* SC90 at 37 °C and 40 °C

Conditions	37 °C			40 °C		
	P	Q_p	E_y	P	Q_p	E_y
Control	83.05 ± 0.02 c	1.73 ± 0.00 d	93.79 ± 2.20	73.00 ± 0.28 a	2.03 ± 0.01 c	94.34 ± 0.21
Yeast extract						
6.0	83.53 ± 0.55 c	2.78 ± 0.02 b	97.83 ± 0.52	67.17 ± 2.78 b	2.24 ± 0.09 b	92.54 ± 4.06
9.0	88.23 ± 1.71 b	2.94 ± 0.06 a	94.07 ± 0.58	67.47 ± 1.81 b	2.25 ± 0.06 b	97.98 ± 0.50
12.0	87.38 ± 0.96 b	2.91 ± 0.03 a	94.68 ± 2.30	69.25 ± 1.16 ab	2.31 ± 0.04 b	98.18 ± 1.50
Urea						
0.25	92.48 ± 0.87 a	2.57 ± 0.02 c	97.48 ± 0.95	72.87 ± 0.72 a	3.04 ± 0.03 a	87.17 ± 1.23
0.50	92.68 ± 0.29 a	2.57 ± 0.01 c	98.03 ± 0.08	72.20 ± 2.15 ab	3.01 ± 0.09 a	91.91 ± 0.60
0.75	86.76 ± 3.35 b	2.89 ± 0.11 a	97.86 ± 0.11	69.90 ± 1.31 ab	2.91 ± 0.05 a	94.36 ± 0.37
1.00	87.39 ± 0.39 b	2.91 ± 0.01 a	98.36 ± 0.84	73.78 ± 1.17 a	2.05 ± 0.03 c	91.65 ± 2.09
$(NH_4)_2 SO_4$						
0.25	57.57 ± 1.67 f	1.20 ± 0.02 g	91.90 ± 0.53	54.03 ± 0.56 d	1.13 ± 0.01 d	92.66 ± 2.92
0.50	71.85 ± 0.36 d	1.00 ± 0.00 h	91.77 ± 0.42	57.82 ± 0.93 cd	1.20 ± 0.02 d	92.23 ± 4.00
0.75	66.54 ± 2.11 e	1.39 ± 0.04 f	97.53 ± 0.01	56.65 ± 5.91 cd	1.18 ± 0.12 d	86.90 ± 5.20
1.00	72.10 ± 0.78 d	1.50 ± 0.02 e	94.84 ± 1.20	59.81 ± 1.34 c	1.25 ± 0.03 d	97.68 ± 1.26

P: ethanol concentration ($g \cdot L^{-1}$); Q_p: volumetric ethanol productivity ($g \cdot L^{-1} \cdot h^{-1}$); E_y: percentage of ethanol production efficiency (%); Mean values ± SD with different letters in the same column are significant different at $p < 0.05$ based on DMRT analysis.

Ethanol Production in 2-L Bioreactor

The time profiles of ethanol production from SSJ using *S. cerevisiae* DBKKU Y-53 and SC90 at 37 °C and 40 °C in a 2-L bioreactor are illustrated in Figure 2. During the first 12 h after fermentation, ethanol concentrations produced by both strains sharply increased and reached their maximum values at 24 h for *S. cerevisiae* SC90 and 48 h for *S. cerevisiae* DBKKU Y-53. Table 8 summarizes the kinetic parameters of ethanol production from the SSJ at high temperatures using *S. cerevisiae*DBKKU Y-53 and SC90. It can be seen from this finding that the newly isolated thermotolerant yeast *S. cerevisiae* DBKKU Y-53 resulted in a greater ethanol concentration as well as sugar utilization capability compared with SC90.

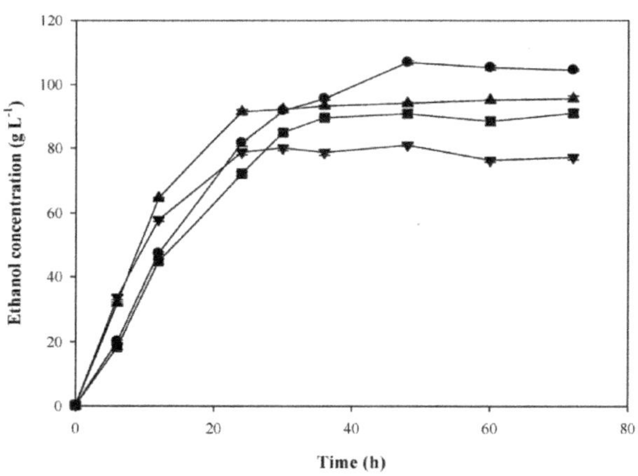

Figure 2: Ethanol production from sweet sorghum juice by *S. cerevisiae* DBKKUY-53 and *S. cerevisiae* SC90 in a 2-L bioreactor at high temperatures. *S. cerevisiae* DBKKUY-53 at 37 °C (●); *S. cerevisiae* DBKKUY-53 at 40 °C (■); *S. cerevisiae* SC90 at 37 °C (▲); *S. cerevisiae* SC90 at 40 °C (▼).

A comparative analysis of the ethanol production from SSJ using the thermotolerant yeast *S. cerevisiae* DBKKU Y-53 with values reported in the literatures using different raw materials and yeast strains was performed, and the results are summarized in Table 9. The ethanol concentration and volumetric ethanol productivity produced by *S. cerevisiae* DBKKU Y-53 were greater than values reported in other works, suggesting that the newly isolated thermotolerant yeast *S. cerevisiae*DBKKU Y-53 was a good candidate for ethanol production from SSJ at high temperatures. Although several isolates of the thermotolerant yeast *K. marxianus* have been reported to be capable of growth and ethanol production at temperatures greater than 45 °C, almost all of these isolates had relatively lower ethanol yields and were less tolerant to ethanol than *S. cerevisiae*. Furthermore, these isolates also required oxygen during ethanol fermentation resulting in an increase in the operating cost. Therefore, the thermotolerant yeast *S. cerevisiae* is more promising for ethanol production at high temperatures on a commercial scale compared with *K. marxianus*.

Table 8: Kinetic parameters of ethanol production from sweet sorghum juice by *S. cerevisiae* DBKKU Y-53 and *S. cerevisiae* SC90 in a 2-L bioreactor at 37 °C and 40 °C

Strains	T (°C)	Parameters (mean ± SD)					
		P	Q_p	$Y_{p/s}$	E_y	t	Sugar Utilized (%)
DBKKU Y-53	37	106.82 ± 0.01 a	2.23 ± 0.01 a	0.45 ± 0.02	88.42 ± 0.11	48	91.84 ± 0.20
	40	85.01 ± 0.03 a	2.83 ± 0.02 a	0.42 ± 0.01	82.06 ± 0.09	30	79.35 ± 0.05
SC90	37	91.59 ± 0.01 b	3.82 ± 0.02 b	0.47 ± 0.01	91.52 ± 0.02	24	82.68 ± 0.02
	40	78.69 ± 0.02 b	3.28 ± 0.03 b	0.46 ± 0.02	89.08 ± 0.01	24	68.19 ± 0.02

T: incubation temperature; P: ethanol concentration ($g \cdot L^{-1}$); Q_p: volumetric ethanol productivity ($g \cdot L^{-1} \cdot h^{-1}$); $Y_{p/s}$: ethanol yield ($g \cdot g^{-1}$); E_y: percentage of ethanol production efficiency (%); t: fermentation time (h); Mean values ± SD with different letters in the same column are significant different at $p < 0.05$ based on DMRT analysis.

Table 9: Comparison of ethanol production by *S. cerevisiae* DBKKU Y-53 and other yeast strains reported in the literatures using different raw materials

Strains	S_0 (g·L^{-1})	C-Source	T (°C)	P (g·L^{-1})	Q_p (g·L^{-1}·h^{-1})	References
S. cerevisiae UV-VS3 100	250	Glucose	30 42	98.0 62.0	2.04 1.29	Sridhar *et al.* [62]
S. cerevisiae VS3	150	Glucose	35 40 42	75.0 60.0 58.0	1.56 1.25 1.21	Kiran Sree *et al.* [12]
S. cerevisiae F111	180	Sugarcane molasses	43	84.0	2.33	Abdel-fattah *et al.* [63]
K. marxianus WR12	180	Sugarcane molasses	43	80.6	2.88	
Pichia kudriavzevii	170	Sugarcane juice	40	71.9	4.00	Dhaliwal *et al.* [64]
K. marxianus DMKU 3-1042	220	Sugarcane juice	37 40	87.0 67.8	1.45 1.13	Limtong *et al.* [13]
Issatchenkia orientalis IPE 100	150	Glucose	37 42	64.3 65.5	1.07 1.37	Kwon *et al.* [65]
S. cerevisiae C3751	100	Glucose	37 41	37.3 38.0	1.55 1.58	Auesukaree *et al.* [20]
S. cerevisiae DBKKU Y-53	250	SSJ	37 40	106.82 85.01	2.23 2.83	This study

S_0: initial sugar concentration; T: incubation temperature; P: ethanol concentration; Q_p: volumetric ethanol productivity.

This work will contribute a significant amount of information on ethanol production at an industrial scale. However, the ethanol production cost at a large-scale should be concerned. There are many techniques which can be employed to reduce the production cost; for example, cell recycling, various fermentation approaches, such as very high gravity fermentation, consolidated bioprocessing, and continuous ethanol fermentation using stirred tank bioreactor coupling with plug flow bioreactor [17,18,43,54].

Real-Time RT-PCR Analysis of Gene Expression

Stressful conditions during ethanol fermentation, such as high temperatures, and high concentrations of ethanol or sugar, have been reported to trigger the expression of several stress-responsive genes including those encoding HSPs, enzymes involved in protein degradation and in the glycolysis pathway, and other proteins involved in the synthesis of compatible solutes and reserve carbohydrates [22,23,25,66,67]. In the present study, the expression levels of *hsp26*, *hsp70*, *hsp90*, *hsp104*,*cdc*, *tps*, *nth*, *gsy*, and *rsp* in the thermotolerant yeast *S. cerevisiae* DBKKU Y-53 and SC90 were determined using real-time RT-PCR. As shown in Figure 3, the expression of all genes was activated in both strains of yeast during ethanol fermentation at 40 °C. Although the expression levels of *tps*, *nth*, *gsy*, and *rsp* in *S. cerevisiae* DBKKU Y-53 and SC90 were not dramatically different, the genes encoding HSPs (*hsp26*, *hsp70*, *hsp90*, *hsp104*) in DBKKU Y-53 were much greater than in SC90. This finding suggests that high growth and ethanol fermentation capabilities of DBKKU Y-53 at a high temperature might be related to increased expression levels of HSP genes. Conversely, the expression level of *cdc* encoding pyruvate kinase, which is involved in ATP production, in SC90 was greater than that in DBKKU Y-53, suggesting that SC90 required more ATP for growth and ethanol production at high temperature. The results in this study were in good agreement with Auesukaree *et al.*[20], who observed high expression levels of genes encoding the small HSP, HSP70, HSP90, and HSP100 family and those genes encoding trehalose-6-phosphate synthase, neutral trehalase, and glycogen synthase in *S. cerevisiae* after heat shock and long-term heat exposure at 37 °C. Piper *et al.* [68] reported that several HSPs are constitutively expressed at appropriate temperatures and play a crucial role in folding and assembling proteins. Many of the HSPs, such as HSP70, HSP90, and HSP104, play an important role as molecular chaperones. These molecular chaperones depend on the energy of ATP hydrolysis for function [69]. Therefore, increasing the expression level of *cdc* during ethanol fermentation at a high temperature may provide sufficient energy not only for growth and ethanol fermentation activity but may also function as molecular chaperones to protect the structural integrity of the proteins in yeast cells. In addition to the HSPs, the accumulation of trehalose has also been reported to be associated with heat stress protection and thermotolerance [70]. Trehalose can stabilize the protein structure, reduce the aggregation of denatured proteins, and cooperate with HSPs to promote protein refolding [71,72,73].

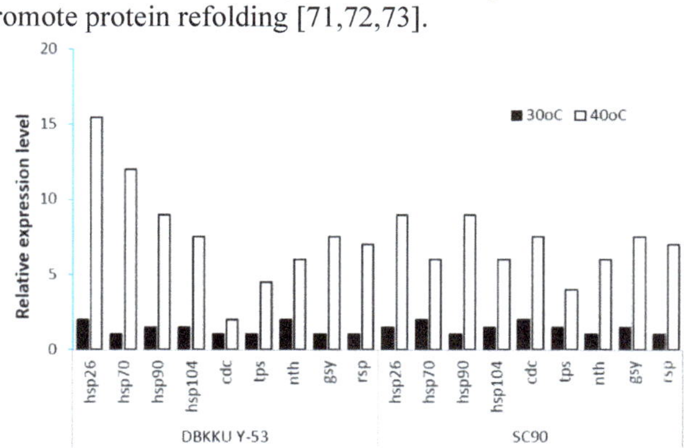

Figure 3: Expression levels of genes encoding HSP26 (*hsp26*), HSP70 (*hsp70*), HSP90 (*hsp90*), HSP104 (*hsp104*), pyruvate kinase (*cdc*), trehalose-6-phosphate synthase (*tps*), neutral trehalase (*nth*), glycogen synthase (*gsy*), and ubiquitin synthase (*rsp*) in the exponential-growth phase of the thermotolerant yeast *S. cerevisiae* DBKKU Y-53 and a reference strain *S. cerevisiae* SC90 during ethanol fermentation at 40 °C. Values presented are the means and relative expression levels of each gene as described in the Materials and Methods section.

In *S. cerevisiae*, ubiquitin ligase plays a key regulatory role in many cellular processes, such as trafficking, sorting, modifying gene expression, DNA repair, RNA transport as well as the degradation of a large number of proteins in multiple cellular compartments [74]. This protein is also involved in the pathways responsible for the regulation of chromatin function and ultimately controls gene expression under limited nutrient conditions [75]. Most recently, Shahsavarani *et al.*[76] also demonstrated that overexpression of *RSP5* encoding ubiquitin ligase improved the ability of *S. cerevisiae* to tolerate high temperatures. An increase in the ubiquitin ligase, which was observed in this study by the high expression level of *rsp*during ethanol fermentation at high temperature, might regulate the transcription of some genes and induce the heat stress response through the ubiquitination process. Therefore, the ability of *S. cerevisiae* DBKKU Y-53 and SC90 to grow and produce a relatively high level of ethanol at a high temperature might be explained by this mechanism.

In the thermotolerant yeast *K. marxianus*, the molecular mechanisms conferring thermotolerance are complicated and are controlled by multiple genes not only for HSPs biosynthesis but also for those genes encoding the proteins involved in DNA replication and repair, RNA processing, ribosome biogenesis, and carbohydrate metabolism process [21]. The expression of*hsp* genes and those functioning to prevent protein denaturation may be insufficient to allow growth and efficient ethanol fermentation of *S. cerevisiae* at high temperatures. To gain a better understanding and provide useful information for fundamental and applied research for innovative applications, further studies are required to clarify the precise mechanism conferring thermotolerance in *S. cerevisiae*.

CONCLUSIONS

Utilization of high-potential thermotolerant ethanol-producing yeast is a promising approach to reduce the energy used in cooling systems and to also reduce the operating cost of ethanol production at high temperatures. In this study, the newly isolated thermotolerant ethanol-producing yeast designated as *S. cerevisiae* DBKKU Y-53 exhibited high growth and ethanol production efficiencies at high temperatures (37 °C and 40 °C) compared with the other isolated strains and the industrial ethanol producer *S. cerevisiae* SC90. The optimum conditions for ethanol production by this thermotolerant yeast using SSJ as a raw material were the following: a pH of 5.5, a sugar concentration of 250 $g \cdot L^{-1}$, a cell concentration of 1.0×10^8 $cells \cdot mL^{-1}$. The SSJ without the addition of an exogenous nitrogen source can be used directly as substrate for ethanol production at high temperatures by the thermotolerant yeast *S. cerevisiae* DBKKU Y-53. During ethanol fermentation at 40 °C, genes encoding HSP26, HSP70, HSP90, HSP104, pyruvate kinase, trehalose-6-phosphate synthase, neutral trehalase, glycogen synthase, and ubiquitin ligase were highly expressed in *S. cerevisiae* DBKKU Y-53 and SC90 compared with the expression levels at 30 °C. This finding suggests that the growth and ethanol fermentation activity of yeast at high temperatures were not only correlated with the expression of genes involved in heat-stress response but were also correlated with genes involved in ATP production, trehalose and glycogen metabolism, and the protein degradation process.

ACKNOWLEDGMENTS

This research was financially supported by the Energy Policy and Planning Office, Ministry of Energy, Thailand and the Research Fund for Supporting Lecturer to Admit High Potential Student to Study and Research on His Expert Program Year 2009, the Graduate School, Khon Kaen University, Thailand. Apart of this work was also supported by the Center for Alternative Energy Research and Development, the Fermentation Research Center for Value Added Agricultural Products (FerVAAP), The New Core to Core Program (CCP) A. Advanced Research Networks on "Establishment of an International Research Core for New Bio-research Fields with Microbes from Tropical Areas World Class Research Hub of Tropical Microbial Resources and Their Utilization", and a grant from Khon Kaen University.

AUTHOR CONTRIBUTIONS

Sunan Nuanpeng performed the experiments; Sudarat Thanonkeo analyzed the data; Mamoru Yamada contributed reagents and analysis tools; Pornthap Thanonkeo conceived and designed the experiments, analyzed the data and wrote the paper.

REFERENCES

1. Farrell, A.E.; Plevin, R.J.; Turner, B.T.; Jones, A.D.; O'Hare, M.; Kammen, D.M. Ethanol can contribute to energy and environmental goals. *Sciences* 2006, *311*, 506–508.
2. Hill, J.; Nelson, E.; Tilman, D.; Polasky, S.; Tiffany, D. Environmental, economic, and energetic costs and benefits of biodiesel and ethanol biofuels. *Proc. Natl. Acad. Sci. USA* 2006, *103*, 11206–11210.
3. Yan, J.; Lin, T. Biofuels in Asia. *Appl. Energy* 2009, *86*, S1–S10.
4. Zhou, A.; Thomson, E. The development of biofuels in Asia. *Appl. Energy* 2009, *86*, S11–S20.
5. Salassi, M.E. *The Economic Feasibility of Ethanol Production from Sugar Crops*; LSU AgCenter: Baton Rouge, LA, USA, 2007.
6. Daniel, P.H.; Lueschen, W.E.; Kanne, B.K.; Hoverstad, T.R. A comparison of sweet sorghum cultivars and maize for ethanol production. *J. Prod. Agric.* 1991, *4*, 377–381.
7. Bvochora, J.M.; Read, J.S.; Zvauya, R. Application of very high gravity technology to the cofermentation of sweet stem sorghum juice and sorghum grain. *Ind. Crops Prod.* 2000, *11*, 11–17.
8. Woods, J. Integrating Sweet Sorghum and Sugarcane for Bioenergy: Modeling the Potential for Electricity and Ethanol Production in SE Zimbabwe. Ph.D. Thesis, University of London, London, UK, 2000.
9. Chan-u-tit, P.; Laopaiboon, L.; Jaisil, P.; Laopaiboon, P. High level ethanol production by nitrogen and osmoprotectant supplementation under very high gravity fermentation conditions. *Energies* 2013, *6*, 884–899.
10. Yu, M.; Li, J.; Chang, S.; Du, R.; Li, S.; Zhang, L.; Fan, G.; Yan, Z.; Cui, T.; Cong, G.; *et al.* Optimization of ethanol production from NaOH-pretreated solid state fermented sweet sorghum bagasse. *Energies* 2014, *7*,

4054–4067.

11. Gnansounou, E.; Dauriat, A.; Wyman, C.E. Refining sweet sorghum to ethanol and sugar: Economic trade-offs in the context of North China. *Bioresour. Technol.* 2005, *96*, 985–1002.
12. Kiran Sree, N.; Sridhar, M.; Suresh, K.; Banat, I.M.; Venkateswar, R.L. Isolation of thermotolerant, osmotolerant, flocculating *Saccharomyces cerevisiae* for ethanol production. *Bioresour. Technol.* 2000, *72*, 43–46.
13. Limtong, S.; Sringiew, C.; Yongmanitchai, W. Production of fuel ethanol at high temperature from sugar cane juice by a newly isolated *Kluyveromyces marxianus*. *Bioresour. Technol.* 2007, *98*, 3367–3374.
14. Kourkoutas, Y.; Bekatorou, A.; Banat, I.M.; Marchant, R.; Koutinas, A.A. Immobilization technologies and support materials suitable in alcohol beverages production: A review. *Food Microbiol.* 2004, *21*, 377–397.
15. Zafar, S.; Owais, M. Ethanol production from crude whey by *Kluyveromyces marxianus*. *Biochem. Eng. J.* 2006, *27*, 295–298.
16. Yuan, W.J.; Zhao, X.Q.; Ge, X.M.; Bai, F.W. Ethanol fermentation with *Kluyveromyces marxianus* from Jerusalem artichoke grown in salina and irrigated with a mixture of seawater and freshwater. *J. Appl. Microbiol.* 2008, *105*, 2076–2083.
17. Yuan, W.J.; Chang, B.L.; Ren, J.G.; Liu, J.P.; Bai, F.W.; Li, Y.Y. Consolidated bioprocessing strategy for ethanol production from Jerusalem artichoke tubers by *Kluyveromyces. marxianus* under high gravity conditions. *J. Appl. Microbiol.* 2011, *112*, 38–44.
18. Hu, N.; Yuan, B.; Sun, J.; Wang, S.A.; Li, F.L. Thermotolerant *Kluyveromyces marxianus* and *Saccharomyces cerevisiae*strains representing potentials for bioethanol production from Jerusalem artichoke by consolidated bioprocessing. *Appl. Microbiol. Biotechnol.* 2012, *95*, 1359–1368.
19. Edgardo, A.; Carolina, P.; Manuel, R.; Juanita, F.; Baeza, J. Selection of thermotolerant yeast strains *Saccharomyces cerevisiae* for bioethanol production. *Enzyme Microb. Technol.* 2008, *43*, 120–123.
20. Auesukaree, C.; Koedrith, P.; Saenpayavai, P.; Asvarak, T.; Benjaphokee, S.; Sugiyama, M.; Kaneko, Y.; Harashima, S.; Boonchird, C. Characterization and gene expression profiles of thermotolerant *Saccharomyces cerevisiae* isolates from Thai fruits. *J. Biosci. Bioeng.* 2012, *114*, 144–149.
21. Lertwattanasakul, N.; Kosaka, T.; Hosoyama, A.; Suzuki, Y.; Rodrussamee, N.; Matsutani, M.; Murata, M.; Fujimoto, N.; Suprayogi; Tsuchikane, K.; *et al.* Genetic basis of the highly efficient yeast *Kluyveromyces marxianus*: Complete genome sequence and transcriptome analyses. *Biotechnol. Biofuels* 2015, *8*.
22. Walter, S.; Buchner, J. Molecular chaperones-cellular machines for protein folding. *Angew. Chem. Int. Ed. Engl.* 2002,*41*, 1098–1113.
23. Borges, J.C.; Ramos, C.H. Protein folding assisted by chaperones. *Protein Pept. Lett.* 2005, *12*, 257–261.
24. Purvis, J.E.; Yomano, L.P.; Ingram, L.O. Enhanced trehalose production improved growth of *Escherichia coli* under osmotic stress. *Appl. Environ. Microbiol.* 2005, *71*, 3761–3769.
25. Li, L.L.; Ye, Y.R.; Pan, L.; Zhu, Y.; Zheng, S.P.; Lin, Y. The induction of trehalose and glycerol in *Saccharomyces cerevisiae* in response to various stresses. *Biochem. Biophy. Res. Commun.* 2009, *387*, 778–783.
26. Parrou, J.L.; Teste, M.A.; Francois, J. Effects of various types of stress on the metabolism of reserve carbohydrates in*Saccharomyces cerevisiae*: Genetic evidence for a stress-induced recycling of glycogen and trehalose. *Microbiology* 1997,*143*, 1891–1900.
27. Unnikrishnan, I.; Miller, S.; Meinke, M.; LaPorte, D.C. Multiple positive and negative elements involved in the regulation of expression of *GSY1* in *Saccharomyces cerevisiae*. *J. Biol. Chem.* 2003, *278*, 26450–26457.
28. Kurtzman, C.; Robnett, C.J. Identification and phylogeny of ascomycetous yeasts from analysis of nuclear large subunit (26S) ribosomal DNA partial sequences. *Antonie van Leeuwenhoek* 1998, *73*, 331–371.
29. Harju, S.; Fedosyuk, H.; Peterson, K.R. Rapid isolation of yeast genomic DNA: Bust n' Grab. *BMC Biotechnol.* 2004, *4*.
30. O'Donnell, K. *Fusarium* and its near relatives. In *The Fungal Holomorph: Mitotic, Meiotic and Pleomorphic Speciation in Fungal Systematics*; Reynolds, D.R., Taylor, J.W., Eds.; CAB International: Wallingford, UK, 1993; pp. 225–233.
31. Tamura, K.; Peterson, D.; Peterson, N.; Stecher, G.; Nei, M.; Kumar, S. Mega 5: Molecular evolutionary genetics analysis using maximum likelihood evolutionary distance, and maximum parsimony methods. *Mol. Biol. Evol.* 2011,*28*, 2731–2739.
32. Saitou, N.; Nei, M. The neighbor-joining method. A new method for reconstructing phylogenetic trees. *Mol. Biol. Evol.* 1987, *4*, 406–425.
33. Livak, K.J.; Schmittgen, T.D. Analysis of relative gene expression data using real-time quantitative PCR and the $2^{-\Delta\Delta C_T}$ method. *Methods* 2001, *25*, 402–408.
34. Zoecklein, B.W.; Fugelsang, K.C.; Gump, B.H.; Nury, F.S. *Wine Analysis and Production*; Chapman & Hall: New York, NY, USA, 1995.
35. Dubois, M.; Gilles, K.A.; Hamilton, J.R.; Robers, P.K.; Smith, F. Colourimetric method for determination of sugar and related substances. *Anal. Chem.* 1956, *28*, 350–356.
36. Banat, I.M.; Nigam, P.; Singh, D.; Marchant, R.; McHale, A.P. Review: Ethanol production at elevated temperatures and alcohol concentrations: Part I—Yeasts in general. *World J. Microbiol. Biotechnol.* 1998, *14*, 809–821.
37. Van Uden, N. Cardinal temperatures of yeast. In *Handbook of Microbiology*; Laskin, A.I., Lechevalier, H.H.A., Eds.; CRC Press: Boca Raton, FL, USA, 1984; Volume VI, pp. 1–8.
38. Lee, C.; Yamakawa, T.; Kodama, T. Rapid growth of a thermotolerant yeast on palm oil. *World J. Microbiol. Biotechnol.* 1993, *9*, 187–190.

39. Ballesteros, I.; Ballesteros, M.; Cabanas, A.; Carrasco, J.; Martin, C.; Negro, M.; Saez, F.; Saez, R. Selection of thermotolerant yeasts for simultaneous saccharification and fermentation (SSF) of cellulose to ethanol. *Appl. Biochem. Biotechnol.* 1991, *28–29*, 307–315.

40. Banat, I.M.; Nigam, P.; Marchant, R. Isolation of thermotolerant, fermentative yeasts growing at 52 °C and producing ethanol at 45 °C and 50 °C. *World J. Microbiol. Biotechnol.* 1992, *8*, 259–263.

41. Tanimura, A.; Nakamura, T.; Watanabe, I.; Ogawa, J.; Shima, J. Isolation of a novel strain of *Candida shehatae* for ethanol production at elevated temperature. *SpringerPlus* 2012, *1*, 27.

42. Kaewkrajay, C.; Dethoup, T.; Limtong, S. Ethanol production from cassava using a newly isolated thermotolerant yeast strain. *ScienceAsia* 2014, *40*, 268–277.

43. Charoensopharat, K.; Thanonkeo, P.; Thanonkeo, S.; Yamada, M. Ethanol production from Jerusalem artichoke tubers at high temperature by newly isolated thermotolerant inulin-utilizing yeast *Kluyveromyces marxianus* using consolidated bioprocessing. *Antonie Van Leeuwenhoek* 2015, *108*, 173–190.

44. Walker, G.M. The roles of magnesium in biotechnology. *Crit. Rev. Biotechnol.* 1994, *14*, 311–354.

45. Roukas, T. Solid-state fermentation of carob pods for ethanol production. *Appl. Microbiol. Biotechnol.* 1994, *41*, 296–301.

46. Torija, M.J.; Beltran, G.; Novo, M.; Poblet, M.; Guillamon, J.M.; Mas, A.; Rozes, N. Effects of fermentation temperature and *Saccharomyces* species on the cell fatty acid composition and presence of volatile compounds in wine. *Int. J. Food Microbiol.* 2003, *85*, 127–136.

47. Tofighi, A.; Assadi, M.M.; Asadirad, M.H.A.; Karizi, S.Z. Bio-ethanol production by a novel autochthonous thermo-tolerant yeast isolated from wastewater. *J. Environ. Health Sci. Eng.* 2014, *12*, 107.

48. Narendranath, N.V.; Power, R. Relationship between pH and medium dissolved solids in terms of growth and metabolism of *Lactobacilli* and *Saccharomyces cerevisiae* during ethanol production. *Appl. Environ. Microbiol.* 2005, *71*, 2239–2243.

49. Ercan, Y.; Irfan, T.; Mustafa, K. Optimization of ethanol production from carob pod extract using immobilized*Saccharomyces cerevisiae* cells in a stirred tank bioreactor. *Bioresour. Technol.* 2013, *135*, 365–371.

50. Singh, A.; Bishnoi, N.R. Ethanol production from pretreated wheat straw hydrolyzate by *Saccharomyces cerevisiae* via sequential statistical optimization. *Ind. Crops Prod.* 2013, *41*, 221–226.

51. Laopaiboon, L.; Thanonkeo, P.; Jaisil, P.; Laopaiboon, P. Ethanol production from sweet sorghum juice in batch and fed-batch fermentations by *Saccharomyces cerevisiae*. *World J. Microbiol. Biotechnol.* 2007, *23*, 1497–1501.

52. Ingledew, W.M. Alcohol production by *Saccharomyces cerevisiae*: A yeast primer. In *The Alcohol Textbook*, 3rd ed.; Jacques, K.A., Lyons, T.P., Kelsall, D.R., Eds.; Nottingham University Press: Nottingham, UK, 1999; pp. 49–87.

53. Ozmichi, S.; Kargi, F. Ethanol fermentation of cheese whey powder solution by repeated fed-batch operation. *Enzyme Microb. Technol.* 2007, *41*, 169–174.

54. Laopaiboon, L.; Nuanpeng, S.; Srinophakun, P.; Klanrit, P.; Laopaiboon, P. Ethanol production from sweet sorghum juice using very high gravity technology: Effects of carbon and nitrogen supplementations. *Bioresour. Technol.* 2009,*100*, 4176–4182.

55. Bai, F.W.; Chen, L.J.; Zhang, Z.; Anderson, W.A.; Moo-Young, M. Continuous ethanol production and evaluation of yeast cell lysis and viability loss under very high gravity medium conditions. *J. Biotechnol.* 2004, *110*, 287–293.

56. Yue, G.; Yu, J.; Zhang, X.; Tan, T. The influence of nitrogen sources on ethanol production by yeast from concentrated sweet sorghum juice. *Biomass Bioenergy* 2012, *39*, 48–52.

57. Ter Schure, E.G.; van Riel, A.A.W.; Verrips, C.T. The role of ammonia metabolism in nitrogen catabolite repression in*Saccharomyces cerevisiae*. *FEMS Microbiol. Rev.* 2000, *24*, 67–83.

58. Magasanik, B.; Kaiser, C.A. Nitrogen regulation in *Saccharomyces cerevisiae*. *Gene* 2002, *290*, 1–18.

59. Magasanik, B. Regulation of nitrogen utilization. In *The Molecular and Cellular Biology of the Yeast Saccharomyces: Gene expression*; John, E.W., Pringle, J.R., Broach, J.R., Eds.; Cold Spring Harbor: New York, NY, USA, 1992; pp. 283–317.

60. Beltran, G.; Esteve-Zarzoso, B.; Rozès, N.; Mas, A.; Guillamón, J.M. Influence of the timing of nitrogen additions during synthetic grape must fermentations on fermentation kinetics and nitrogen consumption. *J. Agric. Food Chem.*2005, *53*, 996–1002.

61. Beltran, G.; Rozès, N.; Mas, A.; Guillamón, J.M. Effect of low-temperature fermentation on yeast nitrogen metabolism.*World J. Microbiol. Biotechnol.* 2007, *23*, 809–815.

62. Sridhar, M.; Kiran Sree, N.; Venkateswar, R.L. Effect of UV radiation on thermotolerance, ethanol tolerance and osmotolerance of *Saccharomyces cerevisiae* VS1 and VS3 strains. *Bioresour. Technol.* 2002, *83*, 199–202.

63. Abdel-fattah, W.R.; Fadil, M.; Nigam, P.; Banat, I.M. Isolation of thermotolerant ethanologenic yeasts and use of selected strains in industrial scale fermentation in an Egyptian distillery. *Biotechnol. Bioeng.* 2000, *68*, 531–535.

64. Dhaliwal, S.S.; Oberoi, H.S.; Sandhu, S.K.; Nanda, D.; Kumar, D.; Uppal, S.K. Enhanced ethanol production from sugarcane juice by galactose adaptation of a newly isolated thermotolerant strain of *Pichia kudriavzevii*. *Bioresour. Technol.* 2011, *102*, 5968–5975.

65. Kwon, Y.J.; Ma, A.Z.; Li, Q.; Wang, F.; Zhuang, G.Q.; Liu, C.Z. Effect of lignocellulosic inhibitory compounds on growth and ethanol fermentation of newly-isolated thermotolerant *Issatchenkia. orientalis*. *Bioresour.*

Technol. 2011, *102*, 8099–8104.

66. Ni, H.T.; LaPorte, D.C. Response of a yeast glycogen synthase gene to stress. *Mol. Microbiol.* 1995, *16*, 1197–1205.
67. Verghese, J.; Abrams, J.; Wang, Y.; Morano, K.A. Biology of the heat shock responses: Building yeast (*Saccharomyces cerevisiae*) as a model system. *Microbiol. Mol. Biol. Rev.* 2012, *76*, 115–158.
68. Piper, P.W. Molecular events associated with acquisition of heat tolerance by the yeast *Saccharomyces cerevisiae*. *FEMS Microbiol. Rev.* 1993, *11*, 339–356.
69. Lindquist, S. Heat-shock proteins and stress tolerance in microorganisms. *Curr. Opin. Genet. Dev.* 1992, *2*, 748–755.
70. Singer, M.A.; Lindquist, S. Thermotolerance in *Saccharomyces cerevisiae*: The Yin and Yang of trehalose. *Trends Biotechnol.* 1998, *16*, 460–468.
71. Hottiger, T.; De Virgilio, C.; Hall, M.N.; Boller, T.; Wiemken, A. The role of trehalose synthesis for the acquisition of thermotolerance in yeast II. Physiological concentrations of trehalose increase the thermal stability of proteins *in vitro*.*Eur. J. Biochem.* 1994, *219*, 187–193.
72. Zancan, P.; Sola-Penna, M. Trehalose and glycerol stabilize and renature yeast inorganic pyrophosphate inactivated by very high temperatures. *Arch. Biochem. Biophys.* 2005, *444*, 52–60.
73. Lindquist, S.; Kim, G. Heat-shock protein 104 expression is sufficient for thermotolerance in yeast. *Proc. Natl. Acad. Sci. USA* 1996, *93*, 5301–5306.
74. Krsmanović, T.; Kőlling, R. The HECT E3 ubiquitin ligase Rsp5 is important for ubiquitin homeostasis in yeast. *FEBS Lett.* 2004, *577*, 215–219.
75. Cardona, F.; Aranda, A.; Del Olmo, L. Ubiquitin ligase Rsp5 is involved in the gene expression changes during nutrient limitation in *Saccharomyces cerevisiae*. *Yeast* 2009, *26*, 1–15.
76. Shahsavarani, H.; Sugiyama, M.; Kaneko, Y.; Chuenchit, B.; Harashima, S. Superior thermotolerance of *Saccharomyces cerevisiae* for efficient bioethanol fermentation can be achieved by overexpression of *RSP5* ubiquitin ligase. *Biotechnol. Adv.* 2012, *30*, 1289–1300.

CHAPTER 19

POTENTIAL OF SORGHUM AS AN ALTERNATIVE TO CORN FORAGE

Girma Getachew, Daniel H. Putnam, Christopher M. De Ben, Edward J. De Peters

Keywords: Sorghum, Climate Change, Drought Tolerance, Salinity

INTRODUCTION

The aim of this review is to summarize the literature related to Sorghum as forage for livestock and assess its role as silage compared with corn silage. Sorghum forage has predominantly been a minor silage forage for livestock in the United States. Due to climate change and predictions for drier conditions (lower rainfall and less water for irrigation) in some regions, for example California with its large dairy and beef cattle industries, there is a search for alternative forages that require less water, fit the production system that integrates forage production and manure management, and maintains agricultural sustainability [1] . In California, In California, water policy is driven by the needs of often competing entities, for example agriculture, human population, and natural animal and plant species. Sorghum is currently of interest as forage because of drought conditions in various dairy regions, but a comprehensive review on Sorghum forage is not available. A goal of this review of Sorghum is that it will serve as a reference guide to livestock and dairy producers, nutritionists, nutrient management planners, and policy makers.

Sorghum is in the subfamily Panicoideae and the tribe of Andropogoneae (the tribe of big bluestem and sugar cane). It is one of the major cereal crops grown in the world, along with wheat, oats, corn, rice, and barley. Sorghum ranks fifth among the world's most important crops [2] . In traditional form, Sorghum is a towering plant that grows over 1.8 m tall, although many recent cultivars designed for cultivation are dwarf varieties, specially developed for ease of harvest. In Africa, however, traditional tall Sorghum is still grown, and the stalks are put to a variety of uses (e.g. fuel or construction). Sorghum belongs to an annual grass group that is drought tolerant and is an excellent choice for livestock feed in arid and relatively dry regions. This tolerance offers potential in areas where water is and will continue in the future to be in decreasing supply, for example the Central Valley of California.

Sorghum varieties are grown for various purposes. In many parts of the world, especially in areas where moisture is a limiting factor, Sorghum is grown for grain for human consumption. In Texas and Oklahoma where annual precipitation is low, sorghum is grown for grain that is used for feeding livestock and poultry. More recently sweet Sorghum is being used for biofuel production. Sorghum grain is also used around the world to brew beer. Broom Sorghum is cultivated for the manufacture of traditional straw brooms.

Sorghums in general can be classified into two broad types: forage types and grain types. The forage Sorghums are further grouped into four types: 1) hybrid forage Sorghum; 2) Sudangrass; 3) Sorghum x Sudan hybrids (also known as Sudan hybrids); 4) sweet Sorghum.

In United States when grain and hay crops are considered, Sorghum ranks in the top 10 crops based on land area of cultivation. When Sorghum area planted for all purposes was considered, in the 2014 cropping season, Kansas was the leading state followed by Texas with an area of 2.8 million and 2.5 million acres, respectively, allocated to Sorghum [3] . In fact, the United States is the largest producer and along with Argentina and Australia, is one of the largest exporters of Sorghum grain. However, in United States, the area under Sorghum cultivation is much smaller (3.1% of the total land area devoted to grain and hay crops) compared with corn, which covers 42% of the area. About 2.5 million hectares were devoted to corn for silage whereas 153,780 hectares were used for Sorghum silage (Table 1). Grain type Sorghum has become a popular rotation crop in the Great Plains [4] . Although Sorghum is able to thrive and produce a reasonable yield in marginal areas, it is surprising that Sorghum forage has not received more interest in areas where water availability is low and soil fertility is marginal. Discussion of the potential of Sorghum forages for livestock feeding is a topic of this review.

Various types of forage Sorghums and their corresponding agronomic and nutritional characteristics were reviewed and are summarized. The reviewed studies covered different

Citation: Girma Getachew, Daniel H. Putnam, Christopher M. De Ben, Edward J. De Peters (2016). Potential of Sorghum as an Alternative to Corn Forage. American Journal of Plant Sciences,07,1106-1121. doi:10.4236/ajps.2016.77106.

genetic materials, various agronomic conditions, and the impact on production performance with the inclusion of forage Sorghums in diets of various classes of livestock.

BMR SORGHUM

Two main types of forage Sorghum hybrids, brown midrib (bmr) and non-bmr, are utilized most often for silage production. Brown midrib phenotypes in corn plants were first described by Jorgensen [5] as having brown pigmentation in the leaf midrib, stem, tassel, cob, and roots. The bmr gene was associated with lower plant lignin [6] . As single-locus mutants identified in a number of grass species, bmr varieties offer a rapid and effective approach to genetically modify the nutritional value of forage plants. As single locus recessive mutations, they can be backcrossed readily into elite lines [7] .

The bmr hybrids are characterized by the expression of a bmr gene associated with lower lignin concentration than comparable non-bmr hybrids. Reduced lignin content was associated with increased overall apparent digestibility of the fiber component by ruminants, and thus the overall nutritive value of the bmr forage was improved [8] . A disadvantage of the bmr hybrid varieties was lodging, particularly if harvest was delayed past the optimum stage.

Table 1: Crop area planted and harvested in the United States during the 2013 crop season

Grain and Hay Crops	Area Harvested (hectares)	% of the Area
Barley	1,214,070	1.35
Corn for grain	35,478,360	39.31
Corn for silage	2,531,740	2.81
Hay, all	23,576,030	26.12
Alfalfa	7,188,510	
Others	16,387,520	
Oats	416,830	0.46
Proso millet	258,190	0.29
Rice	998,770	1.11
Rye	112,500	0.12
Sorghum for grain	2,642,630	2.93
Sorghum for silage	153,780	0.17
Wheat, all	18,274,590	20.25
Winter	13,112,770	
Durum	575,060	
Other spring crop	4,586,760	5.08
Total	90,244,250	100.00

Source: Crop Production 2013 Summary (January 2014), USDA, National Agricultural Statistics Service.

Lodging was influenced more by variety than bmr genotype. When harvested at early heading stage of growth, there was no difference in lodging between bmr-101 and conventional variety FS-5 whereas when harvested soft dough stage lodging was 28% and 58% in FS-5 and bmr101, respectively [9] . However, others reported that lodging was not increased in bmr cultivars compared with the nonbmr cultivars when harvested at the soft-dough stage and there was little correlation between lodging and lignin content, grain yield, or plant height [6] [10] .

Brown midrib mutations both in Sorghum and corn are phenotypically characterized by the presence of brown vascular tissue in the leaf blade and sheath as well as in the stem. The bmr phenotype becomes visually apparent once plants have reached the four-leaf stage and tends to begin to fade as plants approach physiological maturity [11] . Agronomic traits attributed to bmr can be summarized as follows:

- Dwarfing gene (bmr-6) increases leaf to stem ratios
- Adapts to heavier grazing pressure
- Excellent disease resistance
- Rapid growth and regrowth
- Drought stress tolerant
- Improved tillering capacity
- Increased leafiness allows for faster dry-down for hay
- Adaptable to various soil pH ranges including acidic soils

Nutritional attributes of bmr mutants when present in homozygous recessive state include:

- Reduced lignification which contributes to reduced cell-wall concentration, increased digestibility [12] , and increased voluntary intake by ruminants [7]
- Improved animal performance with increased rate of gain [13] and increased milk yield in dairy [14]
- No prussic acid or hydrogen cyanide (HCN) risk

Overall, the bmr trait dramatically improved forage quality of Sorghums as forage for dairy cattle. The lower lignin content in bmr plants resulted increased apparent digestibility of fiber component as well as the whole plant. It is well documented that lignin is for the most part indigestible, and it plays a critical role in forage quality by affecting digestibility of cell wall polysaccharides. However, as lignin contributes to the structural component of plant, it also plays a role in contributing rigidity that reduces lodging in plants. There were wide variations (0% to 60%) in the extent of lodging of Sorghum varieties [15] . The contrasting roles of plant lignin create challenging issues for the development of bmr varieties. Lower lignin content enhances forage quality but can reduce agronomic performance. Lodging is a serious issue when forage is mechanically harvested because of associated costs of labor and time as well as the loss of harvested organic matter. An optimum balance must be found when the bmr gene is introduced into forage varieties of Sorghum, Sudangrass, and Sorghum x Sudangrass hybrids.

The characteristic bmr mutant coloration of the leaf mid-veins was associated with reduced lignin content and altered lignin composition, traits useful to improve dry matter digestibility for livestock based on in vitro analyses [12] (Table 2). Brown midrib phenotype was correlated with two homologous loci in maize (bm1 and bm3) and Sorghum (bmr6 and bmr12) [12] . In corn, the bmr phenotype was associated with increased apparent DM digestibility in ruminants but at a cost of significant reduced forage and grain yields. In Sorghum, yield reductions were apparent in near isogenic lines, but were ameliorated through construction of hybrids that maintain reduced lignin content and increased digestibility [12] . Forage DM yield was lower for bmr Sorghum compared with the wild type, however, with respect to animal performance, digestibility and milk yield were higher in bmr varieties compared with the wild type [12] .

Reduced tillering and plant height of the bmr plants appeared to be mechanisms accounting for reduced forage yield [7] . Predicted net returns from feeding Sudangrass hay were similar for first harvest conventional and bmr lines, but net return was severely depressed for bmr lines in second harvest, due to significant reductions in plant yield [7] . In a yield and lodging study, yield of different varieties of Sorghum ranged from 12 to 16 ton DM/ha in two harvests [15] . In some Sorghum varieties [15] , lodging was as high as 60% (Table 3).

FORAGE SORGHUM

Forage Sorghum is a warm season annual crop and is an excellent choice for a one-cut system (similar to corn for silage). However, in contrast to corn, forage Sorghum is better adapted to production systems under marginal agronomic conditions. For forage production, Sorghum used about 30% less water than corn and also required less nitrogen fertilizer than corn [16] . Forage Sorghums were similar to grain types but were taller and had higher forage quality [16] .

Table 2: Yield, chemical composition, and nutrient digestibility of bmr forage Sorghum silage

	Wild type	bmr6	bmr18	Reference
Forage DM yield (t/ha)*	15.0	12.8	13.5	[6]
Silage				
NDF (%)	58.1	50.2	48.2	[14]
ADF (%)	37.7	33.6	28.5	[14]
ADL (%)	2.9	2.3	2.5	[14]
CP (%)	7.3	7.5	7.8	[14]
Starch (%)	10.9	14.5	16.8	[14]
Total tract digestibility				
DM digestibility (%)	52.5	62.9	69.1	[14]
NDF digestibility (%)	40.8	54.4	47.9	[14]
Lactation performance				
Milk (kg/d)	31.0	34.1	32.2	[14]
Milk fat (%)	3.57	3.89	3.77	[14]
4% Fat-corrected milk (kg/d)	29.1	33.7	31.2	[14]

ADF, Acid Detergent Fiber; NDF, Neutral Detergent Fiber; ADL, Acid Detergent Lignin; CP, Crude Protein.

Table 3: Yields and lodging of forage Sorghums grown in different locations

Varieties	Locations	DM yield from two cuts (ton/ha)			Lodging %		Source
		1st cut	2nd cut	Total yield	1st cut	2nd cut	
Brawley	Imperial Valley, CA*	25.1	11.8	36.9	1	3	[15]
Lindsey 101F	Imperial Valley, CA	14.5	20.0	34.5	0	8	[15]
NK 300	Imperial Valley, CA	11.4	20.8	32.1	0	0	[15]
Dekalb FS1a	Imperial Valley, CA	11.4	19.7	31.1	0	0	[15]
Dekalb FS-22	Imperial Valley, CA	19.6	10.9	30.4	55	49	[15]
Hegari	Imperial Valley, CA	12.0	17.9	30.0	0	0	[15]
Sila King	Imperial Valley, CA	19.2	9.3	28.4	47	55	[15]
Atlas	Imperial Valley, CA	16.7	10.6	27.4	0	60	[15]
Taylor Evans Haygrazer	Davis, CA**	17.9	10.1	28.0	-	-	[15]
Asgrow Grazer	Davis, CA	14.0	11.9	25.9	-	-	[15]
NK 145	Davis, CA	15.3	10.1	25.3	-	-	[15]
Frontier Hydan	Davis, CA	14.2	10.1	24.3	-	-	[15]
DeKalb SX-11	Davis, CA	13.1	9.4	22.5	-	-	[15]
Durrant GX-200	Davis, CA	14.1	9.0	23.1	-	-	[15]
Lindsey 77F	Davis, CA	13.5	8.1	21.6	-	-	[15]
Piper	Davis, CA	6.7	8.9	15.7	-	-	[15]
Greenleaf-bmr	Arlington, VA	9.1	3.2	12.2	-	-	[7]
Greenleaf Conventional	Arlington, VA	10.6	3.9	14.6	-	-	[7]
Piper-bmr	Arlington, VA	8.4	3.0	11.3	-	-	[7]
Piper Conventional	Arlington, VA	12.3	4.4	16.8	-	-	[7]
Greenleaf-bmr	Ithaca, NY	12.4	4.0	16.4	-	-	[7]
Greenleaf Conventional	Ithaca, NY	11.3	5.2	16.5	-	-	[7]
Piper-bmr	Ithaca, NY	9.0	4.3	13.3	-	-	[7]
Piper Conventional	Ithaca, NY	12.4	6.1	18.4	-	-	[7]

*Cut at flowering stage; **Cut at soft dough stage of maturity.

SUDANGRASS

Sudangrass has finer stems, tillers more profusely, and is leafier than hybrid forage Sorghums. It produces few seeds, and its rate of regrowth after cutting or grazing is generally superior to that of hybrid forage Sorghums. For these reasons, Sudangrasses are often used for temporary rotational grazing. Furthermore, Sudangrass accumulates less of the poisonous compound prussic acid (HCN; hydrocyanic acid) than forage Sorghums.

HYBRID FORAGE SORGHUMS

Sorghum and Sudangrass hybrids are crosses between forage-type Sorghums and Sudangrass. The hybrids have the highest yield potential of any of the summer annuals if adequate rainfall is received or irrigation is provided. However, the hybrids possess characteristics that give them advantages when water is in short supply. The hybrids were reported to have less leaf area, more secondary roots, and a waxier leaf surface than corn; traits that favor Sorghum and Sudangrass hybrids under growing conditions of low water availability [17] . Consequently, Sorghum-Sudangrass hybrids offer potential for silage production in drought areas or agronomic conditions with reduced water availability.

Hybrid forage Sorghums are commonly referred to as forage Sorghum and grow from 1.8 to 3 m in height and have relatively large stem diameters. These varieties were selected for a single cut system for either hay or silage. In some instances forage Sorghum produced DM yields similar to silage corn, but the forage quality was generally lower than corn [8] . Forage Sorghums can produce a forage crop containing up to 50% grain by weight, depending upon the hybrid and stage of maturity at harvest. These Sorghum grain yields approach levels associated with corn varieties. Corn forage harvested for ensiling is typically kernel processed to enhance fermentation during ensiling as well as fermentation in the rumen when fed. However, Sorghum's small seed size is a detriment to digestion by ruminants unless it is physically processed. Although technology exists for kernel pro- cessing of corn for silages, there was no information in the literature on kernel processing for Sorghum grain in forage used for silage. Sorghum grain is processed in the feed mill industry, typically steam-flaked (heat to gelatinize the starch and rolling to disrupt the starch granules), prior to feeding to cattle (beef and dairy). Careful selection of bmr hybrids and timing of harvest may help in maximizing total energy yield of forage. Highest crude protein (CP) content and apparent DM digestibility will usually be obtained by harvesting in the vegetative growth stage, whereas DM yield will increase as the plant matures. Even though the maximum total gross energy harvested per hectare occurs with harvesting Sorghums when the grain is in the hard-dough stage, lower digestible energy concentration per unit of DM is achieved by livestock. This relationship between forage digestibility and organic matter yield as maturity progresses is common among most forage species. As a winter cereal forages matured from bloom to hard dough, the NDF concentration decreased but lignin concentration increased [18] . In vivo and in vitro DM digestibility decreased with advancing physiological maturity with plant lignification although empirical equations based on fiber concentration would inaccurately predict a higher digestible energy content. The relationship between digestibility and forage yield will impact agronomic practices and the development of new Sorghum forage cultivars.

In the southeastern United States, Sorghum and Sudangrass hybrids are commonly used as a forage crop for stocker cattle and dry cows (non-lactating cows). Sorghum and Sudangrass forage crops are designed for multiple harvests and can be used as hay, silage, pasture, or green chop. However, the hybrids dry slowly, even if an impeller (flail) or roller-crimper conditioner is used during crop harvest. The waxier leaf surface that favors growth with restricted water possibly hinders field drying of the forage for hay. Consequently, hay production from these species is at greater risk of damage in areas with an unpredictable rain pattern as drying of plant material becomes more difficult, but the agronomic advantage of Sorghum and Sudangrass is forage production under limited water.

The hybrids have an excellent planting period so these forages fit well into many crop rotation systems with winter-grown small grain or vegetable crops. Crop rotation with Sorghum is often practiced in the Central Valley of California to utilize residual fertilizers from previous crops as a nutrient management method. Sorghum also utilizes nutrients associated with manure application as a component of the nutrient management program of a dairy farm. Some vegetable producers use Sorghum as cover crops to increase soil organic-matter content, retain soil nutrients, and reduce weed and pest populations. The hybrids work well in this rotation for dairy and beef cattle operations where high-quality grazing, green chop or silage is needed.

FACTORS AFFECTING FORAGE YIELD

Forage yield of Sorghum is influenced by agronomic factors including soil type, temperature, and availability of irrigation water. When four Sorghum varieties were compared [19] , forage yields ranged from 15.3 and 16.5 tons DM per hectare (Table 4). The DM yield of 14 Sorghum varieties (Table 5) ranged from 10.5 to 14.9 ton/ha with the average of 12.8 ton/ha and similarly the DM yield in 11 varieties of corn ranged from 9 to 15.3 ton/ha with an average of 12.5 ton DM/ha [20] . Under rain fed conditions the agronomic characteristics of corn and Sorghum were similar for DM yield although there were notable variations in yield for both corn and Sorghum varieties (Table 5). Although DM yields were similar, quality measurements favored corn silage for chemical composition (i.e. lower ADF content, higher digestible DM) and animal performance (i.e. greater average daily gain and DM intake) compared with Sorghum silage. Variability in ADF concentration, DM digestibility, DM intake, and ADG was higher for Sorghum silage compared with corn silage. High variability was also observed for ADG for Sorghum (16.3 g/d Mean: 21.0 SD) compared with corn (70 g/d Mean; 18.4 SD). In addition, with three Sorghum silages sheep lost weight during the feeding study. Estimated ADF intake was similar for corn silage (225 g/d) and Sorghum silage (218 g/d). However, estimated digestible DMI of sheep was 516 for corn silage and 360 g for Sorghum, a 30% reduction in intake with Sorghum silages, reflecting the lower performance with Sorghum silage.

Table 4: Yield, lodging and nutritional value of Sorghum varieties

Varieties	DM Yield (t/ha)	Lodging (%)	CP (%)	IVDMD (%)	IVFD (%)	TMR Intake (kg/d)	Milk Production (kg/day)	Milk Fat (%)	Body Weight Change in 5 Weeks (kg/cow)	Source
FS-5	13.0		5.07	64	56	25.4	40.7	3.88	−49	[8]
BMR-101	10.8		5.83	67	60	24.9	41.4	3.81	−33	[8]
Corn	17.8		5.95	67	56	25.8	42.1	3.52	−10	[8]
FS-5	16.4	26.7	5.1	68	59	-	-	-	-	[19]
Silobuster	16.5	56.7	5.2	73	68	-	-	-	-	[19]
Supersile-20	16.4	30.0	5.3	71	65	-	-	-	-	[19]
BMR-101	15.3	43.3	5.8	72	52	-	-	-	-	[19]

CP, Crude Protein; IVDM, In Vitro Dry Matter Digestibility; IVFD, In Vitro Fiber Digestibility; TMR, Total Mixed Ration.

Data summarized from different studies and locations (Table 3) indicate considerable variations (mean and standard deviation of 23.2 t/ha and 7.47 respectively) in forage DM yield of Sorghum varieties (11.3 to 37 t/ha) in agreement with the findings of others [20] (Table 5). A recent three-year multi-location study indicated that corn for silage yielded 23 ton DM/ha with a low variance for yield whereas Sorghum and millet varieties had DM yields of 3 - 5 ton/ha lower than silage corn yields and again there was considerable variability in yields within years and locations for Sorghum [21] . Thus, when considering Sorghum for forage, consideration must be given to the large expected variation in yield of DM, chemical composition, and subsequent animal performance with Sorghum. This variation will also impact nutrient budgeting on dairy farms since large variation in DM yield will result in large variation in estimates of nutrient uptakes from soil application of dairy manure.

SOIL TYPES AND FERTILITY

A positive attribute of Sorghum is that it is adapted to a variety of soil types. Sweet Sorghums, for example, were more tolerant of soil salinity than other forage crops [22] . Though Sorghum grows best on deep, fertile, well-drained loamy soils, Sorghum is also much more tolerant of shallow soil and drought conditions than corn. Sorghum can be successfully grown on clay, clay loam, or sandy loam soils.

Strong crusts, typically 0.64 to 1.27 cm thick, hinder a plant's ability to emerge. Soil crusts typically form after heavy rain combines with warm temperature, low humidity, and high winds to promote rapid soil drying. Soils with high content of fine sand and silt more readily form crusts than coarse sand and clay textured soils. Light tillage with a rotary harrow or sprinkler irrigation can sufficiently weaken crusts to allow the plants to emerge. Cropping systems that leave large amounts of plant residue on the soil surface, such as no-till or conservation tillage, help reduce soil crusting. Traditionally Sorghum has been grown on soils with pH of > 6.5, but soil acidity (pH 5.42) reduced grain yield by 10% [4] .

The time from planting to emergence (usually 5 to 10 days) depends on the growing conditions including soil temperature and moisture, the depth of planting, and to some extent, seed vigor. Slow emergence often results in uneven, skimpy plant stands and decreased forage yield. Before emergence, the plant is dependent on food reserves in the seed from the endosperm for survival. Slow emerging plants risk depleting these reserves, which are important to early plant growth in the days immediately following emergence. In addition to soil type and crusting, temperature influences emergence and early plant growth.

Although Sorghum can grow and yield a reasonable forage biomass from residual soil nutrients when planted after heavily fertilized crops such as corn, it responds well to added nutrients particularly N. In general, forage Sorghum should receive 54 to 68 kg N/ha for a yield goal of 6 to 7.5 ton DM/ha [16] . Irrigated forage Sorghums should receive up to 30% more N [16] . When calculating the amount of N needed, it is always good to consider residual soil N, soil organic matter, N content in irrigation water, and N in manure applications. Application of N can be reduced if Sorghum follows legumes or highly fertilized crops [16] . It has been shown that sweet Sorghum was more environmentally friendly than corn because of the lower N required for Sorghum production [22] .

Table 5: Agronomic and nutritional quality of corn and Sorghum silages ([20])

	Dry matter yield (t/ha)	ADF (%)	DDM (%)	DMI (g)	ADG (g)
Corn					
Morden 77	9.0	23.6	69.1	905	91.0
Golden Jewel	9.9	28.7	61.8	867	75.3
NK Opaque	13.9	29.5	64.1	889	64.3
MN 8803-05	9.9	25.1	66.5	806	78.3
AES 101	11.6	25.7	69.0	783	109.6
Mn Syn 16	9.8	29.1	63.3	811	65.7
Oh 5 x Oh 43	13.0	27.0	65.6	819	72.7
Mn hy 5301	15.0	27.9	62.9	780	63.9
Pioneer 3558	15.3	29.8	62.9	769	58.7
Penn E Syn 2	11.4	32.9	60.3	698	45.4
Cargi 1 HS 50	13.2	32.7	63.0	663	49.1
Average[1]	12.0	28.4	64.4	799.1	70.4
SD[1]	2.22	2.93	2.84	74.00	18.36
Sorghum					
NK 133	11.6	27.2	61.4	836	59.9
Robusto	10.5	29.0	61.2	748	2.1
Duet	12.8	25.7	59.3	729	41.1

Grace 22F	13.4	31.3	60.6	660	−0.5
Trudan 11	10.7	37.1	53.4	723	26.9
Sumax	14.4	33.4	58.9	612	20.0
Pay. 401R	14.9	33.3	56.4	617	32.2
NK 318	13.0	34.3	55.3	610	26.3
Dekalb FS1a	12.1	37.2	55.2	588	12.8
NK 325	11.5	36.2	52.5	598	16.2
NR 367	15.7	42.4	53.9	554	−22.9
MRA FS 500	16.6	36.0	53.5	583	14.6
Sweet Sioux	12.5	39.8	49.2	595	8.3
Pioneer 931	14.4	41.1	51.8	516	−8.7
Average[1]	13.2	34.6	55.9	640.6	16.3
SD[1]	1.85	5.02	3.84	87.71	20.98

ADF, Acid Detergent Fiber; DDM, Digestible Dry Matter; DMI, Dry Matter Intake by Sheep; ADG, Average Daily Gain; [1]Average and Standard Deviation (SD) calculated by the authors.

An advantage for the use of Sorghum as forage rather than corn is the ability of Sorghum to grow in soils moderately saline or alkaline [15] . Although forage yield may decrease under these conditions, the Sorghum forage plant can produce a crop in areas where corn as a silage crop would fail [15] .

WATER-REQUIREMENTS

Although the biomass yield for Sorghum was lower than corn, the water requirement for Sorghum was much less than what corn requires for optimum yield [23] . Nevertheless, although Sorghum is adapted to an arid environment with low water requirement, it responds well to both increased level and frequency of irrigation. Increasing irrigation level from 20 mm to 180 mm increased DM yield from 16.5 t/ha to 30.1 t/ha (Table 6). Protein yield also increased in response to irrigation from 1.13 t crude protein/ha vs. 2.42 t protein/ha; a 114% increase [24] . Even though Sorghum is drought tolerant, forage DM yield was increased by as much as 100% in response to increased irrigation water application ([24]). The increased forage DM yield was associated with increased CP content but reduced in vitro DM digestibility [24] . The decreased in vitro DM digestibility reflected the increased NDF content and the lower in vitro NDF digestibility associated with the great DM yield (Table 6).

Although Sorghum will respond to irrigation like most crops, a significant advantage of Sorghum compared with corn for forage is the water requirement for plant growth. Sorghum was able to survive and produce comparable DM yield with half the amount of irrigation water that was applied to corn (250 vs. 550 mm) [8] .

Water-use efficiency of sweet Sorghum was higher when water stress occurred during the early stage of growth compared with the later phases [25] . These authors suggested that irrigation strategies should consider the growth phase of the plant with respect to points of the growth phase that are more or less sensitive to water stress. Water-use efficiency of sweet Sorghum was reported to be high for C4 crops under well-watered conditions as well as water-stressed conditions [26] . When several forage varieties under different water availability conditions from 2005 to 2009 were compared, a wet year had 120% of normal annual rainfall while a dry year had 80% [27] . During a normal year, water use efficiency (62 kg/ha/mm) and DM biomass yield (23.1 t/ha/yr) for corn exceeded Sorghum hybrid Sudangrass (24 kg/ha/mm and 7.7 t/ha/yr), respectively [27] . However, in dry years the water-use efficiency for Sorghum hybrid Sudangrass was 28 kg/ha/ mm compared with 26 kg/ha/ mm for corn and biomass yield of DM was the same for Sorghum hybrid, Sudangrass and corn (10.6 t/ha/yr) demonstrating the superiority of Sorghum production under low water available conditions. Corn production required water for high yields of DM. However, chemical composition of the forage crops was not reported, and this would be an important aspect to consider with respect to water allocation during the growing season for Sorghum [27] . The frequency of irrigation was found to be more effective than the amount of water in increasing the DM yield; more frequent irrigation was more effective than single or less frequent irrigation. Averaged over two growing seasons, maximum DM yields of 16.3, 11.8, and 10.5 t/ha were reported for frequent (10 irrigations), intermediate (7 irrigations), and infrequent irrigation (5 irrigations), respectively [28] .

Even though Sorghum is able to grow and survive in regions where rainfall is relatively low or where frequent dry periods occur, plant growth and forage yield were still closely linked to available water. Water-use estimates for Sorghum typically range from about 38 - 76 cm of water to produce 25 to 37 tons of forage DM/ha [16] . Across this range of water use, bmr forage Sorghum showed an increase in DM yield of approximately 1.86 t/ha for each additional 2.54 cm of water use, from about 22.4 t/ha at 38 cm of water use to about 49.3 t/ha at 76 cm of water use [16] . Yield of Sorghum forages increased with water availability, DM yields per unit of water use were similar between Sorghum and corn.

Table 6: Effect of irrigation level on yield and forage quality of forage Sorghum [24]

		Irrigation levels (mm)		
	cuts	20	100	180
DM yield (t/ha)	1st cut (early heading)	8.26	12.3	11.1
	2nd cut (soft dough)	8.27	13.9	19.0
	total	16.5	26.2	30.1
CP (%)	1st cut (early heading)	7.36	8.41	9.14
	2nd cut (soft dough)	6.37	6.41	7.41
		6.87	7.41	8.28
IVDM (%)	1st cut (early heading)	0.74	0.72	0.61
	2nd cut (soft dough)	0.65	0.63	0.63
NDF (%)	1st cut (early heading)	61.5	62.5	65.8
	2nd cut (soft dough)	63.3	64.8	65.8
NDFD (%)	1st cut (early heading)	0.72	0.69	0.62
	2nd cut (soft dough)	0.62	0.57	0.62

CP, Crude Protein; NDF, Neutral Detergent Fiber; IVDM, In Vitro Dry Matter Digestibility; NDFD, In Vitro NDF Digestibility.

However, a primary advantage of Sorghum is that it produced higher forage DM yield with less water than corn. Therefore, in regions where water availability is low, because of lack of rainfall or irrigation water is either not available or available only in limited amounts, Sorghum offers an attractive alternative to corn for forage production. But, lower yields for Sorghum than corn for silage must be expected with the genetic base of Sorghums currently available. In California dairy production systems where double and triple cropping are options, Sorghum forage could integrate well in a system that produces corn for silage with Sorghum planting following corn harvest and Sorghum harvesting occurring prior to planting of winter cereals. In areas where water is not available to support corn production then different cropping scenarios will evolve over time and experiences.

TEMPERATURE

A quantitative definition of temperature stress in Sorghum, as in any crop, is difficult to provide since it will depend on a number of factors, which include the duration of exposure of either high or low temperature, the activity or stage of growth of the exposed tissue, and finally the thermal adaptation of the particular Sorghum cultivar [29] . The adaptation productivity of a Sorghum crop will be most affected when temperature stress coincides with critical stages of growth [30] . Heat stress is a major factor influencing the productivity and adaptation of many wild and cultivated plants. Many crop species of tropical and subtropical origin are sensitive to high temperature in the range 30°C - 55°C [29] [30] . However, Sorghum is less tolerant to cold temperature compared with corn. The growth and yield responses of Sorghum to high temperature depend on a number of factors. These factors include the duration of exposure of either high or low temperature, the activity or stage of growth of the exposed tissue at the time of temperature stress, and finally the thermal adaptation of the particular Sorghum cultivar. Plants prefer warm, moist soil for germination and emergence. In contrast cool, wet soils promote the development of diseases. Temperature for Sorghum germination and emergence was 21°C and leaf appearance increased linearly with air temperature from 13°C to 23°C [29] . Although germination can occur at temperatures below 10°C, emergence is delayed. For optimum germination and emergence, delay planting until the 5-day average daily soil temperature at the 5 cm depth reaches 15.6°C. In central and south Texas, predominantly dark color soils and higher frequency of warm spring temperatures allow growers to begin planting when soil temperatures are around 10°C to 12.8°C. If planting in cool, wet soils cannot be avoided, plant shallow and use fungicide seed treatments.

CHEMICAL COMPOSITION

A short coming of Sorghums is their protein content, which is often lower than corn for silage. In one study [8] the highest CP content (DM basis) achieved was 5.8% in bmr and the lowest was 5.1% in variety FS-5 [8] (Table 4). The low CP forage was also associated with low forage digestibility reflecting the relationship between nutrient content and forage digestibility with advancing maturity. The bmr forage had the highest in vitro forage DM digestibility (72%) while FS-5 had the lowest (68%). In one study, Sorghum varieties produced yields of DM similar to corn [20] . However, the corn hybrids have changed dramatically since 1976 with respect to dry matter yield per acre as well as the amount of grain produced in some hybrids used for silage. Indeed, more recent work [8] reported higher yields of DM for corn than two Sorghum varieties (Table 4).

In a total mixed ration, the total tract digestibility was highest for bmr18 lines, intermediate for bmr6 lines, and lowest for wild type lines [12] . Similarly, bmr6 and bmr12 had highest in vitro fiber digestibility compared with the wild type [12] . Brown-midrib lines averaged 9.0% lower lignin content and 7.2% higher in vitro NDF digestibility than the respective conventional lines [7] . Others [31] measured in situ digestibility of NDF at 48 h and found no difference in NDF digestibility for corn silage, whole plant grain Sorghum silage, and forage Sorghum silage. In vitro DM digestibility of 64.7% for corn silage and 51.5% were reported for grain Sorghum silage, but in vitro NDF digestibility was not determined [32] . When grain-Sorghum silage and corn silages were compared, grain-Sorghum yielded 15.5 t/ha whole plant DM and 6.5 t/ha grain whereas corn yielded 7.0 t/ha and 2.5 t/ha whole plant DM and grain, respectively [32] . The authors attributed the lower in vitro DM digestibility of Sorghum silage to a low digestibility of Sorghum stalks and leaves with the assumption that the Sorghum grain starch was digested. The starch contents of the pasture, corn silage, and grain Sorghum silage were 15, 137, and 229 g/kg DM, respectively [32] . In a study that compared Sorghum varieties and corn, total tract digestibility of starch in wild type, bmr6, bmr18, and corn silages were 85.7%, 82.3%, 79.7%, and 91.7 %, respectively [14] .

When corn and Sorghum silages were fed to milking cows, apparent digestibility of dry matter, crude protein and digestible energy were comparable for both Sorghum and corn silages [33] (Table 7). However fat corrected milk yield significantly higher for corn silage than Sorghum silage. In a University of California study that consisted of five Sorghum lines and four corn lines, Sorghum forage was slightly higher in CP content, 7.9% compared with corn at 7.5% although calculated protein yield would favor corn (1.85 vs. 1.72 t/ha) (Table 8). Protein is an expensive component of diets. However, national energy policies for ethanol production from corn grain increased the amount of corn distillers' grains as a feedstuff. In addition, nutrient management to meet regulations for N are incentives for dairy producers not to over feed protein so a lower protein content of Sorghum forages if this occurs is not a critical issue.

When choosing a Sorghum hybrid for silage production both yield and nutritive value potential should be considered. Stage of harvest for Sorghum will depend on a balance between forage DM yield and forage nutrient and energy concentrations, and how these factors fit the overall nutritional needs of the dairy herd as well as the cropping system used. Double and triple cropping can be time critical for planting so harvest time can be impacted by factors other than the quality of forage at harvest time. Each feed commodity in a dairy ration plays a different role. Sorghum's role in the ration will vary for each dairy farm. In recent years there has been a decrease in alfalfa hay fed to dairy cows in California [34] as hay costs have increased and silages have played a more important role in management of nutrients from dairy manure.

Table 7: Apparent digestibility of dry matter and crude protein, and intake energy, yield of fat corrected milk and body weight gain when Sorghum silage or corn silage based diets were fed to lactating cows [33]

Components	Experiment 1		Experiment 2	
	Corn	Sorghum	Corn	Sorghum
Dry matter digestibility (%)	59.1	59.6	65.0	57.3
Crude protein digestibility (%)	43.3	36.7	47.6	48.9
Energy digestibility (%)	58.9	61.3	66.0	57.1
Digestible Energy (kcal/g)	2.82	2.94	2.95	2.54
Daily FCM production (kg/d)	18.7	17.3	15.2	12.4
Daily body weight gain (kg)	0.06	0.35	0.33	0.18

DM, Dry Matter; CP, Crude Protein; FCM, Fat Corrected Milk.

Table 8: Comparison of corn and Sorghum in forage yield and chemical composition in studies conducted at UC Davis in 2012 growing season§

Variety	Type	Yield (t/ha)	Crude Protein*	ADF*	aNDF*
				%	
NK 300	Sorghum Multiple-Use	21.8	8.2	28.3	45.1
Silo 700D	Sorghum Forage type (tall)	19.6	8.3	29.5	49.4
Silo 700D	Sorghum Forage type BMR	13.3	8.1	31.4	51.0
Sorghum 4	Sorghum Evergreen PS	26.3	7.5	37.5	63.5
Sorghum 5	Sorghum Evergreen PS BMR	22.9	7.2	33.8	59.3
	Average for Sorghum	22.0	7.9	32.1	53.7
	S.D.	4.83	0.48	3.67	7.50
TMF2H699	Corn Med. Maturity (110 days)	27.2	7.3	26.2	45.9
F2F622	Corn Med. Maturity BMR (109 days)	22.0	7.9	25.1	44.4
TMF2L871	Corn Late Maturity (117 days)	27.4	7.0	25.7	45.0
F2F622	Corn Late Maturity BMR (115 days)	22.7	7.9	29.2	50.1
	Average for corn	24.7	7.5	26.5	46.4
	S.D.	2.87	0.45	1.82	2.57

§Unpublished data. *Crude protein, ADF, and aNDF values were from fermented corn and Sorghum silage estimated used the near-infrared spectroscopy.

Alfalfa hay played a role in high corn silage based total mixed-rations by supporting fiber mat development in the rumen to support normal rumen function that promotes dry matter digestibility and high production performance. With the decrease in the amount of alfalfa hay fed, straws are often fed at 0.45 to 0.9 kg of DM per cow daily. However, straws are not nutrient and energy dense. Sorghum silage could provide fiber with physical characteristics needed for proper rumen function along with providing higher nutrients and energy than straws. Likewise, Sorghum silage could replace a portion of the cereal silage fed to lactating cows. Sorghum silage may also be fed to dry cows and growing heifers since these animals do not require high intakes of protein and energy. Sorghum silage will likely not replace corn silage in dairy rations since the energy value associated with corn silage's grain content would be difficult to match. But, Sorghum silage could play a role in the nutritional program of animals due to its agronomic characteristics that support the overall sustainability of the dairy farm.

INTAKE, DIGESTIBILITY, AND ANIMAL PERFORMANCE

In general research data show that Sorghum appears to be inferior to corn in total DM digestibility. The mean average daily gain (ADG) for sheep fed corn silage (65.2 g) was more than three times higher than the ADG for sheep fed Sorghum silage (18.1 g). The poor performance of sheep on Sorghum silage was attributed to the presence of compounds such as tannins or HCN [35] [36] that may negatively have affected the utilization of nutrients in the Sorghum silage. Significantly lower in vitro DM digestibility (52%) for Sorghum silage compared with corn silage (65%) was reported [32] . The in vivo DM digestibility of a corn silage based diet (71%) was also higher than a Sorghum based diet (65%) in cattle [32] .

Total mixed-rations containing bmr hybrid Sorghum silage (bmrSS) or corn silage (CS) at either 35% or 45% of dietary DM were fed to Holstein dairy cows to determine the effect on lactation performance and nutrient digestibility [37] . Dry matter intake was greatest when cows were fed the 35% CS (23.4 kg/d) and 45% CS (23.2 kg/d) diets, was least when cows were fed the 45% bmrSS diet (17.6 kg/d), and was intermediate when cows were fed the 35% bmrSS diet (20.1 kg/d). Increasing the amount of Sorghum silage in the diet of lactating cows was associated with decreased DM intake. However, this difference in DM intake was not reflected in solid-corrected milk yield.

Forage intake was influenced by the DM content of silages. Intake of Sorghum silage was increased as DM content of the silage increased. Increasing DM content from 20% to 40% resulted in increased intake from 1.8 to 2.9 kg/100kg [38] . These researchers attributed the lower DM intake of the higher moisture Sorghums to higher concentration of volatile organic acids that once consumed might reduce DDM intake. Dry matter intake of a Sorghum silage based diet was similar with corn silage, although in vitro DM digestibility was higher for the corn based diet [8] . When the average net return from bmr Sorghum and conventional varieties were compared as hays, the average net return ($/ha) from conventional was higher than its brown midrib counter parts [7] (Table 9).

When a bmr Sudangrass hay was compared with a conventional non-bmr variety in dairy cattle, rumen and total tract apparent digestibility of DM, organic matter, NDF, and energy did not differ as the bmr forage replaced the non-bmr forage in the diet [39] .

Table 9: Milk production and net return of four brown-midrib Sudangrass lines and two conventional Sudangrass lines [7]

Sudangrass lines	Average milk production* kg/t hay (DM basis)	Average net return* (land basis, $/ha)
FG96-1-2	1016	934
FG96-1-9	962	1111
FG96-101-3	975	1081
FG96-101-13	1058	996
Greenleaf conventional	793	1180
Piper conventional	518	1141
Brown-midrib average	1002	1030
Conventional average	651	1160

*Average of two harvests.

However, there was a trend for high intake of digestible energy and for increased amounts of total tract organic matter digestion as bmr replaced non-bmr forage [39] .

MILK YIELD AND BODY WEIGHT GAIN

Incorporation of 65% of the diet with standard Sorghum, bmr Sorghum, and corn silage resulted in fat-corrected milk yield of 29.0, 24.5, 23.7, and 20.7 kg/day for corn silage, alfalfa, bmr Sorghum, and standard Sorghum based diets, respectively [40] . When silage of normal Sorghum, bmr Sorghum, corn, and alfalfa based diets were fed to mid-lactation cows, milk yield was not different for bmr Sorghum silage (26.0 kg/d), corn silage (26.4 kg/d), and alfalfa silage (30.1 kg/d), but cows fed normal Sorghum silage produced less milk (20.3 kg/d) [13] . Intake of DM was highest for bmr Sorghum silage (25.3 kg/d) and lowest for normal Sorghum silage (20.4 kg/d) and alfalfa silage (19.6 kg/d) with corn silage intermediate (23.1 kg/d) [14] . When forage Sorghum silage from normal and bmr varieties was compared with corn silage, fat-corrected milk yield was highest for corn silage (33.3 kg/d) and bmr-6 silage (33.7 kg/d) and lowest for normal Sorghum silage (29.1 kg/d) with bmr-16 silage intermediate (31.2 kg/d). Intake of DM was not different for any silage type and averaged across treatments, 24 kg/d [14] .

Brown midrib phenotypes in Sudangrass and/or hybrids could potentially produce positive economic returns compared with non-bmr genotypes [7] . However, determination of possible linkage and/or epistatic relationship between the bmr-6 locus and loci controlling adaptation and agricultural fitness will be a key component for future progress. Genotype by environment interactions will also be important as they reveal adaptive limitations to specific alleles or allele combinations [7] .

When sweet Sorghum silage based TMR was compared with alfalfa silage based TMR for lactating dairy cows, sweet Sorghum silage decreased milk yield, which the authors attributed to the increased NDF content of the TMR. However 4% fat-corrected milk yield (35.0 kg/d for alfalfa silage and 35.3 kg/d for sweet Sorghum silage) was not affected by silage type [41] . When total mixed rations containing corn silage (CS), whole plant grain Sorghum silage (WPGS), or forage Sorghum silage (FSS) based diets were fed to lactating cows, yields of fat-corrected milk 25.6, 25.8, and 24.1 kg/day in CS, WPGS, and FSS, respectively [31] . The DM intake was slightly lower for FSS compared with the other two groups [31] . However, the diets were not equal in forage content. Corn silage was 41.5% of the diet DM, WPGS was 36.7% of the diet DM, and FSS was 28.0% of the diet DM, which complicates interpretation of the production responses. In an earlier study where BMR Sorghum silage was compared with corn silage, the extent of DM degradability was higher for corn silage (49.5%) than BMR Sorghum silage (40.8%), but there were no differences in DM intake (24.4 vs. 25.4 kg/d, respectively) and 4% fat-corrected milk yield (33.4 vs. 33.9 kg/d, respectively) [42] . However, corn silage was fed at a higher proportion of the diet DM than was Sorghum, which again complicates comparison of the forages. To date the scientific literature indicates that Sorghum silage can substitute for corn silage in dairy rations with no significant impact on milk production for lower production potential dairy cows.

However, with respect to the potential of Sorghum silage for high producing dairy cows and for dairy cows in early lactation, there is a paucity of information. There is also a lack of research information for the new bmr type forage Sorghum hybrids fed as silage. In the recent study [31] DM intake for cows fed corn silage was 20.0 kg/d compared with 18.2 kg/d for forage Sorghum silage, a decrease that was significant at a P = 0.07. The same was true for fat-corrected milk yield with 25.5 kg/d for corn silage and 24.1 kg/d for forage Sorghum, different at P = 0.07. However, these levels of performance do not provide sufficient information to predict what would happen if Sorghum silages were fed to high producing dairy cows in light of the fact that DM intakes were low in general for most previously published research and that DM intakes decreased for cows consuming the Sorghum silage diets. The potential for Sorghum silage in the diets of high producing dairy cows has not been adequately studied.

ANTI-NUTRITIONAL FACTORS

Sorghum basically contains two major anti-nutritional factors; tannin, a polyphenolic compound located in the grain and, dhurrin, a cyanogenic glucoside located mainly in the aerial shoot and sprouted seeds. Tannins are high in Sorghum with brown pericarp and no testa, but tannins are low in unpigmented Sorghum grains. The main anti-nutritional effects of tannins are: reduction in voluntary feed intake due to reduced palatability, diminished digestibility and utilization of nutrients, adverse effects upon metabolism, and toxicity. The concentration of tannins in Sorghum grain was reported to be a predominant factor that influenced its nutritional value [43] . From an agronomic perspective, bird resistance in Sorghum varieties was related to the concentration of phenolic compounds in the grain with higher tannin associated with better bird resistance [44] . In many parts of Africa, birds are the main cause for post-harvest loss of grain. To overcome this problem, bird resistant varieties of grain Sorghum have been developed through intensive selection and breeding programs. Bird resistance in Sorghum was associated with higher concentration of tannins in seeds, which in turn was negatively associated with nutritional value of Sorghum for human consumption.

Forage Sorghum can carry some risk of toxicity to livestock, especially when grazed at an early stage of growth. Toxicities can be attributed to the glucoside dhurrin [45] , which releases HCN when plant tissue disruption allows the glucoside found in the epidermal tissue to mix with the enzymes occurring in either the mesophyll tissue [46] or in rumen fluid [47] . This well-known attribute of forage Sorghum may cause fatal poisoning of cattle and, more commonly, the induction of sulfur deficiency [36] . This may be the limitation of utilization of Sorghum for grazing while it is still young and digestible.

SUMMARY

Sorghum forages offer potential for growing regions with low water availability and soils that are somewhat saline as occurs in different areas of the Central Valley in California that is home to 1.8 million dairy cows. Research results indicate that Sorghum, although inferior in forage yield and quality to corn when growing conditions are optimum, has potential for areas of low precipitation or limited irrigation water supply. Generally, the conventional Sorghum varieties yield higher forage DM but are somewhat inferior in forage quality compared with bmr varieties. The choice for a particular variety may depend on the locations and the associated agro- ecological conditions. Although forage DM yield and quality of forage Sorghum is slightly inferior to corn, its ability to thrive and produce reasonable yield in areas of low precipitation or limited access of irrigation water, makes Sorghum more attractive to many farmers in different parts of the world particularly in areas of limited water availability or lower soil quality. Forage Sorghum offers potential in the Central Valley of California where water availability can be restricted. Forage Sorghum is also suited for double and triple cropping systems. Development of new cultivars that are more forage types and higher yielding in digestible DM shows promise for the future of Sorghum forages. However, there is a paucity of information on the feeding value of Sorghum silages in the diets of high producing dairy cows in controlled studies that measure production performance. Until such time that data become available, it is difficult to assess the true potential of Sorghum silages for high producing dairy cows.

REFERENCES

1. Marsalis, M.A. (2011) Advantages of Silage Crops for Quality: What Is Most Important? 2011 Western Alfalfa & Forage Symposium, Las Vegas, 11-13 December 2011, 69-74.
2. Srivastava, A., Naresh Kumar, S. and Aggarwal, P.K. (2010) Assessment on Vulnerability of Sorghum to Climate Change in India. Agriculture, Ecosystems & Environment, 138, 160-169. http://dx.doi.org/10.1016/j.agee.2010.04.012
3. USDA NASS (2015) Crop Production 2014 Summary. National Agricultural Statistics Services.
4. Butchee, K., Arnall, D.B., Sutradhar, A., Godsey, C., Zhang, H. and Penn, C. (2012) Determining Critical Soil pH for Grain Sorghum Production. International Journal of Agronomy, 6, 1-6.
5. Jorgensen, L.R. (1931) Brown Midrib in Maize and Its Linkage Relations. Journal of American Society of Agronomy, 23, 549-557. http://dx.doi.org/10.2134/agronj1931.00021962002300070005x
6. Oliver, A.L., Pedersen, J.F., Grant, R.J. and Klopfenstein, T.J. (2005) Comparative Effects of the Sorghum bmr-6 and bmr-12 Genes: I Forage Sorghum Yield and Quality. Crop Science, 45, 2234-2239.
7. Casler, M.D., Pedersen, J.F. and Undersander, D.J. (2003) Forage Yield and Economic Losses Associated with the Brown-Midrib Trait in Sudangrass. Crop Science, 43, 782-789. http://dx.doi.org/10.2135/cropsci2003.7820
8. Miron, J., Zuckerman, E., Adin, G., Solomon, R., Shoshani, E., Nikbachat, M., Yosef, E., Zenou, A., Weinberg, Z.G., Chen, Y., et al. (2007) Comparison of Two Forage Sorghum Varieties with Corn and the Effect of Feeding Their Silages on Eating Behavior and Lactation Performance of Dairy Cows. Animal Feed Science and Technology, 139, 23-39. http://dx.doi.org/10.1016/j.anifeedsci.2007.01.011
9. Miron, J., Solomon, R., Adin, G., Nir, U., Nikbachat, M., Yosef, E., Carmi, A., Weinberg, Z.G., Kipnis, T., Zuckerman, E., et al. (2006) Effects of Harvest Stage and Re-Growth on Yield, Composition, Ensilage and in Vitro Digestibility of New Forage Sorghum Varieties. Journal of the Science of Food and Agriculture, 86, 140-147. http://dx.doi.org/10.1002/jsfa.2269
10. Bean, B.W., Baumhardt, R.L., McCollum Iii, F.T. and McCuistion, K.C. (2013) Comparison of Sorghum Classes for Grain and Forage Yield and Forage Nutritive Value. Field Crops Research, 142, 20-26. http://dx.doi.org/10.1016/j.fcr.2012.11.014
11. Porter, K., Axtell, J., Lechtenberg, V. and Colenbrander, V. (1978) Phenotype, Fiber Composition, and in Vitro Dry Matter Disappearance of Chemically Induced Brown Midrib (bmr) Mutants of Sorghum. Crop Science, 18, 205-209. http://dx.doi.org/10.2135/cropsci1978.0011183X001800020002x
12. Sattler, S.E., Funnell-Harris, D.L. and Pedersen, J.F. (2010) Brown Midrib Mutations and Their Importance to the Utilization of Maize, Sorghum, and Pearl Millet Lignocellulosic Tissues. Plant Science, 178, 229-

238. http://dx.doi.org/10.1016/j.plantsci.2010.01.001

13. Grant, R.J., Haddad, S.G., Moore, K.J. and Pedersen, J.F. (1995) Brown Midrib Sorghum Silage for Midlactation Dairy Cows1. Journal of Dairy Science, 78, 1970-1980. http://dx.doi.org/10.3168/jds.S0022-0302(95)76823-0

14. Oliver, A.L., Grant, R.J., Pedersen, J.F. and O'Rear, J. (2004) Comparison of Brown Midrib-6 and -18 Forage Sorghum with Conventional Sorghum and Corn Silage in Diets of Lactating Dairy Cows*. Journal of Dairy Science, 87, 637-644. http://dx.doi.org/10.3168/jds.S0022-0302(04)73206-3

15. Prato, J.D., Marble, V.L., Smeltzer, D.G. and Worker, G.F. (1963) Sorghum Forages for Silage in California. Calfonia Agriculture, 17, 10-11.

16. Newman, Y., Erickson, J., Vermerris, W. and Wright, D. (2013) Forage Sorghum (Sorghum bicolor): Overview and Management. University of Florida, Gainesville.

17. Clark, A. (2007) Managing Cover Crops Profitably. 3rd Edition, Sustainable Agriculture Network, Beltsville.

18. DePeters, E.J., Medrano, J.F. and Bath, D.L. (1989) A Nutritional Evaluation of Mixed Winter Cereals with Vetch Utilized as Silage or Hay. Journal of Dairy Science, 72, 3247-3254. http://dx.doi.org/10.3168/jds.S0022-0302(89)79484-4

19. Miron, J., Zuckerman, E., Sadeh, D., Adin, G., Nikbachat, M., Yosef, E., Ben-Ghedalia, D., Carmi, A., Kipnis, T. and Solomon, R. (2005) Yield, Composition and in Vitro Digestibility of New Forage Sorghum Varieties and Their Ensilage Characteristics. Animal Feed Science and Technology, 120, 17-32. http://dx.doi.org/10.1016/j.anifeedsci.2005.01.008

20. Schmid, A.R., Goodrich, R.D., Jordan, R.M., Marten, G.C. and Meiske, J.C. (1976) Relationships among Agronomic Characteristics of Corn and Sorghum Cultivars and Silage Quality. Agronomy Journal, 68, 403-406. http://dx.doi.org/10.2134/agronj1976.0002196 200680002005 1x

21. Wosnitza, A. and Hartmann, S. (2012) Determination of Region-Specific Data of Yield and Quality of Alternatives to Silage Maize in Fodder Crops: Field Trails with Forage Gras and Clover Grass Mixtures, Sorghum as well as Whole Plant Silage of Grain. Proceedings International Conference on the German Diabrotica Research Program, Berlin, 14-16 November 2012, 147-182.

22. Vasilakoglou, I., Dhima, K., Karagiannidis, N. and Gatsis, T. (2011) Sweet Sorghum Productivity for Biofuels under Increased Soil Salinity and Reduced Irrigation. Field Crops Research, 120, 38-46. http://dx.doi.org/10.1016/j.fcr.2010.08.011

23. Bean, B. and Marsalis, M. (2012) Corn and Sorghum Silage Production Considerations. High Plains Dairy Conference, Amarillo, 7-8 March 2012, 1-7.

24. Carmi, A., Aharoni, Y., Edelstein, M., Umiel, N., Hagiladi, A., Yosef, E., Nikbachat, M., Zenou, A. and Miron, J. (2006) Effects of Irrigation and Plant Density on Yield, Composition and in Vitro Digestibility of a New Forage Sorghum Variety, Tal, at Two Maturity Stages. Animal Feed Science and Technology, 131, 121-133. http://dx.doi.org/10.1016/j.anifeedsci.2006.02.005

25. Mastrorilli, M., Katerji, N. and Rana, G. (1999) Productivity and Water Use Efficiency of Sweet Sorghum as Affected by Soil Water Deficit Occurring at Different Vegetative Growth Stages. European Journal of Agronomy, 11, 207-215. http://dx.doi.org/10.1016/S1161-0301(99)00032-5

26. Steduto, P., Katerji, N., Puertos-Molina, H., U¨nlu¨, M., Mastrorilli, M. and Rana, G. (1997) Water-Use Efficiency of Sweet Sorghum under Water Stress Conditions Gas-Exchange Investigations at Leaf and Canopy Scales. Field Crops Research, 54, 221-234. http://dx.doi.org/10.1016/S0378-4290(97)00050-6

27. Pan, G., OuYang, Z., Luo, Q., Yu, Q. and Wang, J. (2011) Water Use Patterns of Forage Cultivars in the North China Plain. International Journal of Plant Production, 5, 181-194.

28. Saeed, I.A.M. and El-Nadi, A.H. (1998) Forage Sorghum Yield and Water Use Efficiency under Variable Irrigation. Irrigation Science, 18, 67-71. http://dx.doi.org/10.1007/s002710050046

29. Peacock, J.M. (1982) Response and Tolerance of Sorghum to Temperature Stress. Sorghum in the Eighties: Proceedings of the International Symposium on Sorghum, Patancheru, 2-7 November 1981, 143-159.

30. Wilson, G.L. and Eastin, J.D. (1982) The Plant and Its Environment. Sorghum in the Eighties: Proceedings of the International Symposium on Sorghum, Patancheru, 2-7 November 1981, 101-119.

31. Colombini, S., Galassi, G., Crovetto, G.M. and Rapetti, L. (2012) Milk Production, Nitrogen Balance, and Fiber Digestibility Prediction of Corn, Whole Plant Grain Sorghum, and Forage Sorghum Silages in the Dairy Cow. Journal of Dairy Science, 95, 4457-4467. http://dx.doi.org/10.3168/jds.2011-4444

32. Abdelhadi, L.O. and Santini, F.J. (2006) Corn Silage versus Grain Sorghum Silage as a Supplement to Growing Steers Grazing High Quality Pastures: Effects on Performance and Ruminal Fermentation. Animal Feed Science and Technology, 127, 33-43. http://dx.doi.org/10.1016/j.anifeedsci.2005.08.010

33. Lance, R.D., Foss, D.C., Krueger, C.R., Baumgardt, B.R. and Niedermeier, R.P. (1964) Evaluation of Corn and Sorghum Silages on the Basis of Milk Production and Digestibility. Journal of Dairy Science, 47, 254-257. http://dx.doi.org/10.3168/jds.S0022-0302(64)88635-5

34. Castillo, A.L., Silva-del-Rio, N., St-Pierre, N. and Weiss, W.P. (2010) Composition of Diets Fed to Different Groups of Lactating Cows on California Dairies. ADSA 2010, Abstract #: T306.

35. Fisk, J. (1980) Effects of HCN, Phenolic Acids and Related Compounds in Sorghum bicolor on the Feeding Behaviour of the Planthopper Peregrinus maidis. Entomologia Experimentalis et Applicata, 27, 211-222. http://dx.doi.org/10.1111/j.1570-7458.1980.tb02968.x

36. Wheeler, J., Mulcahy, C., Walcott, J. and Rapp, G. (1990)

Factors Affecting the Hydrogen Cyanide Potential of Forage Sorghum. Australian Journal of Agricultural Research, 41, 1093-1100. http://dx.doi.org/10.1071/AR9901093

37. Dann, H.M., Grant, R.J., Cotanch, K.W., Thomas, E.D., Ballard, C.S. and Rice, R. (2008) Comparison of Brown Midrib Sorghum-Sudangrass with Corn Silage on Lactational Performance and Nutrient Digestibility in Holstein Dairy Cows. Journal of Dairy Science, 91, 663-672. http://dx.doi.org/10.3168/jds.2007-0521

38. Ward, G.M., Boren, F.W., Smith, E.F. and Brethour, J.R. (1966) Relation between Dry Matter Content and Dry Matter Consumption of Sorghum Silage. Journal of Dairy Science, 49, 399-402. http://dx.doi.org/10.3168/jds.S0022-0302(66)87882-7

39. Ledgerwood, D.N., DePeters, E.J., Robinson, P.H., Taylor, S.J. and Heguy, J.M. (2009) Assessment of a Brown Midrib (BMR) Mutant Gene on the Nutritive Value of Sudangrass Using in Vitro and in Vivo Techniques. Animal Feed Science and Technology, 150, 207-222. http://dx.doi.org/10.1016/j.anifeedsci.2008.10.001

40. Aydin, G., Grant, R.J. and O'Rear, J. (1999) Brown Midrib Sorghum in Diets for Lactating Dairy Cows. Journal of Dairy Science, 82, 2127-2135. http://dx.doi.org/10.3168/jds.S0022-0302(99)75456-1

41. Amer, S., Seguin, P. and Mustafa, A.F. (2012) Short Communication: Effects of Feeding Sweet Sorghum Silage on Milk Production of Lactating Dairy Cows. Journal of Dairy Science, 95, 859-863. http://dx.doi.org/10.3168/jds.2011-4884

42. Colombini, S., Rapetti, L., Colombo, D., Galassi, G. and Crovetto, G.M. (2010) Brown Midrib Forage Sorghum Silage for the Dairy Cow: Nutritive Value and Comparison with Corn Silage in the Diet. Italian Journal of Animal Science, 9, 273-277.

43. Etuk, E.B., Okeudo, N.J., Esonu, B.O. and Udedibie, A.B.I. (2012) Antinutritional Factors in Sorghum: Chemistry, Mode of Action and Effects on Livestock and Poultry. Journal of Animal and Feed Research, 2, 113-119.

44. Reed, J.D., Tedla, A. and Kebede, Y. (1987) Phenolics, Fibre and Fibre Digestibility in the Crop Residue from Bird Resistant and Non-Bird Resistant Sorghum Varieties. Journal of the Science of Food and Agriculture, 39, 113-121. http://dx.doi.org/10.1002/jsfa.2740390204

45. Conn, E.E. (1980) Cyanogenic Compounds. Annual Review of Plant Physiology, 31, 433-451. http://dx.doi.org/10.1146/annurev.pp.31.060180.002245

46. Kojima, M., Poulton, J.E., Thayer, S.S. and Conn, E.E. (1979) Tissue Distributions of Dhurrin and of Enzymes Involved in Its Metabolism in Leaves of Sorghum bicolor. Plant Physiology, 63, 1022-1028. http://dx.doi.org/10.1104/pp.63.6.1022

47. Majak, W., McDiarmid, R.E., Hall, J.W. and Cheng, K.J. (1990) Factors that Determine Rates of Cyanogenesis in Bovine Ruminal Fluid in Vitro. Journal of Animal Science, 68, 1648-1655.

INDEX

O

P

Q

R

S

T

W